The Sweetpotato

Gad Loebenstein · George Thottappilly
Editors

The Sweetpotato

Springer

Editors
Dr. Gad Loebenstein
Agricultural Research
Organization
Dept. Virology
Bet Degan
Israel
Gad-talma@Barak.net.il

Dr. George Thottappilly
Teekay Villa, P.O. Karoor
Via. Thazhekkad-680697
Trichur Dist., Kerala
India
george_thottappilly@rediffmail.com

Book Cover (Collage by Dov Ronen):

Upper right enclosure – *Upper part* – Sweetpotato harvest (2-row mechanical chain harvester) (Courtesy of Tara Smith, Louisiana State University Agricultural Center)
Lower part – Sweetpotato nursery planted with virus-tested plantlets derived from meristem cultures (Courtesy of Gad Loebenstein, ARO, Israel)

Middle enclosure – Sweetpotato production field in California and sweetpotato just prior to harvest (Courtesy of Scott Stoddard, University of California at Merced)

Lower right enclosure – *Right part* – Sweetpotato plant (cv. Lamote) in tissue culture for three months (Courtesy of Victor Gaba, ARO, Israel)
Left part – Sweetpotato soft drink (Courtesy G. Padmaja, Central Tuber Crops Research Institute, Thiruvananthapuram, Kerala, India.

ISBN 978-90-481-8135-3 e-ISBN 978-1-4020-9475-0

DOI 10.1007/978-1-4020-9475-0

Printed on acid-free paper

9 8 7 6 5 4 3 2 1

springer.com

Preface

Root and Tuber crops are basic to the diets of millions in the tropics and subtropics where most of the world's undernourished people live. Among these crops, sweetpotatoes contain substantial levels of protein in addition to carbohydrates. They also provide substantial amounts of vitamins and minerals. Sweetpotato is the seventh most important food crop in the world in terms of production. It produces more biomass and nutrients per hectare than any other food crop in the world, and is one of the crops with a unique role in relief famine. Sweetpotato is grown for both the leaves, which are used as greens, and the tubers, for a high carbohydrate and beta-carotene source. In East Africa's semiarid, densely populated plains, thousands of villages depend on sweetpotato for food security and the Japanese used it when typhoons demolished their rice fields. In the developed world it is a tasty food, rich in nutritional value and served both cooked, fried and in pies.

Once established, apart from occasional weeding, sweetpotatoes need little care. Sweetpotatoes are fast growing and rapidly produce ground cover to prevent soil erosion.

Sweetpotato is genetically a challenging polyploid plant. It can hardly be considered as a model species for studies in molecular biology. Therefore, sequence data on the genomic DNA and expressed gene sequences accumulate slowly. Molecular data and sequence information would be pivotal for improving sweetpotato by engineering its own genes and their expression.

There is considerable interest in sweetpotato biotechnology. Also, there is great interest to grow sweetpotato in space missions. Tissue culture, disease elimination, germplasm storage and distribution are all well developed for sweetpotato. Further research is needed in molecular markers and marker assisted breeding, DNA fingerprinting and germplasm characterization, and more genes and transformation of sweetpotato with these genes.

Recently, a patent was granted for the production of bread from 100% sweetpotato flour. It is hoped that these products will appeal to consumers who are allergic to grain breads and flours. Also scientists at two different institutes in the United States have developed genetically modified sweetpotatoes containing edible vaccines. One of these vaccines works against hepatitis B and the other against the Norwalk virus found in food that has not been handled or stored correctly. Edible vaccines such as these may provide cheap protection for some of the poorest people in the world.

The book has two parts: A. General, including *inter alia* Origin, Botany and Physiology, Cultivars, Genetics, Biotechnology, Pests and diseases, and Part B on the Sweetpotato crop in different areas and regions.

In the last four decades of the twentieth century the use of sweetpotato was diversified beyond their classification as subsistence, food security, and famine-relief crop. It is hoped that millions of subsistence landholders in Africa, Asia, and Latin America will be able to use sweetpotatoes for food, stock food, and processed products and to generate income. It is hoped that by bringing the latest developments on this crop in this book, more research, especially interdisciplinary, will be initiated, as well as international conferences together with funding agencies and policy makers. We sincerely hope that this book will give the motivation to direct our research efforts with the strong commitment to help national governments.

We thank all the contributors, Ms. Zuzana Bernhart and Springer who made this book possible.

Bet Dagan, Israel — Gad Loebenstein
Kodakara, Kerala — George Thottappilly

Contents

Contributors

Akoroda, M. Faculty of Agriculture and Forestry, University of Ibadan, Nigeria, m_akoroda@yahoo.com and M.AKORODA@CGIAR.ORG

Andrade, Maria International Potato Center, Sub-Saharan Africa

Barker, I. International Potato Center, Lima, Peru

Bourke, R. Michael Department of Geography, Research School of Pacific and Asian Studies, The Australian National University, Canberra, ACT 0200, Australia, a145250@anumail.anu.edu.au

Campilan Dindo UPWARD Program, International Potato Center (CIP); c/o IRRI, DA, Metro Manila, Philippines, d.campilan@cgiar.org

Carpena, Azucena L. University of the Philippines Los Baiios, College, Laguna, Philippines, alcarpena@yahoo.com

Chujoy, E International Potato Center (CIP), Apartado 1558, Lima 12, Peru

Clark, Christopher A. Department of Plant Pathology and Crop Physiology, Louisiana State University Agcenter, Baton Rouge, LA 70803, USA, CClark@agcenter.Isu.edu

Cohen, J. Agriculture Research Organization, Bet Dagan, 50–250, Israel, gacovcohen@013.net

Crissman, C. International Potato Center, Peru

Dar, Z. The Agricultural Extension Service, Ministry of Agriculture, Bet Dagan 50–250, Israel, ziondar@shaham.moag.gov.il

Edison, S. Central Tuber Crops Research Institute, Thiruvananthapuram, India, ctcritvm@yahoo.com

Estes, Edmund A. Department of Agricultural and Resource Economics, College of Agricultural and Life Sciences, North Carolina State University, Raleigh, North Carolina, 27695-8109, USA, eaestes@ncsu.edu

Ferrin, Donald M. Department of Agricultural and Resource Economics, College of Agricultural and Life Sciences, North Carolina State University, Raleigh, North Carolina, 27695-8109, USA

Firon Nurit Institute of Plant Sciences, The Volcani Center, Agricultural Research Organization, Bet Dagan, 50250, Israel, vcfiron@volcani.agri.gov.il

Fuentes, S. International Potato Center (CIP), Apartado 1558, Lima 12, Peru, s.fuentes@cgiar.org

Gaba, Victor Department of Plant Pathology and Weed Science ARO Volcani Center, POB 6 Bet Dagan 50250, Israel, vpgaba@volcani.agri.gov.il

Ghislain, Marc International Potato Center (CIP), Applied Biotechnology Laboratory, Germplasm Enhancement and Crop Improvement Division, Lima 12, Peru

Hegde, Vivayaka Central Tuber Crops Research Institute, Trivandrum, India

Holmes, Gerald J. Department of Plant Pathology, North Carolina State University, Raleigh, NC 27695, USA

Kfir, Y. Institute of Plant Sciences, The Volcani Center, Agricultural Research Organization, Bet Dagan, 50250, Israel

Kreuze, Jan F. International Potato Center (CIP), Applied Biotechnology Laboratory, Germplasm Enhancement and and Crop Improvement Division, Lima 12, Peru, J.KREUZE@CGIAR.ORG

LaBonte, D. Louisiana State University AgCenter, School of Plant, Environmental, and Soil Sciences, Baton Rouge, LA 70803, USA

Lemaga, Berga International Potato Center-Sub-Saharan Africa, Nairobi, Kenya

Liu, Qingchang Plant Biotechnology Laboratory, College of Horticulture, Northwest University of Agriculture and Forestry, Yangling 712100, Shaanxi, P.R. China

Loebenstein, G. Agricultural Research Organization, Bet Dagan, 50250, Israel, Gad-talma@barak.net.il

Low, J. Intenzational Potato Center-Sub-Saharan Africa, Nairobi, Kenya, J.LOW@CGIAR.ORG

Lynam, J Independent Agricultural Economist, Nairobi, Kenya

Makeshkumar, T. Central Tuber Crops Research Institute, Trivandrum, India, makeshkumar@yahoo.com

McGregor, C. Louisiana State University AgCenter, School of Plant, Environmental, and Soil Sciences, Baton Rouge, LA 70803, USA

Namanda, S. International Potato Center-Sub-Saharan Africa, Nairobi, Kenya

Overstreet, Charles Department of Plant Pathology and Crop Physiology, Louisiana State University Agricultural Center, Baton Rouge, LA 70803, USA, COverstreet@agcenter.lsu.edu

Padmaja, G. Central Tuber Crops Research Institute, Trivandrum, Kerala, India, g_padmaja1@yahoo.co.in

Pressman, E. Institute of Plant Sciences, The Volcani Center, Agricultural Research Organization, Bet Dagan, 50250, Israel

Schultheis, Jonathan Department of Horticultural Science, North Carolina State University, Raleigh, USA, jonr@ncsu.edu

Shankle, Mark Pontotoc Research and Extension Center, Mississippi State University, Pontotoc, MS 38863, NC 27695, USA

Singer, Sima Department of Plant Pathology and Weed Science, ARO Volcani Center, Bet Dagan 50250, Israel

Smith, Tara P. Sweetpotato Research Station, Louisiana State University Agricultural Center, Chase, LA 71324, USA, TSmith@agcenter.lsu.edu

Sorensen, Kenneth A. Department of Entomology, North Carolina State University, Raleigh, North Carolina 27695, USA, kenneth_sorensen@ncsu.edu

Srinivas, T. Section of Social Sciences, Central Tuber Crops Research Institue, Trivandrum, Kerala, India, srinictcri@yahoo.com

Stoddard, Scott Cooperative Extension Service, University of California, Merced, CA 95340, USA

Suja, G. Central Tuber Crops Research Institute, Trivandrum, Kerala, India

Thiele, G. International Potato, Peru

Thottappilly, G. Sardaya College of Engineering and Technology, Kodakara, Kerala, India, george_thottappilly@rediffmail.com

Valkonen, Jari P.T. Department of Applied Biology, FIN-00014, University of Helsinki, Finland, jari.valkonen@helsinki.fi

Villordon, A. Lousiana State University AgCenter, Sweet Potato Research Station, Chase, LA 71324, USA

Wang, Qingmei Plant Biotechnology Laboratory, College of Horticulture, Northwest University of Agriculture and Forestry, Yangling 712100, Shaanxi, P. R. China

Wang Qiaochun Plant Biotechnology Laboratory, College of Horticulture, Northwest University of Agriculture and Forestry, Yangling 712100, Shaanxi, P.R. China, qiaochunwang@nwsuaf.edu.cn

Wheatley, C. Independent Post-harvest Specialist, New Zealand

Zhang, Liming Institute of Field Crops, Shandong Academy of Agricultural Science, Jinan 250100, Shandong, P.R. China

Color Plates

Plate 1 **a**: Root from a bed affected by sclerotial blight with mycelia of *Sclerotium rolfsii* growing on the surface of the root (courtesy Gerald Holmes, North Carolina State University). **b**: Wilting of sprouts and areas without plants due to sclerotial blight (courtesy Christopher Clark, Louisiana State University AgCenter). **c**: A plant with foot rot growing in a greenhouse showing a lesion at the soil line that girdled the plant causing secondary wilting (photo by Weston J. Martin, Louisiana State University AgCenter). **d**: Yellowing and wilting of lower leaves caused by Fusarium wilt (courtesy Christopher Clark, Louisiana State University AgCenter). **e**: Cross sections of stems of plants with Fusarium wilt (*left*) showing discoloration of the vascular system and stems from healthy plants (*right*) (courtesy Christopher Clark, Louisiana State University AgCenter). **f**: Bacterial stem rot caused by *Dickeya didantii* (courtesy Gerald Holmes, North Carolina State University) (See also Plate 7.1 on page 83)

Plate 2 Root diseases. Symptoms on storage roots: **a**: black rot (caused by *Ceratocystis fimbriata*) (courtesy Gerald Holmes, North Carolina State University), **b**: foot rot (caused by *Plenodomus destruens*) (photo by Weston J. Martin, Louisiana State University AgCenter), **c**: circular spot (caused by *Sclerotium rolfsii*) (courtesy Gerald Holmes, North Carolina State University), **d**: scurf (caused by *Monilochaetes infuscans*) (courtesy Gerald Holmes, North Carolina State University), and **e**: Streptomyces soil rot (caused by *Streptomyces ipomoeae*) (courtesy Gerald Holmes, North Carolina State University). Symptoms of fibrous roots: **f**: Streptomyces soil rot (caused by *Streptomyces ipomoeae*) (courtesy Christopher Clark, Louisiana State University AgCenter) (See also Plate 7.2 on page 86)

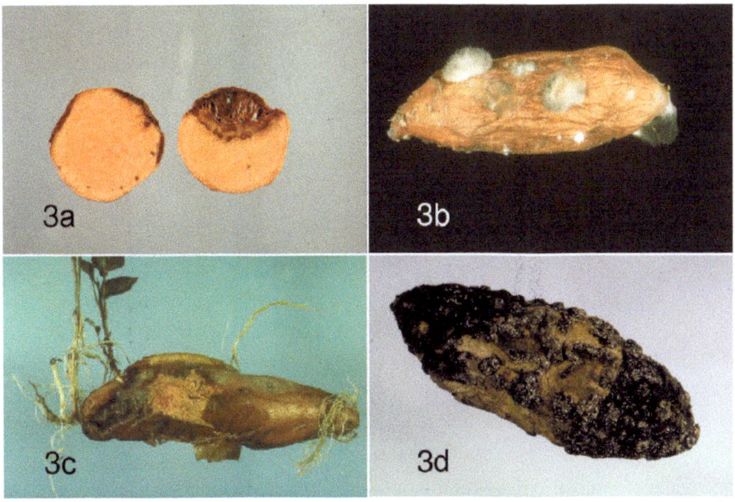

Plate 3 Storage diseases. **a**: Fusarium root rot (*caused by Fusarium solani*) (courtesy Christopher Clark, Louisiana State University AgCenter), **b**: Rhizopus soft rot (caused by *Rhizopus stolonifer*) (photo by Weston J. Martin, Louisiana State University AgCenter), **c**: bacterial soft rot (caused by *Dickeya didantii*) (courtesy Christopher Clark, Louisiana State University AgCenter), and **d**: Java black rot (caused by *Lasiodiplodia theobromae*) (courtesy Gerald Holmes, North Carolina State University) (See also Plate 7.3 on page 91)

Plate 4 Foliar diseases. **a**: stem and leaf scab (caused by *Elsinoe batatas*) (courtesy Christopher Clark, Louisiana State University AgCenter, **b**: chlorotic leaf distortion (caused by *Fusarium denticulatum*) (courtesy Christopher Clark, Louisiana State University AgCenter, **c**: white rust (caused by *Albugo ipomoeae-panduranae*) as seen on adaxial surfaces of leaves (courtesy Gerald Holmes, North Carolina State University), **d**: white rust as seen on abaxial surfaces (courtesy Gerald Holmes, North Carolina State University), **e**: Alternaria blight (caused by *Alternaria spp.*) (courtesy Christopher Clark, Louisiana State University AgCenter, and **f**: Alternaria leaf spot (caused by *Alternaria spp.*) (courtesy Christopher Clark, Louisiana State University AgCenter (See also Plate 7.4 on page 94)

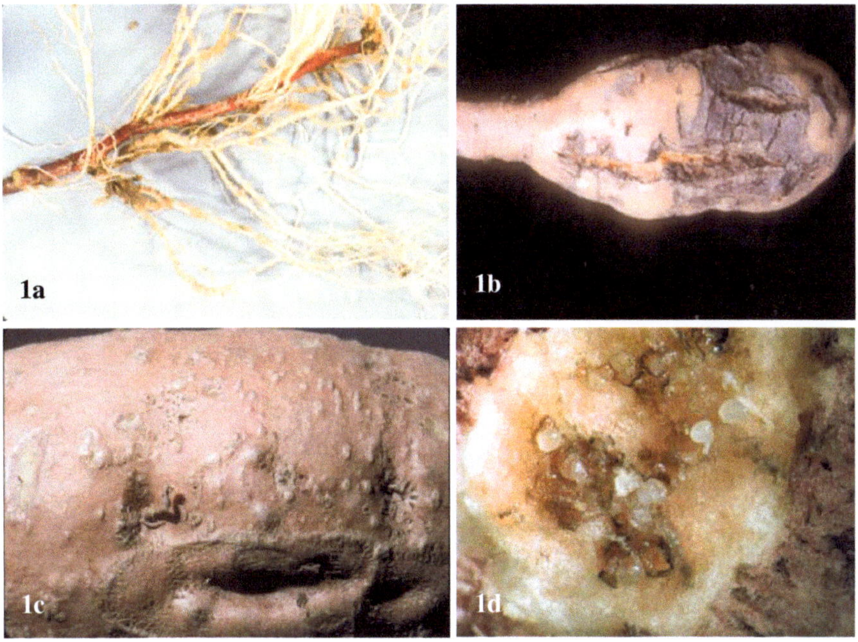

Plate 5 Root-knot nematode *Meloidogyne incognita* (**a**) Root showing galling (courtesy Charles Overstreet, Louisiana State University AgCenter). (**b**) Cultivar 'Centennial' showing cracking and some decay from infection (courtesy Chris Clark, Louisiana State University AgCenter). (**c**) Cultivar 'Beauregard' showing the typical blistering from infection and slight cracking (courtesy Charles Overstreet, Louisiana State University AgCenter). (**d**) Females of root-knot nematode exposed beneath one of the blisters on 'Beauregard'. Several females teased out and laying on the cut area (courtesy Charles Overstreet, Louisiana State University AgCenter) (See also Plate 9.1 on page 139)

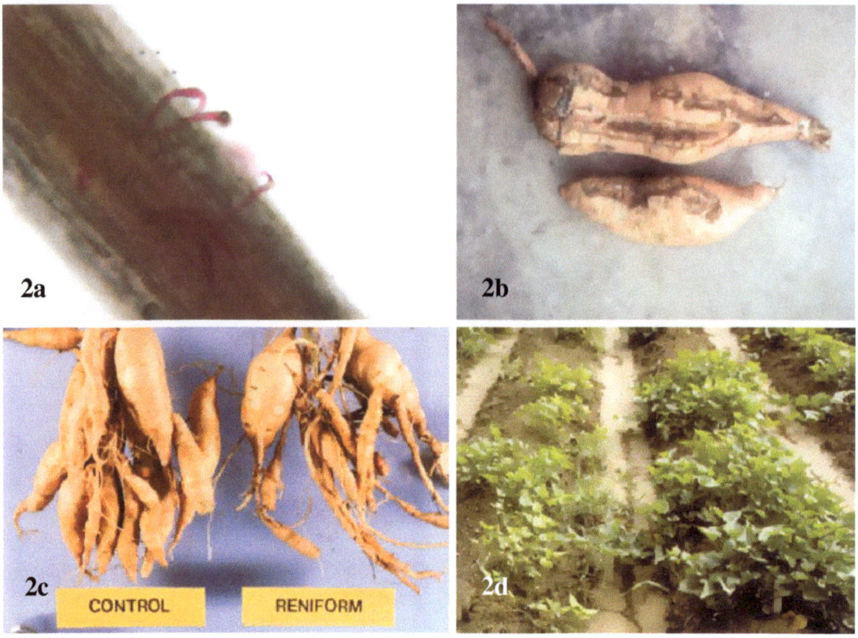

Plate 6 Reniform nematode *Rotylenchulus reniformis* (**a**) Developing females (stained red) inside root (courtesy Charles Overstreet, Louisiana State University AgCenter). (**b**) Cultivar 'Goldrush' showing cracking from infection by this nematode (courtesy Chris Clark, Louisiana State University AgCenter). (**c**) Individual hills of sweetpotato from an untreated area (*left*) compared to an area infested with reniform on the *right* (courtesy Chris Clark, Louisiana State University AgCenter). (**d**) A sweetpotato breeding line in a reniform nursery that is untreated on the left and fumigated on right (courtesy Chris Clark, Louisiana State University AgCenter) (See also Plate 9.2 on page 143)

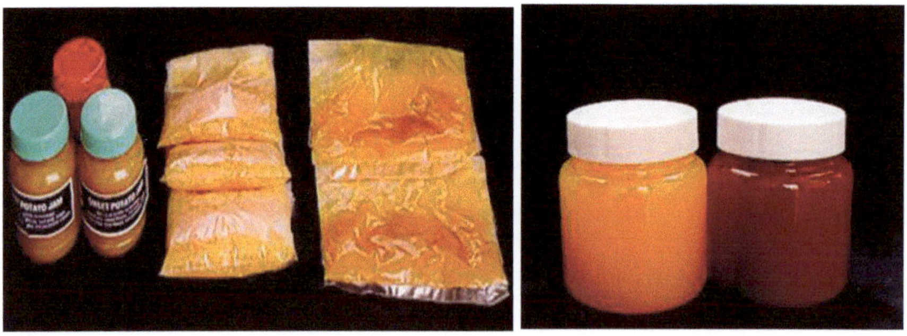

Plate 7 Sweetpotato jam (Courtesy: Padmaja and Premkumar, 2002) (See also Plate 11.1 on page 202)

Plate 8 Sweetpotato soft drink (Courtesy: Padmaja and Premkumar, 2002) (See also Plate 11.2 on page 203)

Plate 9 Sweetpotato based gulab jamuns and instant mix (Padmaja et al., 2005b) (See also Plate 11.3 on page 208)

Plate 10 Sweetpotato grading and packing line located at grower-shipper packing shed in Columbus County, North Carolina (See also Plate 13.1 on page 275)

Plate 11 Commercial grower-shipper bulk bin storage room equipped with environmental management controls, Nash County, North Carolina (See also Plate 13.2 on page 275)

Plate 12 Yams and sweetpotatoes for sale side-by-side at Lowe's Grocery Stores, Wake County, North Carolina (See also Plate 13.3 on page 281)

Plate 13 Cello-wrapping of sweetpotatoes on grading line at grower-shipper packing shed in Columbus County, North Carolina (See also Plate 13.4 on page 281)

Plate 14 Sweetpotato bedding wagon and seed distribution in a hotbed or seed bed (Courtesy of Tara Smith, Louisiana State University Agricultural Center) (See also Plate 14.1 on page 300)

Plate 15 Sweetpotato bed covering machine, covering seed potatoes (Courtesy of Tara Smith, Louisiana State University Agricultural Center) (See also Plate 14.2 on page 300)

Plate 16 Sweetpotato mulch/plastic layer being applied to a seed bed (Courtesy of Tara Smith, Louisiana State University Agricultural Center) (See also Plate 14.3 on page 300)

Plate 17 Sweetpotato hotbeds in California (Courtesy of Scott Stoddard, University of California at Merced) (See also Plate 14.4 on page 300)

Plate 18 Sweetpotato seed beds with transplants (Courtesy of Tara Smith, Louisiana State University Agricultural Center) (See also Plate 14.5 on page 301)

Plate 19 Sweetpotato 2-row mechanical transplanter, (Courtesy of Tara Smith, Louisiana State University Agricultural Center) (See also Plate 14.6 on page 302)

Plate 20 Sweetpotato transplants or slips that will be planted in a commercial production field (Courtesy of Tara Smith, Louisiana State University Agricultural Center) (See also Plate 14.7 on page 302)

Plate 21 Virus-tested sweetpotato plants in test tubes (Courtesy of Scott Stoddard, University of California at Merced) (See also Plate 14.8 on page 305)

Plate 22 Sweetpotato commercial production field (Courtesy of Tara Smith, Louisiana State University Agricultural Center) (See also Plate 14.9 on page 305)

Plate 23 Sweetpotato production field in California just prior to harvest (Courtesy of Scott Stoddard, University of California at Merced) (See also Plate 14.10 on page 306)

Plate 24 Sweetpotato harvest (2-row mechanical chain harvester) (Courtesy of Tara Smith, Louisiana State University Agricultural Center) (See also Plate 14.11 on page 313)

Plate 25 Wooden storage bins used to harvest and store sweetpotatoes (Courtesy of Scott Stoddard, University of California at Merced) (See also Plate 14.12 on page 314)

Plate 26 Sweetpotato harvest (shaker harvester method) (Courtesy of Tara Smith, Louisiana State University Agricultural Center) (See also Plate 14.13 on page 314)

Plate 27 Typical sweetpotato storage facility for a large commercial operation (Courtesy of Scott Stoddard, University of California at Merced) (See also Plate 14.14 on page 316)

Plate 28 Marketing containers (Courtesy of Tara Smith, Louisiana State University Agricultural Center and Scott Stoddard, University of California at Merced) (See also Plate 14.15 on page 317)

Plate 29 Marketing containers (Courtesy of Tara Smith, Louisiana State University Agricultural Center and Scott Stoddard, University of California at Merced) (See also Plate 14.16 on page 317)

Plate 30 Marketing containers (Courtesy of Tara Smith, Louisiana State University Agricultural Center and Scott Stoddard, University of California at Merced) (See also Plate 14.17 on page 318)

Plate 31 **A.** "Guangshu 69", **B.** "Yanshu 5", **C.** "Guangshu 79", **D.** "Jishu 15", **E.** "Jishu 18", **F.** "Jishu 98", **G.** "Xushu 18", **H.** "Xushu 25", **I.** "Jinshan57", **J.** "Nanshu 88", **K.** "Eshu 5", **L.** "Chuanshu 34", **M.** A typical meristem used for virus elimination. **N.** Plantlet regenerated from meristem culture, **O.** Culture room of sweetpotato. **P.** Cross breeding. Bars indicate 5 cm (See also Plate 15.1 on page 351)

Plate 32 **A.** Propagation bed. **B.** Preparation of cuttings. **C.** Ridge planting. **D.** Mulching field. **E.** Large scale production. **F.** Production of vegetable sweetpotato. **G.** Harvesting of vegetable sweetpotato. **H.** Harvesting of sweetpotato. **I.** Sun-drying of sweetpotato. **J.** Sweetpotato noodles. **K.** and **L.** Preserved Sweetpotato chips. **M.** Sweetpotato starch. **N.** Processing of Sweetpotato noodles. **O.** fresh tuber roots for market (See also Plate 15.2 on page 352)

Part I
General

Chapter 1
Introductory Remarks

G. Thottappilly

A tropical American vine, sweetpotato *(Ipomoea batatas)* is in the botanical family *Convolvulaceae* along with common plants, such as bindweed and morning glory. In developing countries, sweetpotato is ranked fifth in economic value production, sixth in dry matter production, seventh in energy production, ninth in protein production, and it has tremendous flexibility of utilization as food, feed and industrial products (Gregory, 1992). Sweetpotatoes should not be confused with potato *(Solanum tuberosum)* which belongs to Solanaceae and are entirely unrelated, although their uses can be similar. Orange-fleshed sweetpotatoes are often known as yams, especially in the southern United States, but they are quite different from true yams (*Dioscorea* sp.) in growth habit and use. Furthermore, unlike *Dioscorea* yams, which is native to Africa and Asia, the greens of sweetpotatoes are edible and provide an important source of food in Africa (Guinea, Sierra Leone and Liberia) as well as in East Asia (Scott et al., 2000a, b).

Sweetpotato is grown from underground tuberous roots. Leaves are variable in shape, size, and color. The single flowers are funnel shaped and white or rose violet. Where stem nodes touch the ground, the edible storage roots develop to usually four to ten storage roots per plant. It was introduced into Europe in the 16th century and later spread to Asia. Sweetpotatoes are used mostly for human consumption but are sometimes fed to swine. They yield starch, flour, glucose, and alcohol and are especially rich in vitamin A. Sweetpotato's role as an important health food is recognised due to high nutrient content with its anti-carcinogenic and cardiovascular disease-preventing properties (Hill et al., 1992). All varieties of sweetpotato are good sources of vitamins C, B2 (Riboflavin), B6 and E as well as dietary fiber, potassium, copper, manganese and iron, and they are low in fat and cholesterol. Despite the name "sweet", it may be a good food for diabetics as it helps to stabilize blood sugar levels and to lower insulin resistance (Answers.com, 2008). It is expected therefore that sweetpotato marketing will increase in the coming years.

G. Thottappilly (✉)
Sardaya College of Engineering and Technology, Kodakara, Kerala, India
e-mail: george_thottappilly@rediffmail.com

G. Loebenstein, G. Thottappilly (eds.), *The Sweetpotato*,
DOI 10.1007/978-1-4020-9475-0_1, © Springer Science+Business Media B.V. 2009

Sweetpotatoes are grown on about 9 million hectares, yielding c $\sim$ 124 million tons, with an average yield of about 13.7 ton/ha (FAOSTAT, 2006). They are mainly grown in developing countries, which account for over 97% of world output. The cultivated area of sweetpotato in China, about 4.7 million ha, accounted for 52% of the total area of sweetpotato cultivation in the world. The rest of Asia accounts for 6 percent, Africa 5 percent, Latin America 1.5 percent, and the United States 0.45 percent. Vietnam is the second largest producer. The largest European producer is Portugal with only .02 percent of world production. These figures clearly illustrate that sweetpotatoes are an important crop in third world countries but are only a secondary foodstuff in developed countries (Scott et al., 2000a, b). Sweetpotato is a 'poor man's crop' in Africa, with most of the production done on a small or subsistence level. This crop plant has a long and unique history of saving lives in relief famine and for providing food security. It matures fast, is rich in nutrients, and is often the first crop planted after a natural disaster, providing abundant food for otherwise starving populations. In eastern Africa, the sweetpotato is known as "the protector of children" or *cilera abana* because it is often the only food that stands between a child's survival and starvation (Answers.com, 2008). Sweetpotatoes were grown in gardens by North American Indians and were an important staple food during the American Revolutionary War and the American Civil War. They were also an essential part of the diet of the slave population in southern states.

Sweetpotatoes vary enormously in taste, size, shape, and texture, although all are smooth-skinned with roots. Sweetpotato flesh can be white, orange, yellow, purple, red, pink and violet, while skin color varies among yellow, red, orange, and brown. Varieties with pale yellow or white flesh are less sweet and moist than those with red, pink, or orange flesh. They also have little or no beta-carotene and higher levels of dry matter, which means their textures are drier and mealier and they stay firmer when cooked. The orange-and red-fleshed forms of sweetpotato are particularly high in beta-carotene, the vitamin A precursor. In developed countries, where the sweetpotato is used more as a vegetable or for sweet dishes, the red-or orange-fleshed types are preferred for their moist flesh and sweet flavor.

It is well suited to survive in fertile tropical soils and to produce tubers without fertilizers and irrigation. Once established, apart from occasional weeding, sweetpotatoes need little care. Sweetpotatoes are fast growing and rapidly produce ground cover to prevent soil erosion (Orno, 1991; Gregory, 1992). They are tolerant to severe weather and can be used as relay crops in various cropping systems (Gregory, 1992), however, the plant does not tolerate frost. Sweetpotatoes can also be grown in the summer in temperate regions as long as they have at least five frost-free months combined with fairly warm days and nights.

Many villages in East Africa depend on sweetpotato for food security and the Japanese used it when typhoons demolished their rice fields. Yields differ greatly in different areas or even fields in the same location. These differences in yields are mainly due to variation in quality of the propagation material. Sweetpotatoes are vegetatively propagated from vines, root slips (sprouts) or tubers, and farmers in African and other countries often take vines for propagation from their own fields year after year. Thus, if virus diseases are present in the field, they will inevitably be transmitted with the propagation material to the newly planted field, resulting in a decreased yield. Often these fields are infected with several viruses, thereby compounding the effect on yields. In China, on average, losses of over 20% due to sweetpotato virus diseases are observed (Gao et al., 2000), mainly due to *Sweet potato feathery mottle virus* (SPFMV) and *Sweet potato latent virus* (SwPLV). The infection rate in the Shandong province reaches 5–41% (Shang et al., 1999). In countries, where care is taken to provide virus-tested planting material as, amongst others in the US and Israel, yields increase markedly, up to 7 times and more. Attempts are being made to incorporate resistance genes by genetic engineering-but so far without success.

Sweetpotato roots can be boiled, steamed, baked, and fried. They are also canned or dried and made into flour, cereal, and noodles. Like pumpkins, sweetpotato roots are often used in sweet dishes, such as pies, puddings, biscuits, cakes, and desserts. In some countries roots are processed to produce starch and fermented to make alcohol. Cooked red-or orange-fleshed sweetpotato roots are sweet, soft, and starchy with a flavor that resembles roasted chestnuts and baked squash. Sweetpotatoes are prepared by scrubbing and cutting into appropriately sized pieces. Once the roots are cooked, they can be served whole or peeled and mashed, pureed, or sieved and served as a sweet or savory vegetable, depending on what is added. In the United States most of the sweetpotato crop is canned. These are usually the smaller roots. Roots of good size are sold fresh, and any that are too large are generally processed into baby food. In the United States sweetpotatoes are probably best known for their use in pies and as a candied vegetable. They are a traditional accompaniment to "Thanksgiving dinner" and often appear on the menu at other festival times, such as Christmas and Easter. Sweetpotatoes can be substituted for potatoes, apples, or squash in almost any recipe. Cooked, mashed sweetpotatoes are also used to replace some of the wheat flour in breads, cakes, muffins, and cookie recipes, as is sweet-potato flour. Sweetpotatoes cooked in their skins can be frozen. Sweetpotatoes-particularly the pale variety-can be substituted for regular potatoes in most recipes. They can be prepared in a variety of ways including baking, boiling and sautéing. Sweetpotato chips can now be found on some restaurant menus. Fresh sweetpotatoes are available sporadically throughout the year, though not as readily during the summer months. Canned and frozen sweetpotatoes are available year-round and are sometimes labeled as yams.

In third world countries sweetpotatoes are processed into starch, noodles, candy, desserts, and flour. This allows the farm household to extend the availability of the crop. In China, for example, sweetpotato starch production has become an important cottage industry, while in Uganda sweetpotatoes are sliced and dried, which allows them to be kept for about five months. The dried pieces are also ground into flour, which is then rehydrated and eaten as a thick porridge known as *atapa*. All parts of the sweetpotato are used as stock feed.

In eastern and southern Africa some 3 million children under the age of five suffer from *xerophthalmia* or dry eye, which causes blindness. Due to lack of vitamin A in the diet, many of the affected children die within a few months of becoming blind. The yellow-and orange-fleshed varieties of sweetpotatoes are high in beta-carotene, which can be converted into vitamin A in the intestines and liver. It has been shown that even small amounts of these sweetpotatoes as a regular part of the diet will eliminate vitamin A deficiency in adults and children. African countries have traditionally grown white-fleshed sweetpotatoes, which are low in vitamin A.

Sweetpotato roots and leaves are used in folk remedies to treat illnesses as diverse as asthma, night blindness, and diarrhea. Easily digestible, they are good for the eliminative system. It is believed they bind heavy metals, so they have been used to detoxify the system.

In South America, the juice of red sweetpotatoes is combined with limejuice to make a dye for cloth; every shade from pink to purple to black can be obtained (Verrill, 1937).

The United States is also exploring the potential of sweetpotato products. A patent was granted for the production of bread made from 100 percent sweetpotato flour. It is hoped that these products will appeal to consumers who are allergic to grain breads and flours. Also scientists at two different institutes in the United States have developed genetically modified sweetpotatoes containing edible vaccines. One of these vaccines works against hepatitis B and the other against the Norwalk virus found in food that has not been handled or stored correctly. Edible vaccines such as these may provide cheap protection for some of the poorest people in the world (Answers.com, 2008).

The inclusion of sweetpotato by NASA as a food for long term space missions-because of its nutrition, versatility and growth habits-has created new opportunities for research in Controlled Ecological Life support Systems (Hill et al.,1992).

Recently the uses of sweetpotatoes were diversified beyond their classification as subsistence, food security, and famine-relief crops. In particular the last decade of the century saw a concentrated, coordinated effort to fully realize the potential of this crop. Hopefully millions of subsistence landholders in Africa, Asia, and Latin America will be able to use sweetpotatoes for food, stock food, and processed products and to generate income.

References

Answers.com. 2008. www.answers.com/sweet potato.

FAOSTAT 2006. FAO Statistical Databases. URL http://apps.fao.org/

Gao, F., Gong, Y.F. and Zhang, P.B. 2000. Production and development of virus-free sweet potato in China. *Crop Protect*. 19: 105–111.

Gregory, P. 1992. Feeding tomorrow's hungry: The role of root and tuber crops. In: *Sweet potato technology for the 21st century*, pp.: xxxvii–xxxviii. Walter A. Hill, Conrad K. Bonsi and Philip A. Loretan (eds), Tuskegee University, Auskegee, Alabama. 607 pp.

Hill, W.A., Bonsi, C.K. and Loretam, P.A. 1992. Sweetpotato research: Current status and future needs. In: *Sweet potato technology for the 21st century*, pp.: xvii–xxv. Walter A. Hill, Conrad K. Bonsi and Philip A. Loretan (eds), Tuskegee University, Auskegee, Alabama, 607 pp.

Orno, J. 1991. Farmer perception on economic and environmental importance of sweet potato. *Proceedings of the 2nd Annual UPWARD International Conference*, University of the Philippines at Los Banos, Laguna, Philippines.

Scott, G.J., Best, R., Rosegrant, M. and Bokanga, M. 2000a. *Roots and tubers in the global food system: A vision statement for the year 2020*. International Potato Center, Lima, Peru, 111 pp.

Scott, G.J., Rosegrant, M.W. and Ringler, C. 2000b. *Roots and tubers for the 21st century. Trends, projections, and policy options. Food, agriculture, and the environment*, Discussion paper 31. International Food Policy Research Institute and Centro Internacuional de la Papa, 64 pp.

Shang, Y.F., Yang, C.L., Zhao, J.H., Li, C.S., Lu, X.B., Wang, S.J. and Xin, X.Q. 1999. Progress in the researches and application of virus-free sweet potato in Shandong province. *Research progress in plant protection and plant nutrition* 1999, pp. 104–108. China Agricultural Press, Beijing, China.

Verrill, A.H. 1937. *Foods America gave the world*. Boston: L.C. Page & Co.

Chapter 2
Origin, Distribution and Economic Importance

G. Loebenstein

Origin

The Sweetpotato (*Ipomoea batatas* (L.) Lam.) (Batatas an Arawak name) was originally domesticated at least 5000 years ago in tropical America (Austin, 1988; Yen, 1982). Based on the analysis of key morphological characters of sweetpotato and the wild Ipomoea species, Austin (1988) postulated that sweetpotato originated in the region between the Yucatan Peninsula of Mexico and the Orinoco River in Venezuela. Using molecular markers the highest diversity was found in Central America, supporting the hypothesis that Central America is the primary center of diversity and most likely the center of origin of sweetpotato (Huang and Sun, 2000; Zhang et al. 2000). Columbus in1492 brought it to Europe and Portuguese explorers of the sixteenth century took it to Africa, India, Southeast Asia and the East Indies. Spanish ships brought sweetpotatoes from Mexico to the Philippines in the 16th century. Introduction of the sweetpotato to the Pacific islands apparently occurred in prehistoric times (Yen, 1982). Fossil carbonized storage roots of sweetpotato found in northern New Zealand have been dated back some 1000 years (Yen, 1991), which strongly supports the theory of prehistoric transfer, probably (though with some controversy) by Peruvian or Polynesian voyagers (Yen, 1982). The linguistic links between the Quechua and Polynesian names for sweetpotato, support the Peruvian origin and human transfer of the Polynesian sweetpotato. However, studies based on molecular markers showed that Peruvian sweetpotatoes are not closely related to those from Papua New Guinea (Zhang et al. 1998) and are also different from those of Mesoamerica (Zhang et al. 2000). It was suggested that the Oceania sweetpotato probably came from Central America, through non-human dispersal (Rossel et al. 2001).

To China the crop was apparently introduced in the late 16th century by ship from Luzon in the Philippines to Fujian of China (O'Brien, 1972). The year 1594 was a famine year, and a huge area of crops was destroyed. The governor of Fujian ordered

G. Loebenstein (✉)
Agricultural Research Organization, P.O. Box 6, Bet Dagan, 50250, Israel
e-mail: Gad-talma@barak.net.il

G. Loebenstein, G. Thottappilly (eds.), *The Sweetpotato*,
DOI 10.1007/978-1-4020-9475-0_2, © Springer Science+Business Media B.V. 2009

farmers to grow sweetpotatoes extensively, in order to stave off famine (Zhang et al. 2009). Other data give information that sweetpotato was introduced to China from Vietnam, India and Burma. Extension of sweetpotato inside China was apparently from south to east along the coast, and from south to north through the Yangtze River and the Yellow River valleys (Anonymous, 2003).

Distribution and Economic Importance

Sweetpotato is the seventh most important food crop in the world in terms of production. They are grown on about 9 million hectares, yielding *c*- 140 million tons, with an average yield of about 14 ton/ha (FAOSTAT, 2001). They are mainly grown in developing countries, which account for over 95% of world output. Roughly 80% of the world's sweetpotatoes are grown in Asia, about 15% in Africa, and only 5% in the rest of the world. The cultivated area of sweetpotato in China, about 6.6 million ha, accounted for 70% of the total area of sweetpotato cultivation in the world. China produces about 100 million tons, circa 70% of the total world production. Vietnam is the second largest producer. In China 70% of sweetpotato output goes to animal feed, principally to pigs. In Vietnam also sweetpotato production is substantially linked to pig production (Bottema, 1992). In addition to China and Vietnam, sweetpotato-pig systems play an important role in the rural economy of many parts of Asia: the Philippines, India, Korea, Taiwan, some eastern islands of Indonesia (Bali and Irian Jaya), and Papua New Guinea, while these systems are also practiced, to a lesser extent, in Latin America and Africa (Scott, 1991). In Brazil, it is the fourth most consumed vegetable, a food of high-energy value, rich in carbohydrates. It also supplies reasonable quantities of vitamins A, C and some of the B complex (Miranda, 2002). In the US about 37 000 hectares are grown, producing ~720 000 tons per year. In Africa sweetpotato is a 'poor man's crop', (and traditionally considered as "women's work") with most of the production done on a small or subsistence level. Sweetpotato produces more biomass and nutrients per hectare than any other food crop in the world. It is well suited to survive in fertile tropical soils and to produce tubers without fertilizers and irrigation and is one of the crops with a unique role in relief famine. Thus, for example, across East Africa's semiarid, densely populated plains, thousands of villages depend on sweetpotato for food security and the Japanese used it when typhoons demolished their rice fields. Sweetpotato is grown for both the leaves, which are used as greens, and the tubers, for a high carbohydrate and beta-carotene source. Yields differ greatly in different areas or even fields in the same location. Thus, the average yield in African countries is about 4.7 tons/ha, with yields of 8.9, 4.3, 2.6 and 6.5 tons/ha in Kenya, Uganda, Sierra Leone and Nigeria, respectively. The yields in Asia are significantly higher, averaging 18.5 tons/ha. China, Japan, Korea, Thailand and Israel have the highest yields with about 20, 24.7, 20.9, 12 and 33.3 tons/ha, respectively. In South America the average yield is 12.2 tons/ha, with Argentina, Peru and Uruguay in the lead with 18, 11 and 10 tons/ha, respectively. For comparison, the average yield in the US is 16.3 tons/ha (all data are averages for 2000 and 2001 from the FAOSTAT 2002).

These differences in yields are mainly due to variation in quality of the propagation material. Sweetpotatoes are vegetatively propagated from vines, root slips (sprouts) or tubers, and farmers in African and other countries often take vines for propagation from their own fields year after year. Thus, if virus diseases are present in the field they will be inevitable transmitted with the propagation material to the newly planted field, resulting in a decreased yield.

In the US consumption of the crop has increased slightly in recent years. The increased consumption is largely due to national advertising campaigns focused on the nutritional aspects of sweetpotatoes aimed at health conscious consumers. The sweetpotato is very high in nutritive value, and merits wider use on this account alone. Sweetpotatoes rank as one of the healthiest vegetables, because of high levels of vitamins A and C, iron, potassium, and fiber. They are also an excellent source of the vitamin A precursor, beta-carotene. One cup of the orange flesh types contains four times the recommended daily allowance of this important nutrient. It is expected that with improved curing and storage procedures and development of fresh shipping methods (wrapping of individual tubers in plastic ready for microwaving) and processed products such as sweetpotato chips, fries, and canned sweetpotatoes will enlarge the market.

References

Anonymous, 2003. Select Markets for Taro, Sweet Potato and Yam. Rural Industries Research & Development Corporation, BIRDC Publication No 03/052, Australia.

Austin, D. F. 1988. The taxonomy, evolution and genetic diversity of sweet potatoes and related wild species. In P. Gregory (ed.) *Exploration, Maintenance and Utilization of Sweet Potato Genetic Resources*. CIP, Lima, Peru, pp. 27–60.

Bottema, T. 1992. Rapid market appraisal, issues and expereince with sweetpotato in Vietnam. In G. Scott, S. Wiersema, and P. I. Ferguson (eds.) *Product Development for Root and Tuber Crops, Vol. 1- Asia*. International Potato Center, Lima, Peru.

FAOSTAT 2001, 20\02, http://faostat.fao.org/

Huang, J. C. and Sun, M. 2000. Genetic diversity and relationships of sweetpotato and its wild relatives in *Ipomoea* series *batatas* (Convolvulaceae) as revealed by inter-simple sequence repeat (ISSR) and restriction analysis of chloroplast DNA. *Theor. Appl. Genet.* 100:1050–1060.

Miranda J. E. C. 2002. Batata-doce. Available from: www.cnph.embrapa.br/cultivares/bat-doce.htm. Date accessed 28 November 2005.

O'Brien P. 1972. The sweet potato: its origin and dispersal. *Am. Anthropol.* 74:342–365.

Rossel, G., Kriegner, A. and Zhang, D. P. 2001. From Latin America to Oceania: The Historic. Dispersal of Sweetpotato Re-examined Using AFLP. *CIP Program report 1999–2000*. pp. 315–321. www.cipotato.org/publications/program_reports/99_00/39historic.pdf

Scott, G. 1991. Sweet potato as animal feed in developing countries: present patterns and future perspectives. Paper presented at the FAO Experts Consultation on *The Use of Roots, Tubers, Plantains and Bananas in Animal Feeding* held at the Centro International de Agricultura Tropical (CIAT), Cali, Colombia, 21–25 January, 1991.

Yen, D. E. 1982. Sweet potato in historical perspective. In R. L. Villareal and T. D. Griggs (eds.) *Sweet Potato: Proceedings of the First International Symposium*. AVRDC, Tainan, Taiwan, pp. 17–30.

Yen, D. E. 1991. Domestication: The lessons from New Guinea. In A. Pawley, (ed.) *Man and a Half: Essaysin Pacific Anthropology and Ethnobiology in Honour of Ralph Bulmer*. The Polynesian Society, Auckland, pp. 558–569.

Zhang, D. P., Ghislain, M., Huaman, Z., Golmirzaie, A. and Hijmans, R. 1998. RAPD variation in sweetpotato (*Ipomoea batatas* (L.) Lam) cultivars from South America and Papua New Guinea. *Genet. Resour. Crop Evol.* 45:271–277.

Zhang, D., Cervantes, J., Huaman, E., Carey, E. and Ghislain, M. 2000. Assessing genetic diversity of sweet potato (Ipomoesa batatas (L.) Lam.) cultivars from tropical America using AFLP. *Genet. Resour. Crop Evol.* 47:659–665.

Zhang, L., Wang, Q., Liu, Q. and Wang, Q. 2009. Sweetpotato in China. In G. Loebenstein and G. Thottappilly (eds.) *The Sweetpotato*. Springer Dordrecht, The Netherlands, This volume.

Chapter 3
Botany and Physiology: Storage Root Formation and Development

N. Firon, D. LaBonte, A. Villordon, C. McGregor, Y. Kfir and E. Pressman

Introduction

Sweetpotato is a member of the family Convolvulaceae, Genus *Ipomoea*, section Batatas. It is the only hexaploid (6x = 90) in this section. Section Batatas continues toundergo revision but contains approximately 12 other species, most of which are diploid (2n = 30), with a few tetraploids (4x = 60) collected in the wild (Bohac et al., 1993). Jarret et al. (1992) supported a hypothesis that *Ipomoea trifida* (H.B.K.) G. Don and wild Mexican tetraploids are putative ancestors of the cultivated sweetpotato. McDonald and Austin (1990) described several section batatas species (*Ipomoea batatas* var. *apiculata*) found in Mexico with putative linkages to domesticated sweetpotato. Genetic diversity studies by Zhang et al. (2001) found that Mesoamerica (Guatemala, Mexico, Nicaragua, Panama, El Salvador) possessed the most allelic diversity and hence warrants consideration as the primary source of genetic diversity in sweetpotato. Species in section Batatas have been shown to contain unreduced gametes (Bohac and Jones, 1994) but the derivation of the hexaploid sweetpotato remains unknown. Different theories of the evolution of *I. batatas* have been advanced. Shiotani (1988) considered it an autopolyploid derivative of *I. trifida* while others favor an allopolyploid origin involving *I. trifida* and an unidentified tetraploid parent. Tetrasomic segregation patterns in microsatellite primers are consistent with an allopolyploid (Buteler et al., 1999). Zhang et al. (2001) also found tetrasomic patterns of inheritance in sweetpotato. The most compelling evidence of an allopolyploid origin is cytological data reported by Magoon et al. (1970). Meiotic patterns showed various pairing configurations among the chromosomes, e.g., duplicate, triplicate and quadrivalents. Their data suggests that *Ipomoea batatas* consists of two related and a third more distant genome.

Despite the complex and still unresolved origin, sweetpotato has become an important global crop and is grown in diverse ecologies. It is considered as a subsistence or security crop in developing countries, where it is often grown under

N. Firon (✉)
Institute of Plant Sciences, The Volcani Center, Agricultural Research Organization, Bet Dagan, 50250, Israel
e-mail: vcfiron@volcani.agri.gov.il

G. Loebenstein, G. Thottappilly (eds.), *The Sweetpotato*,
DOI 10.1007/978-1-4020-9475-0_3,

marginal conditions. In developed countries, the crop is commercially grown as a high value vegetable crop under intensively managed production systems. All producers are challenged to produce high yields and storage roots, which are uniform in size and attractive for markets. Unfortunately, storage root production is inconsistent from plant to plant; some plants have few or no storage roots while others have four or more marketable ones. However, the underlying genetic mechanisms and the factors that promote storage root formation, the most important physiological process in sweetpotato, are not clearly understood.

Storage roots normally arise from the underground portion of a vine cutting that is being used as planting material (Fig. 3.1). Initially, white adventitious roots develop, and some of these roots subsequently undergo sudden changes in their growth pattern and develop into storage roots. Depending on the number of fibrous roots that will be induced to form storage roots, sweetpotato plants will yield either a high number (4 to 6/plant) of uniform and high- grade roots, e.g., U.S. No 1 grade (elliptical roots 8–23 cm in length and 5–9 cm in diameter), or a low number of roots that may even be reduced to one very large storage root per plant or no marketable roots at all. The wide variability in storage root yield among sweetpotato cultivars, and individual plants of the same cultivar, has been attributed to cultivar, propagation material, environmental and soil-factors (Togari, 1950; Lowe and Wilson, 1974). Genetic and environmental factors influence leaf area, leaf production and abscission, leaf photosynthesis, storage root formation and development, total dry matter production and dry matter partitioning (Ravi and Indira, 1999).

The complex interaction between environmental and genetic factors in sweetpotatoes is well documented. Previous reviews by Kays (1985) and Ravi and Indira (1999) covered source and sink relationship, photosynthesis, translocation and respiration, as well as the effect of growth regulators and environmental factors relating to the physiology and yield of sweetpotato. This review will examine anatomical, environmental, and molecular aspects of storage root initiation and development.

Plate 3.1 Adventitious roots development from a 'Georgia-Jet' vine cutting

Induction of Storage Root Formation

Terminology

The economically important portion of the sweetpotato is developmentally and anatomically a true (botanical) root, lacking nodes and associated preformed meristematic tissue. In contrast, potato (*Solanum tuberosum* L.) tubers are thickened underground stems, and therefore possess meristematic buds found at vestigial nodes. Sweetpotato storage roots can initiate adventitious meristematic activity and produce sprouts and roots from the storage organ. The nomenclature that will be used throughout this chapter is: 'fibrous root' for the non-fleshy roots and 'storage roots' for fleshy, edible roots, which arise directly from the node. The term 'tuberous roots' is used by several authors for roots capable of storage root formation (Togari, 1950; Wilson and Lowe, 1973). Two other terms encountered in the literature are 'Lateral root' and 'Pencil root'. 'Lateral root' represents a fibrous root arising adventitiously from another root while 'Pencil roots' are somewhat thickened, lignified, and have a diameter less than 2 cm.

Morphological and Anatomical Cues

Sweetpotatoes develop from adventitious roots that develop from preformed root primordia that are present in sets of 4–10 adjacent to the leaf bases (Belehu et al., 2004). Anatomically, these adventitious roots first develop the primary cambium, between the protophloem and protoxylem. Lignification of the stele is suppressed through the development of the vascular cambium in the parenchymatous zone between the protophloem and protoxylem. Anomalous cambia develop around the central cell and primary xylem elements (primary cambia) and secondary cambia are formed around secondary xylem elements derived from the vascular cambium. Cell division and expansion in these cambia regions lead to rapid thickening of the roots (Wilson and Lowe, 1973). Anatomically, the appearance of anomalous cambia mark the induction phase of storage root formation (Wilson and Lowe, 1973; Wilson et al., 1982). In Fig. 3.2a, a micrograph of a transverse section of a root exhibiting anomalous cambia development, in the leading sweetpotato variety in Israel, 'Georgia-Jet' is presented.

Several types of root including tuberous and non-tuberous roots have been recognized in the adventitious root system of the sweetpotato. Tuberous roots were interpreted by Wilson (1970, as cited in Wilson and Lowe, 1973) as those roots, which were capable of normal tuber (storage root) initiation, i.e. young, thick, primary roots (Togari, 1950) with a pith and centrally located metaxylem element (Artschwager, 1924), pentarch or hexarch steles and enlarged apical meristems (Togari, 1950). Non-tuberous roots, which did not normally develop storage roots included young, thin roots in which the primary tissues presented a central core of xylem with no central pith, a tetrarch stele (Artschwager, 1924) and a small apical

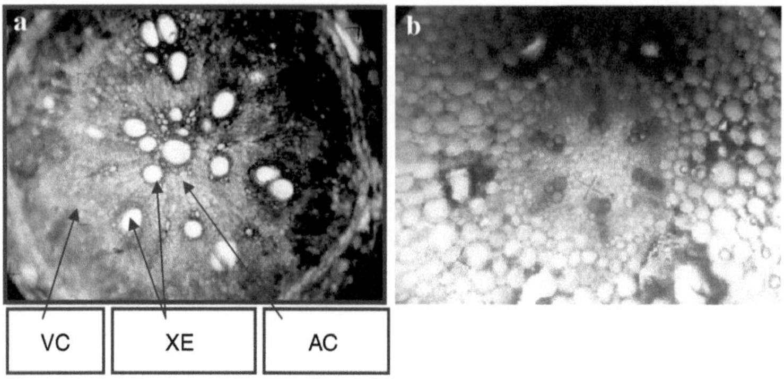

Plate 3.2 Transverse section of a root exhibiting anomalous cambia development (**a**), and the presence of a hexarch stele (**b**) in the leading sweetpotato variety in Israel, 'Georgia-Jet'. VC – vascular cambium; XE – xylem elements; AC – anomalous cambium

meristem (Togari, 1950). In Fig. 3.2b, a micrograph of a transverse section of a root with a hexarch stele in 'Georgia-Jet' is presented. Fibrous roots, i.e. uniformly thickened roots with normal secondary growth leading to complete lignification of the stele (Togari, 1950), were also non-tuberous. Roots in which normal storage root development was halted (Wilson, 1970, as cited in Wilson and Lowe, 1973) lead to the production of intermediate structures, e.g. 'pencil' roots (Togari, 1950).

Comparative analysis of six West Indian sweetpotato cultivars (Lowe and Wilson, 1974) indicated that only roots, which thickened within the first 8 weeks of growth (except in cultivar I62) were capable of realizing their potential for storage root development. Cultivars could be grouped into early storage root initiating types and late initiating types according to the rate of storage root production during the period 4–8 weeks after planting. Differences in yield between the six cultivars resulted largely from differences in bulking rate in the period 16–24 weeks, and particularly in the last 4 weeks of storage root growth. High yields were produced either by a rapid rate of storage root bulking over a short duration or a slower rate of tuber bulking over a longer duration (Lowe and Wilson, 1974).

Togari (1950) proposed a direct link between lignification and storage root initiation, i.e., lignification "prevents" storage root development. The relationship between stele lignification and the inability of pentarch and hexarch adventitious roots to develop into storage roots has also been observed by Wilson and Lowe (1973) and Belehu et al. (2004). Togari (1950) suggested that genetic and environmental factors influence the balance between cambium development and lignification, which in turn will determine to a large degree the final storage root yield. He also suggested that characteristics of the root primordia are altered by environmental factors encountered by the root-tip growing point. If plants are exposed to sun, moisture, nitrogen and potash the number of the primordia with a big root-diameter increases, and moreover, that number is closely related to the number of the storage roots that comes out from this sprout. Adventitious roots from a young healthy sprout have low cambial activity and little lignification. These roots thicken as storage roots.

Old, mature planting stock appears to develop similarly but lignification insues and restricts storage root development (Togari, 1950). High yield varieties tend to have great cambial activity early on and little lignification; low yield varieties do not exhibit this characteristic (Togari, 1950).

Togari (1950) emphasized the influences of environmental conditions during the first 20 days after planting indicating that potash (X4 the amount of standard potash; standard fertilizer being N – 0.5 g, P_2O_5 – 0.5 g, K_2O – 1.2 g per pot. The experiment was conducted in 1/20,000 Wagner's pots) accelerates the cambium activity, optimum temperature (low temperature, maximum of 23.4C during the day) increases it and reduces lignification; both are favorable for the formation of storage roots. Dryness (one third of the water amount supplied) and compactness of soil (25% more soil stuffed tight into pots) increases the cambium activity, but cause an increased lignification leading to a process of development into pencil-like roots. Shade (one sheet of marsh reed screen 60 cm above the plants) reduces both the cambium activity and lignification, and delays development. An overabundance of N (X5 the amount of standard nitrogen) and low soil oxygen (soil was mixed with starch – 10% to the soil weight) decreases cambium activity and increases lignification and make the roots prone to develop into non-storage roots. These changes in the characters of the roots wrought during the 20 days influence the subsequent development of storage roots.

Wilson and Lowe (1973) confirmed Togari's observations that the specialization of the storage root was dependent on the competing processes of lignin biosynthesis or cell division and expansion in the course of storage root ontogenesis. In addition, they indicated that storage root tissue is specialized at the proximal and distal ends. Thus, the tissues of the 'tuber stalk' (proximal end) developed features of a translocatory organ by the differentiation of extensive secondary phloem and phloic vascular bundles. The distal end is responsible for longitudinal expansion of the storage root in early root ontogeny. This later process is poorly understood.

Belehu et al. (2004) suggest the importance of undamaged preformed root primordia on the vines for storage root development. The root primordia originate from the procambium on both sides of the leaf gap (node) and produce adventitious roots, with pentarch, hexarch or septarch steles. Under normal conditions, storage roots will develop from thick roots originating from undamaged preformed root primordia on the nodes of cuttings or nodes of newly formed vines. Wound roots on leaf petioles and stems were also found to have polyarch steles although data were not presented on the percent that become storage roots. Sirju-Charran and Wickman (1988) also note storage root development is not limited to pre-formed root priomordia. As in other clonal cuttings, adventitious roots either arise from preformed root initials on the nodes, or from the cut ends (callus or wound roots). In fields that have been exposed to herbicide damage early during the season, "pencil" or lignified roots can be seen arising from the nodes, while the cut end shows some degree of storage root development. These (mostly anecdotal) observations suggest that once the preformed roots are damaged and the adventitious roots are lignified, there is still some storage root formation possible from the cut ends. From time to time, we also observe these cut ends setting storage roots under "normal" field conditions (A. Villordon, personal communication).

In the case of potato tuber, another well-known storage organ, tuber initiation is induced by short days, high light intensity, and high sucrose levels, and inhibited by high nitrogen levels, high temperatures, and gibberellic acid (Jackson, 1999). However, the fact that the potato tuber develops from the stolon, (underground stem) and not from the root, suggests that distinct induction mechanisms are possibly involved in potato tuber and sweetpotato storage root (You et al., 2003).

Hormones

Past investigations have looked at the influence of hormones on sweetpotato growth and storage root yield. Early work focused on whole plant or organ responses while subsequent research investigated molecular aspects as detailed below.

The effects of benzyl aminopurine (BAP), gibberellic acid (GA3), chloroethyltrimethyl ammonium chloride (CCC) and indol-3yl- acetic acid (IAA) on the growth and storage root production of rooted leaves from a high and low yielding cultivar of sweetpotato were studied by McDavid and Alamu (1980). Enhanced root and storage root growth was detected in response to BAP, while GA3 treatment retarded storage root development.

Cytokinins, particularly t-zeatin riboside (ZR) and abscisic acid (ABA) have been implicated in storage root development in sweetpotato by Nakatani and Komeichi (1991) and Nakatani et al. (2002). A rapid increase in endogenous ZR was found to occur in sweetpotato roots simultaneously with storage root formation. Endogenous ZR was concentrated around the root primary cambium, which plays an important role in storage root formation (McDavid and Alamu, 1980; Nakatani, 1994; Spence and Humphries, 1972). An increase in the level of *t*-ZR, coincided with the greatest rate of storage root formation at 30-to-90 days after planting was demonstrated by Matsuo et al. (1988). The authors suggest that *t*-ZR may play a significant role in storage root development by influencing cell division and subsequent cell enlargement during early storage root development. The longitudinal distribution of *t*-ZR in the developing storage roots showed high levels in the proximal side close to the stem. It is suggested that *t*-ZR localizes at vascular bundles because there are few parenchyma cells in the proximal side. Nakatani et al. (2002) showed that the onset of increased ZR levels was delayed in a late-storage root-forming mutant. Other publications substantiate the role ZR has on storage root ontology (Matsuo et al., 1988; Sugiyama and Hashizume, 1989; Suye et al., 1983). ZR appears associated with the development of primary cambium while ABA is associated with activity of the secondary cambium, after formation of the primary cambium (Nakatani et al., 2002).

Genes coding for Aux/IAA (Auxin/Indole-3-Acetic- Acid) and GH3 family proteins were up-regulated in storage roots, pointing to a possible role for auxin in storage root development (McGregor, 2006). The functions of the different auxin regulated genes in sweetpotato and their possible role in storage root development are unknown. Auxin is known for its acceleration of stem growth however, in roots, auxin strongly inhibits growth over a wide range of concentrations (Tanimoto, 2005). A role for cytokinin and auxin in regulating vascular differentiation was recently demonstrated by Aloni et al. (2006) suggesting that the radial pattern of root protoxylem vs. protophloem strands is induced by altering polar streams of high IAA vs. low IAA concentrations, respectively. Previous studies have reported conflicting results in sweetpotato (Wilson et al., 1982; Nakatani and Komeichi, 1991).

Two gibberellin-responsive GASA protein genes were up regulated in storage roots (McGregor, 2006) in comparison to fibrous roots. Gibberelllin is traditionally associated with cell division and elongation (Harberd et al., 1998). Auxin and GA responsive genes play a role in a modification of the cell wall (Tanimoto, 2005). Other hormones like jasmonic acid and ethylene may be involved in storage root development (McGregor, 2006). Jasmonic acid inhibits root elongation (Staswick et al., 1992; Feys et al., 1994). The stoppage of root elongation has been described during the early stages of storage root development in sweetpotato (Kim et al., 2002a). Until now most research has focused on hormones ABA and cytokinins, however recent results suggest jasmonic acid, ethylene, and auxin deserve attention.

Storage Root Development

Growth of the storage roots occur by continued activity of the vascular cambium and anomalous primary and secondary cambia in storage roots (Wilson and Lowe, 1973). Cambial strips unassociated with vascular tissues also develop within the secondary parenchyma, and contribute to the localized increase in girth in the storage root region of the root. Activity of all cambia, including later-formed tertiary cambia, i.e. cambia developed in tissues derived from secondary cambia, result in the formation of thin-walled, starch-storing parenchyma cells.

Storage root formation is also inextricably linked to the canopy by promoting photoassimilate export and leaf photosynthesis (Keutgen et al., 2002). Storage root growth of sweetpotato clones was shown dependent on sink strength of the developing storage root, the potential of leaves to export photoassimilates, but also on the photosynthetic capacity of leaves based on ^{11}C greenhouse studies. Storage root growth is thus recognized by the canopy and encouraged.

Carbohydrates and Storage Root Development

Recent work has clarified our understanding of storage root carbohydrate physiology and the impact of carbohydrates on storage root formation. Li and Zhang (2003) showed that the sucrose synthase (SuSy) pathway was the predominant pathway for sucrose metabolism in sweetpotato through expression analysis of SuSy, ADP-glucose pyrophosphorylase (AGPase) and invertase. This was later confirmed by McGregor (2006) and extended by two additional genes in the pathway, hexokinase and fructokinase. Fructokinase was elevated in storage roots when compared to fibrous roots, but hexokinase did not differ. An invertase inhibitor like protein was also up regulated in storage roots in comparison to fibrous roots (McGregor, 2006).

Tsubone et al. (2000) investigated the effect of exogenous injection of a sucrose solution into a plant on the activity of AGPase and storage root production. The application of sucrose was found to be effective in increasing storage root production and increasing the ratio of storage root weight to total root weight in a plant. Kwak et al. (2006) showed that two AGPase isoforms are regulated antagonistically in response to endogenous sucrose content in storage roots. The SuSy pathway is also the predominant pathway in sucrose metabolism in potato tubers (Fernie et al., 2002; Geigenberger, 2003; Geigenberger and Stitt, 1991). So, although the storage organs of potatoes and sweetpotatoes have different origins, the stolon and the root respectively, there are similarities between some aspects of their development. This is important since knowledge of potato tuber development is much more advanced (Fernie and Willmitzer, 2001; Jackson, 1999) than that of sweetpotato root development.

In storage roots, the synthesis of granule-bound starch synthase (GBSSI) transcript was found to increase coordinately with storage root expansion: nevertheless, accumulation rates of GBSSI protein in starch granules remained constant regardless of storage root sizes, suggesting an involvement of post-transcriptional regulation for the synthesis of this protein (Wang et al., 1999). Wang et al. (2006) showed that the expression of genes (GBSSI, sporamin, and beta-amylase), starch and sucrose content were highly correlated with the size of storage roots suggesting that fresh weight should be more suitable than growing-day to serve as an indicator for defining developmental stages. Increase in sucrose concentration during development might play an important role in starch accumulation; not only it functions as a substrate for starch synthesis but it can also stimulate transcription of granule-bound starch synthase gene.

Kim et al. (2002b) examined white fibrous roots, thick-pigmented roots and lateral roots for AGPase and chalcone synthase (CHS) expression. These two genes were expressed only in thick-pigmented roots after 3 weeks (Kim et al. 2002b). Based on the analysis of AGPase and CHS expression in different kinds of root tissues and in different developmental stages, these genes were shown to be closely associated with the differentiation of thickening pigmented roots. Expression of both AGPase and CHS was found to start at 2 weeks after planting and highly increase at 3 weeks (Kim et al., 2002b).

An invertase inhibitor like protein was also up-regulated in storage roots in comparison to fibrous roots (McGregor, 2006). Sucrase was also up-regulated in storage roots. The role of this later gene in the conversion of sucrose to starch in storage roots is unclear. The up-regulation of sucrose phosphate synthase, an enzyme that together with sucrose phosphate phosphatase leads to the synthesis of sucrose from UDP-glucose and fructose-6-phosphate, might seem counter-intuitive. However, sucrose synthesis has been reported in potato tubers via both SuSy and sucrose phosphate synthase (Fernie et al., 2002).

After storage root initiation, development of the roots depends on the increase in both the number and size of the cells in the stele and the development of starch granules in the cells (Hahn and Hozyo, 1984). Starch granules themselves vary in size, depending on whether they are located near the anomalous cambium compared to those localized in the outer layer, as reported by Chang et al. (2000). This observation may imply that the starch accumulation signal is triggered in the cells near the anomalous cambium, and subsequently differentiated into the active starch biosynthesizing cells. Therefore anomalous cambium cells around xylem vessels in sweetpotato roots might play a significant role in starch metabolism and further storage root development.

Skin Formation

The periderm in sweetpotato is composed of three layers: phellem, phelloderm, and phellogen. The phellogen originates the phelloderm toward the inside of the root and the phellem toward the outside. The phellem is composed of several layers of cells devoid of starch granules and positioned in a radial manner toward the phellogen. The phelloderm is more difficult to identify as it resembles cortical cell and it usually is distinguished by the radial position of the cells in reference to the phellogen (Kono and Mizoguchi, 1982 as cited by Villavicencio et al. (2007)).

The outer peridermal cells of sweetpotato become partly lignified during growth and are progressively sloughed off. The phellogen layer, however, remains active until harvest, and its activity causes the thickness of the periderm layer to remain constant (Artschwager and Starrett, 1931 as cited by Villavicencio et al. (2007)).

One of the problems in post-harvest handling of sweetpotato roots is the loss of the epidermis/skin from the surface of the root, referred to as "skinning." During skinning, the superficial layers of the periderm separate from underlying tissue; leading to an increased rate of moisture loss, weight loss, shriveling of the root surface, increased susceptibility to pathogen attack, and unattractive appearance. In North Carolina, and in the United States in general, one of the most popular cultivars is 'Beauregard' due to its smooth skin, color, and high yields. This cultivar, however, is very prone to skin loss during harvest and handling.

Villavicencio et al. (2007) suggested that because skin loss occurs due to breakage of the cell walls of peridermal cells, lignin content and polygalacturonase (PG) and pectin methylesterase (PME) activities could be part of the processes or factors that lead to increased or decreased skin adhesion in sweetpotato roots. If this is the case, skin adhesion might correlate with these variables, making it possible to use them as indicators of susceptibility to skinning.

The periderm of roots grown at higher temperatures was thicker and had more layers than that of roots grown at lower temperatures. Histochemical studies of the periderm of sweetpotato showed that the anatomical and structural composition of the cell walls differ depending on growth temperature (Villavicencio et al., 2007).

Pre-harvest canopy removal was demonstrated by LaBonte and Wright (1993) to linearly reduce skinning damage in 'Beauregard' and 'Jewel' varieties combined. Skinning damage was reduced by 62% when the canopy was mowed 10 days pre-harvest, 53% when mowed 8 days pre-harvest and 26% at 4 days, when compared to 0 days pre-harvest.

Environmental Effects

The establishment stage of both sweetpotato and another root crop cassava (*Manihot esculenta* Crantz) (about 4 weeks and 10–12 weeks in sweetpotato and cassava, respectively) is considered to be the most critical phase in the growing cycle of these crops because their state of development during this time largely determines the later growth of the crops and their concomitant yield (Pardales and Esquibel, 1996). Fluctuations in the amount of water in the soil brought about variable development in the individual component roots in a root system, with deficient moisture as the most depressive, even though the plants had been exposed previously to well watered conditions (Pardales and Yamauchi, 2003).

Bhattacharya et al. (1985) demonstrated that elevated CO_2 concentrations stimulated the production of roots from stem cuttings of 'Georgia Jet' in the presence of IAA. This effect was more pronounced at 675 ppm than at 1000 ppm CO_2 atmosphere. The authors suggest that the stimulation in the number and length of roots in water, and more profusely in the presence of IAA in the medium at 675 ppm CO_2, indicates that the level of carbohydrates produced in leaves was adequate to balance most of the applied auxin concentrations this resulted in stimulation of differentiation of cambial derivatives into root primordia and the enlargement of cells.

Thompson et al. (1992) showed that marketable yields increased with applied irrigation amounts until a total water application of 76% of pan evaporation (Epan) was reached and then decreased rapidly with applied irrigation amounts. Weight loss and decay of roots during storage showed quadratic responses to irrigation amounts and were minimal at the irrigation level of maximum yields. Contents of dextrins and maltose increased with irrigation amounts. Glucose content was maximal at a total water amount of 94% Epan and fructose content decreased with increased amounts of irrigation. Sensory ratings for appearance, flavor, texture, and preference, and objective color measurements of cooked flesh also reached their highest values near the irrigation amount of maximum yield.

Genes Involved in Induction of Storage Root Formation

Recently, gene expression studies have been used in an effort to elucidate factors involved in storage root formation. You et al. (2003) constructed a cDNA library of early-stage storage roots, and identified 22 genes differentially expressed between fibrous roots (non-storage and storage stage) and storage roots. Among these were a no apical meristem (NAM)-like and a MADS-box (MCM1, AGAMOUS, DEFICIENS and SRF) protein gene, both of which were down-regulated in storage roots. Kim et al. (2002a; 2005a, b) also identified MADS-box genes that showed high expression in vegetative tissue, particularly root tissue. McGregor identified another MADS-box gene similar to Kim et al. (2005a). Tanaka et al. (2005) identified 10 genes with differential expression among fibrous, thick, and storage roots. One of the genes, SRF6 encoded a receptor-like kinase with high expression around the primary cambium, and xylem meristem. McGregor (2006) also identified several histone H3 genes involved in cell replication and a homeobox gene was found to be homologous to the tomato VAHOX1 gene (Tornero et al. 1996). In root, VAHOX1 is expressed where secondary growth takes place and at the connection point of the secondary roots. In transgenic tobacco, this promoter was activated in the phloem region at the onset of secondary growth.

McGregor (2006) found several NAC family transcriptional regulator proteins that were down regulated in storage roots, similar to a NAM-like protein described by You et al. (2003). One of the up-regulated genes in storage roots is complementary to an *A. Thaliana* gene in the SENU5 subgroup of the NAC family. This gene is found in tomato roots (John et al., 1997). Two NAM like genes were up regulated in storage roots (McGregor, 2006). NAM genes are generally involved in shoot apical meristem development (Souer et al., 1996).

Identification and characterization of genes that are involved in storage root induction and development will provide a better understanding of the pathways involved in this developmental process and has the potential to make it possible in future to manipulate specific genes that will enable enhancement of yield

Conclusions

Sweetpotato storage root formation involves a number of distinct processes including the induction of anomalous cambial cell formation, cell divisions, amylopasts' biogenesis and starch accumulation, is controlled by endogenous factors such as the hormones cytokinins and ABA and was shown to be affected by external factors such as water availability, temperature, and nutrients. External and internal stimuli interact to determine whether an adventitious root becomes a fibrous root, a storage root or an intermediate structure, i.e., a pencil root. Data is lacking on the involvement of cafffeic acid and peroxidase on lignin biosynthesis, a prime anatomical indicator of a root destined to become a fibrous or pencil root. Little information is available, however, with respect to the molecular mechanisms underlying these processes and the cross-talk between the SR- inhibitory processes of stele lignification versus the SR-inducing process of continued anomalous cambial cells proliferation. Identification and characterization of genes that are involved in SR initiation and those involved in the control of lignification will provide data that will enable enhancement of SR formation and yield by improving breeding efficiency, and enable genetic modification, including the generation of the appropriate transgenic plants.

There are still limited data concerning SR initiation-specific genes and overall genome sequence information. The generation of additional sweetpotato sequencing data will enable the generation of a microarrays that will include better representation of SR-expressed genes. Additional experiments, focusing on transcriptome, proteome and metabolome of SR initiation and development will contribute to the identification and characterization of the relevant genes and to our understanding of the factors that control this important process.

Acknowledgments This chapter was written as part of a collaboration between the US and Israeli authors being supported by the United States-Israel Binational Agricultural Research and Development Fund (grant no. US-4015-07).

References

Aloni, R., Aloni, E., Langhans, M. and Ullrich, C.I. 2006. Role of cytokinin and auxin in shaping root architecture: regulating vascular differentiation, lateral root initiation, root apical dominance and root gravitropism. *Ann. Bot.* 97:883–893.

Artschwager, E. 1924. On the anatomy of sweet potato root with notes on internal breakdown. *J. Agric. Res.* 23:157–166.

Belehu, T., Hammes, P.S. and Robbertse, P.J. 2004. The origin and structure of adventitious roots in sweet potato (*Ipomoea batatas*). *Austr. J. Bot.* 52:551–558.

Bhattacharya, N.C. and Strain, B.R. 1985. Rooting of sweet potato stem cuttings under CO2-enriched environment and with IAA treatment. *HortScience* 20:1109–1110.

Bohac, J., Austin, D. and Jones, A. 1993. Discovery of wild tetraploid sweet-potatoes. *Econ. Bot.* 47:193–201.

Bohac, J. and Jones, A. 1994. Unreduced pollen in hexaploid sweet-potato (Ipomoea batatas). *J. Hered.* 85:162–166.

Buteler, M., Jarret, R. and La Bonte, D. 1999. Sequence characterization of microsatellites in diploid and polyploid *Ipomoea*s. *Theor. Appl. Genet.* 99:123–132.

Chang, S-C., Lin, P-C, Chen, H-M, Wu, J-S. and Juang, R-H. 2000. The isolation and characterization of chaperonin 60 from sweetpotato roots – Involvement of chaperonins in starch. *Bot. Bull. Acad. Sinica* 41:105–111.

Fernie, A.R. and Willmitzer, L. 2001. Molecular and biochemical triggers of potato tuber development. *Plant Physi.* 127:1459–1465.

Fernie, A.R., Willmitzer, L. and Tretheway, R.N. 2002. Sucrose to starch: a transition in molecular plant physiology. *Trends Plant Sci.* 7:35–41.

Feys, B.F., Benedetti, C.E, Penfold, C.N. and Turner, J.G. 1994. *Arabidopsis* mutants selected for resistance to the phytotoxin coronatine are male sterile, insensitive to methyl jasmonate, and resistant to a bacterial pathogen. *Plant Cell.* 6:751–759.

Geigenberger, P. 2003. Regulation of sucrose to starch conversion in growing potato tubers. *J. Exp. Bot.* 54:457–465.

Geigenberger, P. and M. Stitt. 1991. Sucrose synthase catalyses a readily reversible reaction in vivo in developing potato tubers and other plant systems. *Planta* 189:329–339.

Hahn, S.K. and Hozyo, Y. 1984. Sweetpotato. *In*: P.R. Goldsworthy and N.M. Fisher (eds.). *The physiology of tropical field crops*. Wiley, New York, pp. 551–567.

Harberd, N.P., King, K.E., Carol, P., Cowling, R.J., Peng, J. and Richards, D.E. 1998. Gibberellin inhibitor of an inhibitor of ? *Bioessays* 20:1001–1008.

Jackson, S.D. 1999. Multiple signaling pathways control tuber induction in potato. *Plant Physiol.* 119:1–8.

Jarret, R.L., Gawel, N. and Whittemore, A. 1992. Phylogenetic-relationships of the sweet-potato [*Ipomoea-batatas* (L.) lam]. *J. Am. Soc. Hort. Sci.* 117:633–637.

John, I., Hackertt, R., Cooper, W., Drake, R., Farrell, A. and Grierson, D. 1997. Cloning and characterization of tomato leaf senescence-related cDNAs. *Plant Mol. Biol.* 33:641–651.

Kays, S.J. 1985. The physiology of yield in the sweetpotato. *In*: J.C. Bouwkamp (ed.). *Sweet potato products: a natural resource for the tropics*. CRC Press. Boca Raton, Fl., pp. 79–132.

Keutgen, N., Mukminah, F. and Roeb, G.W. 2002. Sink strength and photosynthetic capacity influence tuber development in sweet potato. *J. Hort. Sci. Biotechnol.* 77, 106–115.

Kim, S-H., Hamada, T., Otani, M. and Shimada, T. 2005a. Isolation and characterization of MADS box genes possibly related to root development in sweetpotato (*Ipomoea batatas* (L.) Lam.). *J. Plant Biol.* 48:387–393.

Kim, S-H., Hamada, T., Otani, M. and Shimada, T. 2005b. Cloning and characterization of sweetpotato MADS-box gene (*IbAGL17*) isolated from tuberous root. *Plant Biotechnol.* 22:217–220.

Kim, S-H., Mizuno, K. and Fujimura, T. 2002a. Isolation of MADS-box genes from sweet potato (*Ipomoea batatas* (L.) Lam.) expressed specifically in vegetative tissues. *Plant Cell Physiol.* 43:314–322.

Kim, S-H., Mizuno, K. and Fujimura, T. 2002b. Regulated expression of ADPglucose pyrophosphorylase and chalcone synthase during root development in sweet potato. *Plant Growth Regul.* 38:173–179.

Kwak, M.S., Noh, S.A., Oh, M-J., Huh, G.H., Kim, K-N., Lee, S.W., Shin, J.S. and Bae, J.M. 2006. Two sweetpotato ADP-glucose pyrophosphorylase isoforms are regulated antagonistically in response to sucrose content in storage roots. *Gene* 366:87–96.

LaBonte, D.R. and Wright, M.E. 1993. Image analysis quantifies reduction in sweetpotato skinning injury by preharvest canopy removal. *HortScience* 28:1201.

Li, X-Q. and Zhang, D. 2003. Gene expression activity and pathway selection for sucrose metabolism in developing storage root of sweetpotato. *Plant Cell Physiol.* 44:630–636.

Lowe, S.B. and Wilson, L.A. 1974. Comparative analysis of tuber development in six sweet potato (*Ipomoea batatas* (L.) Lam.) Cultivars. 1. Tuber initiation, tuber growth and partition of assimilate. *Ann. Bot.* 38:307–17.

Magoon, M., Krishnan, R. and Vijaya, K. 1970. Cytological evidence on the origin of sweet potato. *Theor. Appl. Genet.* 40:360–366.

Matsuo, T., Mitsuzono, H., Okada, R. and Itoo, S. 1988. Variations in the levels of major free cytokinins and free abscisic acid during tuber development of sweet potato. *J. Plant Growth Regul.* 7:249–258.

McDavid, C.R. and Alamu, S. 1980. The effect of growth regulators on tuber initiation and growth in rooted leaves of two sweet potato cultivars. *Ann. Bot.* 45:363–364.

McDonald, J. and Austin, J. 1990. Changes and additions in *Ipomoea* section Batatas (Convolvulaceae). *Brittonia* 42:116–120.

McGregor, C.E. 2006. Differential expression and detection of transcripts in sweetpotato (*Ipomoea batatas* (L.) Lam.) using cDNA microarrays. Dissertation. Louisiana State University, 143 pp.

Nakatani, M. 1994. *In vitro* formation of tuberous roots in sweet potato. *Jpn. J. Crop Sci.* 63:158–159.

Nakatani, M. and Komeichi, M. 1991. Changes in the endogenous level of zeatin riboside, abscisic acid and indole acetic acid during formation and thickening of tuberous roots in sweet potato. *Jpn. J. Crop Sci.* 60:91–100.

Nakatani, M., Tanaka, M. and Yoshinaga, M. 2002. Physiological and anatomical characterization of a late-storage root-forming mutant of sweetpotato. *J. Am. Soc. Hortic. Sci.* 127:178–183.

Pardales, J.R. and Esquibel, C.B. 1996. Effect of drought during the establishment period on the root system development of cassava. *Jpn. J. Crop Sci.* 65:93–97.

Pardales, J.R. and Yamauchi, A. 2003. Regulation of root development in sweetpotato and cassava by soil moisture during their establishment period. *Plant and Soil* 255:201–208.

Ravi, V. and Indira P. 1999. Crop physiology of sweetpotato. *Hortic. Rev.* 23:277–339.

Shiotani, I. 1988. Genomic structure and the gene flow in sweet potato and related species. *In*: Exploration, maintenance, and utilization of sweet potato genetic resources. Report of the First Sweet Potato Planning Conference 1987. International Potato Center, Lima. pp. 61–73.

Sirju-Charran, G. and Wickman, L.D. 1988. The development of alternative storage sink sites in sweet potato, *Ipomoea batatas*. *Ann. Bot.* 61:99–102.

Souer, E., Van Houwelingen, A., Kloos, D., Mol, J. and Koes, R. 1996. The no apical meristem gene of Petunia is required for pattern formation in embryos and flowers and is expressed at meristem and primordia boundaries. *Cell* 85:159–170.

Spence, J.A. and Humphries, E.C. 1972. Effect of moisture supply, root temperature, and growth regulators on photosynthesis of isolated rooted leaves of sweetpotato (*Ipomoea batatas*). *Ann. Bot.* 36:115–121.

Staswick, P.E., Su, W. and Howell, S.H. 1992. Methyl jasmonate inhibition of root growth and induction of a leaf protein are decreased in an *Arabidopsis thaliana* mutant. *Proc. Nat. Acad. Sci. USA* 89:6837–6840.

Sugiyama, T. and Hashizume, T. 1989. Cytokinins in developing tuberous roots of sweetpotato (*Ipomoea batatas*). *Agric. Biol. Chem.* 53:49–52.

Suye, S., Sugiyama, T. and Hashizume, T. 1983. Mass spectrometric determination of ribosyl trans-zeatin from sweetpotato tubers (*Ipomoea batatas* L. cv. Kohkei No. 14). *Agric. Biol. Chem.* 47:1665–1666.

Tanimoto, E. 2005. Regulation of root growth by plant hormones –roles for auxin and gibberellin. *Crit. Rev. Plant Sci.* 24:249–265.

Tanaka, M., Takahata, Y. and Nakatani, M. 2005. Analysis of genes developmentally regulated during storage root formation of sweetpotato. *J. Plant Physiol.* 162:91–102.

Thompson, P.G, Smittle, D.A. and Hall, M.R. 1992. Relationship of sweetpotato yield and quality to the amount of irrigation. *HortScience* 27:23–26.

Togari, Y. 1950. A study of tuberous root formation in sweet potato. *Bull. Natl. Agric. Exp. Stn.* Tokyo 68:1–96.

Tornero, P., Conejero, V. and Vera, P. 1996. Phloem-specific expression of a plant homeobox gene during secondary phases of vascular development. *Plant J.* 9:639–648.

Tsubone, M., Kubota, F. Saitou, K. and Kadowaki, M. 2000. Enhancement of tuberous root production and adenosine-5w-diphosphate pyrophosphorylase (AGPase) activity in sweet potato (*Ipomoea batatas* Lam.) by exogenous injection of sucrose solution. *J. Agron. Crop Sci.* 184:181–186.

Villavicencio, L.E., Blankenship, S.M., Yencho, G.C., Thomas, J.F. and Raper, C.D. 2007. Temperature effect of skin adhesion, cell wall enzyme activity, lignin content, anthocyanins, growth parameters, and periderm histochemistry of sweetpotato. *J. Am. Soc. Hort. Sci.* 132:729–738.

Wang, S-J., Yeh, K-W. and Tsai, C-Y. 1999. Molecular characterization and expression of starch granule-bound starch synthase in the sink and source tissues of sweet potato. *Physiol. Plantarum* 106:253–261.

Wang, S.J., Chen, M.H., Yeh, K.W. and Tsai, C-Y. 2006. Changes in carbohydrate content and gene expression during tuberous root development of sweet potato. *J. Plant Biochem. Biotechnol.* 15:21–25.

Wilson, L.A. and Lowe, S.B. 1973. The anatomy of the root system in West Indian sweet potato (*Ipomoea batatas* (L.) Lam.) cultivars. *Ann. Bot.* 37:633–643.

Wilson, L.A. 1982. Tuberization in sweetpotato (*Ipomoea batatas* (L.) Lam.). Proc. 1st International Symp. Sweetpotato, AVRDC, Tainan, Taiwan, pp. 79–94.

Zhang, D.P., Carbajulca, D., Ojeda, L., Rossel, G., Milla, S., Herrera, C. and Ghislain, M. 2001. Microsatellite analysis of genetic diversity in sweetpotato varieties from Latin America. *In*: Program Report 1999–2000, International Potato Center, Lima, pp. 295–301.

Chapter 4
Important Cultivars, Varieties, and Hybrids

A.L. Carpena

The many descriptor states possible for morphological traits alone of sweetpotato as seen in the IBPGR'S Descriptors List for Sweet Potato (Huaman, 1991) already indicate that there could exist a number of varieties of this crop species in the world. In a recent workshop co-sponsored by the International Potato Center (CIP), Rossel et al. (2007) reported that over 6000 accessions of landraces, breeding lines, and advanced cultivars of sweetpotato were held by the Center in its genebank, a collection which, based on reports in the same workshop of some of the sweetpotato producing countries, seems just a small fraction of the available varieties in the world.

Although several sweetpotato varieties are maintained by national gene banks all over the world, only a few, up to just two in some cases, predominate the sweetpotato growing areas in each major sweetpotato producing country. The choice of varieties to grow appears to depend largely on how the produce, the roots, are utilized, whether as food either directy or in processed forms, as feed component, or as sources of industrial starch. For food varieties, preferences also seem to vary among and even within countries.

Since the varieties considered important, that is, widely grown, differ from country to country, the dominant varieties in some of the major sweetpotato growing countries are presented below by country.

China

For more than 400 years since the introduction of sweetpotato to China, sweetpotato cultivation in the country relied on local cultivars which, although possessing tolerance/resistance to abiotic/biotic stresses after having survived natural and farmers' selection pressures, were low yielding and with limited ranges of adaptability (Ma et al., 1998). The introduction of the US variety Nancy Hall in 1938 and the Japanese variety Okinawa 100 in 1941 provided the country with two outstanding varieties that were used not just for production of the root crop but also in the genetic

A.L. Carpena (✉)
University of the Philippines Los Baños, College, Laguna, Philippines
e-mail: alcarpena@yahoo.com

G. Loebenstein, G. Thottappilly (eds.), *The Sweetpotato*,
DOI 10.1007/978-1-4020-9475-0_4, © Springer Science+Business Media B.V. 2009

Plate 4.1 Sweetpotato variety 'Xushu 18' of China

development of new varieties. One of the varieties that emerged in the breeding programs involving the two introduced ones is Xushu 18 (Fig. 4.1), a third generation descendant of the two (pedigree given in Xue-qin et al., 2006 and 2007), which to this day is the leading sweetpotato variety in China both in annual hectareage (about 1.5 million ha) and in total root production (about 30 million MT) (data from Li Hongmin, personal communication). Li Hongmin also described Xushu 18 as a widely adapted variety with tolerance to high moisture stress and resistance to root rot, but susceptible to stem nematodes and black rot. It has spreading growth habit, green cordate and slightly toothed leaves, green vines with purple stripes, and elliptic roots with red skin and white flesh with purple rings in some places. The variety is utilized mainly for feed and starch production. Based on the data in the Catalog of Sweet Potato Cultivars in China published in 1993 by the Xuzhou Sweet Potato Research Center in China (cited by Ma et al., 1998) which developed the variety, Xushu 18 had only 25% root dry matter content, which is considered just moderate and 11–15% starch, although data provided by Li Hongmin of the same institution showed the root dry matter and starch contents of the variety to be slightly higher, respectively ranging from 26–32% and 14–20%.

The next leading variety in the country is Nanshu 18, a cross between a local variety (Jinzhuan 7) and an introduced one (America Red), and at present being grown in 0.5 million ha, mainly in middle China where it is more adapted, and producing 12 million MT of fresh roots (Li Hongmin, personal communication). The variety was released in 1988 by the Nanchong Agricultural Research Center, just one of the only two out of 23 varieties released in Chine during the period 1986–1990 that did not trace their ancestries to Nancy Hall and Okinawa 100 (Ma et al., 1998). Nanshu 88 is also a starch and feed variety with root dry matter content of 24–30% and starch content of 12–18%. It is tolerant to drought but susceptible to black rot, root rot, and stem nematodes and with poor tolerance to abiotic stresses. Its vines are short and purple and its lobed leaves are purple when young and turn green as they mature. Other important food varieties according to Li Hongmin, are Guanshu 128 and Yanshu 5, both products of crosses between local varieties and are, respectively, being grown in approximately 80,000 and 60,000 ha, mainly in South China, and producing 2.4 and 1.8 million MT of roots. The former is resistant to scab and tolerant to wet soil while the latter is resistant to *Fusarium oxysporum* and tolerant to drought. Both have spreading growth habit and toothed/lobed leaves which are green at all stages of growth for the former but purple when young for the latter. Root characteristics of the three leading food varieties indicate preferences in China for elliptic root shape (although Beijing 553 is roundish or more ovate), orange (pale or deep) flesh color, and moderate dry matter content (20–26% for the three varieties).

Indonesia

In Indonesia, sweetpotato is utilized mainly as human food, 88% of total production in 1996 (World Bank 1998–99 report summarized by CIP, 1999) although it could be a lower figure if it is true as claimed by Jusuf (2003) that the crop's use as food in recent years in Indonesia has decreased but reliable statistics are lacking. Thus, varieties widely grown are those preferred on the basis of traits related to eating quality such as taste and flavor and root appearance rather than root yield (Jusuf, 2003; Zuraida, 2003).

Sweetpotato is planted in 180,000–200,000 ha annually in Indonesia (Jusuf, 2003; Zuraida, 2003; FAO, 2003) with Java occupying the biggest area (34.8%). The popular varieties in Java are mostly local varieties and differ from area to area with SQ27, an introduction from Puerto Rico and released to farmers way back in the 1950s and was referred to then as Southern Queen, as the most popular in Bogor (West Java) which also grows four other local varieties; Bestak Putih and two others in Karanganyar (Central Java); and Gropak and Kamplong Putih, respectively, in Pacet, Mojokerto and Magetan, both in East Java (Zuraida, 2003). Widodo and Jusuf (2001) listed four varieties grown in Mojokento which did not include Gropak and with varieties IR and IR Melati occupying 45% and 35%, respectively, of the area planted in that province. IR Melati also occupies 50% of the sweetpotato area in Malang, another province in East Java (Jusuf, 2001a). In one production center, Blitar (East Java), two well known varieties in Indonesia, namely, Genjah Rante and Samarinda, occupy 50% and 30% of the production area (Jusuf, 2001b).

Root skin and root flesh are both cream in SQ27 and, respectively, red and white in Gropak. Kamplong Putih is reported to have only medium yielding ability (Jusuf, 2001c; Zuraida, 2003) but with good eating quality (Zuraida, 2003) and early maturity (Jusuf, 2001c). Zuraida (2003) who also reported that Genjah Rante and Samarinda are the varieties widely grown in Blitar, described Genjah Rante to have roots with red skin and yellow flesh and to be drought tolerant and high yielding and Samarinda to also have red root skin but purplish flesh and good eating quality. Genjah Rante seems to be better tasting though than Samarinda based on the report of Jusuf (2001b). Based on the paper of Zuraida (2003), most of the popular local varieties of sweetpotato in Indonesia have colored flesh ranging from cream to orange and at least in one case (Samarinda), purple.

The second largest producer of sweetpotato in Indonesia is Sumatra, 18.8% (Zuraida, 2003). In West Sumatra, the main variety is Tamburin Putih which occupies 30% of the sweetpotato area there as reported by Jusuf (2001) who described this variety as high yielding and with moderate resistance to the sweetpotato weevil but with low dry matter content, a characteristic of most varieties cultivated in Indonesia (Jusuf, 2003).

Vietnam

Before 1980, most sweetpotato varieties cultivated in Vietnam were of local origin and differed among the three geographical regions of Vietnam: Northern, Central, and Southern (Hoanh, 1988). Among the four varieties listed by Hoanh as being grown in Northern Vietnam was Hoang Long, not a local variety but an introduction from China, which along with another variety Chiem Dau was reported to account for 80% of the sweetpotato area in the Red River Delta and Northern coastal provinces (Vu et al., 2000). To this date, Hoang Long (Fig. 4.2) is the most popular variety in that region (Vu, personal communication). Data provided by Vu show that this variety, which is used as food and animal feed, yields from 10 to 15 t/ha and with root dry matter content of 27–28% and starch content of 18–19%. The variety is tolerant to low temperature but susceptible to scab. It has spreading growth habit, purple vines, and triangular leaves with purple green blade and purple ribs/veins. Roots are oblong in shape and with pink skin and creamy yellow flesh.

Plate 4.2 Sweetpotato variety 'Hoang Long' of Vietnam

Another popular variety at present is Da Nang, a local variety that is adapted in Central Vietnam (Vu, personal communication) and is also used both as food and feed. It yields 10–12 t/ha, has 30.5% dry matter in its roots, and is tolerant to drought. It has spreading growth habit, green vines, triangular slightly lobed green leaves, and roots that are oblong and white outside and inside.

More adapted in the southeastern region of the country is Bi Da Lat (Vu, personal communication), a local variety preferred in Ho Chi Minh City for fresh consumption (Vu et al., 2000). It has spreading growth habit, triangular purple leaves, green-purple vines, and oblong roots red outside and orange inside which probably explains why it is usually processed into dehydrated snack items.Another popular variety is KL5, a selection from an open pollinated population (Vu et al., 2000). It is a dual-purpose variety but is more used as animal feed. It has wide adaptation but more recommended in cooler areas because of its tolerance to low temperature. It has spreading growth habit, green deeply lobed leaves, green vines, and long roots with red skin and yellow flesh.

With increasing production in recent years of many food crops including rice, the use of sweetpotato as food has decreased in Vietnam (Hoanh and Diep, 2001, cited by Hoanh et al., 2005). Breeding work to develop sweetpotato varieties for animal feed has resulted in the development and subsequent release in late 2000 of K51, a cross of varieties CN 1028-15 and N8, as a national variety adapted to North and Central Vietnam (Hoanh et al., 2005). K51 has semi-erect to erect growth habit and with heavily branching vines. Yield of roots which show early bulking is 15–30 t/ha depending on the season, but even in winter time, K51 yields higher than other varieties. Because of the absence of tannin in the vines/leaves and the high protein content in the leaves (4.1–7.0%, dry weight basis) of K51, the variety is also used as fresh forage in Vietnam (Hoanh and Diep, 2001, cited by Hoanh et al., 2005).

Japan

Most of the sweetpotatoes produced in Japan are used as food (50%) and as source of starch (20%) (Komaki and Yamakawa, 2006; Katayama et al., 2006). The leading table variety is Beniazuma (Fig. 4.3) which is also the no. 1 cultivar in terms of total production area in Japan according to information from the National Institute of Crop Science (NICS) of Japan Sweet Potato Breeding Laboratory published in its website. An offspring of sweetpotato varieties Kanto No. 85 and Koganesengan, this variety was released by the National Institute of Crop Science (NICS) of Japan in 1981. Tarumoto (1989) described Beniazuma as a high yielding variety that is resistant to diseases. Its roots are long and spindle in shape, have deep red skin and yellow flesh, and tasty and very sweet when cooked (duclong.en.busytrade.com). Other major table varieties are Norin No. 1, the first cultivar produced by Japan's Ministry of Agriculture, Forestry and Fishery modern breeding program using the Norin system described in Tarumoto (1989), released in 1942, and Tamayutaka,

Plate 4.3 Sweetpotato variety 'Beniazuma' of Japan
(Source: National Institute of Crop Science website – address in reference list)

from the cross Kanto No. 33xKanto No. 19 developed by Kanto-Tosan Agricultural Experiment Station and released in 1960. Tamayutaka is the leading cultivar for steamed and dried processing (for chips).

The sweetpotato cultivars currently used in Japan for starch production are Koganesengan, Shiroyutaka, and Shirosatsuma (Komaki and Yamakawa, 2006). The first two were developed by Kyushu National Agicultural Experiment Station in 1966 and 1985, respectively, while the third variety came from the National Agricultural Research Center and was released in 1986. All three varieties have roots with yellow white skin and high starch content (22–26%) and are high yielding (Tarumoto, 1989). Komaki and Yamakawa (2006) noted that although it appears possible to produce sweetpotato varieties with even higher root starch content, say 30%, there is a need to improve certain properties of sweetpotato starch to make it comparable with starch from other crops and, thus, expand its uses which at the present time is only for producing liquid sugar or glucose in Japan. The paper cited studies reporting the development of breeding lines with improved starch characteristics such as low amylose content. One cultivar released from these studies, by NICS in 2002, is Quick Sweet, an offspring of the cross Beniazuma x Kyushu No. 30, which is similar with ordinary cultivars in root appearance and starch and amylose contents but with about 20% lower gelatinization and pasting temperatures and altered starch fine structure (Katayama et al., 2005, 2006). This variety will allow the development of new products from sweetpotato starch which in turn will make it another important starch cultivar.

Some new uses for sweetpotato have emerged relatively recently in Japan resulting in new varieties being developed for such purposes or already existing ones increasing in importance. For example, a purple variety of sweetpotato was released in 1995, namely, Ayamurasaki, for the food colorant industry (Yamakawa et al., 1997, cited by Komaki and Yamakawa, 2006). The variety has a very high anthocyanin content in its roots making it a very economical source of the purple pigment for a variety of food products. The variety also has very high dry matter content in its roots, about 35%, that it is now also used to produce purple sweetpotato flour (Komaki and Yamakawa, 2006).

Papua New Guinea

The sweetpotato is the no. 1 food crop in Papua New Guinea (PNG), accounting for over 60% of the total production for 18 staple food crops in 2000 based on the data presented by Bourke (2006) who estimated that about 5000 cultivars of the crop are grown in the country, that vary greatly in root flesh and skin colors, with no single cultivar dominating production. However, varieties with low moisture contents are preferred.

Based also on the data presented by Bourke (2006), about 75% of sweetpotato produced in PNG come from the highlands. The common varieties grown in these high altitude areas are of long duration, taking 9–12 months to mature, and, thus, do not usually escape crop-damaging or severe frosts when they occur. Preliminary evaluation to identify early maturing varieties from the highland areas resulted in the selection of 10 varieties out of 30 varieties, that yielded at least 6.33 t/ha of marketable size roots in six months of growing period, with the highest yielder, Lian Morea, producing 9.00 t/ha (Sitango and Dopo, 2004). The average yield of the 30 varieties was 5.06 t/ha. Bang (2006) also reported that on farm trials in six sites of 20 varieties selected from 60 varieties screened in three trials at two highland stations of PNG's National Agricultural Research Institute (NARI) resulted in the selection of 12 varieties. The average marketable root yield for all varieties in the on farm trials was 8 t/ha in six months.

Although as mentioned above, more of the sweetpotato in PNG are produced in the highlands, the importance of the crop in the lowlands is increasing as it tends to replace taro as a staple crop in these parts of the country (Wiles and Demerua, 2002). Evaluation of cultivars for the lowlands was done alongside that for the highlands from June 1990–December 1998 and was almost complete by December 1998 while that for the highlands was less advanced (Wiles and Demerua, 2002). During NARI's 75th Anniversary celebration in May 2004, 75 first class lowland varieties were officially released (NARI, 2006). Although the varieties were the results of highly extensive testings: five large screenings, 16 replicated trials over 9 years in which climatic conditions varied from very wet to very dry, researchers at NARI realized that the number of varieties released was still high and should be reduced. Further analysis of the yield data of the released varieties to identify varieties with stable yields over environments resulted in the selection of 16 varieties including four drought tolerant ones, namely, K9, B11, Nuguria 5, and SI 278. All varieties were selected for their high to good yield with good market appeal and/or preferred taste. Several varieties have orange root flesh and high root dry matter content (at least 30%). They are also tolerant to the scab disease owing to their vigorous growth. A number of the varieties are deep-rooted which should make them less vulnerable to sweetpotato weevil and rat infestations. After boiling, texture of root flesh ranges from soft to firm and taste, from non-sweet to sweet.

Uganda

Tanzania is the most widely grown sweetpotato variety in Uganda (Mwanga, 2008; personal communication) and probably in some other African countries such as in Tanzania, Malawi, Kenya, and Zambia where it is known by some other variety names according to the paper of Mwanga et al. (2001) which is also the source of the following information on Tanzania and another popular variety. Tanzania, a selection from Ugandan landrace germplasm, was officially released in 1995 along with three other landrace selections, namely, New Kawogo, Bwanjule, and Wagabolige, and a selection from the progenies of a polycross involving 18 popular farmers' cultivars in Uganda, namely, Sowola (Mwanga et al., 2001). Tanzania and New Kawogo were already dominant cultivars before their release, the former, in drier northeastern districts of Kumi and Soroti, and the latter, in the tall grassland agroecological zone near Kampala.

The cultivar Tanzania has spreading plant type, green vines and green leaves that are deeply lobed and lanceolate in shape, and obovate roots with pale yellow skin and flesh. It has 32% root dry matter and is, thus, somewhat dry but sweet when cooked. Under field conditions, it is resistant to the sweetpotato virus disease (SPVD) and moderately resistant to Alternaria stem blight but susceptible to sweetpotato weevil. Average root yield in various trials was 22.9 t/ha. New Kawogo has also a spreading plant type, green with purple tips veins, and semi-elliptic moderately lobed leaves that are purple when young and green when mature. Veins and that part of the petiole attached to the leaf are also purple. Roots are obovate with shallow horizontal constrictions, have medium red skin with brownish orange tinge and white flesh, and contain 32% dry matter. Roots are somewhat dry when cooked and have moderate sweetness. Fresh root yield averaged 23.3 t/ha in various trials. The cultivar is moderately resistant to sweetpotato weevil and highly resistant to SPVD but highly susceptible to Alternaria stem rot in the field. The lone variety among the five that is not a landrace, Sowola, gave the highest average root yield of 25.6 t/ha and also the highest root dry matter content of 34%.

In 1999, six sweetpotato breeding lines were approved for release by the country's Plant Variety Release Committee, namely NASPOT (Namulonge Sweetpotato) 1 – NASPOT 6 (Mwanga et al., 2003). All were selected from bulked seeds harvested from a polycross nursery of 24 parents which included the first set of varieties released in Uganda listed above, popular farmer landraces, advanced clones from the Ugandan sweetpotato breeding program, and three introduced varieties from the International Institute of Tropical Agriculture (IITA) in Nigeria. With their high root yields (21–29 t/ha), high root dry matter contents (29–35%), at least moderate resistance to SPVD, good root shape (2 round elliptical and 4 obovate, and a variety of root skin (cream and purple) and flesh color (white, cream, pale yellow, and orange), these varieties should soon become important in the country.

The latest set of varieties officially released in Uganda, in 2004, consisted of two orange-fleshed cultivars, namely, Ejumula and SPK004 (also called Kakamega), respectively, Ugandan and Kenyan local landraces which were selected from a batch of 25 sweetpotato clones of various origins, e.g, from the US, Japan, and Taiwan (Mwanga et al., 2007). Both varieties have high root dry matter (respectively, 34.2% and 33.2%) and, thus, dry texture when cooked. By the time of their release, the two cultivars had already spread and had reached 28 districts in Uganda. In communities where orange-fleshed sweetpotatoes (OFSP) are being promoted, OFSP accounted for 22.4% of total production in 2006 compared to only 3.2% in 2004 with the frequency of farmers producing OFSP increasing from 21.7% to 64.3% during this period. Consumption of OFSP also increased from 25% of farm households in 2004 to 69% in 2006 indicating fast acceptance in Uganda of the two new varieties. Thus, these two varieties could also become important ones in the country in the near future.

Brazil

Brazil leads South American countries in sweetpotato hectareage and production, accounting for close to 40% of the total area harvested as well as total root production in these countries annually during the period 1999–2003 based on FAO statistics (FAO, 2003). The sweetpotato is the fourth most important vegetable in Brazil (Miranda, 2002a). If results of studies in the southern region of Sao Paulo state (Bressan et al., 2005; Veasey et al., 2007) are any indication, sweetpotato varieties grown in the country are highly diverse, at least phenotypically and no single varieties seem to dominate cultivation, although according to Bressan et al. (2005), varieties most frequently mentioned in certain municipalities of Sao Paulo were "batata roxa" (purple potato) and "batata branca" (white potato).

In 1979, the national center for horticultural research (CNPH) in Brazil started research on sweetpotato to increase production, which at the time was only 8.7 t/ha (Miranda, 2002a). Out of 36 cultivars collected from different states of Brazil in 1980, four have been selected on the basis of root yield and other agronomic traits and commercial value, namely, Brazlandia Branca, Brazlandia Rosada, Brazlandia Roxa, and Coquinho. The following descriptions of the four varieties are taken from Miranda (2002a).

Brazlandia Branca has creeping growth habit, green vines and leaves, and long oblong roots with white skin and cream flesh which, after cooking, turns to light yellow and has a dry texture but more moist than varieties Coquinho and Brazlandia Roxa. It is an intermediate maturing variety (harvested in 120–150 days). Root yield at CNPH is 37 t/ha after 5 months.

Brazlandia Rosada has also a creeping growth habit, green vines and leaves, long oblong pointed very uniform roots with pink skin and cream flesh which turns yellow after cooking. It is an intermediate variety in maturity, harvested in

120–150 days, but behaves like a late variety producing huge heavy roots when planted with wider spacing. Recommended time to harvest the variety is when roots have reached the commercial size of 150–250 g each. Root yield obtained at CNPH is 33 t/ha in five months. The variety has very high dry matter content in its roots, 39.7%, 81.8% of which are starch and sugar, and is thus suitable for alcohol production.

Brazlandia Roxa is also of creeping growth habit and has green vines and leaves and long oblong pointed roots that vary in shape depending on soil with red skin and cream flesh with low fiber content which turns to creamish yellow after cooking. This is a late variety, harvested in 150 days variety, and does not produce big roots. It has also low resistance to soil pests. Root yield at CNPH is 25 t/ha in five months.

Coquinho is best grown in the spring/summer season where its vegetative development is faster. During this time, its root yield at CNPH is 28 t/ha in four months. Coquinha is, thus, a relatively early variety and rarely produces big roots. Roots vary in shape depending on the type of soil where it is grown, with long and oblong shapes more common. Both vines and leaves are green and roots have pale yellow skin and white flesh with very low fiber content and which, after cooking, becomes off white and of dry texture.

Miranda (2002b) also reported on a sweetpotato variety selected from 132 cultivars in the germplasm collection of CPNH which has outstanding characteristics, namely, Princesa. This variety has a creeping growth habit, green vines and leaves which are lobed and with red petioles and also red veins at the nether side, and uniform long roots which are cream outside and inside and with very high dry matter and starch contents, 39.2% and 30.7%, respectively. It is resistant to "mal-do-pe" or "bad foot" disease caused by the fungus *Plenodomus destruens* Harter and intermediate in resistance to nematodes *Meloidogyne javanica* and *M. incognita.* It is also low in resistance to the herbicide metribuzin. This is a late variety, harvested in 150 days. Root yield at CNPH is 27 t/ha in five months when planted in Nov or Dec. It is used mainly as a table variety. However, its rapid and vigorous plant growth makes its vines ideal for animal feed.

United States

The sweetpotato is used only as food in the northern United States and as both food and livestock feed in the southern US (Huntrods, 2008). Food varieties in the US are classified into two types: dry-flesh and moist-flesh, a classification that is not based on the moisture content of the roots but rather on the characteristics of the solids in cooked roots (USDA ARS, 1971). Moist-flesh varieties are believed to convert starch into sugars more completely during cooking and are, thus, sweeter than the dry-flesh varieties. They are referred to in the US as "yams", rather incorrectly. The leading variety in the US at present in terms of acreage is Beauregard (Fig. 4.4), occupying about 60% (Schultheis, personal communication) to 65% (La Bonte, personal communication) of the sweetpotato production

Plate 4.4 Sweetpotato variety 'Beauregard' of the US

area. Beauregard, a polycross selection with varieties/lines L59-89, L67-29, L70-197, and L78-21 as seed parents (the first two digits refer to the true seed year), was released by the Louisiana State University (LSU) Agricultural Center in 1987 (La Bonte, personal communication). La Bonte (personal communication) provided the following descriptions/characteristics of Beauregard: spreading growth habit but with less vine growth than most varieties (sparse canopy); cordate green leaves with very slight purple coloration at the base above the petiole; elliptic roots with rose skin and orange flesh; resistant to fusarium wilt, fusarium root rot, and rhizopus root rot; intermediate to resistant reaction to soil rot; susceptible to root-knot nematode and bacterial soft rot; with good baking quality and a favorite as a canned product. Beauregard is also one of the main sweetpotato cultivars in New Zealand (Lewthwaite, 2006).

Another important variety in the US is Covington, released by North Carolina Agricultural Research Station in 2005, which is estimated to occupy 30% of the sweetpotato production area (Schultheis, personal communication). The North Carolina Cultivar Improvement Association (NCCIA) described the variety as high yielding with a good percentage of high-grade roots, resistant to Fusarium wilt, and moderately resistant to Streptomyces soil rot and southern root-knot nematode. Its leaves are green and are from heart-shaped to slightly lobed. Roots have rose-colored skin and orange flesh and blocky to fusiform shapes. Other varieties being grown in the US include Jewel (Huntrods, 2008), an old variety.

A new variety that could become important soon in the US is Evangeline, which was also developed by the LSU AgCenter. This variety contains 30% more sugar than Beauregard and has a more consistent root shape and darker orange root flesh color (Huntrods, 2008).

Acknowledgments The author wishes to thank the following people who provided data, information, pictures, and/or references for some of the varieties presented in this chapter: Dr. Vu Dinh Hoa, Hanoi Agricultural University, Vietnam; Dr. Li Hongmin, Xuzhou Sweet Potato Research Center, China; Dr. Robert Mwanga, National Agricultural Research Organization, Namulonge Agricultural and Animal Production Research Institute, Uganda; Ms. Hidelisa de Chavez, UPWARD-CIP, Philippines; Prof. Algerico Mariscal, Philippine Root Crops Research and Training Center, Visayas State University, Philippines; and Mr. Samuel Ordoñez and Dr. Don La Bonte, Louisiana State University Agricultural Center, USA. Thanks are also due to Dr. Catalino Flores and Chili Flores of Filmex Genetics, Mexico for translating from Portuguese into English the two references from Brazil.

References

Bang, S. 2006. Research and development projects on sweet potato by National Agricultural Research Institute (NARI). Retrieved 12 May 2008 from the World Wide Web: http://rspas.anu.edu.au/lmg/sweet_potato_workshop/BangSergie.pdf.

Bressan, E.A., Veasey, E.A., Peroni, N., Felipim, A.P. and dos Santos, K.M.P. 2005. Collecting yam (*Dioscorea* spp.) and sweet potato (*Ipomoea* batatas) germplasm in traditional small-holdings in the Vale do Ribeira, Sao Paulo, Brazil. *Plant Genetic Resources Newsletter* Issue No. 144: 8–13.

Bourke, M. 2006. Sweet potato in Papua New Guinea. Retrieved 6 May 2008 from the World Wide Web: http://rspas.anu.edu.au/lmg/sweet_potato_workshop/BourkeMike.pdf.

CIP (International Potato Center). 1999. Sweetpotato facts: Production, utilization, consumption, feed use.

FAO. 2003. Production Yearbook vol 57.

Hoanh, M.T. 1988. Conditions of sweet potato development in Vietnam: Selection and creation of new varieties. In: Howeler, R.H. (ed.) *Proceedings of 8th International Symposium of the International Society for Tropical Root Crops*. 30 October–5 November 1988. Bangkok, Thailand. 1990. 724 p.

Hoanh, M.T. and Diep, N.T. 2001. Short-growing duration sweetpotato variety K51. *Science & Technology Journal of Agriculture and Rural Development*, 299–300.

Hoanh, M.T., Hoang, T.D., Lan, N.T., Diep, N.T. and Daihung, P. 2005. Breeding of sweetpotato varieties for animal feed in North and Central Vietnam. In: *Concise Papers, Second International Symposium on Sweetpotato and Cassava* held in Kuala Lumpur, Malaysia 14–17 June 2005, pp. 65–66.

Huaman, Z. (ed.) 1991. Descriptors List for Sweetpotato. IBPGR, Rome.

Huntrods, D. 2008 (revision; profile created 2004). Sweet potato profile. Retrieved 18 April 2008 from the World Wide Web: http://www.agmrc.org/agmrc/commodityvegetables/sweetpotatoes/sweetpotatoprofile.htm.

Jusuf, M. 2002a–c. Adoption of sweetpotato varieties in major producing area. Appendix tables b2i, b2ii, b2v (pp. 98, 99, 102). In: Rao, R. and Campilan, D. (eds.) 2002. Exploring the Complementarieties of *In Situ* and *Ex Situ* Conservation Strategies for Asian Sweetpotato Genetic Resources. *Proceedings of 3rd International Workshop of the Asian Network for Sweetpotato Genetic Resources* (ANSWER). Denpasar, Bali, Indonesia. 2–4 October 2001. IPGRI-APO, Serdang, Malaysia.

Jusuf, M. 2003. Breeding improved sweetpotato varieties in Indonesia. In: Fuglie KO (ed.) Progress in Potato and Sweetpotato Research in Indonesia. *Proc CIP-Indonesia Research Review Workshop* held in Bogor Indonesia 26–27 Mar 2002, pp. 177–190. International Potato Center (CIP), Bogor, Indonesia, 2003.

Katayama, K., Komae, K., Kato., T., Tamiya, S., Kuranouchi, T., Komaki, K. and Nakatani, M. 2005. New sweetpotato cultivar, "Quick Sweet" having low gelatinization temperature and altered starch structure. In: *Concise Papers*, *Second International Symposium on Sweetpotato and Cassava* held in Kuala Lumpur, Malaysia 14–17 June 2005.

Katayama, K., Komae, K., Tamiya, S., Khoyama, K., Nakatani, M. and Komaki, K. 2006. Studies in breeding for improving starch properties in sweet potato. *JARQ* 40:115–122.

Komaki, K. and Yamakawa, O. 2006. R & D collaboration with industry – The Japanese sweetpotato story. In: Tan, S.L. (ed) *Proceedings of Second International Symposium on Sweetpotato and Cassava* held in Kuala Lumpur, Malaysia 14–17 June 2005.

Lewthwaite, S.L. 2006. Sweetpotato products in a modern world: The New Zealand experience. In: Tan S.L. (ed). *Proceedings of Second International Symposium on Sweetpotato and Cassava* held in Kuala Lumpur, Malaysia 14–17 June 2005.

Ma, D.F., Li, H.M. and Mok, I.G. 1998. Introduction and use of exotic germplasm in the Chinese sweetpotato breeding program. *CIP Program Report* 1997–98.

Miranda, J.E.C. 2002a (based on date given in Bressan et al., 2005). Batata-doce. Retrieved 15 April 2008 from the World Wide Web: http://www.cnph.embrapa.br/cultivares/bat-doce.htm.

Miranda, J.E.C. 2002b (date patterned after paper directly above). Princesa. Cultivar de batata-doce resistente ao "mai-do-pe". Retrieved 29 April 2008 from the World Wide Web: http://www.cnph.embrapa.br/cultivares/princesa.htm.

Mwanga, R.O.M., Odongo, B., Ocitti p'Obwoya, C., Gibson, R.W., Smit, N.E.J.M. and Carey, E.E. 2001. Release of five sweetpotato cultivars in Uganda. *HortScience* 36:385–386.

Mwanga, R.O.M., Odongo, B., Turyamureeba, G., Alajo, A., Yencho, G.C., Gibson, R.W., Smit, N.E.J.M., and Carey, E.E. 2003. Release of six sweetpotato cultivars ('NASPOT 1' to 'NASPOT 6') in Uganda. *HortScience* 38:475–476.

Mwanga, R.O.M., Odongo, B., Niringiye, C., Alajo, A., Abidin, P.E., Kapinga, R., Tumwegamire, S., Lemaga, B., Nsumba, J., and Carey, E.E. 2007. Release of two orange-fleshed sweetpotato cultivars, 'SPK 004' ('Kakamega') and 'Ejumula', in Uganda. *HortScience* 42:1728–1730.

National Agricultural Research Institute (NARI) (Papua New Guinea). 2006 (deduced from the copyright period). Technology release of sweetpotato: Technology release of sweet potato varieties for the lowlands. Retrieved 6 May 2008 from the World Wide Web: http://www.nari.org.pg/res/wlip/tr-sp.html.

National Institute of Crop Science (NICS) (Japan) Sweet Potato Breeding Laboratory. 2009 Major cultivars released by the sweet potato breeding laboratory. Retrieved 22 April 2008 from the World Wide Web: http://nics.naro.affrc.go.jp/hatasuka/kansyo/cultivare.html.

North Carolina Crop Improvement Association (NCCIA). 2009 Sweet potato varieties. Retrieved 24 April 2008 from the World Wide Web: http://www.nccrop.com/varieties.php/8/Sweet_Potato.

Rossell, G., Reynoso, D., Espinosa, C., Manrique, I., Panta, A. and Roca, W. 2007. The sweetpotato germplasm collection at CIP, Lima-Peru. Paper presented in the *CIP-UPWARD Workshop "Sweetpotato Global Conservation Strategy*. The Global Crop Diversity Trust. Manila, Philippines, 30 April–2 May 2007.

Sitango, K. and Dopo, N. 2004. Early maturing sweet potato varieties for high altitude highlands of Papua New Guinea. Retrieved 9 May 2008 from the World Wide Web: http://www.cropscience.org.au/icsc2004/poster/3/4/7/645_sitangok.htm.

Tarumoto, I. 1989. Sweet potato breeding in Japan: Its past, present and future. In: International Potato Center. 1989. Improvement of Sweet Potato (*Ipomoea batatas*) in Asia. Report of the "*Workshop on Sweet Potato Improvement in Asia*", held at ICAR, India 24–28 October 1988. 256 p.

USDA Agricultural Research Service. 1971. Sweetpotato culture and diseases. *Agricultural Handbook* No. 388. 74p.

Veasey, E.A., Silva, J.R.Q., Rosa, M.S., Borges, A., Bressan, E.A. and Peroni, N. 2007. Phenology and morphological diversity of sweet potato (*Ipomoe batatas*) do Vale do Ribeira. *Sciencia Agricola* (Piracibaba, Brazil) vol. 64 no. 4. Retrieved 1 April 2008 from the World Wide Web: http://www.scielo.br/scielo.php?script_arttex&pid=S0103-90162007000400013.

Vu, D.H., Loc, D.T., Ho, T.V. and Kim, H. 2000. Sweet potato in the postrice areas of Vietnam. In: Rasco Jr ET, Amante VdR (eds.). *Sweet Potato in Tropical Asia*, Chap. 6 pp. 189–220. Philippine Council for Agriculture, Forestry, and Natural Resources Research and Development Book Series No. 171/2000. 230 p.

Widodo, Y. and Jusuf, M. 2002. Adoption of sweetpotato varieties in major producing area. Appendix table 2iv (p.101). In: Rao, R. and Campilan, D. (eds.). 2002. Exploring the Complementarieties of *In Situ* and *Ex Situ* Conservation Strategies for Asian Sweetpotato Genetic Resources. *Proceedings of 3rd International Workshop of the Asian Network for Sweetpotato Genetic Resources* (ANSWER). Denpasar, Bali, Indonesia. 2–4 October 2001. IPGRI-APO, Serdang, Malaysia.

Wiles, G.C. and Demerua, J. 2002. In: Rao, R. and Campilan, D. (eds.). 2002. Exploring the Complementarieties of *In Situ* and *Ex Situ* Conservation Strategies for Asian Sweetpotato Genetic Resources. *Proceedings of 3rd International Workshop of the Asian Network for Sweetpotato Genetic Resources* (ANSWER). Denpasar, Bali, Indonesia. 2–4 October 2001. IPGRI-APO, Serdang, Malaysia.

Xue-qin, H., Qing-Chang, L., Ishiki, K., Hong, Z. and Yu-Ping, W. 2006. Genetic diversity and genetic relationships among Chinese sweetpotato landraces revealed by RAPD and AFLP markers. 2006. *Breeding Science* 56: 201–207.

Xue-qin, H., Qing-Chang, L., Ishiki, K., Zhai, H. and Yu-ping, W. 2007. ISSR analysis of genetic diversity and relationships among sweet potato (*Ipomoea batatas*) landraces in China. *Plant Genetic Resources Newsletter* 150: 35–41.

Yamakawa, O., Yoshinaga, M., Hidaka, M., Kumagai, T. and Komaki, K. 1997. "Ayamurasaki": A new sweetpotato cultivar. *Bulletin of the Kyushu National Agriculture Experimental Sta* 33:49–72.

Zuraida, N. 2003. Sweet potato as an alternative food supplement during rice shortage. *Jurnal Litbang Pertanian* 22: 150–155.

Chapter 5
Genetic Engineering

J.F. Kreuze, J.P.T. Valkonen and M. Ghislain

Genetic transformation through direct gene transfer methods holds promise for introducing novel traits to sweetpotato in cases where no solutions by conventional breeding are available. This may be if the trait is not known in sweetpotato or it is governed by complex inheritance. Sweetpotato is clonally propagated, highly heterozygous, polyploid and out-crossing – in other words, a challenging crop to breed. Combined with the low fertility found in sweetpotato, even introgression of dominant single gene traits may present a challenge.

Transformation of sweetpotato is however not a trivial task. Many protocols have been described for plant regeneration and/or transformation in sweetpotato by either somatic embryogenesis (Al-Mazrooei et al. 1997; Chen et al. 2006; Gama et al. 1996; Liu et al. 1999, 2001; Newell et al. 1995; Otani et al. 1998; Sihachakr et al. 1997; Song et al. 2004; Yu et al. 2007) or organogenesis (Gosukonda et al. 1995a, 1995b; Morán et al. 1998; Porobo Dessai et al. 1995). However, most protocols are highly genotype dependent, with only very few genotypes responding. When coupled to transformation, the numbers of regenerated plants drop dramatically. The lack of reproducibility of many of the described protocols further aggravates the situation.

The International Potato Center (CIP) was the first to report successful development of a transgenic sweetpotato plant using *Agrobacterium rhizogenes* and organogenic plant regeneration from hairy roots (Dodds et al. 1991). Since then various protocols for *A. rhizogenes* (Dodds et al. 1992), and later *A. tumefaciens*, particle bombardment, and protoplast electroporation mediated genetic transformation of several cultivars have been described, coupled to plant regeneration by either somatic embryogenesis (Cipriani et al. 1999, 2001), or organogenesis (Luo et al. 2006).

J.F. Kreuze (✉)
International Potato Center (CIP), Applied Biotechnology Laboratory, Germplasm, Enhancement and Crop Improvement Division, Lima 12, Peru
e-mail: J.KREUZE@CGIAR.ORG

G. Loebenstein, G. Thottappilly (eds.), *The Sweetpotato*,
DOI 10.1007/978-1-4020-9475-0_5,

Target Traits

Efforts to improve sweetpotato by genetic transformation have concentrated on a few major traits and for which solutions do not seem likely via conventional breeding in the near future. The major ones are resistance to weevils and viruses and modified starch properties. Furthermore, some studies have engineered nematode resistance, herbicide resistance, stress tolerance and improved amino-acid composition.

Insect Resistance

Sweetpotato weevils are the most devastating insect pests of sweetpotato (Hartemink et al. 2000). Particularly in Africa, losses due to weevil damage are huge (Smit 1997; Qaim 2001). Screening sweetpotato germplasm has revealed differences in damage caused by weevils, which may be governed by additive genetic factors (Thompson et al. 1994, 1999). However, progress in breeding weevil-resistant cultivars has been slow because the heritability of the trait is extremely low. Therefore, weevil resistance was among the first traits for which genetic transformation was applied in sweetpotato. The early work focused on transformation with proteins that decrease the digestibility of sweetpotato for insects. Newell et al. (1995) transformed sweetpotato with a cowpea (*Vigna unguiculata*) trypsin inhibitor (CTI) and the mannose binding snowdrop lectin. Cipriani et al. (1999, 2001) transformed sweetpotatoes with a soybean (*Glycine max*) Kunitz-type trypsin inhibitor (SKTI) and a rice (*Oryza sativa*) cystein proteinase inhibitor (OCI). Non-choice feeding tests with the West Indian sweetpotato weevil (*Euscepes postfasciatus*) in a screenhouse bioassay (Zhang et al. 2000) revealed a moderate increase of weevil resistance in two transgenic events produced by Newell et al. (1995). These results suggested that trypsin inhibitors could be useful for control of weevils (Zhang et al. 2000). Field trials carried out at two sites in Peru with the same material showed increased resistance to the root-knot nematodes (*Meloidogyne* spp.) in the transgenic plants (Zhang et al. 2000). Despite these initially promising results, the strategy of using proteinase inhibitors was later abandoned due to the relatively small increase of resistance observed and because there were concerns regarding nutritional impact of such proteins on the human diet.

More recently, toxins from *Bacillus thuringiensis* (Bt) were tested against the two African weevil species, *Cylas puncticollis* and *C. brunneus*, and against the American and Asian species *C. formicarius*. A diet incorporation methodology provided reliable toxicity measures of eight Bt proteins, which were chosen based on prior evidence of toxicity (Maingi et al. 2002) or anti-Coleopteran activity. All Bt proteins tested were toxic to both species at less than 1 ppm, with two proteins consistently more toxic than the others (Moar et al. 2007). Four of these Bt proteins were selected for plant expression because of their toxicity and low sequence identity, which is important for the potential of cross-resistance (binding studies are currently ongoing). Several gene construct have been developed using chemically synthesized sequences for the coding and polyA regions and optimized for sweetpotato, and sweetpotato promoters to express high level in the tuberous root.

Transformation with *Agrobacterium* is underway for two African varieties, Tanzania and Wagabolige, at CIP.

Virus Resistance

Viruses constitute a major problem in vegetatively propagated plants because they are transmitted to new crops in infected planting materials. Hence, from the very beginning of growth, the virus is there to interfere with plant physiology and development, which can result in substantial yield losses. The most devastating diseases develop in sweetpotato plants infected with *Sweet potato chlorotic stunt virus* (SPCSV, genus *Crinivirus*; *Closteroviridae*). By a mechanism that is not yet elucidated this virus is able to eliminate effective resistance to many heterologous viruses (Karyeija et al. 2000; Mukasa et al. 2006; Untiveros et al. 2007). Consequently, the greatest yield losses are caused by Sweet Potato Virus Disease (SPVD), which is induced following co-infection of plants with SPCSV and *Sweet potato feathery mottle virus* (SPFMV, genus *Potyvirus*; *Potyviridae*) (Sheffield 1953; Hahn 1979; Gibson et al. 1998; Gutiérrez et al. 2003; Njeru et al. 2004). Sweetpotato genotypes expressing significant levels of resistance to SPVD have not been found although there are differences in symptom severity and the rate of progression of SPVD (Miano et al. 2008). A level of resistance expressed as reduced incidence of the disease in the field has been reported and used in a breeding programme (Mwanga et al. 2002).

Sweetpotato has been transformed with the coat protein (CP) encoding sequence of SPFMV and found to be resistant to SPFMV following experimental inoculation by grafting (Okada et al. 2001; 2002). This type of pathogen-derived resistance (PDR) has been used against a number of viruses in many crop species (Latham and Wilson 2008). However, resistance that works under controlled experimental conditions may not necessarily work out in the field. This was witnessed with the first transgenic sweetpotato lines engineered for resistance to SPFMV using PDR. Their resistance broke down in East Africa (New Scientist, 7 February 2004, p. 7) possibly because the transgene was not from a locally prevalent SPFMV strain or because the plants became infected with SPCSV. Indeed, PDR may be lost following infection of the plants with a virus that is not targeted by PDR (Latham and Wilson 2008). This is explained by suppression of RNA silencing by the untargeted virus. RNA silencing is a fundamental antiviral defense system in plants and other cellular organisms. It becomes activated or primed by transgene-driven over-expression of viral RNA in cells of transgenic plants and by double-stranded structures of the viral RNA during infection of non-transgenic plants (Haasnoot et al. 2007). Hence, PDR actually activates the natural antiviral resistance to be ready for action when the target virus initially infects cells. However, PDR is sequence-specific and therefore not able to target viruses that show less than ca. 87–90% sequence identity with the transgene sequence (Jones et al. 1998). Consequently, the virus that circumvents

PDR will accumulate and produce proteins that suppress RNA silencing (Voinnet 2005), which will convert the plant susceptible also to the virus that was supposed to be the target of PDR. For these reasons, the commonly encountered mixed virus infections in the field and the genetic variability of sweetpotato viruses possess an important challenge that needs to be met before sustainable virus resistance can be obtained (Tairo et al. 2005).

Resistance to potyviruses mediated by a rice cystein proteinase inhibitor (*OCI*) has been reported in tobacco (*Nicotiana tabacum*) (Gutierrez-Campos et al. 1999). *OCI* mediated resistance to potyviruses is thought to act through inhibiting the viral cystein proteinase NIa that processes the potyviral polyprotein. Closteroviruses also encode cystein proteinases to modify some of their proteins. Therefore, it was considered that expression of cystatins in transgenic plants might confer resistance to both viruses involved in SPVD. Cipriani et al. (2001) reported increased resistance to SPFMV in sweetpotato plants of cv. 'Jonathan' transformed with the *OCI*. However, no efficient resistance was obtained and typical symptoms of SPVD developed in plants infected with SPCSV and SPFMV.

Recently, a landrace sweetpotato variety 'Huachano' that is extremely resistant to SPFMV was genetically engineered for resistance to SPCSV (Kreuze et al. 2008). In this cultivar as in many others (Karyeija et al. 2000) the high levels of resistance to SPFMV breaks down following infection with SPCSV and the plants succumb to the severe SPVD. This exemplifies how important the resistance to SPCSV would be in order to protect sweetpotatoes against SPVD and other severe synergistic diseases induced by SPCSV with other viruses. The transgene was designed to express an SPCSV-homologous transcript that forms a double-stranded structure and hence efficiently primes virus-specific resistance. Many transgenic lines accumulated only low concentrations of SPCSV following infection and no symptoms developed. These results showed that sweetpotato could be protected against the disease caused by SPCSV using PDR. However, the low concentrations of SPCSV in the transgenic plants were still sufficient to break down the natural high levels of resistance to SPFMV. Indeed, immunity to SPCSV seems to be required for prevention of the severe virus diseases in sweetpotato.

Starch Modification

Starch is a storage carbohydrate composed by a linear polymer of sugars and ramified chains, amylose and amylopectin, respectively. The use of starch for different purposes requires specific physicochemical properties, which in turn are strongly affected by the content of amylose, amylopectin and their relative amounts (Richardson et al. 2000; Lii and Tsai 1996). The range of amylose content in sweetpotato starch is relatively narrow (10–25%) in comparison to other crops (Noda et al. 1998), which limits its uses for processing purposes. Development of new cultivars with novel properties of starch is therefore one objective of sweetpotato breeding. High-amylose starches as well as amylose-free starches are in demand for their specific uses in food and other industries ranging from paper production to production of renewable, biodegradable and CO_2 neutral packing materials (Richardson et al. 2000; Satin 2005).

Starch is synthesized from glucose by multiple enzymes (Plate 5.1). Cytosolic sucrose is converted to hexose phosphates, which are then transported to the amyloplast via a hexose translocator. The glucose-6-phosphate so formed is then converted to glucose-1-phosphate by the enzyme phosphoglucomutase. Four enzymes play a crucial role in the transformation of glucose-1-phospate into starch. Adenosine diphosphate glucose pyrophosphorylase (AGPase) catalyses the synthesis of ADP-glucose from ATP and glucose-1-phosphate and is rate limiting. The enzyme GBSSI (granule bound starch synthase I) is required to produce the linear polysaccharide amylose by adding ADP-glucose to the non-reducing end of a glucan chain forming α (1,4) linkages. On the other hand, the soluble starch synthases (SS) and starch branching enzymes (SBE) act in concert to produce amylopectin. The starch branching enzymes, responsible for forming α(1,6) linkages in the glucan are divided into two isoforms SBEI and SBEII which differ in their expression pattern and substate specificity (Lea and Leegood 1999).

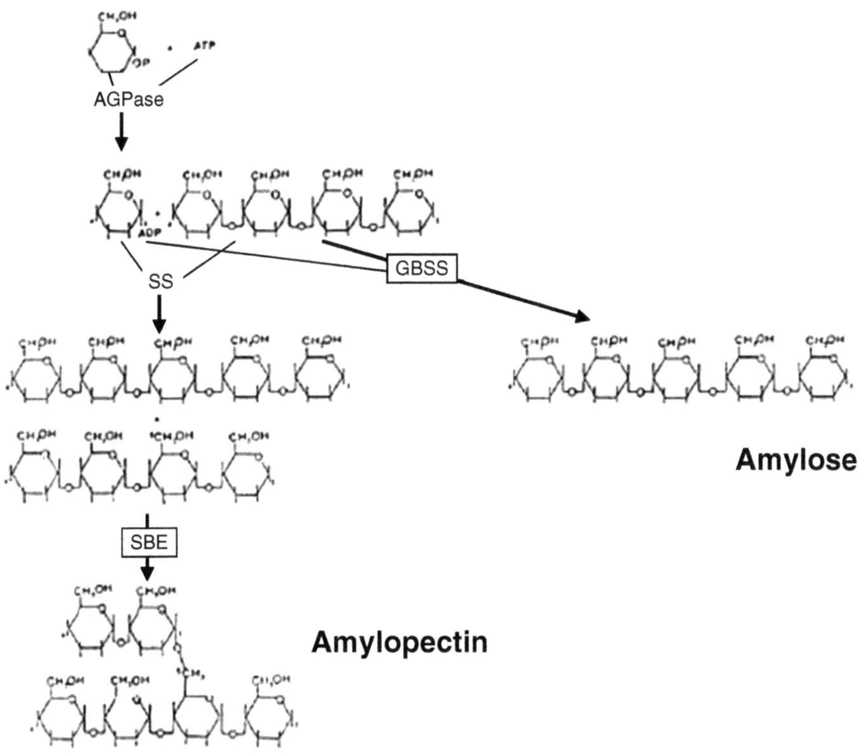

Plate 5.1 Main enzymes involved in starch synthesis in plants. Boxed enzymes have been manipulated through genetic engineering in sweetpotato

Inactivating the key genes responsible for biosynthesis of the two starch components has produced amylose-high and amylose-free sweetpotatoes. Kimura et al. (2001) generated amylose free transgenic sweetpotatoes of their model genotype Kokei 14 by co-suppression of GBSSI, the sole enzyme responsible for the production of amylose in sweetpotato. They also produced transgenic events of the same cultivar with increased amylose content by RNA silencing of sweetpotato SBEII (*IbSBEII*; a class A starch branching enzyme) using inverted repeat constructs (Shimada et al. 2006). Indeed, genes can be silenced at the posttranscriptional level by the same principle that is used in PDR against viruses. Hence, gene -silencing technology allows selective blockage of transcription of certain genes with high efficiency and specificity (Wesley et al. 2001; Brummell et al. 2003). The amylose content increased only modestly by 5–10%, which was expected because experiments on potato have shown that both SBE enzymes need to be inactivated to eliminate efficiently the production of amylopectin (Schwall et al. 2000; Andersson et al. 2006). To achieve this, transgenic sweetpotato plants of the clone CIP199004.2 with high amylose content have been generated using an RNA silencing construct targeting *IbSBEI* and *IbSBEII* simultaneously, but the plants remain to be analyzed (CIP, unpublished).

Besides modification of starch quality, there is a great potential for the modification of starch quantity in sweetpotato. Due to its high productivity, sweetpotato has great potential as a bio-fuel crop for ethanol production. Starch production could probably be increased by 20–35% through genetic engineering, as has been shown with potato (McKibbin et al. 2006) and cassava (Ihemere et al. 2006) by transforming plants with a bacterial AGPase gene under root specific promoter and modified to prevent allosteric regulation. Consequently, AGPase activity is substantially increased in the storage roots, which creates a stronger carbohydrate sink in roots and modulates the flow of assimilates from production of vines and foliage to increased starch production in the storage roots. Manipulation of certain transcription factors can also modify the partition of photoassimilates to the storage roots (Schwachtje et al. 2006). Heterologous thermostable and thermoactive amylases that can be used directly to break starch to sugars for fermentation (Lin et al. 2008) could be introduced to transgenic sweetpotato to improve the suitability of sweetpotato storage roots for bioethanol production.

Other Traits

Another way of increasing the utility of sweetpotato starch is to improve baking quality. Wheat flour contains unique proteins called high-molecular-weight (HMW) glutenins. These glutenins are critiçal in obtaining strong dough and high-quality yeast-raised breads. Dough strength and its ability to contain gas bubbles produced by yeast are known as viscoelasticity. Sweetpotato does not contain HMW-glutenin and is therefore not suitable for bread baking. In an attempt to improve the use of sweetpotato flour as an alternative for bread baking, a HMW glutenin gene of wheat was introduced into sweetpotato variety 'Huachano'. Among the 13 transformed events obtained three expressed glutenin in high amounts but the viscoelasticity of starch has not yet been determined (CIP, unpublished). With the recent rise of wheat prices on the world market, this approach may once again become attractive.

The improvement of protein content was also the aim of the first transformation experiments in sweetpotato where a synthetic peptide (HEAAE), which is rich in essential amino acids, was introduced into sweetpotato (Dodds et al. 1991; 1992). It was however not reported if the plants did contain higher levels of the essential amino acids. Increased lysine content can also be expected in plants transformed with the *dhdps-r1* gene, which is a mutant allele of the dihydrodipicolinate reductase gene of *Nicotiana sylvestris*. This allele is not feedback-inhibited by lysine and thus leads to lysine overproduction. It also confers resistance to the lysine analog AEC toxic to plants. Therefore, Luo et al. (2006) transformed sweetpotato plants with *dhdps-r1* with the aim to replace antibiotics as the selective agent during regeneration of transgenic plants. Although the transgene conferred resistance to AEC (in transgenic plants selected with the antibiotic kanamycin), the use of AEC as a selection agent for transformation never resulted in transgenic plants (CIP, unpublished).

Oxidative stress is one of the major factors causing injury to plants exposed to environmental stress. Transgenic sweetpotato plants expressing the genes of both CuZn superoxide dismutase (CuZnSOD) and ascorbate peroxidase (APX) under the control of an oxidative stress-inducible SWPA2 promoter in the chloroplasts have been reported (Lim et al. 2007). These enzymes are involved in scavenging reactive oxygen species produced in plants during stress. They conferred enhanced resistance to multiple environmental stresses including cold tolerance in the transgenic plants. The increased tolerance to multiple stresses might enable sweetpotato to be cultivated under a wider range of climatic conditions, such as extreme temperature, drought, and salt stress.

Wakita et al. (2001) introduced a ω-3 fatty acid desaturase gene from tobacco (*NtFAD*) into sweetpotato in order to modify the fatty acid composition of the lipids for both functional and nutritional improvement of sweetpotato quality. Modulation of the degree of saturation of fatty acids is an important factor in the metabolic adaptation of higher plants to temperature stress. NtFAD catalyzes the conversion of dienoic fatty acids (16:2 and 18:2) to trienoic fatty acids (16:3 and 18:3) in lipids. An increased content of (18:2 and 18:3) linolenic acid could be obtained but no increase of tolerance to extreme temperatures was reported in the transgenic plants (Wakita et al. 2001).

Besides conferring resistance to potyviruses as mentioned previously (Gutierrez-Campos et al. 1999), cysteine proteinase inhibitors may also confer resistance to nematodes (Atkinson et al. 2004; Urwin et al. 1995). Transgenic sweetpotatoes of the cultivars Jonathan (Cipriani et al. 2001) and Lizixiang (Jiang et al. 2004) expressing the OCI gene have been reported. However, neither article reports on resistance to nematodes in the transgenic plants. However, Zhang et al. (2000) could increase resistance to nematodes in transgenic events of cv. 'Jewel' that was transformed to express the *CTI* (Cowpea tripsin inhibitor) gene or the *CTI* and *GNA* (Snowdrop lectin) gene.

Resistance to the herbicides bialaphos (Otani et al. 2003) and glufosinate (Yi et al. 2007) has been introduced to sweetpotato by genetic transformation. Many sweetpotato cultivars however have vigorous vines that easily smother weeds and weed control is therefore not a big issue.

Transformation Methods

Production of transgenic sweetpotato plants depends on a number of factors that can be divided to two critical processes: transformation and regeneration. Both of these processes have proven to be highly genotype dependent, which has prevented the development of generic protocols, and has limited the number of genotypes that can be transformed to those few that respond well to both processes.

Gene transfer to plants can be achieved by several methods and most of them have been tried also on sweetpotato. *Agrobacterium*-mediated transformation is often preferred if amendable to the crop species in question because it usually results in less complex insertion patterns than the other methods. Therefore, it has been the preferred method also for transformation of sweetpotato. Particle bombardment and protoplast electroporation have also been used by some laboratories. However, for nuclear insertion the latter methods are less suited because incomplete DNA insertions, which are not desirable in the final variety, can not be eliminated by progeny segregation without loosing the original variety.

Agrobacterium-Mediated Transformation

The first success with sweetpotato transformation was achieved by the use of *Agrobacterium rhizogenes* (Dodds et al. 1991, 1992; Otani et al. 1993). The method consisted of wounding and inoculating the stem of *in vitro* plantlets with *A. rhizogenes* and subsequent sub-culturing of the emerging hairy roots until shoots regenerated. *A. rhizogenes* was later replaced with disarmed *A. tumefaciens* strains because they produce less physiological changes in the plants.

The efficiency of transformation can vary significantly between sweetpotato cultivars. For example, transformation with *A. tumefaciens* EHA105 for expression of the marker protein β-glucuronidase (GUS) in kanamycin-resistant calli of two African landrace varieties resulted in 25–27% transgenic calli in cv. 'Wagabolige' as compared to only 0–1% in cv. 'Tanzania' (Moar et al. 2007). Hypervirulent strains of *A. tumefaciens* such as EHA101 and EHA105 can significantly improve transformation efficiency as compared to strains such as LBA4404. Various types of tissues have been transformed using *Agrobacterium*, including embryogenic cell cultures, root disks, stem sections and leaves and combined with various different protocols for plant regeneration (Table 5.1).

Table 5.1 Reports on genetic transformation of sweetpotato

Trait	Gene[1]	Variety	Method[2]	Reference
Marker protein expression (method development)	*GUS*	Chugoku 25, Chugoku 35, Hi-starch, Kyukei 17-3043, Yamakawamurasaki	A.r. direct wounding; stem; organogenesis	Otani et al. 1993
Marker protein expression (method development)	*GUS*	Kokei 14	A.t. EHA101; embryogenic calli from apical meristems; embryogenesis	Otani et al. 1998
Marker protein expression (method development)	*GUS*	Lizixiang	A.t. EHA105 Embryogenic cell suspension from apical meristems; embryogenesis	Yu et al. 2007
Marker protein expression (method development)	*GUS*	Beniazuma	A.t. EHA105; stem, petiole & leaf; embryogenesis	Song et al. 2004
Marker protein expression (method development)	*GUS*	White Star	A.t. EHA101; embryogenic calli from apical meristems; embryogenesis	Gama et al. 1996
Virus resistance (SPCSV)	*hpSPFMV & SPCSV*	Huachano	A.t. EHA105; leaf with petiole; embryogenesis	Kreuze et al. 2008
Virus resistance (SPFMV)	*SPFMV-CP*	Chugoku #25, Chikei 682-11	Electroporation; protoplast; organogenesis/embryogenesis	Okada et al. 2001
Virus resistance (SPFMV)	*SPFMV-CP, SPFMV-NIb*	KSP36, CPT560	A.t.; Leaf with petiole; embryogenesis	Monsanto & KARI, unpublished
Virus resistance (SPFMV)	*SPFMV-CP*	Jewel	A.t. LBA4404; leaf with petiole; organogenesis	Herman et al. 2003
Weevil resistance	*SKTI-4*	Jewel, PI 318846-3, Jonathan, Maria Angola	A.t. LBA4404; leaf with petiole; embryogenesis	Cipriani et al. 1999

Table 5.1 (continued)

Trait	Gene[1]	Variety	Method[2]	Reference
Weevil resistance	*CryIIIa*	Jewel	A.t. C58C1-pGV2260; stems & leaves with petiole; organogenesis	Garcia et al. 2000
Weevil resistance	*CryIIIa*	Jewel	A.t. C58C1-pGV2260; leaf disks; organogenesis	Morán et al. 1998
Amylose-free starch	*GBSSI*	Kokei 14	A.t. EHA101; embryogenic calli from apical meristems; embryogenesis	Kimura et al. 2001
Amylose-high starch	*hpIbSBEII*	Kokei 14	A.t. EHA101; embryogenic calli from apical meristems; embryogenesis	Shimada et al. 2006
Amylose-high starch, marker gene excision	*hpIbSBEI&II* & *Cre-lox*	CIP199004.2	A.t. EHA105; leaf with petiole; Organogenesis	Kreuze et al. unpublished
Baking quality	*HMW-glutenin*	Huachano	A.t. EHA105; leaf & petiole; embryogenesis	Benavides et al. unpublished
Essential amino acids	*HEAAE*	Huachano, Maleno, Ihuanco	A.r. A4 direct wounding, stem, organogenesis	Dodds et al. 1991, 1992
Increased linolenic acid content	*NtFAD3*	Kokei 14	A.t. EHA101; embryogenic calli from apical meristem; embryogenesis	Wakita et al. 2001
Herbicide resistance (bialaphos)	*bar*	Kokei 14	A.t. EHA101; embryogenic calli from apical meristems; embryogenesis	Otani et al. 2003
Herbicide resistance (phos phinothricin ammonium salt, glufosinate; Basta™)	*bar*	Yulmi	Particle bombardment; embryogenic calli from apical meristems; embryogenesis	Yi et al. 2007

Table 5.1 (continued)

Trait	Gene[1]	Variety	Method[2]	Reference
Insect resistance	*GUS, CTI* & *GNA*	Jewel	A.t. LBA4404; tuber disks; Embryogenesis	Newell et al. 1995
Lysine-high amino acid content, non-antibiotic selectable marker	*dhdps-r1*	Jewel	A.t. EHA105; leaf & petiole; Organogenesis	Luo et al. 2006
Nematode and virus resistance	*OCI*	Huachano, Jonathan	A.t. LBA4404; leaf with petiole; embryogenesis	Cipriani et al. 2001
Nematode resistance	*OCI*	Lizixiang	A.t. LBA4404, Embryogenic cell suspension from apical meristems; embryogenesis	Jiang et al. 2004
Stress tolerance	*CuZnSOD* & *APX*	Yulmi	Particle bombardment; embryogenic calli from apical meristems; embryogenesis	Lim et al. 2007

[1] *GUS*: β-glucuronidase reporter gene; *GBSSI*: Granule bound starch synthethase I; *hpIbSBEII*: intron spliced hairpin of Starch Branching enzyme II; *hpIbSBEI&II & Cre-lox*: intron spliced hairpin of starch branching enzyme I and II and Cre-lox marker excision system; HMW-glutenin: high molecular weight glutenin gene from wheat; *HEAAE*: artificial peptide with high essential amino acid content; *NtFAD3*: tobacco microsomal ohgr-3 fatty acid desaturase gene; *bar*: *Streptomyces hygroscopicus* phosphinothricin acetyltransferase gene; *CTI & GNA*: Cowpea tripsin inhibitor and Snowdrop lectin genes respectively; *dhdps-r1*: mutant dihydrodipicolinate synthase gene from *Nicotiana sylvestris*; *OCI*: *Oryza sativa* cystatin I gene; *CuZnSOD & APX*: tobacco CuZn superoxide dismutase and ascorbate peroxidase respecively under oxidative stress inducible promoter; *CryIIIa*: Cry3 crystal protein gene from *Bacillus thuringiensis*; *SKTI-4*: Soybean Kunitz tripsin inhibitor 4; *SPFMV-CP and -NIb*: sweetpotato feathery mottle virus coat protein and Nib gene respectively; hpSPFMV&SPCSV: intron spliced hairpin of polymerases of SPFMV and Sweetpotato chlorotic stunt virus;

[2] Transformation method, explant type and regeneration method used to generate transgenic plants. In case of Agrobacterium mediated transformation the strain is indicated; A.r. :*Agrobacterium rhizogenes*; A.t. :*Agrobacterium tumefaciens.*

Particle Bombardment

Transient or stable expression of marker genes has been reported in sweetpotato following transformation using particle bombardment, including GUS in cv. 'Jewel' and the genotype TIS-70357 following selection with kanamycin (Prakash and Varadarajan 1992) and the green fluorescent protein (GFP) in cv. 'Beauregard' without antibiotic selection (Dhir et al. 1998; Lawton et al. 2000). Lim et al (2007) used particle bombardment to transform embyogenic calli for obtaining transgenic stress-resistant sweetpotatoes. The regeneration procedure included a 6-month long culture of transformed calli in dark on selection medium containing kanamycin. The kanamycin-resistant calli produced somatic embryos on regular culture medium supplemented with kanamycin under low light growth condition. Kanamycin resistant plantlets developed from the embryos on the same medium under increased light. The efficiency of the procedure was not reported in terms of the original number of explants used or the percentage of regenerated transgenic plants. Transgenic status was tested on 11 events by polymerase chain reaction (PCR). Therefore the efficiency of this method cannot be fully evaluated based on the information available.

Electroporation

Protoplasts can be obtained readily from sweetpotato leaves, stems and petioles (Murata et al. 1994; Nakashima 1993; Perera and Ozias-Akins 1991; Sihachakr and Ducreux 1987). Transient and stable expression of GFP has been reported in electroporated sweetpotato protoplasts (Dhir et al. 1998; Lawton et al. 2000) and regeneration of plants from protoplasts has also been reported for many varieties (Murata et al. 1994; Perera and Ozias-Akins 1991; Sihachakr and Ducreux 1987). However, the regeneration efficiencies are low, which is probably the reason why only a single study has reported successful generation of transgenic sweetpotato plants using this technique. Okada et al. (2001) used protoplasts of two cultivars (Chugoku No. 25 and Chikei 682–11) for electroporation with a construct containing the SPFMV coat -protein (CP) encoding region and used hygromycin resistance as the selectable marker. Although equal numbers of hygromycin resistant calli were obtained in both cultivars, only protoplasts of Chikei 682–11 successfully regenerated to plants. Among 449 calli obtained after antibiotic selection, 19 plants from seven independent calli grew to form adventitious shoots. Of these, three independent transgenic lines were obtained based on Southern blot analysis of the CP-transgene and the hygromycin resistance gene.

Regeneration Methods

Regeneration of sweetpotato after transformation can be achieved through two main routes, organogenesis or embryogenesis, of which the latter has been more commonly used. While the response of sweetpotato cultivars to the embryogenesis protocols is better than to organogenesis, the higher efficiency of selection (often 100%) obtained during embryogenesis may be the main factor why this regeneration method is more popular. The disadvantage of embryogenesis protocols, however, is that they are generally lengthy, often requiring 8 months or more from beginning of the experiment before transgenic plants are obtained. The methods reviewed here have not been widely adapted due to low transformation efficiencies and low reproducibility in different laboratories. Most laboratories employ their own unique protocols and sweetpotato genotypes.

Embryogenesis

Regeneration of sweetpotato plants from somatic embryos in semi-solid and liquid media after induction of embryogenic tissues from apical meristems was first reported for cv. 'White Star' (Liu and Cantliffe 1984; Chee and Cantliffe 1989; Chee et al. 1990). However, evidence for the induction of somatic embryogenic tissues at high frequencies was restricted to one or a few cultivars, because most were found to be recalcitrant or to respond poorly (Cavalcante Alves et al. 1994; Desamero et al. 1994; Zheng et al. 1996). Al-Mazrooei et al. (1997) made significant improvements to this protocol by using 2,4,5-T rather than 2,4-D as the primary auxin in the induction medium. They obtained a positive response in 14 out of the 16 tested sweetpotato genotypes. However, some genotypes still responded better to 2,4-D and the optimal concentrations of the hormones varied between genotypes. Hence, optimization was still required for new genotypes.

Embryogenic tissue derived from apical meristems has been used most frequently for transformation purposes (Table 5.1). The first report was on cv. 'White Star' (Gama et al. 1996). Embryogenic calluses were inoculated by co-cultivation with *A. tumefaciens* and regeneration of explants occurred on kanamycin-containing media 7 wk post-inoculation. Up to 12.3% of the regenerants were transgenic as determined by GUS staining. Otani et al. (1998) obtained transgenic sweetpotato plants from embryogenic calluses of apical meristems of cv. 'Kokei 14' on a medium containing 4-FA and hygromycin (Hyg). An average of 10 hygromycin-resistant clusters, of which 53.1% could be regenerated, were obtained from 1 g of embryogenic calli. The same system and cultivar have been used for generating various types of transgenic events (Otani et al. 2003; Kimura et al. 2001; Shimada et al. 2006).

Liu et al. (2001) in collaboration with CIP improved the efficiency of embryogenic tissue formation from apical meristems. They obtained embryogenic tissue from 15 Chinese sweetpotato genotypes. Subculture of the embryogenic tissues in embryogenic suspension cultures enabled subsequent regeneration of plantlets from cell aggregates of all 15 genotypes following their transfer to a solid medium containing ABA. Despite these initially promising results, the establishment of the embryogenic suspension cultures has shown to be technically demanding and reproducibility between laboratories has been low. Another difficulty is experienced by the fact that most cultivars are not easy to regenerate after inoculation with *Agrobacterium* co-culture. Only cv. 'Lizixiang' has been transformed with this system (Jiang et al. 2004; Yu et al. 2007). The use of particle bombardment might circumvent the problem. Lim et al. (2007) used embryogenic calli derived from apical meristems of cv. 'Yulmi' for transformation using particle bombardment and obtained many transgenic events.

Other explant types have also been used for transformation and regeneration of sweetpotato via embryogenesis. CIP routinely uses leaves and petioles as explants for transformation of different cultivars (Figs. 5.2, 5.3, and 5.4) Cipriani et al. 1999; Cipriani et al. 2001; Kreuze et al. 2008). Transgenic plants of cv. Jewel were regenerated through an embryogenic protocol from 2.5 to 10% of freshly harvested storage roots after a series of plant growth regulator changes over 6 months (Newell et al. 1995). The protocol employed by Okada et al. (2001) for transforming protoplasts by electroporation seemed to make use of both the organogenesis and embryogenesis development routes.

Organogenesis

The first success with transforming sweetpotato was obtained using organogenesis without antibiotic selection. Dodds et al. (1991) were the first to demonstrate regeneration of stably transformed sweetpotato plants as confirmed by Southern blot analysis. These plants were obtained by adventitious shoot formation from hairy roots after infection with *A. rhizogenes*. Out of the 24 sweetpotato varieties tested, 21-formed typical hairy roots, but only three varieties regenerated to shoots from the transgenic hairy roots (Dodds et al. 1992). Otani et al. (1993) using *A. rhizogenes* obtained shoots from hairy roots induced on leaf explants of five cultivars among the 14 varieties tested. Both of these laboratories later moved to use disarmed *A. tumefaciens* strains to avoid the morphologically aberrant plants usually produced by *A. rhizogenes* transformation.

Direct organogenesis in sweetpotato explants has been demonstrated in many studies. Medina (1991) used nine different genotypes and showed that the best explants for this purpose were leaves with petioles. Advantageous shoots were obtained in all sweetpotato cultivars using a two-stage protocol involving 3–5 days of culture on medium containing auxin and cytokinin (2,4-D and zeatine riboside) and subsequent transfer to a medium with only cytokinin. Regeneration efficiencies ranged from 2 to 93% depending on the cultivar. Cytokinin in the first-stage medium is however not critical and Porobo Dessai et al. (1995) used a modified protocol with only 2,4-D in the first medium. They achieved 10–83% regeneration efficiency with 15 out of the 25 tested genotypes. The optimal type of auxin and cytokinin used in each stage of the protocol may vary between genotypes. Generally, 2,4-D and zeatine work well in most cases. These protocols have been developed based on the hypothesis that the organogenetic process has two stages. The first is supposed to be an auxin-related dedifferentiation phase in which cells acquire competence for an organogenic response. The second phase is organogenesis during which formation of new shoot meristems can be induced by exogenous cytokinin application (Sugiyama 1999).

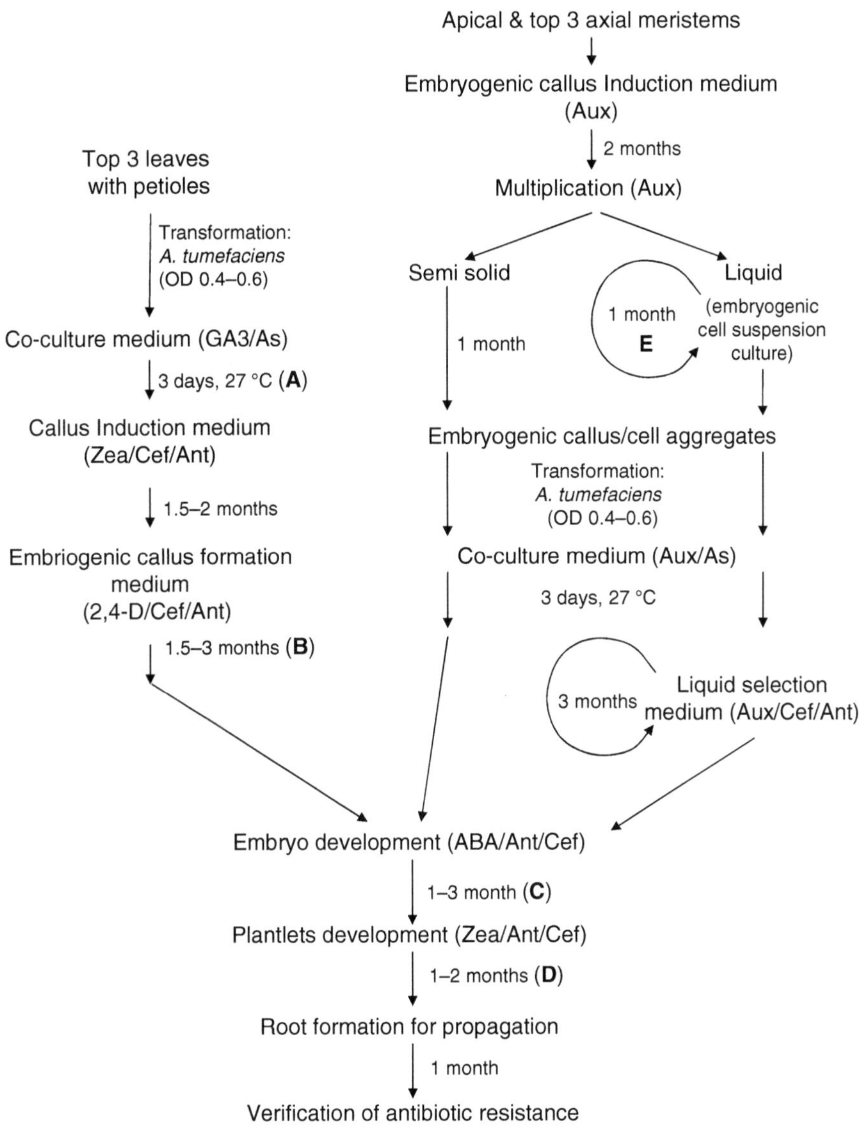

Plate 5.2 Schematic presentation of transformation protocols based on embryogenesis, using leaves with petioles (*left*) or meristems (*right*) as source explants. Approximate durations of each step, and hormones required in each media are indicated. Bold capital letters refer to pictures of the corresponding stage in Plate 5.4. Abbreviations: ABA, abscisic acid; Ant, antibiotic for transgenic selection (kanamycin or hygromycin); As, acetosyringone; Aux: auxin (4-FA, 2,4-D, 2,4,5-T have been used); Cef, Cefotaxime or Carbenicilin to eliminate *Agrobacterium*; Zea, zeatine

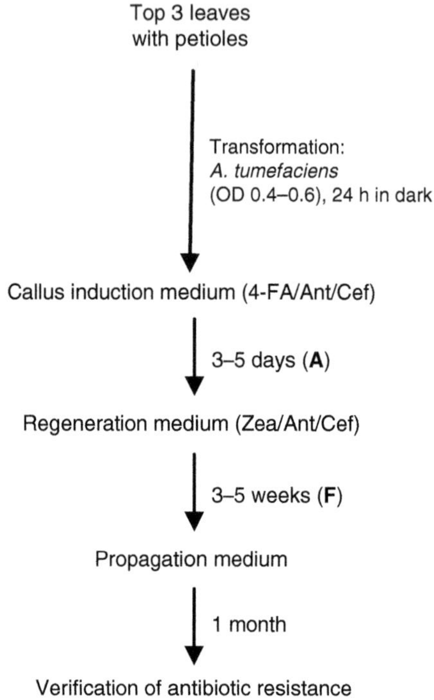

Plate 5.3 Schematic presentation of transformation protocol (from CIP) using whole leaf explants and plant regeneration via organogenesis including approximate durations of each step, and hormones required in media. Bold capital letters refer to pictures of the corresponding stage in Plate 5.4. See Plate 5.2 for abbreviations

However, in spite of the high-frequency regeneration observed in some studies (Song et al. 2004; Garcia et al. 2000; Gosukonda 1995a, 1995b; Medina 1991; Porobo Dessai 1995; Otani et al. 1996) plant recovery drops dramatically when coupled with transformation, which has been the major limitation to this methodology. It is therefore important to assess both transformation and regeneration efficiencies simultaneously. It is also important to verify the exact origin of the regenerated tissue. Although regeneration usually takes place via a callus originating from the cut surface of the petiole, this can vary markedly between genotypes. Some genotypes may regenerate shoots at high frequency directly from the middle of the petiole where *Agrobacterium* cannot penetrate easily upon infection. Wounding of tissue to increase access for *Agrobacterium* infection increases the numbers of transformed calli but is strongly counterproductive in terms of shoot regeneration.

Because of these problems there has been only limited success in transforming sweetpotato using *A. tumefaciens* and an organogenesis protocol. Morán et al. (1998) obtained 32 shoots from 45 leaf disc explants of cv. 'Jewel'. Half of these 32 kanamycin-resistant shoots rooted under kanamycin selection and were considered stable transformants. However, it was not clear whether all shoots originated from different explants, and only one transgenic line was tested by Southern blot analysis. Later, also stem and leaf explants of cv. 'Jewel' were used and it was reported that

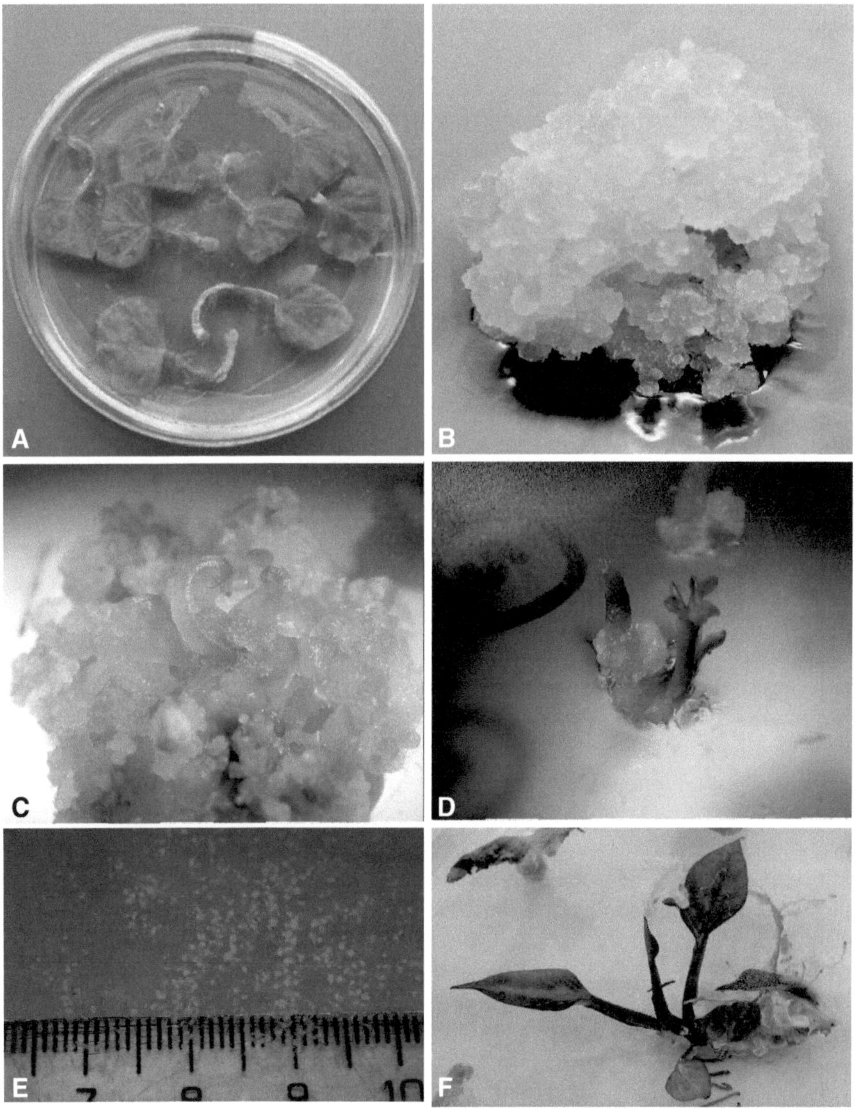

Plate 5.4 Sweetpotato tissues at different stages of development during transformation protocols (as indicated in Figs. 5.2 and 5.3). (**A**) leaves with petioles after infection with *Agrobacterium tumefaciens*. (**B**) Embryogenic callus. (**C**) Mature embryos on embryo maturation media. (**D**) Plant developing from embryo. (**E**) Embryogenic cell suspension. (**F**) Direct regeneration through organogenesis of plant from transgenic callus derived from the petiole of entire leaf

the best results were obtained with leaves and petioles as compared to stem or leaf explants, but no quantitative data were provided (Garcia et al. 2000). Luo et al. (2006) described a two-stage protocol including the first three days of cultivation on a medium with an auxin (4-FA or IAA), followed by culture on medium with the kinetin zeatine. They used leaf-disks, which formed compact callus that did not regenerate. They also tested internodes, which regenerated shoots with a high efficiency but all were non-transgenic. Entire leaves (including petiole) produced 57 regenerants out of the 160 explants tested and 15 regenerants were confirmed to be transgenic. In those two experiments in which the most efficient hormone combination (4FA-zeatin) was used, the transformation efficiency was 10–20% and plantlets were obtained within 6–10 weeks after inoculation with *Agrobacterium*. Low selection efficiency with 60–85% of regenerants representing escapes was however also observed. Experiments at CIP have indicated that results can be extremely variable between experiments and laboratories and the transformation efficiencies may be very low (<1%).

Despite the problems encountered with the organogenic regeneration even in the highly responsive sweetpotato genotypes, the shorter time required to regenerate plants as compared to embryogenesis protocols makes it more attractive. For example, CIP still invests major effort on optimizing organogenic protocols for new sweetpotato genotypes.

Conclusions

Genetic transformation offers possibilities to accelerate improvement of sweetpotato cultivars with traits that are slow or difficult to breed using conventional crossing schemes or for which no good sources have been found in sweetpotato germplasm. Sweetpotato is clonally propagated and a single superior transgenic event could be sufficient for obtaining a significantly improved cultivar for use. However, difficulties to transform and regenerate a broad range of sweetpotato genotypes with the current methods limit the approach. Novel traits cannot be introduced to all those cultivars that would make an elite variety if the weak point, such as susceptibility to viruses or weevils or the less optimal composition of starch or amino acids, was adjusted by introducing the missing trait by genetic transformation. Another challenge is that the knowledge of the mechanism behind viral synergism is not yet sufficiently advanced to make a targeted approach, which could prevent the severe viral diseases of sweetpotato. They are caused by SPCSV that eliminates effective antiviral defense and allows the co-infecting heterologous viruses to cause heavy yield reduction. Finally, sweetpotato is genetically a challenging polyploid plant. It can hardly be considered as a model species for studies in molecular biology. Therefore, sequence data on the genomic DNA and expressed gene sequences accumulate slowly, mainly in studies of relatively few research groups that work on applied aspects of this crop. Enhancement of molecular data and sequence information would be pivotal for improving sweetpotato by engineering its own genes and their expression.

References

Al-Mazrooei, S., Bhatti, M.H., Henshaw, G.G., Taylor, N.J. and Blakesley, D. (1997) Optimisation of somatic embryogenesis in fourteen cultivars of sweet potato [*Ipomoea batatas* (L.) Lam.]. *Plant Cell Rep.* 16: 710–714

Andersson, M., Melander, M., Pojmark, P., Larsson, H., Bülow, L. and Hofvander, P. (2006) Targeted gene suppression by RNA interference: An efficient method for production of high-amylose potato lines. *J. Biotech.* 123: 137–148

Atkinson, H.J., Grimwood, S., Johnston, K. and Green, J. (2004) Prototype demonstration of transgenic resistance to the nematode *Radopholus similis* conferred on banana by a cystatin. *Transgenic Res.* 13: 135–142

Brummell, D.A., Balint-Kurti, P.J., Harpster, M.H., Palys, J.M., Oeller, P.W. and Gutterson, N. (2003) Inverted repeat of a heterologous 30-untranslated region for high-efficiency, high-throughput gene silencing. *Plant J.* 33: 793–800

Cavalcante Alves, J.M., Sihachakr, D., Allot M., Tizroutine, S., Mussio, I., Servaes, A. and Ducreux, G. (1994) Isozyme modifications and plant regeneration through somatic embryogenesis in sweet potato [*Ipomoea batatas* (L.) Lam.]. *Plant Cell Rep.* 13: 437–441

Chee, R.P. and Cantliffe, D.J. (1989) Composition of embryogenic suspension cultures of *Ipomoea batatas* Poir. and production of individualized embryos. *Plant Cell Tissue Organ Cult.* 17: 39–52

Chee, R.P., Schultheis, J.R. and Cantliffe, D.J. (1990) Plant recovery from sweetpotato somatic embryos. *HortScience* 25: 795–797

Chen, L., Bhagsari, A. and Carter, J. (2006). Effects of medium composition and culture duration on *in vitro* morphogenesis of sweetpotato. *Biologia Plant.* 50: 114–117

Cipriani, G., Fuentes, S., Bello, V., Salazar, L.F., Ghislain, M. and Zhang, D.P. (2001) Transgene expression of rice cysteine proteinase inhibitors for the development of resistance against sweetpotato feathery mottle virus. In: *CIP program report 1999–2000.* International Potato Center, Lima, pp. 267–271

Cipriani, G., Michaud, D., Brunelle, F., Golmirzaie, A. and Zhang, D.P. (1999) Expression of soybean proteinase inhibitor in sweetpotato. In: *CIP program report 1997–1998.* International Potato Center, Lima, pp. 271–277

Desamero, N.V., Rhodes, B.B., Decoteau, D.R. and Bridges, W.C. (1994) Picolinic acid induced direct somatic embryogenesis in sweetpotato. *Plant Cell Tissue Organ Cult.* 37: 103–110

Dhir, S.K., Oglesby, J. and Bhagsari, A.S. (1998) Plant regeneration via somatic embryogenesis, and transient gene expression in sweet potato protoplasts. *Plant Cell Rept.* 17: 665–669

Dodds, J.H., Merzdorf, C., Zambrano, V., Sigüeñas, C. and Jaynes, J. (1991) Potential use of agrobacterium-mediated gene transfer to confer insect resistance in sweetpotato. In: *Sweetpotato pest management: A global perspective*, R.K. Jansson and K.V. Raman (Eds), West view Press, Oxford, UK, pp. 203–219

Dodds, J.H., Benavides, J., Buitron, F., Medina, F. and Sigueñas, C. (1992) Biotechnology applied to sweetpotato improvement. In: *Sweetpotato technology for the 21st century.* W.A. Hill, C.K. Bonsi and P.A. Loretan (Eds), Tuskegee Univ., Tuskegee, Ala. pp 7–19

Gama, M., Leite, R.P., Cordeiro, A.R. and Cantliffe, D.J. (1996) Transgenic sweet potato plants obtained by *Agrobacterium tumefaciens*-mediated transformation. *Plant Cell Tissue Organ Cult.* 46: 237–244

Garcia, R., Morán, R., Mena, J., Somontes, D., Pimentel, E., Aldúa, Z., López, A. and Garcia M. (2000) Sweet potato (*Ipomoea batatas* L.) regeneration and transformation technology to provide weevil (*Cylas formicarius*) resistance. Field trial results. In: *Plant Genetic Engineering: Towards the Third Millenium.* A.D. Arencibia (Ed), Elsevier Science BV, Amsterdam, pp . 112–117

Gibson, R.W., Mpembe, I., Alicai, T., Carey, E.E., Mwanga, R.O.M., Seal, S.E. and Vetten, H.J. (1998) Symptoms, aetiology and serological analysis of sweet potato virus disease in Uganda. *Plant Pathol.* 47: 95–102

Gosukonda, R.M., Porobo Dessai, A., Blay, E., Channapatna, S., Prakash, Peterson, C.M. (1995a) Thidiazuron-induced advetitious shoot regeneration of sweetpotato (*Ipomoea batatas*). *in vitro* Cell Devel. Biol. 31: 65–71

Gosukonda, R.M., Prakash, C.S. and Porobo Dessai, A. (1995b) Shoot regeneration *in vitro* from diverse genotypes of sweetpotato and multiple shoot production per explant. *HortScience* 30: 1074–1077

Gutiérrez, D.L., Fuentes, S. and Salazar, L.F. (2003) Sweetpotato virus disease (SPVD): distribution, incidence, and effect on sweetpotato yield in Peru. *Plant Dis.* 87: 297–302

Gutierrez-Campos, R., Torres-Acosta, J.A., Saucedo-Arias, L.J. and Gomez-Lim, M.A. (1999) The use of cysteine proteinase inhibitors to engineer resistance against potyviruses in transgenic tobacco plants. *Nat. Biotech.* 17: 1223–1226

Haasnoot, J., Westerhout, E.M. and Berkhout, B. (2007) RNA interference against viruses: strike and counterstrike. *Nat. Biotech.* 12: 1435–1443

Hahn, S.K. (1979) Effect of virus (SPVD) on growth and yield of sweetpotato. *Exp. Agr.* 15: 253–256

Hartemink, A.E., Poloma, S., Maino, M., Powell, K.S., Egenae, J. and O'Sullivan, J.N. (2000) Yield decline of sweet potato in the humid lowlands of Papua New Guinea. *Agr. Ecosys. Environ.* 79: 259–269

Herman, M., Ambarwati, A.D. and Sisharmini A. (2003) Developing transgenic sweetpotato for resistance to insect pests and diseases. In: *Progress in potato and sweetpotato research in Indonesia: Proceedings of the CIP-Indonesia Research Review Workshop 2002.* K.O. Fuglie (Ed), CIP-ESEAP, Bogor, Indonesia pp. 201–209. http://www.eseap.cipotato.org/MF-ESEAP/Publications/PSP-2003/PSPReport-2003.htm

Ihemere, U., Arias-Garzon, D., Lawrence, S. and Sayre, R. (2006) Genetic modification of cassava for enhanced starch production. *Plant Biotech. J.* 4: 453–465

Jiang, S.Y., Lu., Q.C., Zhai., H., Wu., L.S. and Wang, Y.P. (2004). Regeneration of sweet potato transgenic plants with oryzacystatin-I (*OCI*) gene. *Chinese J. Agric. Biotech.* 1: 99–102

Jones, A.L., Johansen, E.I., Bean, S.J., Bach, I. and Maule, A.J. (1998) Specificity of the resistance to pea seed-borne mosaic potyvirus in transgenic peas expressing the viral replicase (NIb) gene. *J. Gen. Virol.* 79: 3129–3137

Karyeija, R.F., Kreuze, J.F., Gibson, R.W. and Valkonen, J.P.T. (2000) Synergistic interactions of a potyvirus and a phloem-limited crinivirus in sweet potato plants. *Virology* 269: 26–36

Kimura, T., Otani, M., Noda, T., Ideta, O., Shimada, T. and Saito, A. (2001) Absence of amylose in sweet potato [*Ipomoea batatas* (L.) Lam.] following the introduction of granule-bound starch synthase I cDNA. *Plant Cell Rep.* 20: 663–666

Kreuze, J.F., Samolski Klein, I., Untiveros Lazaro, M., Cuellar Chuquiyuri, W., Lajo Morgan, G., Cipriani Mejía, P.G., Ghislain, M. and Valkonen, J.P.T. (2008) RNA silencing mediated resistance to a crinivirus (*Closteroviridae*) in cultivated sweetpotato (*Ipomoea batatas*) and development of sweetpotato virus disease following co-infection with a potyvirus. *Mol. Plant Pathol.* 9: 589–598

Latham, J.R. and Wilson, A.K. (2008) Transcomplementation and synergism in plants: implications for viral transgenes? *Mol. Plant Pathol.* 9: 85–103

Lawton, R., Winfield, S. and Daniell, H. (2000) Expression of green-fluorescent protein gene in sweet potato tissues. *Plant Mol. Biol. Rep.* 18: 139a–139i

Lea, P.J. and Leegood, R.C. (1999) Plant biochemistry and molecular biology, 2nd edition. John Wiley & Sons, Chichester. ISBN 0-471-97683-0

Lii, C.Y., Tsai, K.H. (1996) Effect of amylose content on the rheological properties of rice starch. *Cereal Chem.* 73: 415–420

Lim, S., Kim, Y.H., Kim, S.H., Kwon, S.Y., Lee, H.S., Kim, J.S., Cho, K.Y., Paek, K.Y. and Kwak, S.S. (2007) Enhanced tolerance of transgenic sweetpotato plants that express both CuZnSOD and APX in chloroplasts to methyl viologen-mediated oxidative stress and chilling. *Mol. Breed.* 19: 227–239

Lin, K.H., Fu, H., Chan, C.H., Lo, H.F., Shih, M.C., Chang, Y.M. and Chen, L.F.O. (2008) Generation and analyses of the transgenic potatoes expressing heterologous thermostable β-amylase. *Plant Sci.* 174: 649–657

Liu, J.R. and Cantliffe, D.J. (1984) Somatic embryogenesis and plant regeneration in tissue cultures of sweetpotato (*Ipomoea batatas* Poir.). *Plant Cell Rep.* 3: 112–115

Liu, Q.C., Zhai, H., Lu, D.H., Wang, Y. and Zhang, D.P. (1999) An efficient system of embryogenic suspension cultures and plant regeneration in sweetpotato. In: *CIP Program Report, 1997–1998.* International Potato Center, Lima, Peru, pp. 265–270

Liu, Q.C., Zhai, H., Wang, Y. and Zhang, D.P. (2001) Efficient plant regeneration from embryogenic suspension cultures of sweetpotato. *In Vitro Cell Dev. Biol. Plant* 37: 564–567

Luo, H.R., Santa Maria, M., Benavides, J., Zhang, D.P., Zhang, Y.Z. and Ghislain, M. (2006) Rapid genetic transformation of sweetpotato (*Ipomoea batatas* (L.) Lam) via organogenesis. *Afr. J. Biotech.* 19: 1851–1857

Maingi, D., Sivasupramaniam, S., Brown, G., Wagner, R., and Folk, W.R. (2002) *Bacillus thuringiensis* and cholesterol oxidase activity against the sweet potato weevil (Coleoptera: Curculionidae) in semi artificial diet assays. In Proceedings of the 5th Annual meeting of the Entomological Society of America, 2002

McKibbin, R.S., Muttucumaru, N., Paul, M.J., Powers, S.J., Burrell, M.M., Coates, S., Purcell, P.C., Tiessen, A., Geigenberger, P. and Halford, N.G. (2006) Production of high-starch, low-glucose potatoes through over-expression of the metabolic regulator SnRK1. *Plant Biotech. J.* 4: 409–418

Medina, L.F. (1991) Organogenesis "*in vitro*" a partir de entrenudos, Raices y hojas de nueve cultivares de camote (*Ipomoea batatas* Lam.) aspectos hormonales e histologicos. Universidad Peruana Cayetano Heredia, Lima, Peru

Miano, D.W., LaBonte, D.R. and Clark, C.A. (2008) Identification of molecular markers associated with sweet potato resistance to sweet potato virus disease in Kenya. *Euphytica* 160: 15–24

Moar, W.J., Mwanga, R.O.M., Odongo, B., Ekobu, M., Solera, M., and Ghislain, M. (2007) Progress towards engineering resistance to weevil in sweetpotato using Bt gene technology, In *Biotechnology, Breeding and Seed Systems for African Crops*. Maputo, Mozambique, The Rockefeller Foundation, NY, pp. 162

Morán, R., García, R., López, A., Zaldúa, Z., Mena, J., García, M., Armas, R., Somonte, D., Rodríguez, J., Gómez, M. and Pimentel, E. (1998) Transgenic sweet potato plants carrying the delta-endotoxin gene from Bacillus thuringiensis var. tenebrionis. *Plant Sci.* 139: 175–184

Mukasa, S.B, Rubaihayo, P.R. and Valkonen, J.P.T. (2006) Interations between a crinivirus, an impomovirus and a potyvirus in coinfection sweetpotato plants. *Plant Pathol.* 53: 458–467

Murata, T, Fukuoka, H. and Kishimoto, M. (1994) Plant regeneration from mesophyll and cell suspension of sweet potato, *Ipomoea batatas* (L.) Lam. *Breeding Sci.* 44: 35–40

Mwanga, R.O.M., Kriegner, A., Cervantes-Flores, J.C., Zhang, D.P., Moyer J.W. and Yencho, G.C. (2002) Resistance to sweetpotato chlorotic stunt virus and sweetpotato feathery mottle virus is mediated by two separate recessive genes in sweetpotato. *J. Am. Soc. Hort. Sci.* 127: 798–806

Nakashima, J.T. (1993) Studies on the Peruvian C1 isolate of Sweet Potato feathery Mottle Potyvirus. PhD Thesis, University of Birmingham, UK

Newell, C.A., Lowe, J.M., Merryweather, A., Rooke, L.M. and Hamilton, W.D.O. (1995) Transformation of sweet potato (*Ipomoea batatas* (L.) Lam.) with *Agrobacterium tumefaciens* and regeneration of plants expressing cowpea trypsin inhibitor and snowdrop lectin. *Plant Sci.* 107: 215–227

Njeru, R.W., Mburu, M.W.K., Cheramgon, E., Gibson, R.W., Obudho, E. and Yobera, D. (2004) Studies on the physiological effects of viruses on sweet potato yield in Kenya. *Ann. Appl. Biol.* 145: 71–76

Noda, T., Takahata, Y., Sato, T., Suda, I., Morishita, T., Ishiguro, K., and Yamakawa, O. (1998) Relationships between chain length distribution of amylopectin and gelatinization properties within the same botanical origin for sweet potato and buckwheat. *Carbohydrate Polymers* 37: 153–158

Okada, Y., Nishiguchi, M., Saito, A., Kimura, T., Mori, M., Hanada, K., Sakai, J., Matsuda, Y. and Murata, T. (2002) Inheritance and stability of the virus-resistant gene in the progeny of transgenic sweet potato. *Plant Breed.* 121: 249–253

Okada, Y., Saito, A., Nishiguchi, M., Kimura, T., Mori, M., Hanada, K., Sakai, J., Miyazaki, C., Matsuda, Y. and Murata, T. (2001) Virus resistance in transgenic sweetpotato [*Ipomoea batatas* L. (Lam)] expressing the coat protein gene of sweet potato feathery mottle virus. *Theor. Appl. Genet.* 103: 743–751

Otani, M., Mii, M., Handa, T., Kamada, H. and Shimada, T. (1993) Transformation of sweet potato (*Ipomoea batatas* (L.) Lam.) plants by *Agrobacterium rhizogenes*. *Plant Sci.* 94: 151–159

Otani, M., Masahiro, M. and Shimada, T. (1996) High frequency plant regeneration from leaf calli in sweet potato cv. Chugoku 25. *Plant Tissue Culture Lett.* 13: 23–27

Otani, M., Shimada, T., Kimura, T. and Saito, A. (1998) Transgenic plant production from embryogenic callus of sweet potato (*Ipomoea batatas* (L.) Lam.) using *Agrobacterium tumefaciens. Plant Biotech.* 15: 11–16

Otani, M., Wakita, Y. and Shimada, T. (2003) Production of herbicide-resistant sweetpotato (*Ipomoea batatas* (L) Lam.) plants by *Agrobacterium tumefaciens*-mediated transformation. *Breed. Sci.* 53: 145–148

Perera, S.C. and Ozias-Akins, P. (1991) Regeneration from sweetpotato protoplasts and assessment of growth conditions from flowsorting of fusion mixtures. *J. Amer. Soc. Hort. Sci.* 116: 917–922

Porobo Dessai, A., Gosukonda, R.M., Blay, E., Korsi Dumenyo, C., Medina-Bolivar, F. and Prakash, C.S. (1995) Plant regeneration of sweetpotato (*Ipomoea batatas* L.) from leaf explants *in vitro* using a two-stage protocol. *Scient. Hort.* 62: 217–224

Prakash, C.S. and Varadarajan, U. (1992) Genetic transformation of sweetpotato. In: *Sweetpotato Technology for the 21st Century*. W.A. Hill, C.K. Bonsi and P.A. Loretan (Eds), Tuskegee University, Tuskegee, Ala. pp. 27–37

Qaim, M. (2001). A prospective evaluation of biotechnology in semi-subsistence agriculture. *Agric. Econ.* 25: 165–175

Richardson, P.H., Jeffcoat, R., and Shi, Y-C. (2000) High-amylose starches: from biosynthesis to their use as fFood ingedients. *M.R.S. Bull.* 25: 20–24

Satin, M. (2005) Functional properties of starches. FAO (http://www.fao.org/ag/ags/agsi/starch41.htm)

Schwachtje, J., Minchin, P.E.H., Jahnke, S., van Dongen, J.T., Schittko, U. and Baldwin, I.T. (2006) SNF1-related kinases allow plants to tolerate herbivory by allocating carbon to roots. *Proc. Natl. Acad. Sci. USA* 103: 12935–12940

Schwall, G., Safford, R., Westcott, R.J., Jeffcoat, R., Tayal, A.,Shi, Y-C., Gidley, M.J. and Jobling, S.A. (2000) Production of very-high-amylose potato starch by inhibition of SBE A and B. *Nature Biotechnol.* 18: 551–554

Sheffield, F.M.L. (1953) Virus diseases of sweetpotato in parts of Africa. *Emp. J. Exp. Agric.* 21: 184–189

Shimada, T., Otani, M., Hamada, T. and Kim, S-H. (2006) Increase of amylose content of sweetpotato starch by RNA interference of the starch branching enzyme II gene (IbSBEII). *Plant Biotechnol.* 23: 85–90

Sihachakr, D. and Ducreux, G.C. (1987) Plant regeneration from protoplast culture of sweet potato (*Ipomoea batatas* Lam.) *Plant Cell Rep.* 6: 326–328

Sihachakr, D., Haicour, R., Cavalcante Alves, J.M., Umboh, I., Nzohgé, D., Servaes, A. and Ducreaux, G. (1997) Plant Regeneration in sweetpotato (*Ipomoea batatas* L., Convolvulaceae). *Euphytica* 96: 143–152

Smit, N.E.J.M. (1997). The effect of the indigenous cultural practices of in-ground storage and piecemeal harvesting of sweetpotato on yield and quality losses caused by sweetpotato weevil in Uganda. *Agric. Ecosys. Environ.* 64: 191–200

Song, G-Q., Honda, H. and Yamaguchi, K-H. (2004). Efficient Agrobacterium tumefaciens-mediated transformation of sweet potato (*Ipomoea batatas* (L.) Lam.) from stem explants using a two-step Kanamycin–Hygromycin selection method. *In Vitro Cell Dev. Biol. Plant* 40: 359–365

Sugiyama, M. (1999) Organogenesis *in vitro*. *Curr. Opin. Plant Biol.* 2: 61–64

Tairo, F., Mukasa, S.B., Jones, R.A., Kullaya, A., Rubaihayo, P.R. and Valkonen, J.P.T. (2005) Unravelling the genetic diversity of the three main viruses involved in sweetpotato virus disease (SPVD), and its practical implications. *Mol. Plant Pathol.* 6: 199–211

Thompson, P.G., Schneider, J.C. and Graves, B. (1994) Genetic variance component and heritability estimates of freedom from weevil injury to sweet potato. *J. Am. Soc. Hort. Sci.* 119: 620–623

Thompson, P.G., Schneider, J.C., Graves, B. and Sloan, R.C. (1999) Insect resistance in sweetpotato plant introductions. *HortScience* 34: 711–714

Untiveros, M., Fuentes, S. and Salazar, L.F. (2007) Synergistic Interaction of *Sweet potato chlorotic stunt virus* (*Crinivirus*) with carla-, cucumo-, ipomo-, and potyviruses infecting sweet potato. *Plant Dis.* 91: 669–676

Urwin, P.E., Atkinson, H.J., Waller, D.A. and McPherson, M.J. (1995) Engineered oryzacystatin-I expressed in transgenic hairy roots confers resistance to *Globodera pallida*. *Plant J.* 8: 121–131

Voinnet, O. (2005) Induction and suppression of RNA silencing: Insights from viral infections. *Nature Rev. Genet.* 6: 206–220

Wakita, Y., Otani, M., Hamada, T., Iba, K. and Shimada, T. (2001) A tobacco microsomal ω-3 fatty acid desaturase gen increases the linolénico acid content in transgenic sweet potato (*Ipomoea batatas*). *Plant Cell Rep.* 20: 244–249

Wesley, S.V., Helliwell, C.A., Smith, N.A., Wang, M-B., Rouse, D.T., Liu, Q., Gooding, P.S., Singh, S.P., Abbott, D., Stoutjesdijk, P.A., Robinson, S.P., Gleave, A.P., Green, A.G. and Waterhouse, P.M. (2001) Construct design for efficient, effective and high-throughput gene silencing in plants. *The Plant J.* 27: 581–590

Yi, G., Shin, Y.M., Choe, G., Shin, B., Kim, Y.S. and Kim, K.M. (2007) Production of herbicide-resistant sweet potato plants transformed with the *bar* gene. *Biotechnol. Lett.* 29: 669–675

Yu, B., Zhai, H., Wang, Y., Zang, N., He, S. and Liu, Q. (2007) Efficient Agrobacterium tumefaciens-mediated transformation using embryogenic suspension cultures in sweetpotato, *Ipomoea batatas* (L.) Lam. *Plant Cell Tissue Organ Cult.* 90: 265–273

Zhang, D., Cipriani, G., Rety, I., Golmirzae, A., Smit, N. and Michaud, D. (2000) Expression of protease inhibitors in sweetpotato. In: *Recombinant Protease Inhibitors in Plants.* D. Michaud (Ed), Landes Bioscience, Georgetown, Texas, pp. 167–178

Zheng, Q., Dessai, A.P. and Prakash, C.S. (1996) Rapid and repetitive plant regeneration in sweetpotato via somatic embryogenesis. *Plant Cell Rep.* 15: 381–385

Chapter 6
Propagation of Sweetpotatoes, *In Situ* Germplasm Conservation and Conservation by Tissue Culture

V. Gaba and Sima Singer

Propagation of sweetpotatoes is largely by vegetative propagation, which plays a major role in sweetpotato agriculture, improvement, conservation and diversity. Several topics involving propagation will be covered, including *in situ* germplasm conservation, virus cleaning by tissue culture, mass propagation by tissue culture, germplasm conservation by tissue culture, and the use of tissue culture for the international transport of sweetpotato germplasm.

Propagation of Sweetpotato

Sweetpotato *(Ipomoea batata* var. *batatas*) is an outbreeding, highly heterozygous polyploid (probably hexaploid) crop. As the seeds are heterogeneous, sweetpotato is propagated vegetatively. Sweetpotato seeds are currently used only for the purposes of breeding. Organs that can be used for the asexual propagation of sweetpotato include storage roots, shoot tips, and stem cuttings. Vine cuttings, often only 30 cm from the apex, are used for propagation. Sometime cuttings from lower down a shoot are used. Such sweetpotato cuttings root and establish within a few days, and further cuttings can be made shortly afterwards. Sweetpotato is an annual crop in temperate regions with seed bedded in a nursery, followed by transplanting, growth, harvest, and root storage (Dangler et al., 1994). In temperate zones the storage root is the main organ used for propagation: sprouts from the stored roots of the previous crop are generally used for planting. The storage roots can only be stored for a limited period (a few months). In tropical regions sweetpotato is a perennial plant, growing year-round in the field, reproduced from vine cuttings from the previous crop (Rao and Campilan, 2002).

V. Gaba (✉)
Department of Plant Pathology and Weed Science, ARO Volcani Center,
POB 6 Bet Dagan 50250, Israel
e-mail: vpgaba@volcani.agri.gov.il

G. Loebenstein, G. Thottappilly (eds.), *The Sweetpotato*,
DOI 10.1007/978-1-4020-9475-0_6,

Tissue Culture of Sweetpotato

Naturally, asexual reproduction of sweetpotato either in tropic or temperate zones is liable to infection by diseases and pests, making good quality high yields difficult to obtain. It is possible to use a group of techniques based on plant tissue culture for the crop improvement of sweetpotato. Tissue culture has greatly aided the exploitation of the sweetpotato as an international crop. Many of the issues for the development of the worldwide sweetpotato industry have concerned phytosanitary problems, for which solution tissue culture has been an important tool.

Much work has been done on the biotechnology of sweetpotato, but the most practical and useful to date has been the development of a group of technologies based on the tissue culture of stem node cuttings. These technologies were reviewed by Henderson et al. (1984), Love et al. (1987), Kuo (1991) and Guo et al. (2001).

Virus Cleaning by Tissue Culture

The production of virus-free sweetpotato is achieved almost exclusively by meristem culture *in vitro*. In meristem culture the essence is to take as large a meristem as possible, while excluding virus-infected tissue. Therefore a smaller apical (meristematic) explant might be clean from virus, but not a longer apex subtending some leaf primordia (Henderson et al. 1984). However, it is important to strike a balance, as too small apical explants may not be able to regenerate plants with some genotypes (Kuo, 1991). Survival of the meristem is an important constraint, and is improved by the addition of leaf primordia (Kuo, 1991). Additionally, use of shoot apices has been more successful than the apices from lateral buds (Kuo, 1991). This important methodology is the basis of the production of virus-clean sweetpotato. Virus cleaning of sweetpotato from internal cork virus was first reported by Nielsen (1960). Subsequently Mori (1971) demonstrated removal of three important sweetpotato viruses (internal cork, rugouse mosaic, feather mottle). Afterwards shoot tips 0.4–0.8 mm long of ten sweetpotato cultivars were introduced into tissue culture by Alconero et al. (1975), on Murashige and Skoog (1962) (MS) medium augmented with auxin and cytokinin. Complete plants developed *in vitro*, of which 47% were symptom-free on grafting onto *Ipomoea setosa*, a virus-sensitive indicator plant closely related to sweetpotato. Axenic culture of shoot tips of 0.25–0.4 mm length was utilized by Frison and Ng (1981) to produce plants that were repeatedly indexed negatively (i.e. were found to be not virus-infected by grafting to *I. setosa*). Henderson et al. (1984) report the work of Liao and Chung (1979) who were unable to free 50 mm long sweetpotato apices from virus infection by heat treatment of 38–42 °C for 30–90 days. Virus elimination was also not achieved by the tissue culture of 5 mm apices; eventually virus removal was achieved by culturing apices of 0.3–0.6 mm taken from plants that had been treated with high temperature (38–42 °C for 4 weeks). Again, growth of the apex into a plant was achieved *in vitro* on MS medium, with added auxin and cytokinin, before testing for virus-free status by inoculation of, or grafting onto, *I. nil* indicator plants (Liao and Chung, 1979).

The preparation of virus-clean and indexed sweetpotato is discussed in detail by Love et al., 1987, who describe the procedures for meristem isolation, sweetpotato plant regeneration and testing for the presence of virus. Love et al. (1987) stress the necessity for using a small final explant (0.1 mm) that contains the shoot apical meristem without leaf primordia, which have been removed during the dissection.

Virus cleaning by meristem culture was the basis of the California state program for sweetpotato improvement (Dangler et al., 1994). Virus cleaning by meristem culture has been reported many times since (e.g. Kuo et al., 1985; Marco and Walkey, 1992), without any marked improvements on the techniques noted here, although with much variation of the growth regulator (auxin/cytokinin) regime necessary for cultivar-dependent shoot regeneration from the explanted apex.

Virus cleaning of sweetpotato by meristem culture is considered much more effective than by thermotherapy (Kuo, 1991), with an 80% rate of virus-clean shoots noted by Kuo et al. (1985). Nevertheless, there are reports that virus-free sweetpotato propagation material has been produced by heat therapy alone (e.g. Huett, 1982), who treated apical cuttings for 6 weeks at 38 °C, (and failed to find virus particles by electron microscopy), with a 140% yield increase. A protocol for sweetpotato virus elimination combining meristem culture and thermotherapy is used by the International Potato Center (CIP), Lima, Peru (Panta et al., 2007). In this technique 3–4 week old *in vitro* sweetpotato plantlets are treated with thermotherapy for one month at 34–37 °C. Meristems of 0.2–0.35 mm are excised and regenerated with frequent sub-culture until a rooted plantlet is obtained, with a virus-cleaning efficiency of near 100% (Panta et al., 2007). Three MS-based media are used during the process, with additions of ascorbic acid, calcium nitrate, calcium panthotenate, putrescine, arginine, coconut milk and gibberellic acid (Panta et al., 2007).

Sweetpotato virus indexing is accomplished by grafting sweetpotato cuttings onto indicator plants (*I. setosa* and *I. nil*), and symptoms are evaluated after 4 weeks (Panta et al., 2007). Positive symptom observation is followed by Nitrocellulose Membrane Enzyme Linked Immunosorbent Assay (NCM-ELISA) with available antisera [*Sweetpotato feathery mottle virus* (SPFMV); *Sweetpotato latent virus* (SPLV); *Sweetpotato mild mottle virus* (SPMMV); *Sweetpotato virus G* (SPVG); *Sweetpotato mild speckling virus* (SPMSV); *Sweetpotato chlorotic fleck virus* (SPCFV); C-6 virus; *Sweetpotato chlorotic stunt virus* (SPCSV); *Sweetpotato caulimo-like virus* (SPCaLV); and *Cucumber mosaic virus* (CMV)]. Nucleic acid spot hybridization and PCR are optionally used to confirm the presence of some viruses for which antisera are not available. After the initial plant health check, infected accessions are submitted to the virus elimination process and subsequently re-checked (Panta et al., 2007).

Tissue Culture of Nodes of Sweetpotato

Once the sweetpotato shoot apex has been established in culture, and pathogen-indexed (as above), the next task is to multiply the shoots obtained, both for pathogen testing and for agricultural application.

Micro-propagation *in vitro* of sweetpotato is by the use of stem nodal segments (Love et al., 1987), on MS medium, with MS or Gamborg et al., (1968) vitamins, sucrose, and agar as a gelling agent, at 25–27 °C. The use of growth regulators is usually not necessary for micropropagation of established sweetpotato plants. The plants grow and develop in sterile conditions in test tubes on this semi-solid medium, protected from pathogens and the external environment. In tissue culture single node segments are commonly used for micropropagation, but in the case of sweetpotato the use of double node segments (Panta et al., 2007) is more successful, and leads to a much better rooting *in vitro* (Fig. 6.1). Rooting *in vitro* is important for two reasons: firstly, the explant develops into a plant much more quickly, and secondly, rooted plants are necessary for transfer outside the test tube. After a month the explanted section is generally well rooted, and several healthy leaves have developed (Fig. 6.2). Well-rooted plants with well-developed leaves (i.e. Fig. 6.2) can be transplanted to the greenhouse with >98% success, and after another month in an insect-proof greenhouse could be either used for virus testing, or transferred to farmers.

Alternatively, for testing virus infection, plants such as those in Fig. 6.2 can be transferred to sterile soil plugs in large tissue culture vessels (polycarbonate boxes 10 × 10 × 15 cm) with vented lids, so that the plant can grow in a growth room photoautotrophically under artificial light, but completely enclosed (Fig. 6.3) (Singer and Gaba, unpublished). This method provides the stem sections required

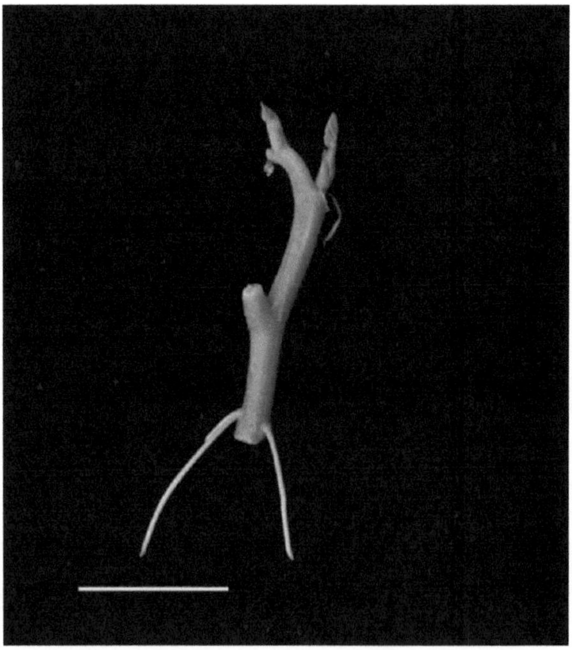

Plate 6.1 Rooting two-node section of sweetpotato (cv. Beauregard), after 2 weeks in culture; the bar = 10 mm

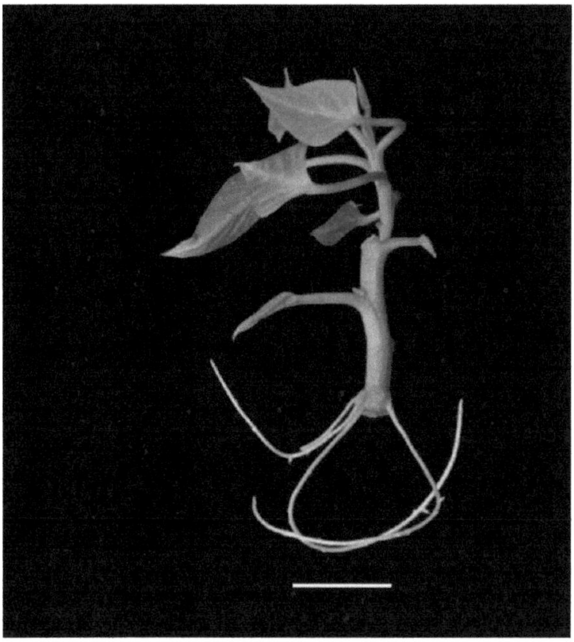

Plate 6.2 Well-developed rooted month-old sweetpotato plant (cv. Jasper); the bar = 10 mm

for the testing of the presence of viruses by graft-transmission to *Ipomoea* indicator plants. Use of such boxes is innovative, as it saves growing plants in a quarantine greenhouse for a month to obtain stem sections adequate for virus testing. Leaf samples from plants at the stage of Fig. 6.2 can also be tested for the presence of the pathogenic bacteria i.e. *Ralstonia solanacerum* and *Erwinia chrysanthemi* pv. Zeae. Plants can grow for 3 months *in vitro* (Fig. 6.4), and then can be subcultured (multiplied) by single or double node cuttings onto fresh medium, thereby achieving good *in vitro* multiplication rates.

Mass Propagation for Horticultural Use

In addition to the simple method of sweetpotato micropropagation outlined above, other methods of mass multiplication of sweetpotato for agricultural purposes have been proposed (a) photoautotrophic micropropagation; (b) somatic embryos for synthetic seed; and (c) bioreactors. The commercial status of these technologies is unclear.

(a) Photoautotrophic micropropagation has been pioneered by Kozai and colleagues. In this technique nodes are micropropagated in a relatively high photon flux density (compared to routine micropropagation) with an enhanced CO_2 level (3–4 times that of atmospheric levels), without sucrose added to the medium, on a variety of substrates (not only agar) producing larger plants than in the controls

Plate 6.3 Sweetpotato plants cv. (Jonathan) grown in large tissue culture vessels to produce stem sections for virus testing by grafting. The vented lid (with a 40 mm vent) has been removed for easier viewing; the box is 10 cm across

(Islam et al., 2004). This method has advantages compared to conventional micropropagation: the resulting plants do not require a hardening stage post *in vitro*, which greatly simplifies and thereby cheapens the micropropagation process. Additionally, as sugar is not used, axenic conditions not required, and the production vessel can be much larger than the usual tissue culture boxes (Heo et al., 2001). However, for this process special equipment is required, and forced introduction of the air fortified with CO_2 is required (Zobayed et al., 2000; Heo et al., 2001).

(b) The use of sweetpotato somatic embryos for synthetic seed production was examined by the group of Cantliffe. The concept was to mass produce somatic embryos in tissue culture, preferably in liquid culture in an air lift fermentor (Bieniek et al., 1995). Subsequently, embryos could be layered with an artificial "seed coat" to be suitable for planting by fluid drilling technology (Schultheis et al., 1990). The automated production of synthetic seeds in combination with fluid drilling technology could also use computer vision analysis of sweetpotato somatic embryos to select those most suitable (Padmanabhan et al., 1998). To add to the attractions of this technology repetitive embryogenesis of sweetpotato is possible, with secondary embryos produced from primary embryos (Zheng et al., 1996). Additionally, there is

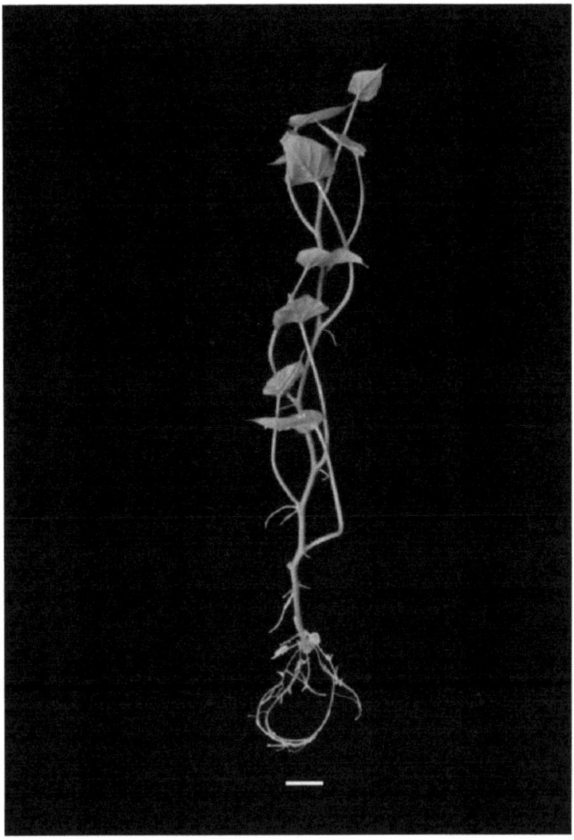

Plate 6.4 Sweetpotato plant (cv. Lamote) in tissue culture for three months. The plant has filled the test tube and ceased elongation. Eventually the leaves will die, but propagation is possible for several weeks more as the axillary buds remain viable; the bar = 10 mm

little cultivar limitation as the production of sweetpotato somatic embryos has been reported in a wide range of cultivars (e.g. Al-Mazrooei et al., 1997; Liu et al., 2001).

(c) Mass multiplication of sweetpotato nodes in liquid culture in a fermentor-type vessel (bioreactor) produced a large number of nodes for micropropagation (Paek et al., 2001), and sweetpotato somatic embryos were grown in an air lift fermentor (Bieniek et al., 1995).

A problem associated with plant tissue culture is the production of "off types", observable deviation from the initial plant material, termed somaclonal variation. Somaclonal variation is obviously a problem in any plant propagated vegetatively. A notable drawback of early studies in sweetpotato somatic embryogenesis was the production of abnormal embryos (DeWald and Cantliffe, 1988), although such problems have been much reduced or eliminated in later studies (e.g. Torres et al., 2001). Generally, micropropagation by node cuttings (from a naturally occurring bud) is the genetically safest procedure: Villordon and LaBonte (1996) determined using a PCR-based method that cloned plants from buds are more genetically uniform than plants propagated from repeated adventitious organogenesis from stored sweetpotato tubers. Somaclonal variation in *in vitro*-regenerated plants from apices during virus cleaning and propagation gave rise to a new cultivar (Moyer and Collins, 1983). There is concern as to the genetic stability of cultures that have been maintained in intermediate-term storage conditions for prolonged periods (Kuo, 1991; Guo et al., 2001), but results appear to be satisfactory with current methods (Guo et al., 2001).

Germplasm Conservation *In Vitro*

There are two major approaches to plant genetic resource conservation: *ex situ* and *in situ. The ex situ* approach with sweetpotato involves maintaining plants in field genebanks or botanical gardens, or *in vitro* as plants in test tubes, cryopreservation of apices, or a few "wild" *Ipomoea* genotypes stored as seeds at low temperatures.

There are two types of *ex situ* germplasm conservation collections. In the first category (the active germplasm collection) plants are maintained, evaluated and distributed as necessary. The second class of clonal collection is a base collection where plant material is not intended for distribution, but is maximally secured against loss (Volk and Walters, 2003).

There are many advantages to the maintenance of a germplasm collection *in vitro*, compared to a green house or field collection: reduction in storage area, year-round accessibility, maintenance of disease-free/-indexed state, rapid *in vitro* propagation, simplicity of shipping, considerable savings in labor and costs (Jarret and Florkowski, 1990; Towill, 2005). The management of field and *in vitro* germplasm collections is elaborated by Reed et al. (2004).

Sweetpotato germplasm is conserved by the United States National Plant Germplasm System by two different *in vitro* methods, at geographically separated sites. The active sweetpotato germplasm collection of the US National Plant Germplasm System is in Griffin, GA, USA. In Griffin, GA, plants are routinely transferred (sub-cultured) to fresh medium for short- to intermediate-term storage, as accessibility for distribution is a major concern. Some plant material is maintained in simple micropropagation (as described above) for ease of distribution worldwide. The active collection is backed up with plants kept in intermediate-term storage at a reduced temperature (21 °C), where plants are maintained for 6–18 months prior to sub-culture (Jarret, 1989). The Griffin, GA, *Ipomoea* collection includes 683 accessions in tissue culture, 423 "wild" accessions maintained as seed, and none in the field or screenhouse (Volk and Walters, 2003).

The CIP sweetpotato collection at CIAT, Cali, Colombia is now wholly maintained *in vitro*, and currently numbers 4,493 accessions (CIP, 2007b). Accessions are maintained in a 16 h photoperiod, for 4 months at 23–27 °C in 2,000 lux, or for 14 months at 18–21 °C in 3,000 lux. The intermediate-term storage medium is based on MS salts with the addition of calcium pantotenate, calcium nitrate, arginine, ascorbic acid, putrescine, sucrose, 3–4% sorbitol, and Phytagel (an agar substitute)

(CIP, 2007b). Prior to transfer, accessions are evaluated according to a number of criteria including the physiological status, possible culture problems (poor rooting due endogenous bacteria) and the phytosanitary status. Post-storage assessments include checks for contamination, rooting ability and viability (and not for genetic stability) (CIP, 2007b). Similar discussions of intermediate-term conservation are made by Kuo (1991), Golmirzaie and Toledo (1998), and Guo et al. (2001). Reduced temperature (18–20 °C), light intensity and an 8–10 h photoperiod, combined with the addition of mannitol (1–1.5%)(or other hyperosmolar material) to the medium are recommended by Guo et al. (2001), as well as reduction in the carbon source (sucrose concentration) and the possible use of a range of growth retardants.

The use of plant growth inhibitors (e.g. ABA and antigibberellins) to limit the sub-culture frequency to years is also possible (Jarret and Gawel, 1991a, b), but some substances (e.g. malaic hydrazide, cycocel, glyphosate) have toxic effects (Golmirzaie and Toledo, 1998). However, in practice, chemical growth retardants are not used in an active or intermediate-term sweetpotato germplasm collection, as such substances hinder the ability to distribute stocks on short notice (Jarret, personal communication).

The base collection of sweetpotato in the US National Plant Germplasm System is now maintained at Fort Collins, CO, in low temperature storage in liquid nitrogen for inactivity, long-term security and cost saving (Volk and Walters, 2003). Cryogenic storage is a back up for the active collection, in case of disaster (NCGRP, 2007). Twenty different sweetpotato lines are maintained in cryopreservation storage at the Fort Collins facility, using a procedure based on that of Hirai and Sakai (2003) where shoot tips are preserved utilizing encapsulation, dehydration and vitrification. There are many publications concerning the cryopreservation of sweetpotato; the theory and practice of cryopreservation is covered by Reed (2008), and she gives protocols for cryopreservation of embryogenic callus (Bhatti et al., 1997) and shoot tips (by Hirai based on Matsumoto et al., 1995).

In Situ Germplasm Conservation

In situ conservation of sweetpotato is carried out in the Asian-Pacific region particularly Indonesia (Irian Jaya, on the island of New Guinea) and the Philippines (see Rao and Campilan, 2002). The *in situ* approach conserves sweetpotato genetic resources in the natural habitat of a traditional crop cultivar (landrace) – a farm or home garden, by farmers (Rao and Sthapit, 2002). Landraces are robustly adapted to their environment, and are an important plant genetic resource, providing resistances for an assortment of climatic, edaphic and biological stresses selected over generations for a particular role and ecological niche. Additionally, *in situ* conservation has the capacity to store large numbers of crop alleles and genotypes. An extra advantage of *in situ* conservation is the ongoing evolution of the farm-conserved material (Sthapit et al., 2002). The correct way to preserve landraces is therefore in their native locale within the traditional agricultural system (Rao and Sthapit, 2002).

The preservation of plant genetic resources in such a manner requires careful building of various levels of community resources (involving public and commercial sectors) over a prolonged period, ensuring that both farmers and the community benefit from *in situ* conservation of their crop. The benefits to farmers include subsistence or livelihood, cash income, cultural use, pride, and crop adaptation to local biotic or abiotic stresses (Rao and Sthapit, 2002). Research and monitoring is needed for successful *in situ* conservation: analysis of genetic variation, registration of species numbers and surveillance of ecological condition and habitat alteration, including farming methods for which documentation of farmers' knowledge of diversity and uses is necessary (Rao and Sthapit, 2002). Identification of the expert farmers who can maintain such a project is often essential. However, Yaku and Widyastuti (2002) felt that exclusion of less up-to-date farmers was problematic as they fell behind, and that site selection and farmer involvement should be arbitrary. Successful *in situ* conservation requires skill in community development. There is a suggestion that *ex situ* collections are more vulnerable to mismanagement and pathogen losses (Sthapit et al., 2002). However, as we will see later the opposite can be true. Challenges such as population increase, land degradation, poverty, ecological change and the introduction of modern sweetpotato cultivars (or cash crops such as rice or cassava) have contributed to the loss of local cultivars in areas where sweetpotato is not a major crop (Sthapit et al., 2002).

Several fascinating case studies of sweetpotato *in situ* conservation have been published (see Rao and Campilan, 2002). In the Philippines a gender division has been observed, with men doing conservation work on farms while the women similarly work in home gardens (Boncodin, 2002). In Bukidnon, teachers and pupils help conserve local traditional sweetpotato varieties by maintaining a school sweetpotato genebank, teaching a culture of conservation, and being a source of good planting stock for the community (Boncodin, 2002). Root crop traders have exceptional knowledge and an influence over conservation. Such traders are typically the source of planting materials and can influence which varieties are planted depending on market demands (Boncodin, 2002). Subsistence growers keep a larger range of genotypes than do commercial growers. Despite being secondary crops in intricate livelihood schemes, sweetpotato is very important for subsistence farmers in a crisis, as a seasonal stable, animal food, for local markets, and money source in an emergency. Multiple uses result in greater crop variety (Boncodin, 2002). Conservation of a cultivar depends mostly on its value, and farmers usually abandon varieties that no longer fulfil certain requirements. For farmers conservation cannot be separated from the use of a crop, a decision rather different from that of a researcher i.e. local people appear to practice a general tactic: conservation through use (Boncodin, 2002). Flexibility in cropping is also important, and aids in responding to market forces e.g. in Baloi, Lanao del Norte, Philippines, by moving preferred local sweetpotato cultivars to upland plots to plant cassava as a cash crop, and subsequently returning sweetpotato to the valley farms as the value of cassava declined (Boncodin, 2002). In Baguio City, The Philippines, women home gardeners have begun sweetpotato-based snack food processing using local cultivars grown in Aringay, La Union, earning a living whilst improving sweetpotato variety conservation (Boncodin, 2002).

An important secondary centre of sweetpotato biodiversity is the island of New Guinea. In the Indonesian (western) portion of the island (Irian Jaya province) surveys found more than 800 local cultivars (see Yaku and Widyastuti, 2002). Sweetpotato has been the staple food on the island since introduction in the 14th century (Yen, 1974). An extended *in situ* conservation project is reported by Yaku and Widyastuti (2002), where they worked with the Dani people in the Baliem valley of Irian Jaya province. The Dani are dependent on sweetpotato for their livelihood, grow a wide range of landraces, and have developed an intensive sweetpotato-based agriculture (Yaku and Widyastuti, 2002). Besides being a staple food, the Dani feed sweetpotato to animals (especially pigs), and use it for religious rites. Widyastuti et al. (2002) discuss the role of Dani women farmers, and their great knowledge of each cultivar: in which agroecozone each should be planted and on what soil type, and where in the mixed sweetpotato beds each cultivar should be placed. Additionally, Dani women understand which cultivars are used for food for babies or infants (containing β-carotene), adults, animal food, or ritual purposes. Other women's knowledge includes the special cultivars to be used for a new field, early-maturing varieties, and the useful differences between traditional and newly introduced (bred) cultivars. The persistent stems of old sweetpotato cultivars do not dehydrate rapidly after harvest and continue branch production until they can be used for planting the next crop of pig feed. New cultivars commonly cease branch production following harvest (Widyastuti et al., 2002). Women's knowledge of sweetpotato culture is learned at different ages and life stages (Widyastuti et al., 2002). The local population has realized the importance of *in situ* conservation to the maintenance of traditional varieties (Yaku and Widyastuti, 2002). As elsewhere, for the reasons already discussed, the number of local cultivars is in decline. However, additionally in the Baliem valley there are other reasons for landraces' decline: replacement with early-maturing cultivars, alteration of cultural standards and the loss of sweetpotato cultivars utilized in abandoned important cultural traditions, and natural disasters (drought, flood, pests) (Yaku and Widyastuti, 2002).

There are several cases of re-introduction, where sweetpotato cultivars lost by local peoples have been returned to their lands from *ex situ* genebanks. Following a severe drought most of the local cultivars were lost in some areas of the Baliem valley. After trying to recover these genotypes from neighbouring villages, or disused plots, eventually the missing cultivars were re-introduced from the Lembang, West Java, *ex situ* collection, co-coordinated by CIP (Yaku and Widyastuti, 2002). Nakatani (2002) reports similar experiences, where local cultivars were re-introduced to two different areas of Japan from an *ex situ* genebank. In one of these areas the local cultivars had fallen into disuse, and in the other were completely destroyed by a tidal wave (Nakatani, 2002). The sweetpotato cultivars used for cultural purposes by New Zealand Maoris were collected by Yen (1974), abandoned in their homeland, maintained *ex situ* in Japan, and later re-introduced from Japan following a Maori cultural revival (Nakatani, 2002). These stories illustrate the necessity for both *ex situ* and *in situ* conservation and for back up at multiple locations to ensure preservation of important genotypes.

The global situation of *ex situ* conservation of sweetpotato has been reviewed recently (CIP, 2007a). Several important *ex situ* sweetpotato collections (e.g. CIP, US National Plant Germplasm System) are now maintained exclusively *in vitro*, as this is a secure and cheap conservation method in an industrialized country. Nevertheless, KONARC (Japan) maintains 1537 accessions (in 2006), with propagation of the storage roots annually in fields. Storage roots are stored during the winter (November-March) (Koji Ishiguro, personal communication).

Much of the diversity of sweetpotato is spread throughout a vast zone, including the Asia-Pacific region, where more than 6,800 native cultivars are maintained in genebanks (Fuglie et al., 2002). A large amount of the Asia-Pacific region endemic sweetpotato cultivars of potential still need to be evaluated and conserved properly, and the use of *ex situ* germplasm conservation is important (Rao and Campilan, 2002; CIP, 2007a), and much cheaper.

For Asian countries (China, India, Indonesia, Philippines, Thailand) using *ex situ* field genebanks, Fuglie et al. (2002) found that the annual cost of maintaining sweetpotato genetic resources varied widely: from \$21.33 per accession in China to \$1.49 per accession in Indonesia. The high cost in China is due to the need for winter storage. Without storage, Asian countries should be able to maintain their field collections for an average of about US\$2.00 per accession per year. The small investments on Asian sweetpotato genetic resource conservation and variety improvement have already begun to pay off, as these programs released 121 improved sweetpotato varieties between 1981 and 2001 (Fuglie et al., 2002).

Other major sweetpotato collections are moving to *in vitro* storage. In India, 373 sweetpotato accessions from the national collections are maintained *in vitro* and 3040 accessions in field genebanks (Naskar et al., 2002). Guo et al. (2001) report on Chinese improvements in *in vitro* culture for intermediate-term storage of sweetpotato genotypes. Other Asia-Pacific countries (Malaysia, Indonesia, Philippines, Papua New Guinea) now maintain *in vitro* part of their important sweetpotato germplasm collections (see Rao and Hermann, 2001; Rao and Campilan, 2002).

Moving Sweetpotato Germplasm Between Countries

Sweetpotato is transported internationally by shipping plants as plantlets in tissue culture under strict quarantine regulations, laid out by Moyer et al. (1989). Moyer et al. (1989) recommendations for vegetative propagating material are as follows. Antibiotics and charcoal should not be added to the media of *in vitro* cultures transported internationally. All *in vitro* material should be virus cleaned by meristem-tip culture, with or without thermotherapy, preferably in the source country or a third country (Frison and Ng, 1981). *In vitro* cultures should be tested for the presence of bacteria (see Moyer et al., 1989). One to four nodes of the meristem tip-regenerated plantlets should be maintained *in vitro* to prevent recontamination. The remainder of the plant should be grown for virus testing in an insect-proof greenhouse. The possibility of systemic bacterial and fungal contamination in cultures should be examined before dispatch and upon receipt. Virus testing of all cultures should be performed in the originating country, in an intermediate quarantine centre, or in quarantine in the receiving country. The movement of sweetpotato seeds is also discussed by Moyer et al. (1989), but occurs rarely for modern cultivars with genetically segregating germplasm.

The above regulations are still followed by major centers such as CIP, which issues both a CIP Phytosanitary Statement and a Peruvian Phytosanitary Certificate with each shipment listing the pathogens for which testing was performed (CIP, 2008).

Much useful information on sweetpotato conservation, tissue culture and control of international movement of sweetpotato germplasm is available from Bioversity International (http://www. bioversity international .org/) and CIP (http://www.cipotato.org/).

Acknowledgments Contribution from the Agricultural Research Organization, The Volcani Center, Bet Dagan, Israel, No. 504/08. The authors thank Drs. Koji Ishiguro and Robert L. Jarret for personal communications.

References

Alconero, R., Santiago, A.G., Morales, F. and Rodriguez, F. 1975. Meristem tip culture and virus indexing of sweet potatoes. *Phytopathology* 65: 769–773.

Al-Mazrooei, S., Bhatti, M.H., Henshaw, G.G., Taylor, N.J. and Blakesley, D. 1997. Optimisation of somatic embryogenesis in fourteen cultivars of sweet potato (*Ipomoea batatas* (L.) Lam.). *Plant Cell Rep.* 16: 710–714.

Bhatti, M.H., Percival, T., Davey, C.D.M., Henshaw, G.G. and Blakesley, D. 1997. Cryopreservation of embryogenic tissue of a range of genotypes of sweet potato (*Ipomoea batatas* (L) Lam.) using an encapsulation protocol. *Plant Cell Rep.* 16: 802–806.

Bieniek, M.E., Harrell, R.C. and Cantliffe, D.J. 1995. Enhancement of somatic embryogenesis of *Ipomoea batatas* in solid cultures and production of mature somatic embryos in liquid cultures for application to a bioreactor production system. *Plant Cell Tissue Organ Cult.* 41: 1–8.

Boncodin, R.M. 2002. Root crop genetic resources R&D in the Philippines: a review of UPWARD experiences. In: Rao, V.R. and Campilan, D (eds), *Exploring the Complementarities of In Situ and Ex Situ Conservation Strategies for Asian Sweetpotato Genetic Resources*, pp. 159–167.

CIP. 2007a. Global strategy for ex-situ conservation of sweetpotato genetic resources. Sweetpotato Genetic Resources Conservation – CIP Genebank November 2007. http://www.croptrust.org/documents/web/SweetPotato-Strategy-FINAL-12Dec07.pdf (link checked 7 April 2008).

CIP. 2007b. Workshop on "*Clonal crops germplasm management*" Lima, Peru, Nov 12–16, 2007. http://research.cip.cgiar.org/confluence/download/attachments/13021/Sweetpotato+and+ARTCs_GPG2a.pdf (link checked 11 March 2008).

CIP. 2008. http://research.cip.cgiar.org/confluence/display/SMTA/Important+information (link checked 11 March 2008).

Dangler, J.M., Scheuerman, R.W., Campbell, R.N. and Clark, C.A. 1994. Meristem-tip culture and California's clean foundation sweetpotato program. *HortTechnology* 4: 227–228.

DeWald, S.G. and Cantliffe, D.J. 1988. Histology of callus initiation and somatic embryogenesis from sweet potato (*Ipomoea batatas* Poir.) shoot apices. *HortScience* 23: 812.

Frison, E.A. and Ng, S.Y. 1981. Elimination of sweet potato virus disease agents by meristem tip culture. *Trop. Pest Manag.* 27: 452–454.

Fuglie, K.O., Mahalaya, S. and Suri, F. 2002. The economics of sweetpotato genetic resource conservation and varietal improvement in Asia. In: Rao, V.R. and Campilan, D (eds), *Exploring the Complementarities of In Situ and Ex Situ Conservation Strategies for Asian Sweetpotato Genetic Resources*, pp. 79–122.

Gamborg, O., Miller, R. and Ojima, K. 1968. Nutrient requirement of suspension cultures of soybean root cells. *Exp. Cell Res.* 50: 151–158.

Golmirzaie, A. and Toledo, J. 1998. *In vitro* conservation of potato and sweet potato germplasm. *CIP Program report 1997–8*, pp. 351–356. http://www.cipotato.org/publications/program_reports/97_98/41invitr.pdf (link checked 7 April 2008).

Guo, X.D., Zhou, M.D., and Wang, Y. 2001. *In vitro* conservation of sweetpotato germplasm. In: Rao V.R. and Hermann M. (eds), *Conservation and Utilization of Sweetpotato Genetic Diversity in Asia*, pp. 16–24.

Heo, J.W., Wilson, S.B. and Kozai, T. 2001. A forced ventilation micropropagation system for photoautotrophic production of sweetpotato plug plantlets in a scaled-up culture vessel: I. Growth and uniformity. *HortTechnology* 11: 90–94.

Henderson, J.H.M., Phills, B.R. and Whatley B.T. 1984. Sweet Potato. In: Sharp, W.R., Evans, D.A., Ammirato, P.V., Yamada, Y. (eds), *Handbook of Plant Cell Culture*, Volume 2. Crop species, Macmillan, New York, pp. 302–326.

Hirai, D. and Sakai A. 2003. Simplified cryopreservation of sweet potato [*Ipomoea batatas* (L.) Lam.] by optimizing conditions for osmoprotection. *Plant Cell Rep.* 21: 961–96.

Huett, D.O. 1982. Evaluation of sources of propagating-material for sweet-potato production. *Sci. Hortic.* 16: 1–7.

Islam, A.F.M.S., Zobayed, S.M.A., Hossain, K.G. and Kozai, T. 2004. Photoautotrophic growth of sweet potato plantlets *in vitro* as affected by root supporting materials, CO_2 concentration, and photosynthetic photon flux. *Trop. Agric.* 81: 80–86.

Jarret, R.L. 1989. A repository for sweetpotato germplasm. *HortScience* 24: 885–8.

Jarret, R.L. and Gawel, N. 1991a. Abscisic acid-induced growth inhibition of sweet potato (*Ipomoea batatas* L.) *in vitro*. *Plant Cell, Tissue Organ Cult.* 24: 13–18.

Jarret, R.L. and Gawel, N. 1991b. Chemical and environmental growth regulation of sweetpotato (*Ipomoea batatas* (L.) Lam.) *in vitro*. *Plant Cell, Tissue Organ Cult.* 25: 153–159.

Jarret R.L. and Florkowski W.J. 1990. *In vitro* active vs. field genebank maintenance of sweet potato germplasm: major costs and considerations. *HortScience* 25: 141–146.

Kuo, C.G., Shen, B.J., Shen, M.J., Green, S.K. and Lee, D.R. 1985. Virus-free sweet potato storage roots derived from meristem-tips and leaf-cuttings. *Sci. Hortic.* 26: 231–240.

Kuo, C.G. 1991. Conservation and distribution of sweet potato germplasm. In: Dodds J.H. (ed), *In Vitro Methods for Conservation of Plant Genetic Resources*. Chapman and Hall, London. Pp 123–149.

Liao, C.H. and Chung, M.L. 1979. Shoot tip culture and virus indexing in sweet potato. *J. Agric. Res. China 28*: 139–144.

Liu, Q.C., Zhai, H., Wang, Y. and Zhang, D.P. 2001. Efficient plant regeneration from embryogenic suspension cultures of sweetpotato. *In Vitro Cell. Develop. Biol. - Plant* 37: 564–567.

Love, S.L., Rhodes, B.B. and Moyer, J.W. 1987. Meristem-tip culture and virus indexing of sweet potatoes. Practical manuals for handling crop germplasm *in vitro* 1. *International Board for Plant Genetic Resources*. IBPGR Headquarters, Rome, 1987. http://www. bioversityinternational.org/publications/Web_version/117 (link checked 11 March 2008).

Matsumoto, T., Sakai, A., Takahashi, C., and Yamada, K. 1995. Cryopreservation of *in vitro*-grown apical meristems of wasabi (*Wasabia japonica*) by encapsulation-vitrification method. *Cryo-Letters* 16: 189–196.

Marco, S. and Walkey, D.G.A. 1992. Rapid propagation of sweet potato (*Ipomoea batatas*) by meristem tip culture in relation to virus eradication. *Phytopathol. Mediterr.* 31: 185–187.

Mori, K. 1971. Production of virus-free plant by means of meristem culture. *J. Agric. Res. Quart.* 6: 1–7.

Moyer, J.W. and Collins W.W. 1983. "Scarlet" sweet potato. *HortScience* 18: 111–112.
Moyer, J.W., Jackson, G.V.H. and Frison, E.A. (eds), 1989. FAO/IBPGR *Technical Guidelines for the Safe Movement of Sweet Potato Germplasm.* Food and Agriculture Organization of the United Nations, Rome/International Board for Plant Genetic Resources, Rome. http://www.bioversityinternational.org/publications/Pdf/503.pdf (link checked 11 March 2008).
Murashige T. and Skoog, F. 1962. A revised medium for rapid growth and bio-assay with tobacco tissue cultures. *Physiol. Plant.* 15: 473–497.
Naskar, S.K., Amma, C.S.E. and Mukherjee, A. 2002. Present status of sweetpotato genetic resources in India. In: Rao, V.R. and Campilan, D (eds), *Exploring the Complementarities of In Situ and Ex Situ Conservation Strategies for Asian Sweetpotato Genetic Resources*, pp. 33–34.
Nakatani, M. 2002. From ex situ to *in situ*: Japan's domestic and international experiences. In: Rao, V.R. and Campilan, D (eds), *Exploring the Complementarities of In Situ and Ex Situ Conservation Strategies for Asian Sweetpotato Genetic Resources*, pp. 166–168.
NCGRP, 2007. http:*//www.ars.usda.gov/Main/docs*.htm?docid=13859 (checked 11 March 2008).
Nielsen, L.W. 1960. Elimination of internal cork virus by culturing apical meristems of infected sweetpotatoes. *Phytopathology* 50: 840–841.
Padmanabhan, K, Cantliffe, D.J., Harrell, R.C. and Harrison, J. 1998. Computer vision analysis of somatic embryos of sweet potato [*Ipomoea batatas* (L.) Lam.] for assessing their ability to convert to plants. *Plant Cell Rep.* 17: 681–684.
Paek, K.Y., Chakrabarty, D. and Hahn, E.J. 2001. Application of bioreactors for large-scale micropropagation systems of plants. *In Vitro Cell. Develop. Biol. - Plant* 37: 149–157.
Panta, A., Espinoza, C. and Roca W. 2007. Health status testing and virus elimination in sweetpotato.http://research.cip.cgiar.org/confluence/download/attachments/13542/4.+SP+Virus+elimination.pdf?version=1 (link checked 11 March 2008).
Reed, B.M. 2008. *Plant Cryopreservation: A Practical Guide*. (editor) Springer.
Reed, B.M., Engelmann, F., Dulloo, M.E. and Engels, J.M.M. 2004. Technical guidelines for the management of field and *in vitro* germplasm collections. *IPGRI Handbooks for Genebanks No. 7.* International Plant Genetic Resources Institute, Rome, Italy. http://www.bioversityinternational.org/Publications/Pdf/1016.pdf (link checked 7 April 2008).
Rao, V.R. and Campilan D. 2002. Exploring the complementarities of *in situ* and ex situ conservation strategies for asian sweetpotato genetic resources. Proceedings of 3rd International Workshop Asian Network for Sweetpotato Genetic Resources (ANSWER) held in Denpasar, Bali, Indonesia on 2–4 October, 2001. *International Plant Genetic Resources Institute Regional Office for Asia, the Pacific and Oceania* (IPGRI-APO), Serdang, Malaysia (editors). (pdf available from http://www.bioversityinternational.org/index.asp).
Rao, V.R. and Hermann, M. 2001. Conservation and utilization of sweetpotato genetic diversity in Asia. Proceedings of the Asian Network for Sweetpotato Genetic Resources Workshop, 3–5 November 1999, Bogor, Indonesia. IPGRI-APO, Serdang, Malaysia (editors). http://www.bioversityinternational.org/Publications/Pdf/739.pdf (link checked 7 April 2008).
Rao, V.R. and Sthapit, B. 2002. *In situ* conservation – a component of complementary conservation strategy. In: Rao, V.R. and Campilan, D (eds), *Exploring the Complementarities of In Situ and Ex Situ Conservation Strategies for Asian Sweetpotato Genetic Resources*, pp. 123–135.
Schultheis, J.R., Cantliffe, D.J. and Chee, R.P. 1990. Optimizing sweet potato (*Ipomoea batatas* (L.) Lam.) root and plantlet formation by selection of proper embryo developmental stage and size, and gel type for fluidized sowing. *Plant Cell Rep.* 9: 356–359.
Sthapit, B., Rao, V.R. and Jarvis, D. 2002. Conceptual basis of *In situ* conservation of sweetpotato agrobiodiversity. In: Rao, V.R. and Campilan D. (eds), *Exploring the Complementarities of in Situ and Ex Situ Conservation Strategies for Asian Sweetpotato Genetic Resources*, IPGRI 2002. pp. 136–141.
Torres, A.C., Ze, N.M. and Cantliffe, D.J. 2001. Abscisic acid and osmotic induction of synchronous somatic embryo development of sweet potato. *In Vitro Cell. Develop. Biol - Plant* 37: 262–267.
Towill, L.E. 2005. Germplasm preservation. In: Trigiano, R.N. and Gray, D.J. (eds), *Plant Development and Biotechnology*. CRC Press, Boca Raton, FL, pp. 277–284.

Villordon, A.Q. and LaBonte, D.R. 1996. Genetic variation among sweetpotatoes propagated through nodal and adventitious sprouts. *J. Am. Soc. Hort. Sci.* 121: 170–174.

Volk, G.M. and Walters, C. 2003. Preservation of genetic resources in the National Plant Germplasm Clonal collections. *Plant Breeding Rev.* 23: 291–344.

Yaku, A. and Widyastuti, C.A. 2002. *In situ* conservation of sweetpotato: an Irian Jaya experience. In: Rao, V.R. and Campilan, D (eds), *Exploring the Complementarities of In Situ and Ex Situ Conservation Strategies for Asian Sweetpotato Genetic Resources*, pp. 142–149.

Yen, D.E. 1974. *The Sweetpotato and Oceania: An Essay in Ethobotany*.Bulletin 236, Bishop Museum Press, Honolulu, Hawaii, pp. 1–389.

Widyastuti, C.A., Prain, G. and Yaku, A. 2002. Dani women's knowledge on and its contribution to maintenance of sweetpotato diversity in Baliem Valley. In: Rao, V.R. and Campilan, D (eds), *Exploring the Complementarities of In Situ and Ex Situ Conservation Strategies for Asian Sweetpotato Genetic Resources*, pp. 150–158.

Zheng, Qi., Dessai, A.P. and Prakash, C.S. 1996. Rapid and repetitive plant regeneration in sweetpotato via somatic embryogenesis. *Plant Cell Rep.* 15: 381–385.

Zobayed, S. M.A., Afreen, F., Kubota, C. and Kozai, T. 2000. Water control and survival of *Ipomoea batatas* grown photoautotrophically under forced ventilation and photomixotrophically under natural ventilation. *Ann. Bot.* 86: 603–610.

Chapter 7
Major Fungal and Bacterial Diseases

C.A. Clark, G.J. Holmes and D.M. Ferrin

Introduction

Although sweetpotato has a reputation as a durable crop able to withstand many adversities, numerous bacterial and fungal diseases have been reported on this crop from different regions of the world. In-depth information on most of these diseases has previously been published in comprehensive monographs. Much of the information in the classic 1929 monograph by Harter and Weimer remains useful. Detailed descriptions of most sweetpotato diseases are presented in the Compendium of Sweet Potato Diseases by Clark and Moyer published in 1988. This chapter will highlight some of the most important bacterial and fungal diseases of sweetpotato and research developments on them since 1988.

The 1929 and 1988 monographs are most thorough in reflecting their authors' experiences in North America, where less than 2% of the world's sweetpotatoes are produced, but where a considerable amount of research has been conducted on diseases of local importance. Unfortunately, less research information has been published on some of the important diseases in the tropics, such as stem and leaf scab and Alternaria stem and petiole blight. Lenné (1991) provided the most extensive information on the geographic distribution of sweetpotato diseases based on data from published literature and herbarium samples. Ames et al. (1997) published a handbook with brief descriptions and excellent photographs of numerous worldwide pests, diseases, and nutritional disorders. Nevertheless, there is less information on the geographic distribution of sweetpotato diseases than for most crops and it is difficult to determine how widespread many of the diseases are. This reflects the fact that sweetpotato in many countries is a poor man's crop that therefore receives very little research support and there are few plant pathologists who work on sweetpotato. It is therefore quite likely that some of the diseases discussed have a wider geographic distribution than is currently recognized in the literature.

The occurrence of sweetpotato diseases is also affected by the local crop production system. Sweetpotato is an indeterminant plant that is usually grown as an

C.A. Clark (✉)
Department of Plant Pathology and Crop Physiology, Louisiana State University Agcenter, Baton Rouge, LA 70803, USA
e-mail: CClark@agcenter.lsu.edu

G. Loebenstein, G. Thottappilly (eds.), *The Sweetpotato*,
DOI 10.1007/978-1-4020-9475-0_7,

annual. In tropical regions, it can be grown all year long and is propagated mostly from vine cuttings taken from other production fields. In temperate and sub-tropical regions, where winters are too harsh for year-long production, storage roots are held overwinter and are used to produce slips for transplanting the following season. It is not always clear how local environments, differences in susceptibility of local cultivars, or differences in occurrence of pathogens contribute to differences in diseases that occur from location to location. Several diseases are primarily carried from one season to the next by their ability to infect storage roots and are thus more problematic in temperate production systems. Some diseases also occur principally in plant beds used in temperate production systems to produce slips from storage roots. In this chapter, diseases will be discussed in the following groups: plant bed diseases, root-borne diseases, soil-borne diseases, foliar diseases, and postharvest diseases. These reflect the primary association of these diseases with stages in the production of the crop, but should not be considered exclusive as, for example, some soil-borne diseases may sometimes also be root-borne and vice versa.

Plant Bed Diseases

In temperate regions, storage roots are usually stored for 4–7 months after they are harvested until they are bedded in the field for production of slips. There are many variations as to how these plant production beds are prepared, but generally, the ‘seed’ roots are removed from storage and placed in the field and covered with field soil or some other medium. They may be heated artificially, covered with mulch to increase the bed temperature, or simply covered with soil. Since the bedding process requires handling of the roots, new wounds often occur on them, providing opportunities for postharvest pathogens to occur in the plant bed that also occur in the transition from storage to market (e.g., Rhizopus soft rot). Unfortunately, there is little information on the incidence or importance of these diseases in sweetpotato production as they occur on sweetpotatoes that are buried under ground and thus are not normally visible. These diseases will be discussed under the postharvest disease grouping below.

Sclerotial Blight (caused by **Sclerotium rolfsii***)*

Sclerotium rolfsii is a common soil-borne fungus that infects a wide array of crops. It is a common disease of sweetpotato in the USA, but there are few reports from elsewhere. In sweetpotato beds, foci of infection can occur throughout the beds (Plate 7.1b). These foci enlarge and on susceptible cultivars cause circular areas in which plants are first wilted and stunted and then die. When conditions are humid, course white mycelia grow out around the base of plants and over the soil and lower stems of the plants. Numerous sclerotia that look similar to mustard seeds are soon

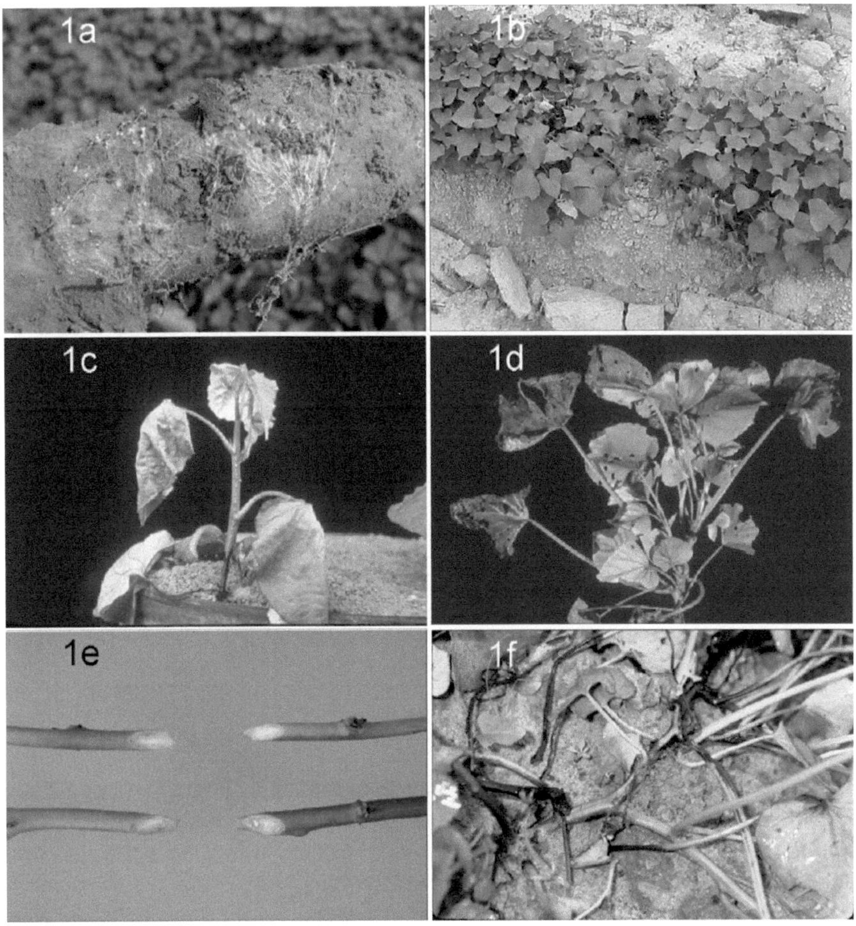

Plate 7.1 **a**: Root from a bed affected by sclerotial blight with mycelia of *Sclerotium rolfsii* growing on the surface of the root (courtesy Gerald Holmes, North Carolina State University). **b**: Wilting of sprouts and areas without plants due to sclerotial blight (courtesy Christopher Clark, Louisiana State University AgCenter). **c**: A plant with foot rot growing in a greenhouse showing a lesion at the soil line that girdled the plant causing secondary wilting (photo by Weston J. Martin, Louisiana State University AgCenter). **d**: Yellowing and wilting of lower leaves caused by Fusarium wilt (courtesy Christopher Clark, Louisiana State University AgCenter). **e**: Cross sections of stems of plants with Fusarium wilt (*left*) showing discoloration of the vascular system and stems from healthy plants (*right*) (courtesy Christopher Clark, Louisiana State University AgCenter). **f**: Bacterial stem rot caused by *Dickeya didantii* (courtesy Gerald Holmes, North Carolina State University) (See also Plate 1 on page xv)

formed on the mycelia. The same pathogen causes a disease known as circular spot that develops in the production field and is discussed below.

It appears that many different stresses can trigger outbreaks of sclerotial blight, especially warm temperatures and water-saturated soil. In addition, disease is favored by the presence of decaying vegetation in the beds that may stimulate eruptive germination of the sclerotia of the pathogen in the soil (Clark, 1989). Thus, decaying seed roots in the beds or decaying leaves on the surface of the beds may stimulate disease development. There are two potential sources of inoculum for sclerotial blight: sclerotia of the fungus that survive in soil for several years, or infected 'seed' roots. Although it is difficult to isolate *S. rolfsii* from within old circular spot lesions, sclerotia may occasionally be produced on the outer surface of these lesions and serve as a source of inoculum. Typically, when seed roots are excavated from within and around foci of infection, they are found to be covered with mycelia of *S. rolfsii* even though roots may show no symptoms of internal infection. Mycelia of the pathogen grow superficially over the surface of the 'seed' roots and enter mostly where the sprouts emerge from the roots (Plate 7.1a). They can then infect both the sprout, causing wilting and death, and the 'seed' root, which then rapidly develops a soft rot.

Several practices can contribute to effective management of sclerotial blight: avoiding use of cultivars that are particularly susceptible, avoiding fields with a known history of diseases caused by *S. rolfsii*, covering the surface of the seed roots with an effective fungicide at the time of bedding, keeping beds well aerated (such as by punching holes in plastic mulches), and using only roots free of disease for 'seed'.

Other Plant Bed Diseases

Sclerotinia rot, caused by *Sclerotinia sclerotiorum* is an important disease in New Zealand (Wright et al., 2003; Lewthwaite and Wright, 2005). Symptoms are wilting of leaves followed by rotting of stems and can occur in both the plant bed and the production field. White mycelia grow out on the surface of infected tissue and later black sclerotia are formed on the mycelia. The local New Zealand cultivar, Toka Toka Gold, is more susceptible than cultivars introduced from the USA or Japan (Broadhurst et al., 1997; Lewthwaite and Wright, 2005).

Although they only grow superficially and do not truly cause a disease, slime molds growing on the surface of plants in plant beds often cause concerns. Several different genera of slime molds are found in sweetpotato beds during periods of warm, wet weather, but other than being unsightly, they do not have a measurable effect on plant production or growth (Clark and Moyer, 1988).

Root-Borne Diseases

Where storage roots are used for propagation of the sweetpotato crop, pathogens that infect those roots have only a short distance to travel to infect the sprouts produced on those roots and hence the succeeding crop. Pathogens that infect plants systemically, such as viruses, or motile organisms such as insects or nematodes, can move from the infected root into the attached sprout with relatively little difficulty. Those pests are discussed in other chapters in this book. Several fungal and bacterial pathogens can also follow this route of infection. Several of these pathogens are not thought to persist in soil for longer than 1–2 years, and thus root to sprout transmission is the principal means of primary infection.

Black Rot (Caused by **Ceratocystis fimbriata***)*

Black rot has been reported on sweetpotato in China, Japan, Kenya, and the USA and is apparently widely distributed (Clark and Moyer, 1988; Farr et al., 2007: Kihurani et al., 2000; Lenné, 1991). Black lesions develop on storage roots (Plate 7.2a) that gradually enlarge and may encompass a large portion of the root surface but do not normally penetrate beneath the cortex. Infected roots taste bitter. Black cankers also develop on lower portions of the stems of sprouts that are below the soil line when the sprouts are on or near infected 'seed' roots, and infected sprouts may be stunted, wilt and occasionally die. If infected slips are transplanted to the field, cankers may continue to develop on the stem, causing stunting, yellowing, and transient wilting of the vine. In later stages of infection, black perithecia may be produced on lesions. The perithecia have long necks ($\sim$ 900 uM) with fimbriae at the apex and often with a cream to pink mass of ascospores exuding from the apex. The ascospores are hat-shaped and hyaline. The fungus also produces numerous endoconidia from phialides that are hyaline and cylindric in shape. The fungus also produces a fruity aroma both in infected tissue and in culture. These characteristics help distinguish black rot from charcoal rot (caused by *Macrophomina phaseolina*), which is characterized by production of microsclerotia in the cortex of infected roots, and Java black rot (caused by *Lasiodiplodia theobromae*, see postharvest diseases), which is characterized by eruption from infected roots of black stroma containing pycnidia filled with two-celled dematiaceous conidia.

Ceratocystis fimbriata has been described as a pathogen of several plants, also causing wilts on several species of trees. However, strains from each host are host-specific and intersterile and recently it has been suggested that they actually represent distinct species (Engelbrecht and Harrington, 2005), with the name *C. fimbriata* being reserved for the sweetpotato pathogen on which the fungus was first described. *C. fimbriata* does not normally persist in soil for longer than 1–2 years. It is unclear whether wounds are required for infection, but wounds such as caused by mice or insects certainly increase the incidence of infection. *C. fimbriata* produces both ascospores within the perithecia and endoconidia from phialides in the mycelia in great abundance and these spores can contaminate equipment, packing lines, and containers that come in contact with infected roots. This pathogen has great potential for spreading during any handling operation such as packing or bedding.

Black rot is also significant historically. From 1958 through 2003, professor Ikuzo Uritani and his co-workers in Nagoya, Japan published numerous articles on the physiology and biochemistry of the interaction between *C. fimbriata* and sweetpotato root tissue. This work was instrumental in improving general understanding of the role of phytoalexins, such as ipomeamarone, and other factors in determining the basis for resistance and host specificity in plant disease (Uritani, 1999).

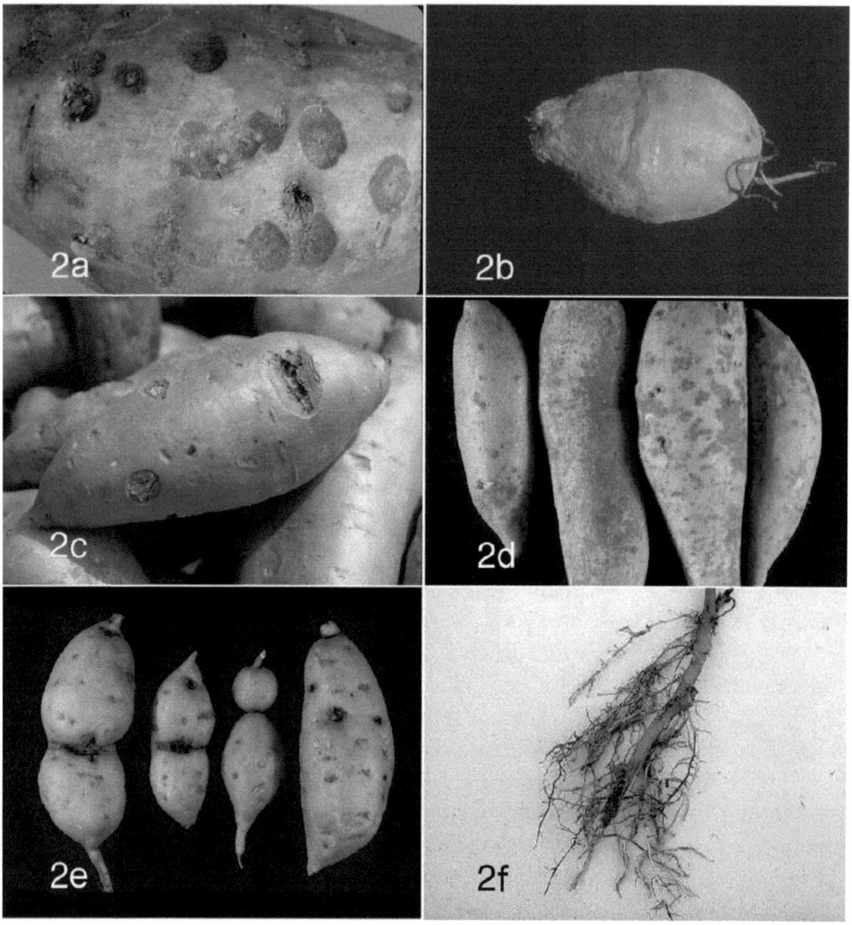

Plate 7.2 Root diseases. Symptoms on storage roots: **a**: black rot (caused by *Ceratocystis fimbriata*) (courtesy Gerald Holmes, North Carolina State University), **b**: foot rot (caused by *Plenodomus destruens*) (photo by Weston J. Martin, Louisiana State University AgCenter), **c**: circular spot (caused by *Sclerotium rolfsii*) (courtesy Gerald Holmes, North Carolina State University), **d**: scurf (caused by *Monilochaetes infuscans*) (courtesy Gerald Holmes, North Carolina State University), and **e**: Streptomyces soil rot (caused by *Streptomyces ipomoeae*) (courtesy Gerald Holmes, North Carolina State University). Symptoms of fibrous roots: **f**: Streptomyces soil rot (caused by *Streptomyces ipomoeae*) (courtesy Christopher Clark, Louisiana State University AgCenter) (See also Plate 2 on page xvi)

Scurf (Caused by **Monilochaetes infuscans***)*

Scurf has been reported from Africa, Asia, Australia, Europe, North and South America, and New Zealand (Farr et al., 2007; Lenné, 1991). This disease is restricted to the below-ground parts of the plant and primarily affects the appearance of storage roots, causing superficial purplish brown spots that are limited to the periderm. Lesion appearance (Plate 7.2d) is similar to that of boron deficiency, however, boron deficiency also causes the affected area to be raised and usually develops only in storage. Scurf lesions develop on the roots before harvest and continue to enlarge and coalesce during storage until most of the root surface is affected. Tate (1996) estimated that in New Zealand, infection levels of 50% at harvest could reach 100% after 3 months storage. Most of the apparent increase in scurf during storage probably represents enlargement of lesions incurred in the field, but Taubenhaus (1916) indicated that scurf could also spread between contacting roots in storage. When lesions cover a large portion of the root surface, increased water loss through them may cause roots to shrivel in storage, but otherwise the disease has little effect on production or quality. The causal fungus, *M. infuscans*, occurs only on sweetpotato and a few related species. It produces a simple dark conidiophore with a bulbous base on which is produced a chain of hyaline conidia that can be disseminated in water or during handling of infected roots to initiate new infections.

The pathogen grows from infected 'seed' roots up onto the lower portion of sprouts. When slips are pulled from beds, the below-ground portion of the stem carries the pathogen to the field, where it can grow back down onto the daughter roots. Consequently, scurf can be eliminated from sprouts simply by cutting them above the soil surface without allowing the knife to make contact with soil. Scurf is often more prominent on the proximal end of storage roots. Cultivars in which the slender attachment between the stem and the enlarged portion of the storage root is short tend to develop more scurf than those with long attachments because the slow-growing pathogen does not grow onto the storage roots of the latter as readily. Soil factors (e.g., greater organic matter content from the presence of animal manure) may enhance survival of *M. infuscans*.

Because the pathogen grows slowly and infects only the periderm, it can be difficult to isolate in culture. However, sporulation can usually be induced on infected roots by placing them in moist chambers.

There has been one preliminary report from New Zealand of fungicide resistance in *M. infuscans* (Broadhurst, 1996). Resistance was indicated to benzimidazole fungicides, which were the only fungicides used for scurf control in that production system.

Foot Rot (Caused by **Plenodomus destruens***)*

Foot rot has been reported as a problem in Argentina, Brazil, Burundi, New Zealand, Rwanda, South Africa, Tanzania, and the USA (Farr et al., 2007; Lenné, 1991). Foot rot can develop in plant beds, production fields, and storage. In plant beds, necrotic lesions extend up the stem from infected 'seed' roots. Sometimes lower leaves of plants may become yellow and in severe cases plants can wilt and die (Plate 7.1c). Symptoms appear in the field as yellowing, stunting, and/or transient wilting of the vines when necrotic lesions that develop at or near the soil line girdle the stem. Daughter roots produced on infected plants may be infected by the fungus growing down from the stem. Infected roots develop a slow rot in storage (Plate 7.2b). If infected roots survive storage and are used as 'seed' roots, the sprouts produced on them may become infected. In the beds, cankers develop on the stems of sprouts below the soil line, which can cause the plants to be stunted and in some cases die. Surviving infected plants can carry the pathogen with them to the field if slips are pulled rather than cut. Lopes and Silva (1993) found that yield losses in southern Brazil could be as great as 80% when symptomless cuttings of a susceptible cultivar obtained from an infested field were used for planting.

Pycnidia of *P. destruens* are produced on stems and in a layer beneath the periderm of infected storage roots. The pycnidia produce biguttulate, hyaline conidia. Conidia produced on storage roots can contaminate healthy roots during the bedding process or conidia produced on stems in plant beds can contaminate healthy plants during the transplanting process. However, the predominant mode of dissemination of the pathogen is by vegetative propagation of an infected crop. The pathogen is not thought to persist in soil for long, though this may differ from location to location and may also depend on whether Convolvulaceous weeds that are alternative hosts for *P. destruens* are present.

Management of Root-Borne Diseases

Crop rotation can greatly reduce risk from infection by soil-borne inoculum with each of these root-borne pathogens since they do not survive as long as true soil-borne pathogens. However, some of them may also infect common weed species within the Convolvulaceae and the role of these alternative hosts in survival in the field is unknown. Implementation of an integrated program of crop rotation and use of only disease-free roots for seed, treating 'seed' roots with an effective fungicide at the time of bedding, and cutting slips or transplants at least 2–3 cm above the soil line, has made black rot and foot rot rare, and greatly reduced the incidence of scurf in the USA. Black rot remains an important problem in some regions where sweetpotato production is very intensive and 1 or 2 sweetpotato crops are planted in the same land every year.

Soil-Borne Diseases

Streptomyces Soil Rot (Caused by **Streptomyces ipomoeae***)*

Where it occurs, Streptomyces soil rot is one of the most destructive diseases of sweetpotato. It has been a serious problem in the USA and Japan, but its geographic distribution outside of those countries is uncertain. The pathogen survives in soil for many years and can infect the sweetpotato root system at any stage of its development. It causes necrotic lesions on feeder roots that effectively prune the root system and reduce yield (Plate 7.2f). It is difficult to excavate the infected portions of the fibrous roots from the soil because they are so thoroughly decayed. It also causes necrotic lesions on the storage roots (Plate 7.2e), which are often centered on the point of emergence of lateral fibrous roots. Tissue within the lesion is black and corky and the infected periderm often has cracks radiating from the center of the lesion. These lesions enlarge slowly, but prevent enlargement of the storage root at the site of infection, causing constrictions, indentations and distortions that reduce quality. Disease development is greater in dry soil. It is also greater when the soil pH is 6.0–7.0 and does not develop at soil pH below 5.2. It is common for susceptible cultivars grown in heavily infested soil under conducive conditions (i.e. dry soil at pH of 6.0–7.0) to fail to yield any marketable roots.

Streptomyces ipomoeae is unique among plant pathogens. It differs from other prokaryotic pathogens, including other pathogenic *Streptomyces* species, in that it is able to penetrate directly into the fibrous roots (Clark and Matthews, 1987; Loria et al., 1997). It does not penetrate the intact periderm of storage roots, but invades these roots from infected fibrous roots that emerge from the storage root. Unlike the other plant pathogenic *Streptomyces* species, it induces necrotic lesions, not the hypertophic erumpent type of lesions produced by *S. scabiei, S. acidiscabiei*, or *S. turigidiscabies* on other root and tuber crops (Loria et al., 1997). Although *S. ipomoeae* is an aggressive pathogen on sweetpotato, it can be difficult to isolate as it grows slowly in culture and necrotic tissue is rapidly colonized by other actinomycetes and bacteria (Clark and Moyer, 1988). Thaxtomin toxins, predominantly thaxtomin A, are produced by *S. scabiei, S. acidiscabiei*, and *S. turigidiscabies* in several growth media. *Streptomyces ipomoeae* does not produce thaxtomins under these conditions (King et al., 1994), but instead produces predominantly thaxtomin C only on sweetpotato storage root tissue.

The origin and distribution of *S. ipomoeae* is uncertain. It has been known in the USA since at least the late 1800s and the only other confirmed report is from Japan, where it was first reported in the late 1940s. Various analyses of strains from the USA and Japan suggest that they are quite similar and may share a common origin (Clark et al., 1998a; Labeda and Lyons, 1992). However, three groups were distinguished among these isolates based on their ability to inhibit each other during pairewise cocultivation on agar, and there was some correlation with rep-PCR analyses (Clark et al., 1998a). One group of strains of *S. ipomoeae* was found to produce a bacteriocin-like protein, ipomicin, that is one of the inhibitory principles (Zhang et al., 2003).

Sweetpotato breeding programs in both the USA and Japan have developed a number of cultivars with resistance to Streptomyces soil rot. To date, these cultivars have provided the most effective and economic control for the disease. Disease severity can also be greatly reduced by maintaining soil pH below 5.2, but productivity of the crop may be reduced by the acidic conditions. Fumigating soil with products containing chloropicrin also has an effect, especially on yield, but so few infections are required to affect storage roots, that it does not completely alleviate quality loss. Incorporating tissue of *Geranium carolinianum* reduced Streptomyces soil rot severity to a similar extent as chloropicrin fumigation (Ooshiro et al., 2006).

Fusarium Wilt (Caused by **Fusarium oxysporum** *f. sp.* **batatas***)*

Fusarium wilt was once an important disease of sweetpotato in the USA and has been reported in other countries. Symptoms first appear as yellowing of older leaves and transient wilting of the vines (Plate 7.1d). Later, vines may become permanently wilted and die. Infected vascular elements become brown (Plate 7.1e) and in late stages the entire stem may rot and light pinkish sporodochia bearing micro- and macroconidia appear on the surface of the stem. In some cases, infected plants may survive and produce storage roots that are infected, usually showing vascular discoloration near the proximal end of the root. If such roots are used for 'seed', they can transmit the fungus to slips that may wilt in the plant bed or later in the production field.

The pathogen may persist in soil for many years, but crop rotations can reduce disease pressure. In places where Fusarium wilt has been a problem, resistant cultivars have been developed that are very effective in providing economic control of this disease. Susceptible plants can also develop induced resistance to Fusarium wilt when inoculated with nonpathogenic strains of *F. oxysporum* (Ogawa, 1989; Ogawa and Komada, 1986) or with the chlorotic leaf distortion pathogen, *F. denticulatum* (Clark, 1994).

Although this disease is generally considered to be caused by *F. oxysporum* f. sp. *batatas*, there are also some strains of the tobacco pathogen, *F. oxysporum* f. sp. *nicotianae* that are capable of causing wilt in susceptible sweetpotato (Clark et al., 1998b). In addition, a third lineage was isolated from Fusarium wilt-resistant sweetpotatoes grown in California that was designated as a new race of *F. oxysporum* f. sp. *batatas* (Clark et al., 1998b).

Fusarium Root and Stem Canker (Caused by **Fusarium solani***)*

Fusarium solani is a common soil-borne fungus that causes cortical rots on many plants. On sweetpotato, the pathogen can affect the crop at any stage of production. It is more commonly noticed late in storage as a dry rot. Externally, lesions have a target appearance with concentric rings of light and darker brown color. Cutting open the storage roots reveals lesions that in later stages can extend into the center of the root often with lens-shaped or biconvex cavities (Plate 7.3a) within the lesion that are lined with white mycelia of the fungus. This decay seems to develop more aggressively late in storage and/or on 'seed' roots used for bedding. The fungus can grow from infected 'seed' roots up into sprouts and cause cankers on the lower portions of the stem. A similar disease, known as surface rot, is caused by the related *Fusarium oxysporum*. Surface rot lesions usually have a uniform brown external color and are restricted to the cortex of the storage root. Unlike *F. solani*, the surface rot pathogen does not seem to invade sprouts.

Fusarium species survive for long periods in soil, primarily as resistant chlamydospores. Storage roots are most frequently infected through wounds incurred during harvesting. Fusarium root rot is often severe when harvest occurs after

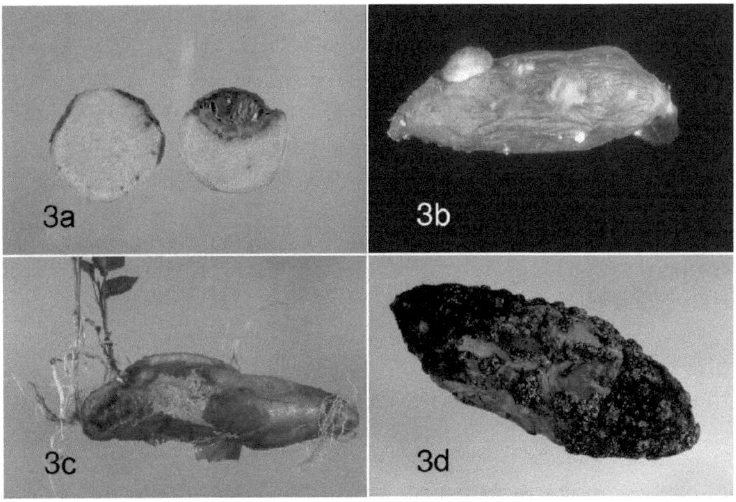

Plate 7.3 Storage diseases. **a**: Fusarium root rot (*caused by Fusarium solani*) (courtesy Christopher Clark, Louisiana State University AgCenter), **b**: Rhizopus soft rot (caused by *Rhizopus stolonifer*) (photo by Weston J. Martin, Louisiana State University AgCenter), **c**: bacterial soft rot (caused by *Dickeya didantii*) (courtesy Christopher Clark, Louisiana State University AgCenter), and **d**: Java black rot (caused by *Lasiodiplodia theobromae*) (courtesy Gerald Holmes, North Carolina State University) (See also Plate 3 on page xvii)

soils have been cold and wet. Anecdotal observations also suggest that drought conditions, which predispose sweetpotatoes to greater skinning during harvesting may also predispose roots to greater infection by *F. solani*. Infections probably also occur through wounds incurred on 'seed' roots during the bedding process. *Fusarium solani* can also follow the "root-borne" cycle of infection, progressing from infected 'seed' roots into slips in which it may grow in advance of the cankers that appear on lower portions of the stem. Transplanting infected slips results in the appearance of a proximal end rot of storage roots that is already well established at harvest. In many cases, *F. solani* is found co-infecting plants with *Dickeya dadantii* (*Erwinia chrysanthemi*) and dual infection results in a more aggressive disease than with either pathogen alone (Duarte and Clark, 1993).

Circular Spot (Caused by Sclerotium rolfsii*)*

Circular spot develops on sweetpotato storage roots in the field shortly before harvest. Lesions are circular, 1–2 cm in diameter and usually shallow (1–4 mm). The surface of the lesion is brown and the decayed tissue below is yellowish brown. Under most circumstances, development of the lesions ceases as soon as the roots are harvested and it is often difficult to isolate the causal fungus, *Sclerotium rolfsii*.

However, if roots are placed in a warm, saturated environment immediately after harvest, typical course white, mycelia of *S. rolfsii* grow out on the surface. On rare occasion, one to a few sclerotia may be found on the surface of the lesion. In storage, abscission zones often form beneath the circular spot lesions and the necrotic tissue may separate from the remainder of the sweetpotato (Plate 7.2c).

Incidence of circular spot is highly variable and some cultivars are more susceptible than others. The factors that affect incidence of circular spot are not fully known, but prolonged periods of flooding in the field, especially when soil is warm, can stimulate much more aggressive development by *S. rolfsii* and lesions may be much wider and deeper or even consume entire roots with soft rot.

Other Soil-Borne Diseases

Although the bacterial wilt pathogen, *Ralstonia solanacearum*, occurs widely in areas around the world where sweetpotatoes are grown, it is only known as a pathogen of sweetpotato in some parts of China where it can cause substantial losses (Clark and Moyer, 1988). The pathogen is soil-borne, surviving longer in upland fields than in flooded paddies. It can also be carried through the cycle of vegetative propagation of the crop. Plants can be infected through wounds at transplanting. Affected plants, either in plant beds or in the field, develop a water-soaked rot progressing upwards from the base of the plant. Terminal portions of the vine may wilt and die. Less severely affected plants may have vascular discoloration in the stem. Storage root symptoms vary from mild yellowish-brown streaking in the vascular tissue to a soft grayish-brown decay that may consume the entire root. It appears that the strains of *R. solanacearum* in China are unique in their host range. Four biotypes of *R. solanacearum* have been described, and biotype IV-1 is composed of strains from sweetpotato, ground nut and ginger (Zheng and Dong, 1995). Resistance has been found to bacterial wilt, but there are at least three races of the bacterium that differentially infect different cultivars (Fang and Yu, 1991; Fang et al., 1994; Fang et al., 1995; Zhang et al., 1999).

Foliar Diseases

A variety of leaf spots and other foliar diseases occur on sweetpotato around the world, but in most cases, they have very little effect on production of the crop. However, two diseases, stem and leaf scab and Alternaria leaf blight, are capable of causing some yield reduction in the limited geographic areas in which they occur. Two other diseases, white rust and chlorotic leaf distortion (CLD), will be discussed because at times they may induce symptoms that are prominent and provoke questions.

Stem and Leaf Scab (Caused by **Elsinoë batatas** *[anamorph =* **Sphaceloma batatas***])*

Stem and leaf scab has been reported in Australia, Asia, many Pacific Islands, the Caribbean (Puerto Rico), Mexico, South America (Brazil), but is absent from the USA except Hawaii (Farr et al., 2007; Lenné, 1991). The range in which it has a significant impact on sweetpotato production is probably more restricted. It is most prominent at elevations in the tropics where conditions are wet and cool to warm (Lenné, 1991). It causes small brown to black, raised necrotic lesions on leaves, petioles, and stems. On leaves, most lesions are produced on major veins and cause the veins to shrink and distort, and if there are many lesions, the leaves become severely distorted (Plate 7.4a). Acervuli are produced below and break through the epidermis and hyaline, spherical-to-ovoid conidia are produced in them. Globose asci, each containing four to six septate ascospores are sometimes produced in dark gray stroma beneath the host epidermis.

The effect of scab on sweetpotato production depends on how early in the growing season the disease develops. Different studies showed that yield losses ranged from 27 to 57% when plants were inoculated at 2 weeks after planting (Lenné, 1991). Ramsey et al. (1988) found a negative correlation between scab severity at 55 and 82 days, but not at 111 and 195 days after planting. Other species of Convolvulaceae are also susceptible to scab, including *Ipomoea aquatica* which is often grown near sweetpotato in much of the range of this disease, and many wild species that may provide inoculum to infect sweetpotato.

There are significant differences in susceptibility of sweetpotato genotypes to stem and leaf scab and use of resistance is one means of managing the disease (Ramsey et al., 1988; Smit et al., 1991). On resistant cultivars, the incubation period is longer, infection frequency is reduced, and lesion size is smaller. The resistant cultivars tend to have thicker cuticles, fewer stomates in leaves, and fewer lenticels on petioles and stems (Bajit and Gapasin, 1987). However, there are preliminary indications that there may be some specialization among strains of the pathogen. Other control practices include use of fungicides and sanitation measures, such as destruction of crop debris and use of disease-free planting material.

Alternaria Stem and Petiole Blight (Caused by **Alternaria** *spp.)*

Alternaria leaf spots, with typical target-like appearance, may be found on sweetpotato from time to time in many parts of the world and several species of *Alternaria* have been reported to cause them (Lenné, 1991). In most cases, the leaf spots are minor, usually developing to a limited extent on older leaves (Plate 7.4f) and do not affect production. However, in some parts of East Africa, including Ethiopia, Kenya, and Uganda, a more aggressive blight was first reported in 1984 (Bruggen, 1984) that develops primarily on stems and petioles (Plate 7.4e). Lesions are gray to black

Plate 7.4 Foliar diseases. **a**: stem and leaf scab (caused by *Elsinoe batatas*) (courtesy Christopher Clark, Louisiana State University AgCenter, **b**: chlorotic leaf distortion (caused by *Fusarium denticulatum*) (courtesy Christopher Clark, Louisiana State University AgCenter, **c**: white rust (caused by *Albugo ipomoeae-panduranae*) as seen on adaxial surfaces of leaves (courtesy Gerald Holmes, North Carolina State University), **d**: white rust as seen on abaxial surfaces (courtesy Gerald Holmes, North Carolina State University), **e**: Alternaria blight (caused by *Alternaria spp.*) (courtesy Christopher Clark, Louisiana State University AgCenter, and **f**: Alternaria leaf spot (caused by *Alternaria spp.*) (courtesy Christopher Clark, Louisiana State University AgCenter (See also Plate 4 on page xviii)

and if they girdle the stem may cause wilting or dieback of terminal portions of the vine. Both *Alternaria bataticola* and *A. alternata* have been isolated from infected plants, but *A. bataticola* is the more aggressive species (Anginyah et al., 2001; Osiru et al., 2007). An Alternaria stem and petiole blight also occurs in Papua New Guinea (Lenné, 1991). Subsequently, *A. bataticola* was reported from the Brazilia-DF area of Brazil, but the Brazilian isolates of the fungus differed from the African isolates in that they lacked the forked beak on conidia that characterizes some of the African strains (Lopes and Boiteux, 1994). Symptoms included necrotic lesions on petioles and stems and necrotic spots on leaves with chlorotic halos. When lesions girdled stems, wilting and dieback of the vine ensued. Lopes and Boiteux (1994) indicate that the disease had been endemic in Brazil for a long time but had recently increased in importance, possibly because of an increase in a susceptible cultivar, Brazlandia Roxa.

Although Alternaria stem and petiole blight is widespread within the countries of East Africa where it occurs, disease severity varies greatly within the region from minimal levels in less humid areas up to 25–50% of the plant infected in other areas, though it is not clear what is responsible for the variation in disease severity (Anginyah et al., 2001; Osiru et al., 2007; Skoglund et al., 1993). In each location where it has been reported, it has been observed that cultivars differ in susceptibility. As Lenné (1991) indicated, there is a dearth of information on the causes, variation, and control of Alternaria diseases of sweetpotato.

White Rust (Caused by **Albugo ipomoeae-panduratae***)*

White rust is a common disease distributed widely around the world on sweetpotato and related species of the Convolvulaceae. Symptoms appear as chlorotic spots on leaves that are delimited by veins and appear angular (Plate 7.4c). Pustules containing white masses of sporangia erupt from within the lesions on the lower sides of leaves (Plate 7.4d). In some places, sweetpotato leaves affected by white rust become severely blistered, but in other locations this is not seen.

Incidence and severity of white rust on sweetpotato is correlated with rainfall and during extended rainy periods a large proportion of the leaf area may be affected. Differences in susceptibility/resistance have been reported for sweetpotato cultivars. However, the few studies that have investigated effects on yield suggest that white rust does not impact yield of sweetpotato storage roots (Cavalcante et al., 2006).

White rust also occurs on many other species of *Ipomoea* and *A. ipomoeae-panduratae* is also cited as the pathogen on these species. In the southern USA white rust is very common on the ivy-leafed morning glory, *I. hederacea* var. *hederacea*, and it is common to find white rust with blistering on this species growing in fields of sweetpotato that are free of white rust symptoms. It is not clear why symptoms vary from host to host or location to location, or why white rust develops on *I. hederacea* even when weather does not favor development on sweetpotato. Cifferi (1928) suggested that there were differences in host specialization within *A. ipomoeae-panduratae*. Extensive systematic studies, including cross inoculations, will be needed to determine the relationship among strains of *A. ipomoeae-panduratae* that occur on different species of Convolvulaceae to determine if differences in disease development reflect differences in pathogen populations or host reactions.

Chlorotic Leaf Distortion (Caused by **Fusarium denticulatum***)*

Chlorotic leaf distortion (CLD) is caused by *Fusarium denticulatum* (originally identified as *F. lateritium* [Clark et al., 1990; Nelson et al., 1995] and later renamed by Nirenberg and O'Donnell [1998]). Although CLD was not recognized until the late 1980s (Clark et al., 1990) it was quickly reported thereafter in Brazil, Kenya, and Peru (Kihurani et al., 1993; Icochea et al., 1994) and the causal fungus, *F. denticulatum* has been isolated from true seed produced in many different countries (Clark and Hoy, 1994b). Thus, *F. denticulatum* is apparently widely distributed with sweetpotato. Using isolates obtained from true seed, Hyun and Clark (1998) found that sweetpotato isolates were relatively homogeneous, but Clark et al. (1995) concluded that isolates from Africa showed greater diversity for pathogenicity and vegetative compatibility grouping. Phylogenetic analysis by O'Donnell et al. (1998) associated sweetpotato CLD isolates of *F. denticulatum* with a clade that originated in Africa.

Symptoms of CLD range in severity from almost inconspicuous to dramatic and are most pronounced, almost exclusively, at the growing tip of the vine. Symptom development is greatest during periods of warm, humid weather with bright sunshine. Under these conditions, leaves nearest the vine tip develop a bright general chlorosis and are often twisted or distorted. They also have a white substance on the upper leaf surface that resembles a wax deposit on young leaves that have just unfolded or resembling a salt deposit near the edges of older leaves (Plate 7.4b). However, microscopic examination reveals that this white substance is composed of mycelia and conidia of the causal fungus. As leaves age, they regain some of the normal green color with only a little diffuse chlorosis near the center of the leaf, even though newly emerging leaves may continue to show dramatic chlorosis. When weather turns cloudy, affected plants may appear to recover. When weather is not conducive for CLD, the only sign of the pathogen may be small clumps of mycelia on the upper leaf surface.

The relationship of *F. denticulatum* with the sweetpotato plant is highly unusual. The fungus does not normally penetrate the vegetative shoot system, but instead colonizes secretions released by the sweetpotato onto the apical meristem and leaf primordia and also grows actively between halves of young leaves that have not yet unfolded (Clark, 1992). As a result, the fungus often contaminates meristems that are excised in efforts to produce virus-free plants. Despite its epiphytic relationship on the vegetative shoot, the fungus can be found inside a relatively high proportion of true seed produced on affected mother plants (Clark and Hoy, 1994b).

Although CLD can induce dramatic symptoms, and cultivars can vary in their susceptibility, in studies by Kim et al. (1996), CLD did not have a measurable effect on yield of storage roots, regardless of symptom severity or susceptibility of different cultivars. Research is needed to determine if CLD affects quality or safety of shoot tips for human or animal consumption. CLD does provide some degree of cross protection against Fusarium wilt development (Clark, 1994).

Other Leaf Diseases

Phytophthora ipomoeae was recently reported from the Central Highlands of México, where it causes a late blight-like disease on *Ipomoea longipedunculata* and *I. purpurea* (Flier et al., 2002; Grünwald et al., 2004). It is suggested that it evolved from a common ancestry with the late blight pathogen of potato and tomato, *P. infestans*. The initial reports indicated that one cultivar of sweetpotato, A26/7, was included in inoculation studies with this pathogen and that it was not susceptible. Although this disease does not appear to pose an immediate threat to sweetpotato production, the potential for *P. ipomoeae* to spread to lower areas of México where sweetpotatoes are grown or to other geographic regions and become established on wild species of Convolvulaceae growing in proximity to sweetpotato and/or adaptation of these strains to sweetpotato should be considered and monitored (Flier et al., 2002).

Postharvest Diseases

Several diseases previously described in this chapter may also be considered as postharvest diseases (e.g., scurf, Fusarium root and stem canker) because they are often detected after harvest. The three diseases discussed below develop after harvest and often after packaging for long-distance transport.

Rhizopus Soft Rot (Caused by **Rhizopus stolonifer** *and* **R. arrhizus***)*

Rhizopus soft rot is probably the most important postharvest disease of sweetpotatoes. Although two species of *Rhizopus* can cause this disease, *R. stolonifer* occurs more frequently and is generally more virulent than *R. arrhizus*. The pathogen can reduce an entire root to a soft, watery mass within 3–4 days at 13 °C. The fungus produces masses of black spores at the tips of fungal growth 1–2 cm above the root surface or as a black, prostrate growth 1–2 mm thick at the root surface (Plate 7.3b). If infection begins at the mid-section of a root, it often produces ring rot, which if left to progress further will quickly involve the entire root. The decay produces a sweet, fermentation odor that quickly attracts fruit flies.

Rhizopus is an ubiquitous fungus, present in air and soils. However, a wound is required in order for infection to occur. Such wounding occurs during harvest as roots are removed from soil, placed into buckets and bins and moved into curing and storage facilities. Wounds that crush or soften root tissue are much more likely to become infected than wounds resulting from clean cuts, surface abrasions or root ends that break (Holmes and Stange, 2002). Another opportunity for wounding occurs when roots are removed from storage, washed, graded and packaged for shipment. This final stage of handling is when Rhizopus soft rot can be most destructive. The short disease cycle and rapid progression of soft, watery rot results in conspicuous and messy losses even when shipping relatively short distances and especially over longer distances.

Rhizopus soft rot is controlled by the integration of several management tactics: use of resistant cultivars, curing, gentle handling to minimize wounding and application of an effective fungicide. Cultivars vary greatly in their susceptibility to Rhizopus soft rot (Clark and Hoy, 1994a). In general, white fleshed cultivars tend to be more susceptible than orange-fleshed cultivars, but there is considerable variation in susceptibility within cultivars of a given flesh color. Cultivar Beauregard is widely grown in the USA and is highly resistant to Rhizopus soft rot. However, significant losses due to this disease may occur even when Beauregard is grown. Thus, additional measures must be taken to manage Rhizopus soft rot.

Curing roots at elevated temperature (30 °C) and relative humidity (95%) for 3–5 days immediately following harvest is a standard practice in developed countries and results in wound "healing" or suberization of wounded root surfaces. This makes host tissue resistant to pathogen invasion and is an effective means of controlling Rhizopus soft rot. Curing also produces the added physiological benefit of increasing sugar content in roots.

Much can be done to minimize injuries during harvest and packing operations. Of particular concern are the distances that roots are dropped during various handling practices. Minimizing the number and the distance of such drops as well as cushioning the fall is key to reducing Rhizopus soft rot.

Finally, fungicides can be used to protect roots from decay after packing. They are typically applied as a 10–20 second dip in aqueous suspension or sprayed onto roots as they pass along a brushbed during the packing process. Dicloran (trade name = Botran) has been used for decades and remains an effective method of controlling the disease.

Bacterial Root and Stem Rot (Caused by **Dickeya dadantii** *[syn.* **Erwinia chrysanthemi***])*

Bacterial root and stem rot was first reported in 1974 (Martin and Dukes, 1977; Schaad and Brenner, 1977) and much remains unknown about the disease. Symptoms may occur at any stage of plant growth from transplanting in the field to postharvest. In the field, the disease typically manifests itself as dark brown, water-soaked tissue on the lower stem and dark streaks in the vascular tissue (Plate 7.1f). In mildly infected storage roots symptoms may not be apparent. In more severe infections, yellowish-brown, longitudinal streaks may develop and water-soaked lesions may appear on the surface. In such cases, the disease progresses rapidly, producing slimy masses of the bacterium, which often ooze out from cracks in the epidermis. The entire root may rot, leaving the outer skin intact (Plate 7.3c). The decaying tissue initially is the same color as healthy tissue but may quickly become cream or gray-brown in color and gives off a foul odor produced by secondary invading bacteria. Transplants, vines, and storage roots each may have latent infections by

D. dadantii that can later develop into active infections under the appropriate environment (Duarte and Clark, 1992).

Disease development is favored by conditions of low oxygen and high temperature and may remain latent under less favorable conditions. After harvest, roots may be exposed to such conditions if they are not continuously refrigerated or inadequate gas exchange is allowed during storage, transit, or in specialized packaging (e.g., plastic wrap).

The pathogen infects the plant primarily through wounds and does not reside in soil except in association with crop debris or weed hosts. Inoculum sources include infected mother roots, contaminated wash water, and harvesting equipment contaminated by infected roots and adhering infested soil.

Management of bacterial stem and root rot relies on several tactics. Wounding during any stage of production should be minimized. Disease-free planting material should be used and vine cuttings used for transplants should be cut above the soil surface to avoid contaminating the knife blade. Since disease development is favored by anaerobic or near anaerobic conditions, adequate gas exchange should be provided to plants and roots during production, storage and transit. Submersion of roots in water should be avoided as this may contaminate the water and inoculate sound roots. If water is used, it should be treated with a sanitizing agent (e.g., chlorine) to kill the bacterium. Cultivars vary in their susceptibility, most being resistant (Clark et al., 1989). Beauregard and Georgia Jet are susceptible while many older cultivars such as Centennial and Porto Rico are less susceptible.

Java Black Rot (Caused by **Botryodiplodia theobromae***)*

Although Java black rot can be one of the most destructive postharvest diseases of sweetpotato, severe losses are much more common in warmer regions. The disease gets its name because it was first reported on sweetpotatoes shipped to the USA from Java.

Symptoms typically appear at one or both root ends. The decay is firm, but moist. Apparently healthy tissue surrounding the lesion gradually turns brown if exposed to air. Infected roots initially produce white cottony growth and the entire root will become completely decayed within 1–2 weeks. The white cottony growth is replaced by a mummified, very hard root that is black. A positive diagnosis is based on the presence of black stromatic masses of the fungus that break through the outermost layers of the root in the form of domes or cushions on the root surface (Plate 7.3d). The stromatic masses are fungal fruiting bodies called pycnidia, which are filled with dark two-celled conidia. As the fruiting bodies break down, spores are released that coat nearby surfaces, including storage containers and adjacent roots.

Spores of *B. theobromae* survive free in the soil for at least one year (Lo and Clark, 1988), and may also survive between seasons on the surface of storage or handling containers (e.g., wooden crates). The pathogen may also survive on host debris in the field.

Roots become infected from soilborne inoculum through wounds. Since roots are severed from the mother plant at harvest, at least one major wound occurs at this time. Often a second major wound occurs at the distal end of the root, as this often snaps off. Both of these wounds are major sites for infection, but can be reduced or eliminated by prompt curing. However, many new wounds are created during the packing process and this secondary disease cycle can be far more destructive.

Susceptibility to Java black rot increases after 5–8 months in storage. This is compounded by the potential for greater inoculum levels over time. The disease is favored by the same conditions that are optimal for curing. Despite this, curing is still one of the best methods of controlling the disease. Chilling injury, flooding and poor ventilation all contribute to increased susceptibility.

Java black rot is managed by the integration of several practices. Previously used storage containers should be washed and sanitized prior to harvest. Wounding during harvest should be minimized and roots should be cured immediately after harvest. Diseased roots should not be used for bedding and it is good practice to cut vines used for transplants above the soil surface. Cultivar resistance exists, but has not been thoroughly studied.

References

Ames, T., Smit, N.E.J.M., Braun, A.R., O'Sullivan, J.N., and Skoglund, L.G. 1997. *Sweetpotato: Major Pests, Diseases, and Nutritional Disorders*. International Potato Center (CIP), Lima, Peru. 152 pp.

Anginyah, T.J., Narla, R.D., Carey, E.E., and Njeru, R. 2001. Etiology, effect of soil pH and sweetpotato varietal reaction to Alternaria leaf petiole and stem blight in Kenya. *African Crop Sci. J.* 9:287–292.

Bajit, G.B., and Gapasin, R.M. 1987. Relationship between morphological characteristics and varietal resistance of sweet potato to scab infection caused by *Sphaceloma batatas* Saw. *Ann. Trop. Res.* 9:75–83.

Broadhurst, P. 1996. Fungicide resistance in Northland. *Commercial Grower* 51:35.

Broadhurst, P.G., Lewthwaite, S.L., and Triggs, C.M. 1997. Field evaluation of sweetpotato cultivars for resistance to Sclerotinia rot. *Proc. 50th N.Z. Plant Prot. Conf.* 1997:89–92.

Bruggen, A.H.C. van. 1984. Sweet potato stem blight caused by *Alternaria* sp.: a new disease in Ethiopia. *Netherlands J. Plant Pathol.* 90:155–164.

Cavalcante, J.T., Ferreira, P.V., Soares, L., Borges, V., Silva, P.P. da, and Silva, J. da. 2006. Análise de trilha em caracteres de rendimento de clones de batatadoce (*Ipomoea batatas* (L.) Lam). *Acta Scientiarum - Agronomy* 28:261–266.

Cifferi, R. 1928. Osservazioni sulla specializzazione dell' 'Albugo Ipomoeae-paaduratae' (Schw.) 8w. *Nuovo Giornale Botanico Italiano* 35:112–134.

Clark, C.A. 1989. Influence of volatiles from healthy and decaying sweet potato storage roots on sclerotial germination and hyphal growth of *Sclerotium rolfsii*. *Can. J. Bot.* 67:53–57.

Clark, C.A. 1992. Histological evidence that *Fusarium lateritium* is an exopathogen on sweetpotato with chlorotic leaf distortion. *Phytopathology* 82:656–663.

Clark, C.A. 1994. The chlorotic leaf distortion pathogen, *Fusarium lateritium*, cross protects sweetpotato against Fusarium wilt caused by *Fusarium oxysporum* f. sp. *batatas*. *Biolo. Control* 4:59–66.

Clark, C.A., Chen, C., Ward-Rainey, N., and Pettis, G.S. 1998a. Diversity within *Streptomyces ipomoeae* based on inhibitory interactions, rep-PCR, and plasmid profiles. *Phytopathology* 88:1179–1186.

Clark, C.A., and Hoy, M.W. 1994a. Identification of resistance in sweetpotato to Rhizopus soft rot using two inoculation methods. *Plant Dis*. 78:1078–1082.
Clark, C.A., and Hoy, M.W. 1994b. Isolation of *Fusarium lateritium* from sweetpotato seed. *Plant Dis*. 78:585–587.
Clark, C.A., Hoy, M.W., and Nelson, P.E. 1995. Variation among isolates of *Fusarium lateritium* from sweetpotato for pathogenicity and vegetative compatibility. *Phytopathology* 85:624–629.
Clark, C.A., Hyun, J-W., and Hoy, M.W. 1998b. Relationships among wilt-inducing isolates of *Fusarium oxysporum* from sweetpotato and tobacco. *Plant Dis*. 82:530–536.
Clark, C.A., and Matthews, S.W. 1987. Histopathology of sweet potato root infection by *Streptomyces ipomoea*. *Phytopathol*. 77:1418–1423.
Clark, C.A., and Moyer J.W. 1988. *Compendium of Sweet Potato Diseases*. APS Press, The American Phytopathological Society, St. Paul, MN, USA. 74 pp.
Clark, C.A., Valverde, R.A., Wilder-Ayers, J.A., and Nelson, P.E. 1990. *Fusarium lateritium*, causal agent of sweetpotato chlorotic leaf distortion. *Phytopathology* 80:741–744.
Clark, C.A., Wilder-Ayers, J.A., and Duarte, V. 1989. Resistance of sweet potato to bacterial root and stem rot caused by *Erwinia chrysanthemi*. *Plant Dis*. 984–987.
Duarte, V., and Clark, C.A. 1992. Presence on sweetpotato through the growing season of *Erwinia chrysanthemi*, cause of stem and root rot. *Plant Dis*. 76:67–71.
Duarte, V., and Clark, C.A. 1993. Interaction of *Erwinia chrysanthemi* and *Fusarium solani* on sweetpotato. *Plant Dis*. 77:733–735.
Engelbrecht, C.J.B., and Harrington, T.C. 2005. Intersterility, morphology and taxonomy of *Ceratocystis fimbriata* on sweet potato, cacao and sycamore. *Mycologia* 97:57–69.
Fang, S-M., Chen, Y-S., and Wei, G-R. 1995. The concurrent resistant reaction of sweet potato varieties to *Pseudomonas solanacearum* and *Fusarium oxysporum* f. sp. *batatas*. *Plant Prot*. 21:34–35.
Fang, S-M., Wu, J-Y., and Chen, Y-S. 1994. Study on resistance test of sweet potato varieties against *Pseudomonas solanacearum*. *J. Fujian Agric. Univ*. 23:154–159.
Fang, S-M., and Yu, H-Q. 1991. Pathogenic groups of sweet potato *Pseudomonas solanacearum* and its distribution and development. *Acta Phytophylacica Sinica* 18:127–132.
Farr, D.F., Rossman, A.Y., Palm, M.E., and McCray, E.B. 2007. Fungal Databases, Systematic Mycology and Microbiology Laboratory, ARS, USDA. Retrieved November 22, 2007, from http://nt.ars-grin.gov/fungaldatabases/
Flier, W.G., Grünwald, N.J., Kroon, L.P.N.M., Bosch, T.H.M. van den, Garay-Serrano, E., Lózoya-Saldaña, H., Bonants, P.J.M., and Turkensteen, L.J. 2002. *Phytophthora ipomoeae* sp. nov., a new homothallic species causing leaf blight on *Ipomoea longipedunculata* in the Toluca Valley of central Mexico. *Mycol. Res*. 106:848–856.
Grünwald, N.J., Garay-Serrano, E., and Lozoya-Saldana, H. 2004. First report of blight on *Ipomoea purpurea* caused by *Phytophthora ipomoeae*. *Plant Dis*. 88:1283.
Harter, L.L., and Weimer, J.L. 1929. A mongraphic study of sweet- potato diseases and their control. *USDA Tech. Bull*. 99: 118pp.
Holmes, G.J., and Stange, R.R. 2002. Influence of wound type and storage duration on susceptibility of sweetpotatoes to Rhizopus soft rot. *Plant Dis*. 86:345–348.
Hyun, J-W., and Clark, C.A. 1998. Analysis of *Fusarium lateritium* using RAPD and rDNA RFLP techniques. *Mycol. Res*. 102:1259–1264.
Icochea, T., Duarte, V., and Clark, C.A. 1994. Occurrence of sweetpotato chlorotic leaf distortion in Brazil and Peru caused by *Fusarium lateritium*. *Plant Dis*. 78:754.
Kihurani, A.W., Carey, E.E., and Narla, R.D. 2000. First report of black rot disease of sweetpotato in Kenya. *African Potato Assoc. Conf. Proc*. 5:415–419.
Kihurani, A.W., Skoglund, L.G., and Gatumbi, R.W. 1993. Occurrence of sweetpotato chlorotic leaf distortion caused by *Fusarium lateritium* in Kenya. *Plant Dis*. 77:1260–1261.
Kim, B-K., La Bonte, D.R., and Clark, C.A. 1996. Studies on sweetpotato with chlorotic leaf distortion, caused by *Fusarium lateritium*. *Inter. J. Pest Manag*. 42:67–70.

King, R.R., Lawrence, C.H., Calhoun, L.A., and Ristaino, J.B. 1994. Isolation and characterization of thaxtomin-type phytotoxins associated with *Streptomyces ipomoeae*. *J. Agric. Food Chem.* 42:1791–1794.

Labeda, D.P., and Lyons, A.J. 1992. DNA relatedness among strains of the sweet potato pathogen *Streptomyces ipomoea* (Person and Martin 1940) Waksman and Henrici 1948. *Appl. Environ. Microbiol.* 58:532–535.

Lenné, J.M. 1991. Diseases and Pests of Sweet Potato. National Resources Institute, Bull. 46. 116 pp.

Lewthwaite, S.L., and Wright, P.J. 2005. Sweetpotato vine resistance to Sclerotinia rot. In: Nichols, M.A., ed., Proceedings of the First International Conference on Sweetpotato, Food and Health for the Future. *Acta Horticulturae* 670:63–69.

Lo, J-Y., and Clark, C.A. 1988. Sources of inoculum and infection courts of *Diplodia gossypina* on sweet potato. *Phytopathology* 78:1442–1446.

Lopes, C.A., and Boiteux, L.S. 1994. Leaf spot and stem blight of sweetpotato caused by *Alternaria bataticola*: A new record to South America. *Plant Dis.* 78:1107–1109.

Lopes, C.A., and Silva, J.B.C. 1993. Management measures to control foot rot of sweet potato caused by *Plenodomus destruens*. *Inter. J. Pest Manag*. 39:72–74.

Loria, R., Bukhalid, R.A., Fry, B.A., and King, R.R. 1997. Plant pathogenicity in the genus *Streptomyces*. *Plant Dis.* 81:836–846.

Martin, W.J., and Dukes, P.D. 1977. A bacterial wilt and root rot of sweet potato caused by *Erwinia chrysanthemi. Plant Dis. Rep*. 61:158–161.

Nelson, P.E., Juba, J.H., and Clark, C.A. 1995. An unusual strain of *Fusarium lateritium* affecting sweetpotato. *Mycologia* 87:507–509.

Nirenberg, H.I., and O'Donnell, K. 1998. New *Fusarium* species and combinations within the *Gibberella fujikuroi* species complex. *Mycologia* 90:434–458.

O'Donnell, K., Cigelnik, E., and Nirenberg, H.I. 1998. Molecular systematics and phylogeography of the *Gibberella fujikuroi* species complex. *Mycologia* 90:465–493.

Ogawa, K. 1989. Studies on biocontrol of Fusarium wilt of sweet potato with cross-protection. *Ann. Phytopath. Soc. Japan* 55:385–387.

Ogawa, K., and Komada, H. 1986. Induction of systemic resistance against Fusarium wilt of sweet potato by non-pathogenic *Fusarium oxysporum*. *Ann. Phytopath. Soc. Japan* 52:15–21.

Ooshiro, A., Kawano, S., Takushi, T., Oomi, N., Tokeshi, M., and Ooshiro, N. 2006. Control of soil rot of sweet potato by the incorporation of the fresh aerial tissue of *Geranium carolinianum* L. into the soil. *Weed Biol. Manag*. 6:241–244.

Osiru, M., Adipala, E., Olanya, O.M., Lemaga, B., and Kapinga, R. 2007. Occurrence and distribution of Alternaria leaf petiole and stem blight on sweetpotato in Uganda. *Plant Pathol. J. (Faisalabad)* 6:112–119.

Ramsey, M.D., Vawdrey, L.L., and Hardy, J. 1988. Scab (*Sphaceloma batatas*) a new disease of sweet potatoes in Australia: fungicide and cultivar evaluation. *Australian J. Exp. Agric.* 28:137–141.

Schaad, N.W., and Brenner, D. 1977. A bacterial wilt and root rot of sweet potato caused by *Erwinia chrysanthemi. Phytopathology* 67:302–308.

Skoglund, L.G., Gatumbi, R.W., and Kihurani, A.W. 1993. Non-viral foliar pathogens and disorders of sweet potato in Kenya. *Inter. J. Pest Manag*. 39:452–458.

Smit, N.E.J.M., Holo, T., and Wilson, J.E. 1991. Sweet potato seedling test for resistance to leaf scab disease (*Elsinoe batatas*). *Trop. Agric. (Trinidad)* 68:263–267.

Tate, K.G. 1996. Kumara scurf & its control. *Commercial Grower* 51:34.

Taubenhaus, J.J. 1916. Soilstain, or scurf, of the sweet potato. *J. Agric. Res*. 5:995–1001.

Uritani, I. 1999. Biochemistry on postharvest metabolism and deterioration of some tropical tuberous crops. *Bot. Bull. Acad. Sin*. 40:177–183.

Wright, P.J., Lewthwaite, S.L., Triggs, C.M., and Broadhurst, P.G. 2003. Laboratory evaluation of sweetpotato (*Ipomoea batatas*) resistance to Sclerotinia rot. *N. Z. J. Crop Hort. Sci*. 31:33–39.

Zhang, L-S., Lu, T., Xie, C-S., Huang, H-C., Qian, A-P., and Xu, J. 1999. A study on resistance to different strain groups of *Pseudomonas solanacearum* of sweet potato varieties in south China. *Acta Agriculturae Universitatis Jiangxiensis* 21:347–350.
Zhang, X., Clark, C.A., and Pettis, G.S. 2003. Interstrain inhibition in the sweet potato pathogen *Streptomyces ipomoeae*: Purification and characterization of a highly specific bacteriocin and cloning of its structural gene. *Appl. Environ. Microbiol.* 69:2201–2208.
Zheng, X-M., and Dong, C. 1995. Biotypes of *Pseudomonas solanacearum* (Smith) Smith from various hosts in Guangdong Province. *J. South China Agric. Univ.* 16:50–53.

Chapter 8
Virus and Phytoplasma Diseases

G. Loebenstein, G. Thottappilly, S. Fuentes and J. Cohen

Sweetpotatoes are vegetative propagated from vines, root slips (sprouts) or tubers, and farmers often take vines for propagation from their own fields year after year. Thus, if virus diseases are present in the field they will inevitable be transmitted with the propagation material to the newly planted field, resulting often in a marked decrease in yields. Yields differ greatly in different areas or even fields in the same location. Thus, the average yield in African countries is about 7.02 tons/ha, with yields of 9.4, 4.4, 2.5 and 3.2 ton/ha in Kenya, Uganda, Sierra Leone and Nigeria, respectively. The yields in Asia are significantly higher, averaging 12.41 tons/ha. China, Japan, Korea and Israel have the highest yields with about 21.6, 25.8, 16.4 and 44.4 tons/ha, respectively. In South America the average yield is 10.74 tons/ha, with Argentina, Peru and Uruguay in the lead with 17.2, 16.35 and 13.68 tons/ha, respectively. For comparison, the average yield in the USA is 20.1 tons/ha (all data are averages for 2005 from the FAOSTAT 2007).

These differences in yields are mainly due to variation in quality of the propagation material, often taken from the previous season of farmer's fields. Often these fields are infected with several viruses, thereby compounding the effect on yields. In China, on average, losses of over 20% due to sweetpotato virus diseases were observed (Gao et al. 2000), mainly due to *Sweetpotato feathery mottle virus* (SPFMV) and *Sweetpotato latent virus* (SPLV). The infection rate in the Shandong province reaches 5–41% (Shang et al. 1999). In countries were care is taken to provide virus-tested planting material as, amongst others in the USA and Israel, yields increase markedly, up to 7 times and more. In some countries, as in Uganda, Kenya and Tanzania virus diseases are a major constraint for sweetpotato production.

G. Loebenstein (✉)
Agricultural Research Organization, Bet Dagan 50-250, Israel
e-mail: Gad-talma@barak.net.il

This chapter is based to a large extent on one published in "Virus and Virus-like Diseases of Major Crops in Developing Countries" (eds) G. Loebenstein and G. Thottappilly; Kluwer Academic Publishers, 2003.

G. Loebenstein, G. Thottappilly (eds.), *The Sweetpotato*,
DOI 10.1007/978-1-4020-9475-0_8, © Springer Science+Business Media B.V. 2009

The Main Viruses

Sweetpotato feathery mottle virus Genus *Potyvirus* (SPFMV) is the most common sweetpotato virus worldwide. Certain isolates in the USA and Japan cause much economic damage by inducing cracking or internal corkiness in some cultivars (Fig. 8.1). In Africa, SPFMV causes a severe sweetpotato virus disease (SPVD) in a complex infection with the whitefly-transmitted *Sweetpotato chlorotic stunt virus* Genus *Crinivirus* (SPCSV) (Syn. *Sweetpotato sunken vein virus*, SPSVV).

Most sweetpotato cultivars infected by SPFMV alone show no or only mild circular spots on their leaves (Fig. 8.2) or light green patterns along veins. However, when infected together with the whitefly-transmitted SPCSV stunting of the plants, feathery vein clearing and yellowing of the plants are observed (Fig. 8.3). In controlled experiments, SPFMV-infection alone did not reduce yields compared to virus-free controls, while the complex infection with SPCSV reduced yields by 50% or more (Gutiérrez et al. 2003; Milgram et al. 1996). In Brazil, yield increases of 118% were observed when virus-tested cuttings derived from meristems of heat-treated plants were compared to yields from cuttings taken from fields (Pozzer et al. 1995). This is a unique report where SPFMV alone markedly decreased yields, and it might be that another virus could have been present.

SPFMV is transmitted in a nonpersistent manner by aphids, including *Aphis gossypii*, *Myzus persicae*, *A. craccivora* and *Lipaphis erysimi*. Aphid transmissibility is dependent on virus encoded helper proteinase and a DAG triplet in the protein coat. The virus can be transmitted mechanically to various *Ipomoea* spp. as *I. batatas, I. setosa, I. nil, I. incarnata* and *I. purpurea*, and some strains to *Nicotiana benthamiana, N. clevelandii, Chenopodium amaranticolor* and *C. quinoa* (Brunt

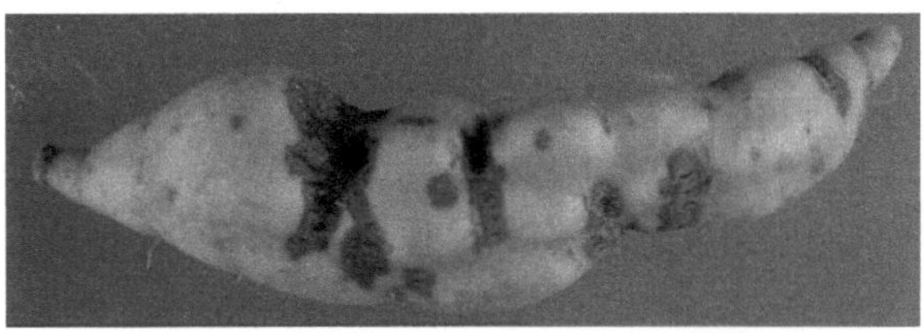

Fig. 8.1 Internal cork disease caused by sweetpotato feathery mottle virus (Courtesy J.W. Moyer)

Fig. 8.2 SPFMV circular spots

Fig. 8.3 Sweetpotato infected by both SPFMV and SPSVV causing stunting of the plants, feathery vein clearing and yellowing of the plants

et al. 1996). The virus is transmitted by grafting but not by seed or pollen or by contact between plants. In Uganda SPFMV was found in 22 *Ipomoea* spp. Hewittia sublobata and *Lepistemon ovariensis* (Tugume et al. 2008).

The virus can best be diagnosed by grafting on *I. setosa*, causing vein clearing (Fig. 8.4) followed by remission, or on *I. incarnata* and *I. nil* inducing systemic vein clearing, vein banding and ringspots. SPFMV can be diagnosed by ELISA, and antisera are commercially available.

However, ELISA reliably detects SPFMV only in leaves with symptoms and when coinfected with SPCSV (Gutiérrez et al. 2003). It is best to sample several leaves from a plant, as the virus seems to be unevenly distributed, especially for meristem-derived plantlets. This has to be followed up by indexing on a susceptible indicator (Green et al. 1988). SPFMV can also be detected by membrane immunobinding (MIBA, also termed NCM-ELISA) and by using a riboprobe (Abad and Moyer, 1992). The latter although being very sensitive, detecting 0.128 pg of RNA, compared with 179 pg of capsid protein, was still somewhat less sensitive than grafting on a sweetpotato indicator (cv. Jewel). The use of a riboprobe also requires a well-equipped laboratory and radioactive materials or digoxigenin for labeling the probe. SPFMV has also been identified by reverse transcription and polymerase chain reaction (RT- PCR), utilizing degenerate genus-specific primers,

Fig. 8.4 *I. Setosa* infected with *Sweetpotato feathery mottle virus* (SPFMV), showing vein clearing

designed to amplify the variable 5′ terminal region of the potyvirus coat protein gene (Colinet et al. 1998).

SPFMV can be purified from infected *I. nil* leaf material (Cali and Moyer, 1981), by rather complex purification procedures, and directly from sweetpotatoes (infected also with SPCSV), by a relatively simple method (Cohen et al. 1988). This method yielded 50–100 mg virus/kg tissue, 5–10 times more when compared with the previous methods. This high yield of virus is due to the synergistic interaction between SPFMV and SPCSV, where SPFMV titer attains more than 600-fold increase (Karyeija et al. 2000) It should be mentioned that virus yields from plants infected by SPFMV alone, without SPCSV, are markedly lower.

Virions are filamentous, not enveloped, usually flexuous, with a modal length of 830–850 nm. The genome consists of single stranded linear RNA, unipartite of 11.6 Kbp, with a poly(A) region (Moyer and Cali, 1985). The complete nucleotide sequence of a sweetpotato feathery mottle potyvirus severe strain (SPFMV-S) genomic RNA was determined (Sakai et al. 1997). The viral RNA genome (S strain) is 10,820 nucleotides (nts) long (GenBank accession D86371) excluding the poly(A) tail, and contained one open reading frame, potentially encoding 3,493 amino acids. Except in the regions of P1 and P3, the polyprotein has a high level of amino acid identity with those of other potyviruses (Sakai et al. 1997). The helper component-proteinase of SPFMV facilitates systemic infection spread of *Potato virus X* in *I. nil* (Sonoda et al. 2000).

Many strains of SPFMV (Moyer, 1986), isolates, variants and serotypes of SPFMV have been reported, as the russet crack (RC), C, S, C1 (from Peru), strain 835 (from Guatemala) and others (Colinet and Kummert, 1993; Karyeija et al. 2001). By comparing coat protein gene sequences of isolates it was shown that isolates from East Africa (EA) form a separate cluster (Kreuze et al. 2000). SPFMV can be divided into four phylogenetic groups (strains) according to the analysis of coat protein (CP) encoding sequences. Thus, strain EA contained the East African, Spanish, and Peruvian isolates of SPFMV. In contrast, strain RC contained isolates from Australia, Africa, Asia and North and South America. Strain O contained isolates from Africa, Asia and South America. The strain C isolates from Australia, Africa, Asia, and North and South America formed a group that was genetically distant from the other SPFMV strains (Tairo et al. 2005, Untiveros et al. 2007; Valverde et al. 2004b). It was suggested that SPFMV-C might be a distinct virus (Lyerly et al. 2003; Tairo et al. 2005). Evidences on recombination events between viruses from different strain groups of SPFMV have been recently reported (Untiveros et al. 2007).

Though SPFMV alone generally causes only minor damage, its control is imperative as in combination with other viruses its effect on plant growth and yields may become substantial. The proven approach is to prepare virus-tested plants from meristems. The plants are then grown under insect-proof conditions, and propagation is continued by cuttings. The farmer plants a certain number of vines in the open field and continues to propagate until the stand of the field is complete. The following year he will again start from virus-tested plants. This scheme has been in operation for more than 15 years in Israel, and presently it is difficult to find SPFMV in the country.

Breeding of SPFMV resistant plants was initiated by CIP (Mihovilovich et al. 2000). Several clones that were resistant to SPFMV in CIP's tests were found to be susceptible, when exposed to Israeli (unpublished) and Ugandan isolates (Karyeija et al. 1998). Apparently, strain diversity requires that breeding and selection have to be done in various locations. On the other hand a substantial number of African sweetpotato landraces have resistance to this virus (Carey et al. 1997).

Another approach was the development of transgenic sweetpotatoes with coat-mediated (CP) resistance to SPFMV (Okada et al. 2001). Also, CP-mediated resistance introduced into several African varieties and cultivar CPT-560 was evaluated in Kenya in a cooperative project between Monsanto Co., USA and Kenya Agriculture Research Institute (KARI) (Odame et al. 2001) However, these transgenic lines were not resistant to the 'complex' infection with SPCSV, causing the SPVD (Wambugu, 2004).

An interesting idea for controlling SPFMV involves the introduction of a rice cysteine-inhibitor gene. This gene inhibits the proteolysis of the viral polyprotein, thereby interfering with viral replication. Improved resistance to SPFMV was observed in 18 of the 25 transgenic lines after challenge-inoculation with the russet crack strain of SPFMV (Cipriani et al. 2001). However, when stem cuttings were prepared from the tolerant transgenic plants and grafted with healthy *I. setosa* scions, virus symptoms appeared on the scions. Apparently, the transgenic sweetpotato plants still contained some virus.

Sweetpotato chlorotic stunt virus Genus *Crinivirus* (SPCSV). {Possible synonym: *Sweetpotato sunken vein virus* (SPSVV)}. Although SPCSV is the name recognized by the International Committee of Taxonomy of Viruses (ICTV), they might be considered as separate viruses for the following reasons: (i) SPCSV and SPSVV differ in their nucleotide and protein sequences by more than 20% (Cuellar et al. 2008); (ii) comparing nucleotide sequences of the Israeli SPSVV with the Ugandan isolate of SPCSV in the heat shock 70-like protein region, revealed a homology of only 0.69 (Cuellar et al. 2008); (iii) symptoms induced by SPSVV on *I. setosa* differ markedly from those by SPCSV; (iv) infection of sweetpotato by SPSVV alone produced on cv. Georgia Jet mild symptoms consisting of slight yellowing of veins, with some sunken secondary veins on the upper sides of the leaves and swollen veins on their lower sides (Fig. 8.5) (at 28–29 °C). Upward rolling of the 3–5 distal leaves was also observed. Similar symptoms were also obtained on graft-inoculated plants of cv. Papel and Camote Negro (Cohen et al. 1992). The disease could easily be detected in the field, where plants had erect branches, with some upward rolling leaves. In some cultivars, SPCSV isolates cause no symptoms or mild stunting combined with slight yellowing or purpling of old leaves (Gibson et al. 1998; Gutiérrez et al. 2003).

Effects on yields by SPSVV or SPCSV alone are minor or close to 30% reduction, but in complex infection with SPFMV or other viruses yield losses of 50% and more are observed (Gutiérrez et al. 2003; Milgram et al. 1996; Untiveros et al. 2007.

SPCSV and/or SPSVV are transmitted by the whitefly *Bemisia tabaci* biotype B, *Trialeurodes abutilonea*, and *B. afer* (Gamarra et al. 2008; Ng and Falk, 2006; Schaefers and Terry, 1976; Sheffield, 1957; Sim et al. 2000; Valverde et al. 2004b)

Fig. 8.5 *Sweetpotato* sunken *vein virus* – swollen veins on the lower side of the leaf; left leaf –healthy control

in a semipersistent manner, requiring at least one hour for acquisition and infection feeding and reaching a maximum after 24 h for both of them. The virus is graft transmissible, but not by mechanical inoculation. The virus was transmitted by whiteflies to *I. setosa, N. clevelandii, N. benthamiana* and *Amaranthus palmeri* from *I. setosa* and by grafting to other Ipomoeas.

The virus is best being diagnosed on a pair of sweetpotato plants- one healthy, the other infected by SPFMV. On the healthy plants hardly any symptoms will become apparent, while (if carrying SPFMV) severe symptoms of SPVD will appear. The virus can also be diagnosed by immunosorbent electron microscopy (ISEM) and by ELISA.

SPSVV can be purified from infected *I. nil* or *N. clevelandii*, maintained at 28 °C (Cohen et al. 1992). Normal length of virus particles from purified preparations was 850 nm, the diameter ~12 nm, with an open helical structure typical for closteroviruses. In leaf dip preparations from *I. setosa* infected with an isolate of SPCSV from Nigeria a modal length of 950 nm was obtained (Winter et al. 1992). The coat protein of the Israeli and Nigerian isolates had a MW of ~ 34, 000 and ~ 29, 000, respectively (Cohen et al. 1992; Hoyer et al. 1996). Double-stranded RNA from the Israeli isolate consisted of one major band of M_r 10.5 Kbp, and 2 minor bands of 9.0 and 5.0 Kbp. Northern blot analysis of the Nigerian isolate revealed the presence of a large dsRNA with an estimated size of ~9.0 Kbp and several smaller ones. Several Kenyan cDNA clones revealed an open reading frame (ORF) of 774 nts coding for the coat protein (Hoyer et al. 1996).

SPCSV can be serologically divided into two major serotypes, which correlate to two genetically distantly related strains/groups based on the coat protein (CP) and the heat shock protein 70 homologue (Hsp70h) genes similarities. The East Africa (EA) group was first identified in East Africa, and also occurs in Peru, while the West Africa (WA) was first identified in West Africa and occurs additionally in the Americas and the Mediterranean, but not in East Africa (Hoyer et al. 1996; Tairo et al. 2005). In East Africa two serotypes (S_{EA1} and S_{EA2}) were distinguished using a panel of monoclonal antibodies (MAbs) to a Kenyan isolate of SPCSV. S_{EA1} serologically resembled the Kenyan isolate of SPCSV whereas S_{EA2} has not previously been reported. S_{EA1} was predominant in eastern Uganda whereas S_{EA2} was predominant in southern and western Uganda. The Hsp70h and CP genes were generated by RT-PCR. Sequence analyses revealed substitutions at two nucleotide positions in the Hsp70h gene, although neither affected deduced amino-acid sequences. Nucleotide substitutions in the CP gene region, which led to 11 amino-acid substitutions, revealed two major groupings plus other minor variants (Alicai et al. 1999). SPCSV isolates from East Africa were phylogenetically distant to SPCSV isolates from elsewhere (Tairo et al. 2005).

The complete nucleotide sequences of genomic RNA1 (9,407 nts) and RNA2 (8,223 nts) were determined, revealing that SPCSV possesses the second largest identified positive-strand single-stranded RNA genome among plant viruses after *Citrus tristeza virus* (Kreuze et al. 2002). RNA1 contains two overlapping open reading frames (ORFs) that encode the replication module, consisting of the putative papain-like cysteine proteinase, methyltransferase, helicase, and polymerase domains. RNA2 contains the *Closteroviridae* hallmark gene array represented by an Hsp70h, a protein of 50 to 60 kDa depending on the virus, the major coat protein, and a divergent copy of the coat protein. The two genomic RNAs of SPCSV contained nearly identical 208-nt-long 3′ terminal sequences, and the ORF for a putative small hydrophobic protein was found in SPCSV RNA1. Furthermore, unlike any other plant or animal virus, SPCSV carried an ORF for a putative RNase III-like protein (ORF2 on RNA1). Several subgenomic RNAs (sgRNAs) were detected in SPCSV-infected plants, indicating that the sgRNAs formed from RNA1 accumulated earlier in infection than those of RNA2. Recently it was shown that the 3-proximal part of RNA 1 and the partial sequence of the Hsp70h gene of the Israeli SPSVV differ markedly in sequence from most of the SPCSV isolates (Cuellar et al. 2008; see also accession numbers EU124491 and EU124487). Additionally, the phylogenetic analysis of the partial Hsp70h, CP, and p60 genes of the Ea and the WA strains indicate that they may belong to different species in the genus *Crinivirus* (Abad et al. 2007; Tairo et al. 2005).

The disease has been reported from East and Southern Africa, Nigeria, Niger, Indonesia, Israel, Egypt, Spain, Argentina, Brazil, Peru (Gutiérrez et al. 2003) and recently from the USA (Abad et al. 2007).

In a recent research a Peruvian sweetpotato landreace was transformed with an introduced hairpin construct targeting the replicase encoding sequences of SPFMV and SPCSV (Kreuze et al. 2008). A high level of resistance to SPCSV was observed in the transgenes, which however did not prevent development of SPVD when plant became infected also with SPFMV.

For control of SPCSV/SPSVV it is best to start from meristems and test the plants on a pair of sweetpotato plants as described above.

Sweetpotato virus disease (SPVD) is caused by the interaction of SPFMV and SPCSV/SPSVV. Characteristic symptoms of the disease include veinclearing (Fig. 8.3), chlorosis and stunting. The disease was described by Schaefers and Terry (1976) in Nigeria and is the most important virus (complex) disease in East Africa, where sweetpotato is often the main food staple (Karyeija et al. 1998). The disease was described in Israel by Loebenstein and Harpaz (1960), Peru (Gutiérrez et al. 2003), the USA (Abad et al. 2007), Spain (Trenado et al. 2007), and occurs probably in Italy (Parrella et al. 2006).

It can cause losses over 50%, especially in Uganda and Kenya, though in another study from Uganda, losses were much smaller, probably due to relatively high levels of virus resistance in their landraces (Gibson et al 1997). In a 3-year field study in Cameroon SPVD reduced root yields by 56–90% in susceptible varieties (Ngeve and Bouwkamp, 1991); and yield reductions of 78% due to SPVD were reported from field trials in Nigeria (Hahn, 1979). In Israel in a 2-year field experiment yield reductions of ~50% were observed in plots planted with SPVD-infected cuttings (Milgram et al. 1996). In Peru in a field study, SPVD reduced root yield by 65–72% (Gutiérrez et al. 2003).

As stated above the optimal approach for controlling SPVD is to supply virus-tested planting material. In practice, however, in Uganda and other African countries this has sofar not been achieved. In Uganda, where sweetpotatoes are grown on a large scale the most important control practice, as perceived by farmers, is to plant cuttings from symptomless parents, and destroying diseased plants (Gibson et al. 2000). Although SPVD-resistant sweetpotato landraces are available, these were perceived by farmers to have poor and late yields. Cultivars Tanzania and New Kawogo were relatively resistant to SPVD in low pressure of whiteflies. SPFMV in these cultivars was difficult to detect, because of low virus titer, and was seldom acquired by aphids. However, resistance to SPFMV was not apparent once infected also with SPCSV (Aritua et al. 1998). In SPVD-infected plants SPFMV titers were much higher than in sweetpotato plants infected with SPFMV alone (Cohen et al. 1988; Karyeija et al. 2000), facilitating the spread of SPFMV by aphids (Rossel and Thottappilly 1988). This increase is not associated with enhanced virus movement but is due to a synergistic interaction with the crinivirus (Karyeija et al. 2000).

Sweetpotato mild mottle virus Genus *Ipomovirus*, (SPMMV). Synonyms – sweetpotato B virus (Sheffield, 1957). SPMMV has so far been reported from West Africa (Kenya, Uganda, Burundi and Tanzania), South Africa, Indonesia, China, Philippines, Papua New Guinea, India, New Zealand, and Egypt.

SPMMV can cause leaf mottling, stunting and loss of yields. Cultivars differ greatly in their reaction to the virus, some being symptomlessly infected, others apparently immune. The virus is transmitted semipersistently by *B. tabaci*, by grafting and by mechanical inoculation. It is not transmitted by seed or by contact between plants. The virus was transmitted to plants in 14 families (Hollings et al. 1976). Diagnostic species: *N. tabacum, N. glutinosa*- vein clearing, leaf puckering, mottling and distortion; *C. quinoa* – local lesions, not systemic; *I. setosa* – conspicous systemic vein chlorosis (Fig. 8.6). Diagnosis can be confirmed by serological tests of *I. setosa*. Commercial antisera are available. After 3–4 weeks, new growth is almost symptomless. Sap transmission from sweetpotato to test plants is often difficult. The virus is best maintained in *N. glutiniosa, N. clevelandii and N. tabacum*, and can be assayed quantitatively on *C. quinoa*, where SPMMV induces local lesions.

The virus can be purified from systemically infected *N. tabacun* (Hollings et al. 1976). Virions are flexous rod shaped particles, 800–950 nm in length, containing 5% RNA and 95% protein. The genome consists of single stranded RNA. The viral RNA was cloned and the assembled genomic sequence was 10,818 nts in length with a polyadenylated tract at the 3′ terminus. The sequence accession code is Z73124. Almost all known potyvirus motifs are present in the polyprotein of SP-MMV, except some motifs in the putative helper-component and coat protein, which

Fig. 8.6 Symptoms of *Sweetpotato mild mottle virus* on *I. setosa* leaves. Healthy leaf is at top left (Courtesy Dr. J. Moyer)

are incomplete or missing. This may account for its vector relations (Colinet et al. 1998). The coat protein has a MW of 37,700.

SPMMV isolates showed a high level of variability with no discrete strain grouping. Sequences of several SPMMV isolates revealed nt sequence identities of 88.0% and 899% or higher for the CP-encoding region and 3′-UTR, respectively, while CP aminoacid (aa) sequences were 93.0–100% identical. Analysis of the CP-encoding nt sequences did not revealed phylogenetically distinguishable groups of SPMMV isolates. Rather, analysis indicated high genetic variability (Tairo et al. 2005).

A synergism was observed in sweetpotato doubly infected by SPMMV and SPCSV (but not by SPFMV) (Untiveros et al. 2007).

Sweetpotato latent virus Genus *Potyvirus* (SPLV) is widespread in China and has been reported also from Taiwan, Korea, Indonesia, Japan, Philippines, Uganda, Kenya, South Africa, India, Egypt and New Zealand.

SPLV may cause mild chlorosis but in most cultivars the infection is symptomless. Symptoms often disappear after infection, but the plants remain infected. Crystal inclusions are observed in the nucleus and pinwheels in the cytoplasm.

SPLV isolates from Japan and China were transmitted by the aphid *Myzus persicae* (Usugi et al., 1991) and the virus can be transmitted by mechanical inoculation and by grafting. It is not transmitted by seed.

Diagnostic species: *N. benthamiana* – systemic mosaic and stunting; *N. clevelandii*- systemic pin-prick chlorotic lesions; *C. quinoa, C. amaranticolor* – brown necrotic local lesions, no systemic infection; *I. setosa* – systemic mottle. Diagnosis can be confirmed by serological tests of *I. setosa.*

The virus is best maintained in *N. benthamiana* or *N. clevelandii* and can be assayed on *C. quinoa* or *C. amaranticolor.*

The virus can be purified according to Liao et al. (1979). Virus particles are flexous rods, 750–790 nm in length. The capsid protein has a MW of 36,000. By using MAbs and polyclonal antibodies some epitopes common to SPLV and SPFMV are found. These can easily be differentiated when potyvirus cross-reactive MAbs are used, indicating a distant relationship (Hammond et al. 1992).

Combining RT-PCR with degenerate oligonucleotide primers derived from the conserved regions of potyviruses, it was possible to identify SPLV, as well as SPFMV and *Sweetpotato G virus* (Colinet et al. 1998), and to differentiate between 2 strains of SPLV (Colinet et al. 1997).

Sequence data accession codes are: X84011 SPLV (Chinese) mRNA for coat protein; X84012 SPLV (Taiwan) mRNA for coat protein. According to Nishiguchi et al. (2001), SPLV has 58% homology to SPFMV-S.

The best way to control this virus, as well as other viruses infecting sweetpotato is by establishing propagation nurseries derived from virus-tested mother plants.

Sweetpotato virus G Genus *Potyvirus* (SPVG) is widespread in China, (Colinet et al. 1994, 1998), and was reported also from Egypt (IsHak et al. 2003), the USA (Souto et al. 2003; Kokkinos and Clark, 2006), Peru (Untiveros et al. 2007), Spain (Trenado et al. 2007), Tanzania (Ndunguru and Kapinga, 2007), Peru (Untiveros et al. 2007), Spain (Trenado et al. 2007), Japan, Ethiopia, Nigeria, and Barbados. Recently, the virus was also found in areas of the Pacific Ocean and their molecular characteristics compared to other isolates (Rannalli et al. 2008). The virus is transmitted mechanically and by aphids *Myzus persicae* and *Aphis gossypii* in a non-persistent manner (Souto et al. 2003). SPVG causes mottling in *I. nil* and chlorotic spotting in *I. setosa* and *I. tricolor* (Souto et al. 2003). Cylindrical inclusions bodies, which consisted of pinwheels and scrolls, were observed in the cytoplasm of epidermal, mesophyll, and vascular cells of infected *I. nil* and *I. setosa* (Souto et al. 2003). Isolates LSU-1 and -3 obtained from sweetpotato plants in Louisiana, USA (Souto et al. 2003) reacted with MAb PTY-1 (Jordan and Hammond, 1991).

A partial sequence of SPVG (X76944) has been obtained after RT-PCR, showing an identity of around 70% and 80% in the amino acid sequence between the complete and conserved core of the coat protein of SPFMV, respectively (Colinet et al. 1998). The SPVG coat protein has 355 amino acids while that of SPFMV has 316 amino acids (Colinet et al. 1994). Comparison with coat protein sequences of known potyviruses indicates that SPVG is a member of the genus *Potyvirus*. Strains of SPVG seem to occur in China (Colinet et al. 1998). SPVG-CH2 shared 89.2% and 90.6% amino acid sequence identities with SPVG-CH in the N-CP and the N-terminal region of the coat protein core (N-CP core), respectively. Sequence identity was much lower with SPFMV and even more so with SPLV and other potyviruses. No biological properties (including host range and symptomatology) or viral characteristics have come to our attention.

Sweetpotato virus 2 Tentative member Genus *Potyvirus* (SPV2). Synonyms: Sweetpotato virus II, Ipomoea vein mosaic virus and Sweetpotato virus Y (Moyer et al. 1989; Souto et al. 2003; Ateka et al. 2004).

The virus was first isolated from sweetpotato plants from Taiwan (Rossel and Thottappilly, 1988), then isolates were obtained from sweetpotato clones from China, Portugal, South Africa, China, USA (Souto et al. 2003), Spain (Trenado et al. 2007), Australia (Tairo et al. 2006), and Peru (Untiveros et al. 2007). SPV2 has filamentous particles of 850 nm in length and induces cytoplasmic cylindrical inclusions consisting of pinwheels and scrolls (Souto et al. 2003; Ateka et al. 2004). The virus was nonpersistently transmitted by *M. persicae* and mechanically transmitted to several species of genera *Chenopodium*, *Datura*, *Nicotiana*, and *Ipomoea*. SPV2 causes vein clearing and leaf distortion on *N. benthamian*, chlorotic local lesions on *Chenpodium* spp., vein mosaic on *I. nil*, *I. setosa* and *I. tricolor* (Souto et al. 2003; Ateka et al. 2007). Isolates LSU-2 and -5 obtained from sweetpotato plants in Louisiana, USA (Souto et al. 2003) reacted with MAb PTY-1 (Jordan and Hammond, 1991).

Although SPV2 has been isolated from sweetpotato plants showing mild symptoms consisting of leaf mottle, vein yellowing and / or ringspots (Rossel and Thottappilly, 1988; Tairo et al. 2006) the significance of this virus to sweetpotato production is not clear as similar symptoms may be caused by other viruses, and sweetpotato cultivars inoculated with SPV2 under greenhouse conditions failed to produce obvious symptoms (Ateka et al. 2004; Souto et al. 2003). However, SPV2 interacts synergistically with SPCSV, increasing symptoms severity on sweetpotato (Kokkinos and Clark, 2006; Tairo et al. 2006), suggesting that SPV2 night be economically important in areas where SPCSV occurs.

It seems that biologically and genetically diverse strains of SPV2 occur. Some differences in test plant reactions and host range appeared to correlate to some extent with the geographic origin and molecular distinctness of the SPV2 isolates (Ateka et al. 2007). Comparison of the CP gene sequences of several isolates revealed nt and aa sequences identities ranging from 81 to 99% and from 86 to 99%, respectively. Phylogenetic analysis of sequences distinguished several groups, which partially correlated with the geographic origin of the isolates (Ateka et al. 2007).

Sweetpotato mild speckling virus Genus *Potyvirus* (SPMSV) was isolated from plants cv. Morada INTA from Argentina showing symptoms of "sweetpotato chlorotic dwarf disease" (Di Feo et al. 2000). These plants show chlorosis, dwarfing, vein clearing, and leaf distortion (Alvarez et al. 1997; Di Feo et al. 2000). The severity of the disease depends on the presence of SPMSV in the complex. Symptoms in plants infected with SPFMV and SPCSV were milder than those in plants infected with the three viruses (Di Feo et al. 2000). SPMSV is synergistic in plants infected with SPCSV but not to SPFMV (Untiveros et al. 2006).

SPMSV has been found in Peru, Argentina, Philippines, China, Indonesia, Egypt, South Africa, Nigeria, and New Zealand. SPMSV is transmitted mechanically and by *M. persicae* in a nonpersistent manner. However, it does not react with the MAb PTY-1 that recognises a cryptotope conserved in most aphid-transmitted potyviruses (Di Feo et al. 2000; Jordan and Hammond, 1991). Its host range is restricted to *Convolvulaceae, Chenopodiaceae*, and *Solanaceae*. *I. setosa* and *I. nil* react with vein clearing, blistering, leaf deformation and mosaic; *N. benthamiana* with vein clearing, and reduction, deformation and down rolling of leaves, and *C. quinoa*, and *N. tabacum* "Samsun" with local infections (Fuentes et al. 1997). SPMSV can easily be detected by ELISA.

This virus has flexuous particles of c. 800 nm. The coat protein sequences of SPMSV showed 63% identity with SPFMV, 68 to 70% with SPLV, 57% with SPVG, and 73% with *Potato virus Y* (Alvarez et al., 1997). Cylindrical inclusion (bundles, laminate aggregates, and pinwheels, neither circles nor scrolls) were observed in the cytoplasm of infected *I. setosa* and *I. batatas* (Nome et al. 2006). This high resemblance favors inclusion of SPMSV within the *Potyvirus* genus.

Sweetpotato leaf speckling virus Possible Genus *Polerovirus* (SPLSV) was isolated from a sweetpotato accession DLP 1541 from CIP's germplasm collection, showing slight leaf curl and whitish speckling symptoms (Fuentes et al. 1996) (Fig. 8.7). It is found occasionally in farmers' materials. The virus was also found in Cuba. Infected plants of cv. Jewel yielded c. 20% less under greenhouse conditions. Virus causes dwarfing and leaf curling in addition to chlorotic and necrotic spotting in *I. nil* and *I. setosa.* SPLSV is transmitted by grafting and by *Macrosiphon euphorbiae* in a persistent manner, but not by *M. persicae, A. gossypi*i or mechanically. Diagnosis can be confirmed by nucleic acid spot hybridization (NASH) from *I. setosa* or *I. nil* (Fuentes et al. 1977). Concentration of SPLSV increases towards the bottom of the plant. On leaves, the virus was detected in the basal part of the leaves and petioles, but not in the apical halves.

SPLSV can be purified from leaves or roots of infected plants (Fuentes et al. 1996). Virions are isometric c. 30 nm in diameter. RT-PCR amplification of RNA from SPLSV-infected plants, using degenerate primers designed to amplify sequences in Luteovirus- or PLRV-RNA, yielded c. 500 bp and c. 600 bp DNA fragments, respectively. The nucleotide sequence of these fragments encoded two polypeptides (coat protein and 17K) characteristic of luteo- and poleroviruses sequences (GenBank accession DQ655700). The coat- and the 17K protein of SPLSV have similarities to of 70% and 63% those of PLRV, respectively. NASH using a PLRV probe can therefore detect SPLSV in sap. However, the heterologous probe gives a lower signal than the homologous reaction with PLRV. SPLSV was not detected in ISEM or ELISA tests using PLRV antibodies.

Sweetpotato chlorotic flecks virus (SPCFV) (tentatively member in Genus *Carlavirus*) was isolated from sweetpotato accession DLP 942 from the CIP's germplasm collection, showing fine chlorotic spots (flecks) (Fuentes and Salazar, 1992). SPCFV was detected in samples from Peru, Japan, China, Korea, Taiwan,

Fig. 8.7 Sweet potato cv Jewel infected with SPLSV showing whitish speckling

Cuba, Panama, Bolivia, Colombia, Brazil, Uganda, Philippines, India, Indonesia, Australia and New Zealand (Aritua et al. 2007; Fuentes and Salazar, 1992; Gibson et al. 1997; Usugi et al. 1991; Jones and Dwyer 2007; Fletcher et al. 2000). This virus is transmitted mechanically but not by aphids and not by seed and has a limited host range in the families *Convolvulaceae* and *Chenopodiaceae*. SPCFV induces in *I. nil* fine chlorotic spots and vein clearing on the first and second true leaf. The virus can be recovered and detected from symptomatic leaves, but not from symptomless ones. No clear symptoms are observed on infected *I. setosa* (only transient vein clearing on 1–2 leaves). In *C. murale* and *C. quinoa* react with local necrotic lesions. Graft inoculated cvs. Nemañete, Jewel, Luby 3074, IITA TIS 2498 did not show symptoms. Infectivity of SPCFV is lost at dilution between 10^{-2} and 10^{-3}, and after storage between 1h and 1 day. The thermal inactivation point is between 50 and 60 °C. Some cytological alterations (hypertrophy of chloroplasts) were observed in *I. nil* but no inclusion bodies ("pinwheels").

Synergism was observed in sweetpotato doubly infected by SPCFV and SPCSV (Untiveros et al. 2006).

SPCFV seems to be serologically related to an uncharacterised Japanese isolate (sweetpotato symptomless virus – SPSV). Although SPCFV and SPSV are serologically related, they differ in some characteristics. SPSV has flexuous short and long particles c. 710 to 760 nm and 1,430 to 1,510 nm in length, and it infects only *Ipomoea* spp., while SPCFV particles are flexuous rod c. 750–800 in length. Also, *I. nil* infected with SPSV showed stronger symptoms (vein clearing, curling and vein necrosis) than those caused by SPCFV. The thermal inactivation point of SPSV (70–80 °C) is higher than that of SPCFV.

Virus can be purified from cotyledons and leaves of infected *I. nil* (12 day after inoculation). The virus particles are flexuous rods c. 750–800 in length with a capsid polypeptide of Mr 34.5 kDa (Fuentes and Salazar, 1992), Its genome (GenBank accession NC_006550) is 9,104 nts long (excluding the poly(A) tail) and potentially includes six open reading frames (ORFs) (Aritua et al. 2007). Its genomic organization and similarity resembles that of genus *Carlavirus* but is the least related to typical carlaviruses. SPCFV genome differed from that of typical carlaviruses in being comparatively large (9,104 vs. 6,480–8,739 nts). The remarkable size of the SPCFV genome results from a large ORF1 (237.5 vs. 200–235 kDa) and the presence of an abnormally long untranslated region (233 vs. <50 nts) between ORF4 and ORF5. Its close relative is melon yellowing-associated virus, a proposed carlavirus from Brazil, which is transmitted by whiteflies (*B. tabaci* biotype B) (Aritua et al. 2007). Comparison of the CP gene of geographically diverse SPCFV isolates showed a wide variation both in their CP nucleotide and the corresponding aa sequences ranging from 75.1–99% and 88.3–99.7%, respectively. At aa level, the East African isolates were 94.0–99.7% similar while isolates of the non- East African origin were more divergent (88.3–96.7%). The phylogenetic analysis of the CP aa sequences separated the SPCFV isolates into various clusters that corresponded to their geographical origin (Aritua et al. 2007).

Sweetpotato caulimo-like virus (SPCaLV) has been found together with SPFMV in widely scattered geographical locations – Madeira, New Zealand, Papua New

Guinea and Solomon Islands (Atkey and Brunt, 1987), China (Gao et al. 2000) and Uganda (WISARD Project Information, 1991 Aritua et al. 2007). Also reported from Kingdom of Tonga, Kenya, Nigeria, Egypt, Philippines, Australia, and Puerto Rico.

Sweetpotato plants infected by SPCaLV are usually without symptoms, though a few leaves occasionally have chlorotic or purple spots, probably induced by SPFMV (Atkey and Brunt, 1987). Virus is not transmitted by mechanical inoculation, not by contact between plants or by seed.

Graft inoculation of *I. setosa* seedlings with SPCaLV resulted in appearance of chlorotic flecks along minor veins or circular interveinal spots on several leaves. Such leaves sometimes become completely chlorotic, wilt and die. Fifty nm caulimo-like particles can be detected in negatively stained sap of *I. setosa* (but not from *I. batatas*), and c. 10 times more by ISEM. In the cytoplasm of infected *I. setosa* leaf cells numerous isometric particles and intracellular inclusions are readily detected. The inclusions are spherical or ovoid, up to 4μm in diameter with a large central lacuna and usually several smaller peripheral lacunae; they do not contain virions (Atkey and Brunt, 1987).

The viral genome consists of double-stranded DNA and the coat protein has a Mr 42–44 kDa (Brunt et al. 1996).

Sweetpotato leaf curl virus Genus *Begomovirus* (SPLCV) was first reported by Shinkai (1979) and Liao et al. (1979) from Japan and Taiwan. The virus has also been reported from USA, Far East and samples from Brazil, Mexico, China, Korea, Puerto Rico (Lotrakul et al. 2002), Kenya (Miamo et al. 2006), and Peru (Fuentes and Salazar 2003). In infected young sweetpotato plants upward curling of leaves and vein swelling on young plants are observed, though later only few symptoms remain and plants become symptomless. SPLCV can cause up to 30% reductions in yield. The virus is transmitted by *B. tabaci* biotype B in a persistent manner and by grafting, but not mechanically or by seeds. Under experimental conditions, the virus is transmitted by its vector at relatively low levels (Valverde et al. 2004a). Various *Ipomoea* species were susceptible to SPLCV, as I. purpurea causing leaf curl and stunt (Osaki and Inouye, 1991), *I. aquatica* – yellow vein symptoms, *I. nil, I. setosa* and *N. benthamiana* – leaf curl symptoms (Lotrakul et al. 1998). The disease is caused by a gemini virus belonging to the *Begomovirus* group (Onuki et al. 2000). Cytopathic changes in the nucleoplasm, as fibrillar rings and crystalline arrays of virus-like particles, typical of gemini-infected tissue were also observed (Osaki and Inouye, 1991). The virus was partially purified yielding typical geminate particles with a size of c. 18 × 30 nm, and Western blot analysis revealed serological relationships between SPLCV and *Bean golden mosaic virus* (Onuki et al. 2000). A complete sequence of SPLCV (AF104036, 2,828 nts) has been determined by Lotrakul and Valverde (1999). Its genomic DNA and organization is similar to that of monopartite begomoviruses. The partial nucleotide sequence identity (from 87% to nearly 100%) of the AC1 gene of different SPLCV isolates and the phylogenetic analysis of them suggest that there may be more than one species of the SPLCV (Lotrakul et al. 2002).

Coinfection of SPLCV and SPFMV in *I. nil* and *I. setosa* induce severe leaf distortion, general chlorosis and stunting. SPLCV titers/replication were significantly greater in plants co-infected with SPFMV compared to plants inoculated with SPLCV alone (Kokkinos and Clark, 2004).

Sweetpotato leaf curl Georgia virus Genus *Begomovirus* (SPLCGV). This virus, formely known as *Ipomoea leaf curl virus* (Lotrakul et al. 2003), is transmitted by *B. tabaci* biotype B. It occurs in USA, Puerto Rico, and India (Prasanth and Hegde, 2008). The virus caused leaf curl symptoms in several *Ipomoea* species. Unlike SPLCV, SPLCGV did not cause yellow vein symptoms on *I. aquatica* and *I. cordatotriloba*. Its genome of 2,773 nts (AF326775) has an organization typical of other Old World monopartite begomoviruses. Sequence comparisons showed that SPLCGV was similar to SPLCV and to another *Ipomoea*-infecting begomovirus, *Ipomoea* yellow vein virus (IYVV) from Spain with about 76 and 78% nucleotide sequence identity, respectively. Although the coat protein of SPLCGV was nearly identical to that of SPLCV and IYVV, the sequence of the common region and the AC1, AC2, AC3 and AC4 ORFs were different (Lotrakul et al. 2003).

Ipomoea yellow vein virus Genus *Begomovirus* (IYVV) was first found infecting *I. indica* plants showing yellow vein symptoms in Spain (Banks et al. 1999) and Italy (Briddon et al. 2005), then found infecting cultivated sweetpotato plants. The complete nucleotide sequence (AJ132548) confirmed its begomovirus nature. Contrary to SPLCV, IYVV was not transmitted by *B. tabaci* biotype B (or Q and S). Phylogenetic analysis of the three *Ipomoea*-infecting begomoviruses species (SPLCV, SPLCGV, and IYVV) recognised by the ICTV revealed that these viruses form a separate cluster that place them apart from all other begomovirus.

Ipomeoa crinkle leaf curl virus (ICLCV). In 1992, Cohen et al. (1997) observed on *I. setosa*, grafted with scions from sweetpotato cv. Georgia Jet plants, introduced from an unknown source in North America, atypical symptoms of little leaf and crinkle symptoms (Fig. 8.8). Geminate particles were observed in crude sap preparations. The virus was transmitted by *B. tabaci* in a persistent manner and by grafting, but not mechanically. The virus was transmitted to several *Ipomoea* species, including *I. hederacea, I. trifida, I. nil, I. littoralis* and *I. setosa* and induced symptoms on them, including on some cultivars of *I. batatas*, but not on cv. Georgia

Fig. 8.8 Symptoms of *Ipomoea crinkle leaf virus* on *I. setosa*. Left- healthy plant

Jet. It might be mentioned that SPLCV did not infect *I. hederacea* and symptoms induced by ICLCV on *I. setosa* and *I. nil* differed from those described for SPLCV (Chung et al. 1985). Based on the host range ICLCV is considered to be distinct from SPLCV, but its exact relationship to other identified viruses remains unclear.

Cucumber mosaic virus Genus *Cucumovirus* (CMV) is one of the most widespread plant viruses, recorded in more than 190 species, belonging to more than 40 families (Francki et al. 1979). CMV has been isolated from *I. setifera* (Migliori et al. 1978) and Martin (1962) succeeded in transmitting CMV by mechanical inoculation to *I. nil, I. purpurea, I. lacunosa*, and *I. trichocarpa* but not to *I. batatas* cv. Puerto Rico. Cohen and Loebenstein (1991) failed in transmitting CMV to healthy sweetpotato plants. However, sweetpotatoes carrying the whitefly-transmitted SPSVV can easily be infected by CMV by aphid, mechanical or graft inoculations (Cohen and Loebenstein, 1991). Untiveros et al. (2007) found that CMV was able to infect sweetpotatoes without the assistance of SPCSV. It appears that CMV strains are nonspecific for infection in sweetpotato. CMV isolated from cucumber (Cohen and Loebenstein, 1991) or *Arracacia xanthorrhiza* (Untiveros et al. 2007) were able to infect sweetpotato plants assisted by SPSVV or no, respectively. In some fields in Israel during the 80es heavy infections together with SPFMV and SPSVV caused severe yellowing and stunting (Fig. 8.9). Later, when farmers used certified planting material such symptoms were hardly found. Apparently, the presence of another virus (SPSVV) facilitates replication or translocation of some CMV strains in sweetpotato. It may be that there is a gene silencing mechanism that inhibits replication of such CMV strains in healthy sweetpotato and is supressed by SPSVV, allowing CMV to replicate and/or move in the sweetpotato plant.

It is interesting to note that although CMV occurs worldwide, in sweetpotato it has been reported so far only from Israel, Japan, New Zealand (Fletcher et al. 2000), Spain, West Africa (Clark and Moyer, 1988) and Egypt. CMV was not found in Kenya (Ateka et al. 2004) and Tanzania (Ndunguru and Kapinga, 2007) even though SPCSV is very widespread SPCSV strains do not support infection of sweetpotatoes with *Cucumber mosaic virus* (CMV), while SPSVV is needed to infect sweetpotatoes with CMV.

Fig. 8.9 Sweet potato plants (cv. Georgia Jet) severely affected by yellowing and stunting caused by double infection with *Cucumber mosaic virus* and *Sweet potato sunken vein virus*

Sweetpotato yellow dwarf virus Genus *Ipomovirus*, Family *Potyviridae* (SPYDV) in *I. batatas* was first reported from Taiwan by Liao et al. (1979), Far East, and Brazil (www.abam.com.br/livroscargil/Capitulo%202/Capitulo%2022.pdf). The virus in sweetpotato causes severe systemic leaf chlorosis and stunting. The virus was transmitted by *B. tabaci* in a persistent manner, as well as by grafting and by mechanical inoculation. Susceptible hosts include in addition to sweetpotato, *Chenopodium* spp., *Gomphrena globosa, Datura stramonium, Cassia occidentalis, Sesamum indicum* (Chung et al. 1986). SPYDV induces in *I. setosa* stunting, general leaf chlorosis, small distinct chlorotic spot, and vein chlorosis. Virions are filamentous, not enveloped, flexuous, with a clear modal length of 750 nm. The viral coat protein has a MW of 33,000. Pinwheel inclusions are present in the cytoplasm of infected cells.

The virus can be elininated by meristem tip culture. Stemtips up to 2.5 cm below the meristems of infected plants which had received a one or two-month treatment of heat (37 °C) were found to be free of SPYDV (Green and Luo, 1989).

Sweetpotato vein mosaic virus Possible Genus *Potyvirus* (SPVMV) was first reported from Argentina by Nome (1973). In sweetpotato, the virus causes a disease called "batata crespa" and infected plants shows vein clearing, mosaic, twisting of leaves, stunting, and noticeable reduction in number and size of the fleshy roots. The virus can be transmitted in a nonpersistent manner by *M. persicae*, as well as by grafting and by mechanical inoculation (Nome et al. 1974). The virus was transmitted to several *Ipomoea* species, including *I. alba, I. nil, I. setosa, I. lacunosa, I. hederacea, I. tricolor, I. trichocarpa, I. kurtziana, I. fistulosa*, and *I. angulata*, generally inducing systemic vein clearing and mosaic. Twisting, chlorosis and size reduction of leaves were marked in *I. setosa, I. nil, I. hederacea*, and *I. batatas*. Virus particles are flexuous rod, 761 nm in length. Virions are found in the cytoplasm of all parts of the host plant. Pinwheels and scrolls and laminated aggregates inclusions are present in the cytoplasm of infected cells (Brunt et al. 1996). The virus could be eliminated by heat treatment (Nome and Salvadores, 1979). Unfortunately the original culture and antiserum are not available.

Sweetpotato C-6 virus. The virus was isolated from sweetpotato Sosa 29, from the Dominican Republic, showing chlorotic spots (Fuentes, 1994). In Louisiana it was found in most of the black ornamental sweetpotatoes (Blackie, Ace of Spades, and Black Beauty) (Clark and Valverde, 2000) Its host range is restricted to C*onvolvulaceae*. The virus induces on *I. nil* and *I. setosa* fine chlorotic spots and vein clearing (Fig. 8.10) but on sweetpotato cv. Paramonguino, Costanero, and Jonathan chlorotic spots. This is followed by yellowing and leaf drop. The virus is transmitted by grafting and by mechanical inoculation (with a low efficiency), using sap from *I. nil* roots but not from leaves. Attempts to transmit the virus with *M. persicae* were unsuccessful. The virus was detected in samples from Peru, Uganda, Cuba, USA, Dominican Republic, Philippines, Indonesia, Egypt, Kenya, South Africa, New Zealand, and Puerto Rico. C-6 virus gave a mild reaction with Potato virus S (PVS) antiserum.

C-6 virus was purified from infected *I. setosa* plants (Fuentes, 1994). Virus particles are flexuous rods, 750–800 nm in length. No "pinwheels" were observed in infected *I. nil*, but chloroplasts and mitochondria showed hypertrophy.

Fig. 8.10 *Ipomoea setosa* infected with C-6 virus showing fine chlorotic spots and vein clearing

Sweetpotato C-3 virus, a suspected flexuous virus, was isolated from a sweetpotato co-infected with SPFMV and showing leaf deformation, vein clearing and mosaic, from Brazil (Fuentes and Salazar, 1989). The virus is transmitted by grafting but neither mechanically nor by aphid *M. persicae*. It induces mosaic, leaf deformation and vein clearing in *I. setosa*. Infected sweetpotato cv. Paramonguino showed mottle symptoms and cv. Georgia Red interveinal mottling. *N. benthamiana* graft-inoculated with C-3 showed yellowing vein, mosaic and leaf deformation, but *I. nil* did not show symptoms.

Sweetpotato ringspot virus Genus *Nepovirus* (SPRSV) was observed in *I. batatas*, imported from Papua New Guinea into the UK. (Brown et al. 1988). The virus on sweetpotato occasionally causes chlorotic ringspots, which disappear soon after infection. The virus can be transmitted by grafting and mechanical inoculation. The virus infects a wide range of experimental hosts, including *I. setosa* – faint systemic chlorotic leaf mottling; *N. megalosiphon* – necrotic rings and systemic mottling; *N. benthamiana* – systemic leaf deformations; *C. quinoa, C. murale* – transient systemic chlorosis (www.ncbi.nlm.nih.gov/ICTVdb/). Virions are not enveloped, 28 nm in diameter, which sediment as three components (T, M and B) with sedimentation coefficients of 47, 81 and 130 S and buoyant densities in cesium chloride of 1.32, 1.45 and 1.52 g cm^{-3}, respectively. All components contain a single polypeptide of Mr 56.6×10^3. T component particles contain only protein but those of M and B components contain ss-RNA 8,848 nt (Mr 2.93×10^6) and 6,670 nt (Mr 2.31×10^6), respectively. Virions can be found in leaves and roots. Although the virus has physico-chemical properties resembling those of nepoviruses, it showed no serological relationship to any of 13 recognised nepoviruses (including *Arabis mosaic virus, Cassava green mottle virus, Cherry leaf roll virus, Raspberry ringspot virus, Strawberry latent ringspot virus* and *Tobacco ringspot virus*) (Brunt and Brown, 1990).

New Potyviruses

Two previously undescribed viruses affecting sweetpotato in North Carolina, USA have been reported by Moyer et al. (2002). These two isolates gave no reaction with antiserum for SPFMV, SPMMV, SPLV, SPCFV, SPMSV, C-6, SPCaLV, and SPCSV, but showed novel dsRNA profiles. The first is a 9–10 Kbp single species and may correspond to a potyvirus, the second is a 7.5 Kbp dsRNA with two putative sub-genomic fragments. Both isolates tested negative in ELISA with a universal antiserum for potyviruses. From the first profile one isolate, GW Beauregard, was selected for dsRNA purification, cloning and sequencing. This isolate when graft inoculated on cv. Tanzania induced severe mosaic, vein banding and leaf deformation The CP gene sequences were used for the phylogenetic analysis. Sequence of the putative potyvirus was more closely related to SPMMV than SPLV, SPVG, and SPFMV.

Sweetpotato chlorotic stunt virus is a wrongly name given to a tentatively *Potyvirus* isolated from sweetpotatoes from the Caribbean area showing stunting and chlorotic leaves (Brown et al. 1988). This virus should not be confused with SPCSV (genus *Crinivirus*) previously described here. We are including some information on this virus, which could be useful when characterizing new potyviruses. This virus is not transmitted by aphids, but is readily sap-transmissible to a wide range of herbaceous species of which *N. benthamiana* is a good propagation host and *C. amaranticolor* and *C. quinoa* are convenient local lesion assay hosts. The virus was also detected in Kenya, Uganda (Wambugu, 1991), and Zimbabwe (Chavi et al. 1997). Virus particles are flexuous rod, 850–950 nm in length, which contain ss-RNA and a single capsid polypeptide of Mr 43 kDa. The virus induces "pinwheels".

Viruses Infecting Wild Ipomoeas or Sporadically Sweetpotato

A virus infecting *I. aquatica* was reported by Brunt et al. (1996). *Tobacco streak virus* and *Tobacco mosaic virus* (Clark and Moyer, 1988) have also been reported infecting sweetpotato.

Little-leaf Phytoplasma Disease

Little-leaf Phytoplasma Disease also referred to as witches' broom, was first described by Van Velsen (1967) from Papua and New Guinea. It has been reported also from the Ryuku islands (Yang, 1969), Tonga (Kahn and Monroe, 1969), Solomon Islands Dabek and Sagar, (1978), Taiwan (Yang, 1969) and Northern Australia (Gibb et al. 1995). This disease has an exceptionally long incubation period in sweetpotato, ranging from 50 to 186 days following graft transmission (Clark and Moyer, 1988). First symptoms are veinclearing of otherwise normal leaves. Subsequent symptoms include large reductions in the size of leaves and internodes and an erect habit of the tips of the vines. Plants which become infected at an early stage yield poorly, producing only one or two pencil-thin tubers (Pearson et al. 1984). Typical phytoplasma bodies were observed in phloem sieve cells. Meristem tip culture was found to be effective in eliminating the phtytoplasma from the infected sweetpotato tissue (Green et al. 1989). The phytoplasma could be diagnosed by PCR, and the full-length chromosome was determined as 600 Kbp, which is one of the smallest phytoplasma genome sizes (Gibb et al. 1995). Symptoms on graft-inoculated *I. setosa* may take 6 months or more to develop (Moyer et al. 1989). The disease is lethal in *I. ericolor*, with a relatively short incubation period of 35 to 49 days, making this species a candidate for an indexing host (Clark and Moyer, 1988).

Control

At present the best way to control virus diseases in sweetpotato is to supply the grower with virus-indexed propagation material. Such plantlets can be obtained from meristems shoot tip cultures, also in combination with cryotherapy (Wang and Valkonen, 2008). Such programs are operating in Israel and in the Shandong province of China (Gao et al. 2000). In Israel, as a result of planting virus-tested material, yields increased at least by 100%, while in China increases ranged between 22–92%. The payoff to the farmer has been high and in Israel use of certified material is common practice, while in China the use of pathogen-free material is being extended. In African countries such programs are operating only on a limited scale, because sweetpotatoes are grown mainly as a food security crop, and not as a commercial one.

Some cultural/phytosanitation practices may facilitate control of viral diseases. Examples of such cultural practices include selection of disease-free planting material, destroying (roguing) of diseased plants and wild *Ipomoea* spp, especially in joung crops, isolating new crops (15–20 m far) from old diseased crops, destroying crop residues, and protecting crops with barriers or intercropping with maize (Gibson and Aritua, 2002).

Breeding programs might be a future answer and such programs are in operation in Uganda, combining SPVD resistance with desirable agronomic traits such as yield, earliness and acceptable culinary quality (Karyeija et al. 2000; Mwanga et al. 2002). Progress has been made and several cultivars were released (Turyamureeba et al. 1998). It will have to be seen if these cultivars will retain their resistance in other places where different strains of virus components of SPVD may be present. Thus, several clones that were resistant to SPFMV in CIP's tests were found to be susceptible, when Israeli (unpublished) and Ugandan isolates were tested (Karyeija et al. 1998).

So far transgenic approaches did not result in cultivars resistant to SPVD. Apparently the small residues of SPCSV were in the plant were sufficient to cause together with SPFMV the severe SPVD.

Table 8.1 A list of recognized viruses known to infect sweetpotato

Virus	Shape and size	Vector	Distribution	References
SwPLV-***Potyvirus***	Flexuous, 750–790 nm	Aphid	Uganda, Kenya, Taiwan, Japan, China, Indonesia, Philippines, India	10, 11, 25
SPMSV-*Potyvirus*	Flexuous, 800 nm	Aphid	Argentina, Peru, Indonesia, Philippines, Egypt	1, 12, 18
SPLSV-*Polerovirus*?	Isometric, 30 nm	Aphid	Peru, Cuba	17
CMV- *Cucumovirus*	Isometric, 30 nm	Aphid	Israel, Uganda, Kenya, Japan, Egypt.	7, 13
SPVMV-Potyvirus?	Flexuous, 761 nm	Aphid	Argentina	29, 30
SPSVV-Crinivirus	Flexuous, 850–950 nm	Whitefly	Africa (Nigeria, Uganda, Kenya, Zaire, Zambia, Madagascar), Asia (Taiwan, China, Indonesia, Philippines), America (USA, Brazil, Argentina, Peru), Israel, Egypt	6
SPMMV-Ipomovirus	Flexuous, 800–950 nm	Whitefly	West Africa (kenya, Uganda, Burundi, Tanzania), Indonesia, Papua New Guinea, India, Egypt	11, 21, 32
SPYDV-Ipomovirus	Flexuous, 750 nm	Whitefly	Taiwan	25
SPLCV-Begomovirus	Geminate	Whitefly	Taiwan, Japan, Korea, China, USA, Puerto Rico, Peru	25, 26, 33
ICLCV-Geminivirus?	Geminate	Whitefly	USA?	8

Table 8.1 (continued)

Virus	Shape and size	Vector	Distribution	References
SPVG-Potyvirus	Flexuous	Unknown	China	9, 11
SPCFV	Flexuous, 750–800 nm	Unknown	Peru, Japan, Brazil, China, Cuba, Panama, Colombia, Bolivia, Indonesia, Philippines, Uganda, India, Austria	16, 20, 35
C-6	Flexuous 750–800 nm	Unknown	Peru, Uganda, Cuba, Dominican Republic, Indonesia, Philippines, USA, Puerto Rico, Egypt	14
SPCV-Caulimo-like	Isometric, 50 nm	Unknown	South Pacific region (Kingdom of Tonga, Papua New Guinea, New Zeland, Salomon Islands, Austria), Madeira, kenya, Uganda, China, Egypt, Puerto Rico,	2, 19
SPCSV-Potyvirus	Flexuous, 850–950 nm	Unknown	Caribbean Region, Kenya, Uganda, Zimbabwe	3
SPRSV-Nepovirus	Isometric, 28 nm	Unknown	Papua New Guinea, Kenya	3, 4
TSV-Ilar-like	Isometric, 30 nm	Unknown	Guatemala	5
C-3	Flexuous?	Unknown	Brazil. Unknown in other countries	15
TMV-Tobamovirus	Rod, 300 nm	None	USA	5

1) Alvarez *et al.* 1997; **2)** Atkey and Brunt, 1987; **3)** Brown *et al.* 1988; **4)** Brunt *et al.* 1996; **5)** Clark and Moyer, 1988; **6)** Cohen *et al.* 1992; **7)** Cohen and Loebentein, 1991; **8)** Cohen *et al.* 1997; **9)** Colinet *et al.* 1994; **10)** Colinet *et al.* 1997; **11)** Colinet *et al.* 1998; **12)** Di Feo *et al.* 2000; **13)** Francki *et al.* 1979; **14)** Fuentes, 1994; **15)** Fuentes and Salazar, 1989; **16)** Fuentes and Salazar, 1992; **17)** Fuentes *et al.* 1996; **18)** Fuentes *et al.* 1997; **19)** Gao *et al.* 2000; **20)** Gibson *et al.* 1997; **21)** Hollings *et al.* 1976; **22)** Karyeija *et al.* 1998; **23)** Karyeija *et al.* 2001; **24)** Kreuze *et al.* 2000; **25)** Liao *et al.* 1979; **26)** Lotrakul *et al.* 2002; **27)** Moyer, 1986; **28)** Moyer *et al.* 1989; **29)** Nome, 1973; **30)** Nome *et al.* 1974; **31)** Pozzer *et al.* 1995; **32)** Sheffield, 1957; **33)** Shinkai, 1979; **34)** Sim, 2001; **35)** Usugi *et al.* 1991.

As *Ipomoea* spp. are susceptible to most sweetpotato viruses it is of importance to survey weeds as hosts and potential reservoirs of viruses.

Concluding Remarks

The demand for root and tuber crops in developing countries to ensure a sustainable food supply is predicted to grow steadily over the next decades. Sweetpotato is one of the major crops, and much of it is grown in developing countries, for supplying carbohydrates and beta-carotene. In Sub-Saharan Africa it serves as a food security crop, and is important also in relief famine. As sweetpotato is propagated vegetativley control of virus diseases is an important factor in obtaining high yields. Over 20 viruses have so far been reported to infect sweetpotato (Table 8.1), but several of them are still not characterised. Antisera are available only for a limited number of viruses. Broad-spectrum PCR for detecting new uncharacterised viruses has been demonstrated especially with potyviruses and geminiviruses. It is important to determine whether the new isolated viruses are strains of previously reported viruses or represent a new virus species. Their identification, economic impact and effect in mixed infections, especially with Poty-, Geminiviruses and SPCSV/SPSVV are an important task for the future.

Knowledge of sweetpotato viruses and their genetic variability that provides information on the components of virus population is the base for designing appropriate control measures that deploy host resistance in sweetpotato crops. Damages caused by sweetpotato viruses are mostly through synergistic mixed infections. SPCSV can synergise with viruses member in potyvirus, ipomovirus, carlavirus, and cucumovirus. It means that development of durable resistance to synergistic diseases, as SPVD, should be resistance to SPCSV/SPSVV.

It was demonstrated that transgenic sweetpotato containing the CP gene from SPFMV or RdRp genes for both SPFMV and SPCSV did not confer resistance (Kreuze et al. 2008) to 'complex' infection with SPCSV, causing the SPVD. SPCSV encodes two unique proteins: RNase 3 and p22, which are expressed early in infection and are co-operatively able to suppress RNA silencing. As not all isolates of SPCSV contains the p22 gene (Cuellar et al. 2008) but are able to synergise with different viruses, it is hoped that transgenes containing the RNase3 gene will achieve resistance to the SPVD.

Recently, a landrace sweetpotato variety 'Huachano' that is extremely resistant to SPFMV was genetically engineered for resistance to SPCSV (Kreuze et al. 2008). In this cultivar as in many others (Karyeija et al. 2000) the high levels of resistance to SPFMV breaks down following infection with SPCSV and the plants succumb to the severe SPVD.

Research on sweetpotato viruses has not received the attention it deserves and the international community and its organizations should increase their efforts to strengthen sweetpotato research in general and sweetpotato virus focused research in particular.

References

Abad, J.A. and Moyer, J.W. 1992. Detection and distribution of sweetpotato feathery mottle virus in sweetpotato by *in vitro*-transcribed RNA probes (riboprobes), membrane immunobinding assay, and direct blotting. *Phytopathology* 82: 300–305.

Abad, J.A., Parks, E.J., New, S.L., Fuentes, S., Jesper, W., and Moyer, J.W. 2007. First Report of *Sweetpotato chlorotic stunt virus*, a component of sweetpotato virus disease, in North Carolina. *Plant Dis.* 91: 327.

Alicai, T., Fenby, N.S., Gibson,R.W., Adipala, E., Vetten, H.J., Foster, G.D., and Seal, S.E. 1999. Occurrence of two serotypes of sweetpotato chlorotic stunt virus in East Africa and their associated differences in coat protein and HSP70 homologue gene sequences *Plant Pathol.* 48: 718–726.

Alvarez, V., Ducasse, D.A., Biderbost, E., and Nome, S.F. 1997. Sequencing and characterization of the coat protein and 3' non-coding region of a new sweetpotato potyvirus. *Arch. Virol.* 142: 1635–1644.

Aritua, V., Alicai, T., Adipala, E., Carey, E.E., and Gibson, R.W. 1998. Aspects of resistance to sweetpotato virus disease in sweetpotato. *Ann. Appl. Biol.* 132: 387–398.

Aritua, V.; Bua, B., Barg, E., Vetten, H.J., Adipala, E., and Gibson, R.W. 2007. Incidence of five viruses infecting sweetpotatoes in Uganda; the first evidence of Sweetpotato caulimo-like virus in Africa *Plant Pathol.* 56: 324–331.

Ateka, E.M., Barg, E, Njeru, R.W., Lesemann, D.-E., and Vetten, H.J. 2004. Further characterization of "sweetpotato virus 2'. *Arch. Virol.* 149: 225–239.

Ateka, E.M., Barg., E., Njeru, R.W., Thompson, G., and Vetten, H.J. 2007. Biological and molecular variability among geaographically diverse isolates of sweetpotato virus 2. *Arch. Virol.* 152: 479–488.

Ateka, E.M., Njeru, R.W., Kibaru, A.G., Kimenju, J.W., Barg, E., Gibson, R.W., and Vetten, H.J. 2004. Identification and distribution of viruses infecting sweetpotato in Kenya. *Ann. Appl. Biol.* 144: 371–379.

Atkey, P.T. and Brunt, A.A. 1987. Electron microscopy of an isometric caulimo-like virus from sweetpotato (*Ipomoea batatas*). *J. Phytopath.* 118: 370–376.

Banks, G.K., Bedford, I.d., Beitia, F.J., Rodrigues-Cerezo, E., and Markham, P.G. 1999. A Novel Geminivirus of *Ipomoea indica* (*Convolvulaceae*) from Southern Spain. *Plant Dis.* 83: 486.

Briddon, R.W., Bull, S.E., and Bedford, I.D. 2005. Occurrence of *Sweetpotato leaf curl virus* in Sicily. URL http://www.bspp.org.uk/ndr/july2005/2005-44.asp.

Brown, J.D., Brunt, A.A., and Hugo, S.A. 1988. Studies on viruses isolated from sweetpotato (*Ipomoea batatas) Rep. Glasshouse Crops Res. Inst. for 1986–1987*, pp. 104–108.

Brunt, A.A. and Brown, J.D. 1990. Sweetpotato ringspot, a newly recognised Nepo-like virus from Ipomoea batatas. (Abstract). *Int. Virol. Congress*, Berlin.

Brunt, A.A., Crabtree, K., Dallwitz, M.J., Gibbs, A.J., and Watson, L. (eds.), 1996. Viruses of Plants: Descriptions and Lists from the VIDE Database. CAB INTERNATIONAL, Cambridge, UK. 1484 pp.

Cali, B.B. and Moyer, J.W. 1981. Purification, serology, and particle morphology of two russet crack strains of sweetpotato feathery mottle virus. *Phytopathology* 71: 302–305.

Carey, E.E., Mwanga, R.O.M., Fuentes, S., Kasule, S., Macharia, C., Gichuki, S.T., and Gibson, R.W. 1997. Sweetpotato viruses in Uganda and Kenya: results of a survey. *Proc. Sixth Triennial Symp. Inter. Soc. Tropical Crops – Africa Branch* pp. 22–28.

Chavi, F., Robertson, A.I., and Verduin, B.J.M. 1997. Survey and characterization of viruses in sweetpotato from Zimbabwe. *Plant Dis.* 81: 1115–1122.

Chung, M.L., Hsu, Y.H., Chen, M.J., and Chiu, R.J. 1986. Virus diseases of sweetpotato in Taiwan. In: *Pl. Virus Diseases of Horticulture Crops in the Tropics and Subtropics*, No. 33, p. 84–90. FFTC Book Series.

Chung, M.L., Liao, C.H., Chen, M.J., and Chiu, R.J. 1985. The isolation, transmission and host range of sweetpotato leaf curl disease agent in Taiwan. *Plant Prot. Bull.* (Taiwan) 27: 333–342.

Cipriani, G., Fuentes, S., Bello, V., Salazar, L.F., Ghislain, M., and Zhang, D.P. 2001. Transgene expression of rice cysteine proteinase inhibitors for the development of resistance against sweetpotato feathery mottle virus. *Scientists and farmer: Partners in research for the 21 st Century*, pp. 267–271.

Clark, C.A. and Moyer, J.W. 1988. Compendium of Sweetpotato Diseases. The American Phytopathological Society. Minnesota, USA. 74 pp.

Clark, C.A. and Valverde, R.A. 2000. Viruses and Sweetpotato Cultivar Decline in Louisiana, USA. Pages 62–69 in: Nakazawa, Y., and Ishiguro, K., eds. Proceedings of International Workshop on Sweetpotato Cultivar Decline Study. Kyushu National Agricultural Experiment Station (KNAES), Miyakonjo, Japan.

Cohen, J., Franck, A., Vetten, H.J., Lesemann, D.E., and Loebenstein, G. 1992. Purification and properties of closterovirus-like particles associated with a whitefly-transmitted disease of sweetpotato. *Ann. Appl. Biol.* 121: 257–268.

Cohen, J. and Loebenstein, G. 1991. Role of a whitefly-transmitted agent in infection of sweetpotato by cucumber mosaic virus. *Plant Dis.* 75: 291–292.

Cohen, J., Milgram, M., Antignus, Y., Pearlsman, M., Lachman, O., and Loebenstein, G. 1997. Ipomoea crinkle leaf curl caused by a whitefly-transmitted gemini-like virus. *Ann. Appl. Biol.* 131: 273–282.

Cohen, J., Salomon, R., and Loebenstein, G. 1988. An improved method for purification of sweetpotato feathery mottle virus directly from sweetpotato. *Phytopathology* 78: 809–811.

Colinet, D. and Kummert, J. 1993. Identification of a sweetpotato feathery mottle virus isolate from China (SPFMV-CH) by the polymerase chain reaction with degenerate primers. *J. Virol. Methods* 45: 149–159.

Colinet, D., Kummert, J., Lepoivre, P., and Semall, J. 1994. Identification of distintict potyvirus in mixedly-infected sweetpotato by polymerase chain reaction with degenerate primers. *Phytopathology* 84: 65–69.

Colinet, D., Kummert, J., and Lepoivre, P. 1994. The complete nucleotide sequences of the coat protein cistron and 3′ non-coding region of a newly-identified potyvirus infecting sweetpotato, as compared to those of sweetpotato feathery mottle virus. *Arch. Virol.* 139: 327–336.

Colinet, D., Kummert, J., and Lepoivre, P. 1997. Evidence for the assignment of two strains of SPLV to the genus Potyvirus based on coat protein and 3′ non-coding region sequence data. *Virus Res.* 49: 91–100.

Colinet, D., Nguyen, M., Kummert, J., Lepoivre, P., and Xia, F.Z. 1998. Differentiation among potyviruses infecting sweetpotato based on genus- and virus-specific reverse transcription polymerase chain reaction. *Plant Dis.* 82: 223–229.

Cuellar,W.J., Tairo,F., Kreuze, J.F., and Valkonen, J.P.T. 2008. Analysis of gene content in sweet potato chlorotic stunt virus RNA1 reveals the presence of P22 protein RNA silencing. suppressor in only few isolates: implications to viral evolution and synergism. *J. Gen. Virol.* 89: 573–582.

Dabek, A.J. and Sagar, C. 1978. Witches' broom chlorotic little leaf of sweetpotato in Guadalcanal, Solomon Islands, possibly caused by mycoplasma-like organisms. *Phytopath. Z.* 92: 1–11.

Di Feo, L., Nome, S.F., Biderbost, E., Fuentes, S., and Salazar, L.F. 2000. Etiology of sweetpotato chlorotic dwarf disease in Argentina. *Plant Dis.* 84: 35–39.

FAOSTAT 2007. FAO Statistical Databases. URL http://apps.fao.org/

Fletcher, J.D., Lewthwaite, S.L., Fletcher, P.J., and Dannock, J. 2000. Sweetpotato (kumara) virus disease surveys in New Zealand. Pages 42–47 in: Nakazawa, Y., and Ishiguro, K., eds. Proc. International Workshop on Sweetpotato Cultivar Decline Study.Kyushu National Agricultural Experimental Station (KNAES), 8–9 September 2000, Miyakonojo Japan.

Francki, R.I.B., Mossop, D.W., and Hatta, T. 1979. Cucumber mosaic virus. No. 213. *Descriptions of Plant Viruses*. Commonwealth Mycol. Inst. Assoc. Appl. Biol. Kew, Surrey, UK.

Fuentes, S. 1994. Preliminary identification of a sweetpotato virus (C-6). *Fitopatologia* 29: 38 (Abstr., in Spanish).

Fuentes, S., Arellano, J., and Meza, M.A. 1977. Preliminary studies of a new virus, C-8, affecting sweetpotato. *Fitopatologia* 32: 9–10 (Abstr. in Spanish).

Fuentes, S., Mayo, M., Jolly, C.A., Nakano, M., Querci, M., and Salazar, L.F.. 1996. A novel luteovirus from sweetpotato, sweetpotato leaf speckling virus. *Ann. Appl. Biol.* 128: 491–504.
Fuentes, S., Querci, M., Salazar, L.F., and Mayo, M. 1997. Detection of SPLSV by Nucleic Acid Spot Hybridization (NASH) Test. *Fitopatología* 32: 204 (Abstr. in Spanish).
Fuentes S. and Salazar, L.F. 1989. Identification of sweetpotato [*Ipomoea batatas* (L.) Lam.] viruses. *Fitopatologia* 24: 43 (Abstr. in Spanish).
Fuentes, S. and Salazar, L.F. 1992. Identification of a new sweetpotato virus. *Fitopatologia* 27: 50 (Abstr., in Spanish).
Fuentes, S. and Salazar, L.F. 2003. First report of Sweetpotato leaf curl virus in Peru. *Plant Dis.* 87: 98. (Abstr.)
Gamarra, H.A., Fuentes, S., Morales, F.J., and Barker, I. 2008. *Bemisia afer*, a vector of *Sweetpotato chlorotic stunt virus*. Plant Dis. (submitted).
Gao, F.,Gong, Y.F. and Zhang, P.B. 2000. Production and development of virus-free sweetpotato in China. *Crop Protect.* 19: 105–111.
Gibb, K.S., Padovan, A.C., and Mogen, B.D. 1995. Studies on sweetpotato little-leaf phytoplasma detected in sweetpotato and other plant species growing in Northern Australia. *Phytopathology* 85: 169–174.
Gibson, R.W. and Aritua, V. 2002. The perspective of sweetpotato chlorotic stunt virus in sweetpotato production in Africa: a review. *Afr. Crop Sci. J.* 10: 281–310.
Gibson, R.W., Jeremiah, S.C., Aritua, V., Msabaha, R.P., Mpembe, I., and Ndunguru, L. 2000. Sweetpotato virus disease in sub-Saharan Africa: evidence that neglect of seedlings in the traditional farming system hinders the development of superior resistant landraces. *J. Phytopath.* 148: 441–447.
Gibson, R.W., Mawanga, R.O.M., Kasule, S., Mpembe, I., and Carey, E.E. 1997. Apparent absence of viruses in most symptomless field-grown sweetpotato in Uganda. *Ann. Appl. Biol.* 130: 481–490.
Gibson, R.W., Mpembe, I., Alicai, T., Carey, E.E., Mwanga, R.O.M., Seal, S.E., and Vetten, H.J. 1998. Symptoms, aetiology and serological analysis of sweetpotato virus disease in Uganda. *Plant Pathol.* 47: 95–102.
Green, S.K., Kuo, Y.J., and Lee, D.R. 1988. Uneven distribution of two potyviruses (feathery mottle virus and sweetpotato latent virus) in sweetpotato plants and ins implication on virus indexing of meristem derived plants. *Trop. Pest Manage.* 34: 298–302.
Green, S.K and Luo, C.Y. 1989. Elimination of sweetpotato yellow dwarf virus SPYDV) by meristem tip culture and by heat treatment. Z. Pflanzenkr. Pflanzenschutz 96: 464–469.
Green, S.K., Luo, C.Y., and Lee, D.R. 1989. Elimination of mycoplasma-like organisms from witches' broom infected sweetpotato. *J. Phytopathol.* 126: 204–212.
Gutiérrez, D.L., Fuentes, S., and Salazar, L.F. 2003. Sweetpotato virus disease (SPVD): Distribution, incidente, and effect on sweetpotato yield in Peru. *Plant Dis.* 87: 297–302.
Hahn, S.K. 1979. Effects of viruses (SPVD) on growth and yield of sweetpotato. *Exp. Agric.* 15: 253–256.
Hammond, J., Jordan, R.L., Larsen, R.C., and Moyer, J.W. 1992. Use of monoclonal antisera and monoclonal antibodies to examine serological relationships among three filamentous viruses of sweetpotato. *Phytopathology* 82: 713–717.
Hollings, M., Stone, O.M., and Bock, K.R. 1976. Purification and properties of sweetpotato mild mottle virus, a whitefly-borne virus from sweetpotato (Ipomoea batatas) in East Africa. *Ann. Appl. Biol.* 82: 511–528.
Hoyer, U., Maiss, E., Jelkmann, W., Lesemann, D.E., and Vetten, H.J. 1996. Identification of the coat protein gene of a sweetpotato sunken vein closterovirus isolate from Kenya and evidence for a serological relationship among geographically diverse closterovirus isolates from sweetpotato. *Phytopathology* 86: 744–750.
IsHak, J.A., Kreuze, J.F., Johansen, A., Mukasa, S.B., Tairo, F., Abo El-Abbas, F.M., and Valkonen, J.P.T. 2003. Some molecular characteristics of three viruses from SPVD-affected sweetpotato plants in Egypt. *Arch. Virol.* 148: 2449–2460.

Jones, R.A.C. and Dwyer, G.I. 2007. Detection of Sweetpotato chlorotic fleck virus and *Sweetpotato feathery mottle virus* – strain O in Australia. *Australia. Plant Pathol.* 36: 591–594.
Jordan, R., and Hammond, J. 1991. Comparison and differentiation of potyvirus isolates and identification of strain-, virus-, subgroup-specific and potyvirus group-common epitopes using monoclonal antibodies. *J. Gen. Virol.* 72: 25–36.
Kahn, R.P., and Monroe, R.L. 1969. Detection of sweetpotato little leaf agent (witches' broom) associated with a mycoplasma-like organism. *Pl. Prot. Bull.* FAO 17: 104–106.
Karyeija, R.F., Gibson, R.W., and Valkonen, J.P.T. 1998. Resistance in sweetpotato virus disease (SPVD) in wild East African Ipomoea. *Ann. appl. Biol.* 133: 39–44.
Karyeija, R.F., Kreuze, J.F., Gibson, R.W., and Valkonen, J.P.T. 2000a. Two serotypes of *Sweetpotato feathery mottle virus* in Uganda and their interaction with resistant sweetpotato cultivars. *Phytopathology* 90: 1250–1255.
Karyeija, R.F., Kreuze, J.F., Gibson, R.W., and Valkonen, J.P.T. 2000b. Synergistic interactions of a potyvirus and a phloem-limited crinivirus in sweetpotato plants. *Virology* 269: 26–36.
Karyeija, R.F., Kreuze, J.F., Gibson, R.W., and Valkonen, J.P.T. 2001. Variability of sweetpotato feathery mottle virus in Africa. *Afr. Crop Sci. J.* 9: 293–299.
Kokkinos, C.D. and Clark, C.A. 2004. Assessment of interactions among important sweetpotato viruses using real-time quantitative PCR. *Phytopathology* 94: S54.
Kokkinos, C.D. and Clark, C.A. 2006. Interactions among *Sweetpotato chlorotic stunt virus* and different potyviruses and potyvirus strains infecting sweetpotato in the United States. *Plant Dis.* 90: 1347–1352.
Kreuze, J.F., Karyeija, R.F., Gibson, R.W., and Valkonen, J.P.T. 2000. Comparisons of coat protein gene sequences show that East African isolates of sweetpotato feathery mottle virus form a genetically distinct group. *Arch. Virol.* 145: 567–574.
Kreuze, J.F., Savenkov, E.I., and Valkonen J.P.T. 2002. Complete genome sequence and analyses of the subgenomic RNAs of sweetpotato chlorotic stunt virus reveal several new features for the genus Crinivirus. *J. Virol.* 76: 9260–9270.
Kreuze, J.F., Samolski, I., Untiveros, M., Cuellar, W.J., Lajo, G., Cipriani, P.G., Ghislain, M., and Valkonen, J.P.T. 2008. RNA silencing mediated resistance to a crinivirus (*Closteroviridae*) in cultivated sweetpotato (*Ipomoea batatas* L.) and development of sweetpotato virus disease following co-infection with a potyvirus. *Mol. Plant Pathol.* 9: (in press).
Liao, C.H., Chien, K., Chung, M.L., Chiu, R.J., and Han, Y.H. 1979. A study of a sweetpotato virus disease in Taiwan. I. Sweetpotato yellow spot virus disease. *J. Agric. Res. China* 28: 127–137.
Loebenstein, G. and Harpaz, I. 1960. Virus diseases of sweetpotatoes in Israel. *Phytopathology* 50: 100–104.
Lotrakul, P. and Valverde, R.A. 1999. Cloning of a DNA-A-like genomic component of sweetpotato leaf curl virus: nucleotide sequence and phylogenetic relationships. *Mol. Plant Pathol.* On-line publication/(http://www.bspp.org.uk/mppol/1999/0206LOTRAKUL.
Lotrakul, P., Valverde, R.A., and Clark, C.A. 2003. Properties of a begomovirus isolated from sweetpotato[Ipomoea batatas (L.) Lam.] infected with Sweetpotato leaf curl virus. *Revi. Mexicana Fitopatologia* 21: 128–136.
Lotrakul, P., Valverde, R.A., Clark, C.A., Hurt, S., and Hoy, M.W. 2002. Sweetpotato Leaf Curl Virus and Related Geminiviruses in Sweetpotato. Proceedings of 1st IS on Sweetpotato. T. Ames (ed.), ISHS. *Acta Hort.* 583: 135–141.
Lotrakul, P., Valverde, R.A., Clark, C.A.,Sim, J., and De La Torre, R. 1998. Detection of a geminivirus infecting sweetpotato in the United States. *Plant Dis.* 82: 1253–1257.
Lyerly, J.H., New, S.L., Abad, J.A., and Moyer, J.W. 2003. Identification of sweetpotato viruses using an RT-PCR based method. *Phytopathology* 93: S55.
Martin, W.J. 1962. Susceptibility of certain Convolvulaceae to internal cork, tobacco ringspot and cucumber mosaic viruses. *Phytopathology* 52: 607–611.
Miamo, D.W., LaBonte, D.R., Clark, C.A., Valverde, R.A., Hoy, M.W., Hurt, S., and Li, R. 2006. First report of a begomovirus infecting sweetpotato in Kenya. *Plant Dis.* 90: 832.
Migliori, A., Marchoux, G., and Quiot, J.B. 1978. Dynamique des populations du virus de la mosaique du concombre en Guadelupe. *Ann. Phytopathol.* 10: 455–466.

Mihovilovich, R., Mendoza, H.A., and Salazar, L.F. 2000. Combining ability for resistance to sweetpotato feathery mottle virus. *HortScience* 35: 1319–1320.
Milgram, M., Cohen, J., and Loebenstein, G. 1996. Effects of sweetpotato feathery mottle virus and sweetpotato sunken vein virus on sweetpotato yields and rate of reinfection on virus-free planting material in Israel. *Phytoparasitica* 24: 189–193.
Moyer, J.F. 1986. Variability among strains of sweetpotato feathery mottle virus *Phytopathology* 76: 1126. (Abstr.).
Moyer, J.W., Abad, J.A., New, J, and Bell, J. 2002. Isolation, identification and detection of undescribed RNA sweepotato viruses. Proceedings of 1st IS on Sweetpotato. T. Ames (ed.), ISHS. *Acta Hort.* 583: 121–127.
Moyer, J.W. and Cali, B.B. 1985. Properties of sweetpotato feathery mottle virus RNA and capsid protein. *J. Gen. Virol.* 65: 1185–1189.
Moyer, J.W., Jackson, G.V.H., and Frison, E.A. (eds.), 1989. *FAO/IBPGR Technical Guidelines for the Safe Movement of Sweetpotato Germplasm.* Food and Agriculture Organization of the United Nations, Rome/International Board For Plant Genetic Resources, Rome.
Mwanga, R.O.M., Yencho, C.G.C., and Moyer, J.W. 2002. Diallel analysis of sweetpotatoes for resistance to sweetpotato virus disease. *Euphytica* 128: 237–248.
Ndunguru, J. and Kapinga R. 2007. Viruses and virus-like diseases affecting sweetpotato subsistence farming in southern Tanzania. *African J. Agric. Res.* 2: 232–239.
Ng, J.C.K. and Falk. B.W. 2006. Virus-vector interactions mediating nonpersistent and semipersistent transmission of plant viruses. *Annu. Rev. Phytopathol.* 44: 183–212.
Ngeve, J.M. and Bouwkamp, J.C. 1991. Effects of sweetpotato virus disease (SPVD) on the yield of sweetpotato genotypes in Cameroon. *Exp. Agric.* 27: 221–225.
Nishiguchi, M., Okada, Y., Sonoda, S., Mori, M., Kimura, T., Hanada, K., Sakai, J., Murata, T., Matsuda, Y., Fukuoka, H., Miyazaki, T., Nakano, M., Usugi, T., and Saito A. 2001. Sweetpotato feathery mottle virus derived resistance: CP mediated resistance and gene silencing. In: Proceedings of International Workshop Sweetpotato Cultivar Decline Study (ed by Kyushu Natl Agric. Expt. Sta.) pp. 120–124.
Nome, S.F. 1973. Sweetpotato vein mosaic in Argentina. *Phytopatho Z.* 77: 44–45.
Nome, C.F. Laguna, I.G., and Nome, S.F. 2006. Cytological alterations produced by Sweetpotato mild speckling virus. *J. Phytopathol.* 154: 504–507.
Nome, S.F. and Salvadores, M.C. 1979. Hot-air treatment for the elimination of sweetpotato vein mosaic virus from sweetpotato plants [Ipomoea batatas]. Argentine meetings on plant protection], Universidad Nacional de Tucuman (Argentina). Facultad de Agronomia y Zootecnia.- San Miguel de Tucuman (Argentina), 1979. v. 2 p. 779–787.
Nome, S.F., Shalla, T.A., and Peterson, L.J. 1974. Comparison of virus particles and intracellular inclusions associated with vein mosaic, feathery mottle, and russet crack diseases od sweetpotato. *Phytopathol. Z.* 79: 169–178.
Odame, H., Kameri-Mbote, P., and Wafula, D. 2001. Innovation and policy process: The case of transgenic sweetpotato in Kenya. www.acts.or.ke/publications/Sweetpotato and innovation process.pdf
Okada, Y., Saito, A., Nishiguchi, M., Kimura, T., Mori, M., Hanada, K., Sakai, J., Miyazaki, C., Matsuda, Y., and Murata, T. 2001. Virus resistance in transgenic sweetpotato [Ipomoea batatas L. (Lam)] expressing the coat protein gene of sweetpotato feathery mottle virus. *Theor. Appl. Genet.* 103: 743–751.
Onuki, M., Honda, Y., and Hanada, K. 2000. Geminate particle morphology of sweetpotato leaf curl virus in partially purified preparation and its serological relationship to two Begomoviruses by western blotting. *J. Gen. Plant Pathol.* 2: 182–184.
Osaki, T. and Inouye, T. 1991. Transmission characteristics and cytopathology of a whitefly-transmitted virus isolated from sweetpotato leaf curl disease. *Bull. Univ. Osaka Prefecture, Series B, Agric. Biol.* 43: 11–19.
Parrella, G., De Stradis, A., and Giorgini, M. 2006. Sweetpotato feathery mottle virus is the casual agent of Sweetpotato Virus Disease (SPVD) in Italy. *Plant Pathol.* 55: 818.

Pearson, M.N., Keane, P.J., and Thagalingham, K. 1984. Little leaf: A disease of sweetpotato in Papua New Guinea probabky caused bt mycoplasma-like organisms. *Phytopath. Z.* 109: 269–276.

Pozzer, l., Dusi, A.N., Lima, M.I., and Kitajima, E.W. 1995. Characterization of a Brazilian isolate of sweetpotato feathery mottle virus infecting sweetpotato. *Fitopat. Brasil.* 20: 65–71.

Prasanth, G. and Hegde, V. 2008. Occurrence of Sweetpotato *feathery mottle virus* and *Sweetpotato leaf curl Georgia virus* on sweetpotato in India. *Plant Dis.* 92: 311.

Rannalli, M., Czekaj, V., Jones, R.A.C., Fletcher, J.D., Davis, R.I., Mu, L., Dwyer, G.I., Coutts, B.A., and Valkonen, J.P.T. 2008. Molecular genetic characterization of *Sweet potato virus G* (SPVG) isolates from areas of the Pacific Ocean and southern Africa. *Plant Dis.* 92: 1313–1320.

Rossel, H.W. and Thottappilly, G. 1988. Complex virus diseases of sweetpotato. Pages 291–302 in: Exploration, Maintenance and Utilization of Sweetpotato Genetic Resources. Report of 1st Sweetpotato Planning Conference, International Potato Center, Lima, Peru.

Sakai, J., Mori, M., Morishita, T., Tanaka, M., Hanada, K., Usugi, T., and Nishigushi, M. 1997. Complete nucleotide sequence and genome organization of sweetpotato feathery mottle virus (S strain) genomic RNA: the large coding region of the P1 gene. *Arch. Virol.* 142: 1553–1562.

Schaefers, G.A. and Terry, E.R. 1976. Insect transmission of sweetpotato agents in Nigeria. *Phytopathology* 66: 642–645.

Shang, Y.F., Yang, C.L., Zhao, J.H., Li, C.S., Lu, X.B. Wang, S.J., and Xin, X.Q. 1999. Progress in the researches and application of virus-free sweetpotato in Shandong province. *Res. Progress Plant Protect. Plant Nutrit.* 1999, p. 104–108. China Agric. Press, Beijing, China.

Sheffield, F.M.L. 1957. Virus diseases of sweetpotato in East Africa. I. Identification of the viruses and their insect vectors. *Phytopathology* 47: 582–590.

Shinkai, A. 1979. Sweetpotato leaf curl disease. *Agric. Hortic.* 54: (1).

Sim, J. 2001. Cytopathology, detection, and identification of viruses infecting sweetpotato. PhD. Thesis. The Departemnet of the Plant Pathology and Crop Physiology. Louisiana State University, U.S.A. 137 pp.

Sim, J., Valverde, R.A., and Clark, C.A. 2000. Whitefly transmission of *Sweetpotato chlorotic stunt virus*. *Plant Dis.* 84: 1250.

Sonoda, S., Koiwa, H., Kanda, K., Kato, H., Shimono, M., and Nishiguchi, M. 2000. The helper component-proteinase of sweetpotato feathery mottle virus facilitates systemic spread of potato virus X in Ipomoea nil. *Phytopathology* 90: 944–950.

Souto, E.R., Sim, J., Chen, J., Valverde, R.A., and Clark, C.A. 2003. Properties of strains of Sweetpotato feathery mottle virus and two newly recognized potyviruses infecting sweetpotato in the United States. *Plant Dis.* 87: 1226–1232

Tairo, F., Jones, R.A.C., and Valkonen, J.P.T. 2006. Potyvirus complexes in sweetpotato: Occurrence in Australia, serological and molecular resolution, and analysis of the Sweetpotato virus 2 (SPV2) component. *Plant Dis.* 90: 1120–1128.

Tairo, F., Musaka, S.B., Jones, R.A.C., Kullaia, A., Rubaihayo, P.R., and Valkonen, J.P.T. 2005. Unravelling the genetic diversity of the three main viruses involved in Sweetpotato Virus Disease (SPVD), and its practical implications. *Mol. Plant Pathol.* 6: 199–211

Trenado, H.P., Lozano, G., Valverde, R.A., and Navas-Castillo, J. 2007. First report of *Sweetpotato virus G* and Sweetpotato virus 2 infecting sweetpotato in Spain. *Plant Dis.* 91: 1687 (Abstr.)

Tugume, A.K., Mukasa, S.B., and Valkonen, J.P.T. 2008. Natural wildhosts of Sweet potato feathery mottle virus show spatial differences in virus incidence and virus-like disease in Uganda. *Phytopathology* 98: 640–652.

Turyamureeba, G., Mwanga, R.O.M., Odongo, B., Ocitti p'Obwoya, C., and Carey, E.E. 1998. Application for inclusion of a crop/variety in the National Cultivar List. Submission to the Variety Release Committee for release of sweetpotato varieties. The Ugandan Ministry of Agriculture, Animal Industries & Fisheries, Uganda.

Untiveros, M., Fuentes, S., and Kreuze, J. 2007. Molecular variability of sweetpotato feathery mottle virus and other potyviruses infecting sweetpotato in Peru. *Arch. Virol.*Online (DOI 10.1007/s00705-007-0019-0)

Untiveros, M., Fuentes, S., and Salazar, L.F. 2006. Synergistic interaction of *Sweetpotato chlorotic stunt virus* (*Crinivirus*) with Carla-, Cucumo-, Ipomo-, and Potyviruses infecting sweetpotato. *Plant Dis.* 91: 669–676

Usugi, T., Nakano, M., Shinkai, A., and Hayashi, T. 1991. Three filamentous viruses isolated from sweetpotato in Japan. *Ann. Phytopathol. Soc. Jpn* 57: 512–521.

Valverde, R.A., Lozano, G., Navas-Castillo, J., Ramos, A., and Valdés, F. 2004a. First report of *Sweetpotato chlorotic stunt virus* and *Sweetpotato feathery mottle virus* infecting sweetpotato in Spain. *Plant Dis.* 88: 428.

Valverde, R.A., Sim, J., and Lotrakul, P. 2004b. Whitefly transmission of sweetpotato viruses. *Virus Res.* 100: 123–128.

Van Velsen, R.J. 1967. "Little leaf", a virus disease of *Ipomoea batatas* in Papua and New Guinea. *Papua New Guinea Agric. J.* 18: 126–128.

Wambugu, F.M. 1991. *In vitro* and epidemiological studies of sweetpotato (*Ipomea batatas*)(L.) Lam. virus diseases in Kenya. PhD Thesis. University of Bath. 271 pp.

Wambugu, F. 2004. Notes in New Scientist, 181 No. 2433, 7 February 2004.

Wang, Q.C. and Valkonen, J.P.T. 2008. Elimination of two viruses which interact synergistically from sweetpotato by shoot tip culture and cryotherapy. *J. Virological Meth.* 154: 135–145.

Winter, S., Purac, A., Leggett, F., Frison, E.A., Rossell, H.W., and Hamilton, R.I. 1992. Partial purification and molecular cloning of a closterovirus from sweetpotato infected with the sweetpotato virus disease complex from Nigeria. *Phytopathology* 82: 869–875.

Yang, I.L. 1969. Studies on witches broom in sweetpotato in Taiwan. *J. Taiwan Agric. Res.* 18: 50–60.

Chapter 9
Nematodes

C. Overstreet

Introduction

One of the major pest groups that cause considerable damage to sweetpotato is that of plant-parasitic nematodes. Nematodes are animals commonly referred to as roundworms belonging to the Phylum Nematoda. There are numerous types of nematodes including those that feed on bacteria, fungi, plants, and animals. Plant-parasitic nematodes are those types which depend on plants as their food source. There are a number of different genera and species that feed on sweetpotato. However, there appear to be only a few types that are associated with any serious damage to sweetpotato. Nematodes are a problem everywhere that sweetpotatoes are grown primarily because of the wide host range of many nematodes to the crops that are grown by man for food and fiber as well as spread of various nematodes over time from infected storage roots of sweetpotato.

Major Nematodes

The most important nematodes throughout the world belong to the genera *Meloidogyne*, *Pratylenchus*, *Heterodera*, *Ditylenchus*, *Globodera*, *Tylenchulus*, *Xiphinema*, *Radopholus*, *Rotylenchulus*, and *Helictotylenchus*, ranked in this order (Sasser and Freckman, 1987). Most of these genera have species that have been found to attack sweetpotato with the exception of *Heterodera*, *Globodera*, and *Tylenchulus*. Other genera and species of nematodes that may have worldwide or regional importance such as *Hirschmanniella oryzae* (parasite of rice in many areas of the world) or *Hoplolaimus columbus* (hosts include cotton, soybean, sweet corn, and lima bean in the southern U.S.) have been reported on sweetpotato as a nonhost (Lewis and Smith, 1976; Babatola, 1979). Sweetpotato is certainly grown in rotation with many

---

C. Overstreet (✉)
Department of Plant Pathology and Crop Physiology, Louisiana State University Agricultural Center, Baton Rouge, LA 70803
e-mail: COverstreet@agcenter.lsu.edu

G. Loebenstein, G. Thottappilly (eds.), *The Sweetpotato*,
DOI 10.1007/978-1-4020-9475-0_9,

different types of crops that may be hosts to similar or even completely different nematode species.

Nematodes of Sweetpotato

There are a number of genera and species of nematodes that have been found in association with sweetpotato. A listing of different types by Goodey et al. (1965) indicated at that time ten genera and a total of seventeen species of plant-parasitic nematodes were found associated with sweetpotato. A report from the French West Indies listed thirteen genera of nematodes from sweetpotato (Massese, 1969). Since there were a number of wild species belonging to the same genus as sweetpotato (*Ipomoea*) including *I. liliacea, I. rubra*, and *I. nil* found in the same fields, some of this diversity in nematode types was considered to possibly be associated with the weed hosts. *Rotylenchulus reniformis* was the most common nematode found. In the Philippines, thirteen genera and twenty-two species of nematodes were associated with sweetpotato (Gapasin, 1979). *Rotylenchulus* sp. was present in 80% of the samples and *Meloidogyne* sp. in 15%. A survey of sweetpotato fields conducted in Korea found that 61% in Haenam, 40% in Iksan, and 31% in Yeju were infested with *Meloidogyne* sp. (Dongro et al., 2006). An area with the greatest diversity of nematodes on sweetpotato was in Uganda including 25 genera and 55 species. (Coyne et al., 2003). Surveys conducted in Kyushu Island, Japan from 85 sweetpotato fields indicated that incidence of *M. incognita* was very high (96% of the fields) and the rest with two other root-knot species, *M. arenaria* or *M. javanica* (Iwahori et al., 2000). A second survey at Kyushu, Japan indicated that 55% of the sweetpotato fields had root-knot nematode present and 94% were *M. incognita* (Iwahori and Sano, 2003). Sweetpotato fields in Northwest Frontier Providence, Pakistan were found to be infested by both *M. incognita* and *M. javanica* (Musarrat et al., 2006). The major nematodes of sweetpotato in New Guinea, Fiji, Samoa, Tonga, and several other Pacific Islands were *M. arenaria*, *M. incognita*, and *M. javanica*. (Bridge, 1988). *Rotylenchulus reniformis* was found in Fiji, Western Samoa, and Tonga where it greatly reduced root weight of sweetpotato when present at high levels (Vilsoni and Hienlein, 1982). *M. javanica* is a major problem of sweetpotato in Africa (Kistner et al., 1993). A report in 1978 of the types of nematodes associated with various crops in Louisiana listed only 10 genera of plant-parasitic nematodes occurring on sweetpotato (Birchfield et al., 1978). The reniform nematode *Rotylenchulus reniformis* was not included at that time. An examination of samples submitted by producers to the Nematode Advisory Service indicates either a change in incidence or spread of this nematode. Of the 565 samples submitted by producers from sweetpotato fields during 1985–2006, *Rotylenchulus reniformis* was by far the dominant pest nematode occurring in 63% of the samples compared to *Meloidogyne* found in only 6% of the fields. The next most frequent nematode type was the spiral nematode, *Helicotylenchus* occurring in 28% of the samples. Apparently, the reniform nematode is continuing to spread into new areas within the southern U.S. (Overstreet and McGawley, 2000; Koenning et al., 2004).

Losses

Plant-parasitic nematodes cause considerable losses to sweetpotato each year. Losses caused by nematodes to sweetpotato around the world were estimated to 10.2% in 1987 causing losses of at least 2.6 billion U.S. dollars (Sasser and Freckman, 1987). The production losses caused by nematodes on sweetpotatoes and the nematodes associated with these losses were reported by several states (Koenning et al., 1999). California reported a 10% loss to sweetpotatoes from *Meloidogyne*, Hawaii and Louisiana with between 5–10% losses from *Meloidogyne* and *Rotylenchulus*, North Carolina and South Carolina with 5–10% loss due to *Meloidogyne*, Oklahoma with 1–5% by *Meloidogyne*, and Texas with a 5–7% loss from *Meloidogyne*. The root-knot nematodes belonging to the genus *Meloidogyne* are generally considered to be the most important nematodes in the production of sweetpotato around the world. Since sweetpotato is an underground crop and nematodes may feed directly on the marketable portion, sometimes it doesn't take very many nematodes to cause problems.

Root-Knot Nematodes

One of the first reports of root-knot nematode as being a pest of sweetpotato was by Bessey in 1911. At that time, the species had not been clearly differentiated for *Meloidogyne*. The original name of root-knot was *Heterodera radicicola*. It was renamed again by Goodey in 1932 to *Heterodera marioni*. Chitwood in 1949 differentiated the root-knot nematodes into five different species and placed them in the genus *Meloidogyne*. Today there are over 92 species identified for *Meloidogyne*. Considering that the original number was one and only much later even five, it's easy to see some of the problems with the taxonomy of these nematodes.

There are four species of root-knot nematode that are most often found throughout the world. These are *M. incognita*, *M. javanica*, *M. hapla*, and *M. arenaria* (Sasser and Carter, 1982). The most common species was *M. incognita*, which occurred in 52% of the 914 root-knot populations collected from around the world. The second most important species was *M. javanica* with a frequency of 31%. The other two species, *M. hapla* and *M. arenaria*, were much less common having a frequency of only 8 and 7%, respectively (Sasser and Carter, 1982). The species of root-knot nematode that are primary pests of sweetpotato include *M. incognita* and *M. javanica*, with some areas having problems with *M. hapla* or *M. arenaria*. These species of *Meloidogyne* are fairly cosmopolitan and are found throughout the tropics, sub-tropics, and even the warmer areas of the temperate zones including most areas where sweetpotatoes are grown. Like a number of species belonging to the genus *Meloidogyne*, these four species that cause problems on sweetpotato are known to have a wide host range. The southern root-knot nematode, *M. incognita*, is known to attack thousands of different host plants including most vegetable crops, agronomic crops such as soybean, cotton, corn, upland rice, sugarcane, and numerous ornamental, native, and weed hosts. This species appears to be the primary root-knot species in most areas where sweetpotato is produced. The second most common root-knot species on sweetpotato is *M. javanica* or Javanese root-knot. This nematode is found in tropical and subtropical regions of the world on sweetpotato such as the Philippines (Gapasin, 1979), Japan (Iwahori et al., 2000), Pakistan (Musarrat, 2006), New Guinea, Fiji, Samoa, Tonga, and several other Pacific Islands (Bridge, 1988), and South Africa (Kistner et al., 1993). *Meloidogyne arenaria* or peanut root-knot is known to cause necrosis to the roots of sweetpotato but apparently doesn't complete its life cycle and produce eggs on most sweetpotato cultivars (Clark and Moyer, 1988). The northern root-knot nematode, *M. hapla*, can attack sweetpotato but generally is considered a minor problem because it is limited to cooler temperate areas (Jatala, 1991). This nematode, even when introduced, does not survive climates with hot summers. Apparently, all of the major species of root-knot nematode are temperature dependent. *Meloidogyne incognita* and *M. arenaria* come from areas with mean temperatures of 15–27°C, *M. javanica* from 18 to 30°C, and *M. hapla* from 3 to 15°C (Sasser and Carter, 1982). Sweetpotato is reported as being a host to a least four other root-knot species, *M. konaensis*, *M. megadora*, *M. cruciani*, and *M. mayaguensis*. There is very little evidence that the first three species are serious pests on sweetpotato (Zhang and Schmitt, 1994; García-Martinez et al., 1982; Almeida, 1997). The root-knot nematode *M. mayaguensis* was described in 1988 and is often found in association with *M. incognita, M. javanica*, and *M. arenaria*. This species is morphologically variable, often with perineal patterns of females similar to *M. incognita* and males and second-stage juveniles similar to *M. javanica* and *M. arenaria* (De Waele and Elsen, 2007). This species has a wide host range including many vegetable crops, ornamentals, potato, and sweetpotato, very similar to that of *M. incognita*. It has now been reported in Africa, the Caribbean, Brazil, and Florida in the U.S. (Brito et al., 2004). It has very likely been around for some time and misidentified as *M. incognita*. It has been reported in West Africa as breaking resistance to a number of crops which are normally resistant to *M. incognita* including the sweetpotato cultivar 'CDH' (Fargette, 1988). It is not clear as to how much threat this nematode species is to sweetpotato at this time.

Symptoms

Root-knot nematode species generally produce distinctive galls on the roots of plants that are attacked and become suitable hosts for the nematode (Plate 9.1a). Some plants such as sweetpotato can still be severely damaged by root-knot species that cannot successfully complete their life cycle on the roots. A great majority of the other hosts that are attacked by root-knot nematodes show characteristic stunting, wilting, off-color foliage, and even early death. A wide range of symptoms have been observed on sweetpotato cultivars over time. During earlier times, there were varying reports of injury and resistance of sweetpotatoes (Kushman and Machmer, 1947; Weimer and Harter, 1925). Heavy infestation of *M. incognita acrita*(now recognized as only *M. incognita*) was reported by Krusberg and Nielsen (1957) to cause reduced plant and root growth on the variety Porto Rico. Root cracking was severe under these high levels of nema-

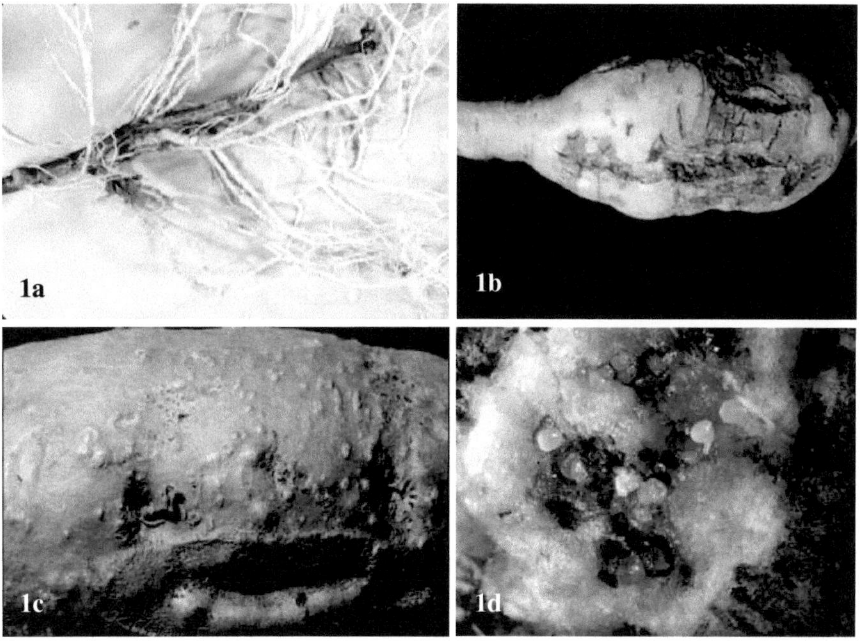

Plate 9.1 Root-knot nematode *Meloidogyne incognita* (**a**) Root showing galling (courtesy Charles Overstreet, Louisiana State University AgCenter). (**b**) Cultivar 'Centennial' showing cracking and some decay from infection (courtesy Chris Clark, Louisiana State University AgCenter). (**c**) Cultivar 'Beauregard' showing the typical blistering from infection and slight cracking (courtesy Charles Overstreet, Louisiana State University AgCenter). (**d**) Females of root-knot nematode exposed beneath one of the blisters on 'Beauregard'. Several females teased out and laying on the cut area (courtesy Charles Overstreet, Louisiana State University AgCenter) (See also Plate 5 on page xix)

tode pressure (Plate 9.1b). The soil fumigant D-D (dichloropropane-dichloropropene mixture) was applied to soil to infer that the effects observed was actually caused by root-knot nematode. Plants from the treated soil had more enlarging roots with fewer cracks than the untreated. Cracking as high as 50% was observed in the untreated but only 15% where a fumigant had been applied. Cracking of the storage roots of sweetpotato has been one of the most recognizable symptoms of damage by root-knot particularly on earlier cultivars (Nielsen and Sasser, 1959; Martin, 1962; Tateishi, 2003; Clark et al., 1980). Gapasin and Valdez (1979) reported cracking of sweetpotato by both *M. incognita* and *M. javanica*, as well as reduced tops, roots, and storage roots. Cultivars react quite differently to the various root-knot species and symptoms reflect this. *Meloidogyne incognita* on both a moderately resistant cultivar 'Jasper' and susceptible cultivar 'Centennial' both caused cracking at even low population levels (Lawrence et al., 1986). However, the moderately resistant 'Jasper' was able to tolerate much greater levels of the root-knot nematode than 'Centennial' and resulted in higher yields and improved root quality. Another common symptom associated with *Meloidogyne* is root necrosis resulting from damage during the initial infection of the root system (Giamalva et al., 1963). These symptoms occur on cultivars of sweetpotato infected by other species of root-knot such as *M. arenaria*, *M. hapla*, or *M. javanica* even when they cannot successfully reproduce and complete their life cycle. Agu (2004a) compared injury of the sweetpotato cultivar 'Koka-6' in soil that was inoculated with either *M. incognita* or the lesion nematode *Pratylenchus brachyurus*. The variables that were measured were leaf area index, leaf area growth rate, leaf area per unit weight, net assimilation, and relative growth. Plants that were infected with *M. incognita* were more damaged than those infected by the lesion nematode. In another study, Agu (2004b) found that the root-knot nematode *M. incognita* reduced shoot growth, shoot-to-storage root ratio, days to maturity, storage roots yield, and the proportion of marketable roots on the cultivar 'Cuba-2'. Symptoms of root-knot nematode are not always easy to recognize. Bridge (1988) indicated that large numbers of nematodes of *Meloidogyne* can be found in roots without any obvious symptoms. Root-knot nematodes can also be embedded deep within necrotic tissues associated with small or distorted storage roots that often have cracks along the surface. Kistner et al. (1993) described the reaction of the susceptible cultivar 'Blesbok' to *M. incognita* and *M. javanica*. There was no cracking or other observable effect on the quality of storage roots but marketable yields were decreased by 11.4%. In Louisiana, 'Beauregard' is the dominant cultivar produced by growers. The only symptoms that are recognized from *M. incognita* are small raised area on the storage roots (Plate 9.1c). If a horizontal cut is made through these raised areas, females and egg masses of root-knot nematode can be seen (Plate 9.1d). Unfortunately, only a low level of root-knot damage to storage roots is acceptable before quality is severely impacted.

Species Reactions on Sweetpotato and Pathotypes

After the species identification of *Meloidogyne* by Chitwood in 1949, a better understanding of the reaction to the various species was possible. Giamalva et al. (1963) was one of the first to look at the reaction of these newly identified species on sweetpotato and their reactions on some of the cultivars that ranged from resistant to susceptible that had been observed at that time. The five species of root-knot nematodes included *M. arenaria, M. javanica, M. hapla, M. incognita* and *M. incognita acrita*. The sweetpotato cultivars evaluated were five cultivars considered susceptible to non-identified species of root-knot nematode included 'Goldrush', 'Acadian', 'Centennial', 'Allgold', and 'La1-80'. The cultivar 'Porto Rico' was considered susceptible to intermediate in its reaction. Two cultivars, 'Heartogold' and 'Nemagold' were considered to be resistant. The peanut root-knot *M. arenaria* did not cause any galling or develop any mature females in any of the sweetpotato cultivars. Only one cultivar 'Allgold' appeared to be very susceptible to *M. javanica* and had very low or no galling on any of the other selections. The northern root-knot *M. hapla* did not appear to be a very aggressive pathogen and caused only light galling on six of the eight selections and none on the other two. The two cultivars which had previously been rated as resistant, 'Heartogold' and 'Nemagold', were found to be resistant to *M. incognita* and *M. incognita acrita*. Giamalva et al. (1963) also found highly significant differences among 14 isolates of *M incognita acrita* on susceptible 'Goldrush' and 'Acadian', moderately resistant 'Porto Rico' and resistant 'Heartogold' suggesting variability within a species. There were two early reports of unusual variability in cultivars to sweetpotato in Louisiana (Martin and Birchfield, 1973). A race of *M. incognita* that was found to be extremely pathogenic on soybeans failed to reproduce on the susceptible cultivar 'Centennial' normally very susceptible to this nematode. The second report was from that of a resistant selection of sweetpotato to *M. incognita* 'La. 4-73' developed in Louisiana being completely susceptible to a population of *M. incognita* from Maryland (Martin and Birchfield, 1973). It was designated the "resistance breaking" (RB) race and a breeding line 'W-51' was developed and released which had resistance against it in 1978 by Dukes et al. The identification of a pathotype that could break resistance to sweetpotato cultivars was described in Japan (Nishizawa, 1974). A population of *M. incognita* identified as N-2 was pathogenic on three cultivars 'Norin No. 2', 'Norin No. 5' and 'Taikaku' which were resistant to another population (K-14) of *M. incognita*. A short time later, five resistance breaking races of *M. incognita* were reported in Japan (Okamoto, 1979). These races were differentiated by their reactions to cultivars of tomato, tobacco, and sweetpotato that had *M. incognita* resistance. Lawrence and Clark (1986) reported that a number of populations of *M. incognita* isolated from Louisiana soils had variable virulence on sweetpotato cultivars ranging from resistant 'Jewel' or breeding line 'L4-73', intermediate 'Jasper', to susceptible 'Centennial'. Several populations were considered to be capable of overcoming resistance. Sano and Iwahori (2005) identified nine races of *M. incognita* in Japan using sweetpotato differential cultivars. These nine races were identified as SP1-SP9 based on reactions to sweetpotato cultivars 'Norin-1', 'Norin-2', 'Tanegashimamurasaki-7', 'Elegant Summer', and 'J-red'. There did appear to be some regional specific distribution of these various races of sweetpotato in Japan. The identification of species and races of *Meloidogyne* were developed by Taylor and Sasser in 1978 based on their reactions to a number of host plants including tobacco, cotton, bell pepper, watermelon, and peanut. At that time races 1–4 were characterized based on host preferences on cotton and tobacco. The frequency of each of these races based on worldwide surveys at that time was determined at 72% for race 1, 13% for both races 2 and 3, and 2% for race 4 (Sasser and Carter, 1982). Obviously, pathotypes or races that can break resistance to sweetpotato cultivars will have to be dealt with as these problems develop over time.

Temperature Effects

Temperature may also be a factor in development of *Meloidogyne* in sweetpotato. Jatala and Russell (1972) found that as temperature increased from 24°C, 28°C, and 32°C, there was increased penetration by *M. incognita* in both the susceptible 'Allgold' and resistant 'Nemagold'. However, there were fewer juveniles found in the resistant cultivar. As the temperature increased, more nematodes reached maturity in the resistant 'Nemagold'. Temperature may be an important factor in expression of resistance against *M. incognita* in sweetpotato. In California, *M. incognita* reproduced on all eight resistant line or cultivars tested when soil temperature was high, over 30°C at the 15-cm depth (Roberts and Scheuerman, 1984).

Reniform Nematode

Reniform nematodes belong to the genus *Rotylenchulus* and are commonly found throughout tropical and subtropical areas around the world. There are ten named species of this genus but only two have been found associated with sweetpotato plants (Robinson et al., 1997). *Rotylenchulus reniformis* has a worldwide distribution occurring throughout the Americas, Africa, India, China, Japan, Philippines, and the Pacific (Robinson et al., 1997). This nematode species also has a wide host range and attacks hundreds of crops or native plants. The other species of *Rotylenchulus* attacking sweetpotato is *R. borealis* (Vovlas, N. and Inserra, R.N., 1982). This species has a limited range of occurrence being found only in Europe and Africa. It also has a very limited host range and has been reported on bean, corn, green pea, potato, sorghum, and sweetpotato. Limited information about this nematode species and the lack of reports as a problem suggest that it is probably a minor pest on sweetpotato.

Symptoms

Martin (1960) was one of the first to report that severe damage could occur to sweetpotato by *R. reniformis*. Infected plants had sparse, necrotic, and discolored roots. Birchfield and Martin in 1965 indicated that damage may range from light to severe by *R. reniformis* depending on initial populations (Plate 9.2a). A significant yield increase of marketable roots was obtained by fumigation with dichloropropene. Vilsoni and Hienlein (1982) found a quadratic response to yield with increasing population levels of *R. reniformis* and a significant reduction in storage root weights at higher levels of this nematode. In the Philippines, Gapasin and Valdez (1979) reported that increasing levels of *R. reniformis* resulted in decreased root, storage-root, and top weights. Storage roots were also deformed, cracked, and smaller compared to the control. Clark and Wright (1983) found that *R. reniformis* causes cracking under field conditions (Plate 9.2b). These cracks were reported to be qualitatively different from those associated with other causes. Cracks from reniform infection were deeper and the exposed surfaces healed over by the formation of callus and periderm. There were no juveniles or adults of *R. reniformis* on or within cracked sweetpotatoes unlike what is found with *M. incognita*. One line 'P-104' was reported as being resistant to cracking. Differences in population development of the reniform nematode have also been reported among sweetpotato selections with most

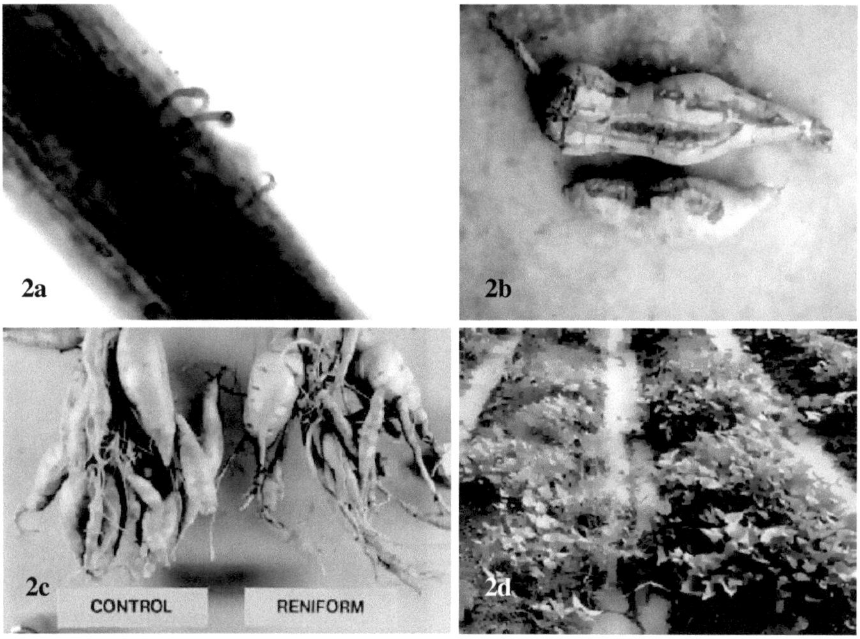

Plate 9.2 Reniform nematode *Rotylenchulus reniformis* (**a**) Developing females (stained red) inside root (courtesy Charles Overstreet, Louisiana State University AgCenter). (**b**) Cultivar 'Goldrush' showing cracking from infection by this nematode (courtesy Chris Clark, Louisiana State University AgCenter). (**c**) Individual hills of sweetpotato from an untreated area (*left*) compared to an area infested with reniform on the *right* (courtesy Chris Clark, Louisiana State University AgCenter). (**d**) A sweetpotato breeding line in a reniform nursery that is untreated on the left and fumigated on right (courtesy Chris Clark, Louisiana State University AgCenter) (See also Plate 6 on page xx)

being very good hosts (Clark et al., 1980). The application of a fumigant showed the different responses and impacts of the nematode to cultivars. The cultivar 'Goldrush' was severely reduced (78%) by *R. reniformis* compared to 'Porto Rico' at 23%. However, even though it sustained the greatest damage, 'Goldrush' allowed the least reproduction by the nematode. In Louisiana, the presence of reniform nematode is considered to slow down development of storage roots on the cultivar 'Beauregard' (C. Clark, personal communication). A delay of even a week or more in sweetpotato development can have serious consequences to producers when hurricanes or heavy rainfall events develop prior to harvest (Plate 9.2c). Yik and Birchfield (1982) showed that infection by the reniform nematode on sweetpotatoes caused malformation of xylem, cambium, and phloem cells which could result in smaller and reduced quality of storage roots.

Occasionally, both *R. reniformis* and *M. incognita* can be found together. In Louisiana, these two nematodes occur together only about 4% of the time based on 16,040 nematode samples processed during the past 20 years by the Nematode Advisory Service. A competitive interaction exists between *M. incognita* and

R. reniformis. When both nematodes were together at low populations levels, there were no differences in final populations of *R. reniformis* compared to the absence of *M. incognita* (Thomas and Clark, 1983a). However, soil counts of juveniles of *M. incognita* were reduced at the end of the season by *R. reniformis*. In fields with a high level of infestation of *R. reniformis*, the artificial infestation with high levels of *M. incognita* was found to inhibit *R. reniformis* populations but not *M. incognita* levels. Either nematode was capable of becoming the dominant population. Thomas and Clark (1983b) also indicated that *R. reniformis* was inhibited and *M. incognita* became the dominant nematode. Root-knot nematode, *M. incognita*, is very destructive to the root system and a self-limiting effect may be responsible for reducing *R. reniformis* population. Two races of *R. reniformis* have been described (Dasgupta and Seshadri, 1971a; 1971b). These races were based on population development on castor or Upland cotton in India. In the U.S., differences in *R. reniformis* populations from across the country or even within a region have been observed (McGawley and Overstreet, 1995; Agudelo et al., 2005). Walters and Barker (1993) looked at two populations of *R. reniformis* from two different states on the sweetpotato cultivar 'Beauregard'. Both populations reproduced well and no differences were observed in fecundity of the nematodes. Although both populations suppressed storage roots, some differences were observed based on the different populations. The Georgia population of *R. reniformis* suppressed yields linearly with increasing population levels of the nematode. There was not a relationship between storage roots and initial populations with the North Carolina population of *R. reniformis*. Root necrosis did increase linearly with increasing levels of nematode for both populations.

Other Nematodes

There are a few other types of plant-parasitic nematodes that are known to cause serious problems on sweetpotato. These nematodes may cause localized or regional problems but on a worldwide basis, are usually considered less of a problem than those caused by either the root-knot nematodes or reniform nematode.

Lesion Nematode

The lesion nematode genus, *Pratylenchus*, has several species that are known to attack and cause some damage to sweetpotato. *Pratylenchus brachyurus* caused slight damage to the cultivar 'Koka-6' with moderate necrotic root lesions and slightly reduced leaf growth in tests conducted in Ethiopia (Agu, 2004a). However, in the same experiment *M. incognita* caused much greater damage than the lesion nematode. Jatala (1991) described some of the damage that lesion nematode causes. Storage roots may have brown or black necrotic lesions affecting quality. In Brazil, Anguiz and Canto-Sáenz (1991) evaluated twenty cultivars of sweetpotato against *P. flakkensis* to see if it might be a good rotation crop for Irish potato, which is seriously damaged by this species. None of the sweetpotato cultivars evaluated was considered a good host and supported very little reproduction of the nematode. Six of the cultivars that were selected for further testing actually showed an increase in storage root weights after being inoculated with this species. Japan appears to be one of the areas that have some of the greatest problems with lesion nematodes. Surveys conducted in sweetpotato fields from central and Southern Kyushu, Japan indicated the lesion nematode *P. coffeae* to be present in 22% of the fields (Iwahori et al., 2000). Other surveys conducted in several islands around Kyushu and the Nansei Islands, Japan found *P. coffeae* in 12.9% of the fields (Iwahori et al., 2001). A third survey conducted in the northern part showed a lower incidence of *Pratylenchus* (Iwahori and Sano, 2003). The dominant species was still *P. coffeae* but *P. penetrans* was found in two fields. Lesion nematodes are such serious pests in Japan that *P. coffeae* is included in the selection process for new cultivars (Yoshida, 1985). Cultivars such as 'Beniotome' have been selected for resistance to both *M. incognita* and *P. coffeae* (Kukimura et al., 1992). In China, Feng et al. (2000) proposed that the cleaning up of viruses from sweetpotatoes by using shoot tip cultures could restore a cultivars yield and quality as well as improving resistance against *P. coffeae*.

Stem Nematodes

Stem nematodes, *Ditylenchus dipsaci* and *D. destructor* are known to cause a brown ring disease in sweetpotato. China appears to be the major area where this disease is a problem (Lin et al., 1993; Zhang et al., 2006). The disease is primarily a storage problem. Storage roots that are infected with this nematode have a brown to blackish layer internally often leading to complete decay (Clark and Moyer, 1988; Scurrah et al., 2005). Scanning electron microscopy revealed that in slight or medium infection by *D. destructor*, starch grains were smaller and distorted with cell shrinkage and large gaps due to loss of water (Jianhua et al., 1998). In severely infected storage roots, starch grains were either greatly reduced or absent, with severe cell shrinkage. Since both nematode species have been reported in China, considerable effort in their breeding program has been addressed with both species. Interspecific F_1 hybrids of sweetpotato from crosses of *I. trifida, I. batatas*, and *I. littoralis* were reported to be more resistant to *D. dipsaci* than intervarietal hybrids (Xie et al., 1992). Wang et al. (1995) evaluated the status of 508 *I. batatas* cultivars and 119 accessions of related wild species and progeny from interspecific hybrids against *D. destructor*. High levels of resistance were found in 68 cultivars, 12 accessions of *I. trifida*, and 52% of progeny lines between *I. trifida* and *I. batatas*. Lin et al. (1999) reported that the cultivars 'Lushu 78066' and 'Shengli-100' were resistant to *D. destructor*. A comprehensive screening of 1496 entries of sweetpotato was conducted during 1991–95 for resistance in China (Zhang et al., 1998). There were 324 entries rated as highly resistant to resistant to *D. dipsaci*. A field study was conducted in China during 2001–2003 to evaluate 100 sweetpotato cultivars against *D. destructor* (Zhou et al., 2005a). From 21 cultivars selected from this group, 10 were highly resistant, two were resistant, and nine either susceptible or highly susceptible. Zhou et al. (2005b) evaluated 200 accessions of a F_1 population of crosses between the highly resistant 'Xu 781' and highly susceptible 'Xushu 18' to determine resistance to *D. destructor*. A molecular marker linked to the resistant gene, OPD01-700, was obtained and evaluated from the range of resistance by various clones (Zhou et al., 2005b). Maosong et al., 1996 looked at the morphological structures of sweetpotato in resistant and susceptible cultivars against *D. destructor*. The resistant cultivars 'Lushu-78066' and 'Shegli-100' had thicker cell walls in xylem parenchyma, more lignification, and more vessels at the top of the root linking it to the subterranean stem. There also may be an interaction between *D. destructor* and *Fusarium* spp. on sweetpotato in China (Zhang and Zhang, 2007).

Burrowing Nematode

Burrowing nematodes belong to the genus *Radopholus*. At least two species of this genus have been reported on sweetpotato, *R. williamsi* (renamed to *Achlysiella williamsi*) and *R. similis* (Bridge, 1988; O'Bannon, 1976). *Radopholus similis* is worldwide occurring in tropical and semitropical areas. Like many nematodes, this species has an extensive host range occurring on over 250 plant species. Banana, one of the most susceptible plants, suffers tremendous losses from this species. The other species, *A. williamsi*, is probably limited in distribution and has only been reported on sweetpotato in Fiji (Bridge, 1988). Two races of *R. similis*, the banana and citrus races, have been reported based on occurrence on these hosts. Sweetpotato is apparently not a host for the banana race of *R. similis* (Milne and Keetch, 1976; Tarte et al., 1981). Koshy and Jasy (1991b) evaluated 17 host differentials and 28 populations of *R. similis*. A total of ten different races could be identified based on the multiplication on these various hosts. The impact of *R. similis* on sweetpotato was studied in India (Koshy and Jasy, 1991a). Increasing populations of *R. similis* reduced growth parameters of sweetpotato including roots and storage roots. The burrowing nematode is well known for causing lesions and rotting of feeder roots and these symptoms were expressed on sweetpotato. Massese (1969) did indicate that *R. similis* was a serious crop of sweetpotato in the French West Indies.

Nematode Management

Sweetpotatoes are grown in diverse regions of the world, with production ranging from subsistence farms to large corporate farms that may be found in the U.S. There are many areas where chemical nematicides are either not available or not affordable by producers. Although resistant cultivars are available in many areas, cultivars that are not resistant may continue to be used because of specific shape, color, taste, or other personal preferences. A wide range of crops are grown in association with sweetpotato around the world. Unfortunately, many of the crops grown in rotation with sweetpotatoes are hosts to the major nematodes impacting this crop. There are several options that can be utilized by producers regardless of where sweetpotatoes are grown in the world. Unfortunately, the sweetpotato industry in Louisiana as well as the U.S. in general tends to rely on a single, dominant cultivar at any one time (Clark et al., 1997). At this time the cultivar 'Beauregard' is by far the dominant cultivar in Louisiana which is susceptible to both root-knot and reniform nematodes. This limitation necessitates the use of either nematicides or rotation rather than the incorporation of resistance into a management plan by producers.

Resistance

The use of resistant cultivars to manage root-knot nematodes has received considerable attention during the past 100 years. Cultivars were selected that were tolerant or exhibited resistance even before the species of *Meloidogyne* were clearly understood (Weimer and Harter, 1925; Tyler, 1941; Kushman and Machmer, 1947). Gentile et al. (1962) indicated that resistance to *M. incognita* and *M. javanica* could be transferred to breeding lines of sweetpotato. Sweetpotato lines that had resistance were reported for combinations of resistant and susceptible parents by Struble et al. (1966). Of the older cultivars of sweetpotato, Jersey lines were considered to be fairly resistant and Porto Rico considered being intermediate against root-knot nematode (Morrison, 1970). A listing of root-knot resistant cultivars for various crops including sweetpotato by Sasser and Kirby in 1979 included the known reactions of 46 cultivars and breeding clones to *M. incognita, M. javanica, M. hapla*, and *M. arenaria*. The genotypes varied considerable in their reactions to the four species of root-knot nematodes. Twenty-two genotypes were considered resistant or highly resistant to *M. incognita*. Shiga and Takemata (1981) found 67 clones of sweetpotato highly resistant and 66 with some resistance of the 408 clones tested. The most resistant selections were found in Papua New Guinea and resistance decreased with distance away from the island. In the Philippines, Gapasin (1984) evaluated 52 cultivars for resistance to *M. incognita* and found 28 cultivars resistant. Only one of the cultivars was susceptible to *M. javanica*. Galls were usually small on most cultivars and an egg mass index was used to assess resistance. In India, 90 accessions of sweetpotato were screened for resistance to *M. incognita*. Thirty-four were resistant and 14 considered highly resistant (Ramakrishnam et al., 1997). Cervantes et al. (2002b) in the U.S. evaluated 26 sweetpotato genotypes that are of economic importance or present in the pedigree of modern cultivars against *M. incognita* races 1–4, *M. javanica*, and *M. arenaria* races 1 and 2. Several additional genotypes including 'Beauregard', 'L86-33', 'PDM P6', 'Porto Rico', and 'Pelican Processor' were also included for additional studies against the various root-knot nematodes. Two landraces 'Tanzania' and 'Wagabolige' were included from Africa because they were considered resistant to all the *Meloidogyne* species. The host status of each of these genotypes depended on the species of *Meloidogyne* or race. The peanut root-knot *M. arenaria* race 1 was able to lightly infect and reproduce on only 'L86-33'. Race 2 of *M. arenaria* reproduced well on 'L86-33' but poorly on 'Nancy Hall' and none of the others. The cultivar 'Beauregard' was susceptible to all four races of *M. incognita* as well as *M. javanica*. Several cultivars such as 'Excel', 'Triumph', 'Regel', 'Tinian', 'Resisto', 'Tanzania', 'Wagabolige', and 'Carogold' were considered very resistant to all the *Meloidogyne* spp. or races. Also, seven genotypes were evaluated against three populations of root-knot nematode from each of the four *M. incognita* races that had been collected from different areas around the world. The cultivar 'Beauregard' was susceptible to all twelve *M. incognita* populations and could be considered a universal susceptible control. Several reports of cultivar resistance against root-knot nematode have come out of Brazil. Silveira and Maluf (1993) evaluated 31 clones and five cultivars against *M. javanica* and races 1–4 of *M. incognita*. Twenty-one clones were rated as resistant to *M. javanica* as well as the cultivars 'Brazlândia Roxa', 'Brazlândia Rosada', and 'Coquinho'. None of the selections had resistance to all four races of *M. incognita*. Maluf et al. (1996) did a more extensive evaluation and included 226 sweetpotato clones of diverse origin against root-knot nematodes. The majority of the clones were found to be resistant or moderately resistant to *M. javanica*. Over 50% of the clones had at least moderate resistance to races 1–4 of *M. incognita*. Maluf et al. (1996) also reported that resistance to the different *Meloidogyne* sp. isolates (*M. javanica* and *M. incognita* races 1–4) were not highly correlated indicating that there may be independent selection for resistance. Fortunately, many cultivars apparently have this wide range of resistance against *Meloidogyne* species and races. Mcharo et al. (2005) reported that resistance to root-knot nematode may be qualitatively as well as quantitatively controlled. A F_1 population of sweetpotato half-sib genotypes developed at Louisiana State University AgCenter from 'Beauregard', 'Excel', 'L94-96', 'L89-110', 'L86-33', and 'L96-117' showed a bimodal response to *M. incognita* infection. This would indicate that a major gene may be controlling the resistance trait. A second population of sweetpotatoes tested against *M. incognita* was from full-sibs developed by the National Agricultural Research Organization, Kampala Uganda and International Potato Center, Lima, Peru. These crosses were 'Beauregard x Wagabolige', 'Kyukei No. 63. x Jonathan W218', 'Jonathan W154 x Wagabolige', 'CN1732-4 x Jonathan W218', 'Tanzania x Wagabolige' and open pollinated 'Tanzania'. Genotypes from this population showed a quantitative response indicating that a few major genes may be involved with nematode resistance. Komiyama et al. (2006) indicated that as the result of studies with the resistant cultivar 'Minamiyutaka' and the diploid *Ipomoea trifida* against *M. incognita*, resistance mechanisms may be controlled by two dominant genes. The Brazilians evaluated 70 sweetpotato genotypes of which 58 were clones coming from polycrosses against two root-knot species (Peixoto et al., 1998). Additionally, three genotypes were provided by growers in the State of Minas Gerais and seven other genotypes including 'Brazlandia Rosada', 'Coquinho', 'Rio Doce', 'Morena Roxa', 'Arroba', 'Pira' and 'O42', a resistant control 'Surpresa' and susceptible control 'Brazlandia Branca'. The nematodes evaluated were *M. incognita* races 1, 2, and 3, *M. javanica*, and field populations of *Meloidogyne* sp. Twenty-one of the clones were resistant to all the nematodes tested. The cultivars 'Surpresa', 'Arroba', 'Pira 1', and 'Coquinho' showed at least some resistance to all the nematodes. Wanderly and Santos (2004) evaluated 35 sweetpotato cultivars against

M. incognita and 15 were considered resistant. In Nigeria, Hahn et al. (1989) evaluated 414 clones of sweetpotato for resistance to *M. incognita* and *M. javanica*. Fifty-five of these clones were rated as resistant to both root-knot nematode species. Crozzoli et al. (1994) in Venezuela looked at ten sweetpotato selections against *M. incognita*. Three of the selections 'UCV-2', 'UCV-7', and 'Catemaco' were not damaged by *M. incognita*. However, all three supported infection and reproduction as well as the other susceptible cultivars or lines. A second evaluation in Venezuela by Montes et al. (1998) looked at 28 selections of sweetpotato against *M. incognita* race 2. All selections except 'UCV-8' were considered susceptible. In Korea, none of the cultivars evaluated by Dongro et al. (2006) were resistant against *M. incognita*. Three cultivars 'Jimmi', 'Jeungmi' and 'Borami' were found to be resistant against *M. javanica, M. hapla*, and *M. arenaria*. In the U.S., Thies (2000) looked at several breeding clones for resistance to *M. incognita* race 3. Four regional lines were rated as being highly resistant. Some of the benefits of using resistance were reported by Dukes and Bohac (1994). Ten sweetpotato cultivars were planted in the same exact areas for several years in a field infested with *M. incognita*. Cultivars that were very susceptible such as 'Georgia Jet' and 'Red Jewel' had marketable yield reductions each year. Highly resistant cultivars or lines such as 'Sumor' and 'W-279' performed well each year with high yields and good quality of storage roots. Nematode reproduction was kept low on these highly resistant selections over time indicating durability of resistance to *Meloidogyne*. However, Tateishi (2003) found that alternating a sweetpotato cultivar 'Beniotome' which is resistant to *M. incognita* and a susceptible cultivar 'Kokei 14' had some negative consequences. After the fifth cropping season, populations of the nematode began increasing on the resistant cultivar to 50–80% levels of the susceptible. Cultivars selected for resistance should be able to successfully keep nematode populations down, show few if any damage symptoms, and yield well. Trudgill (1991) indicates that tolerance to nematode infection is a necessary adjunct to resistance. The cultivar 'Beauregard' is a good example of tolerance to reniform nematode since it doesn't show signs of cracking or root necrosis and may yield reasonably well even in the presence of this nematode. Even where resistance is not available, tolerance to nematode attack is important. Tolerance has always been used in breeding programs or even where landraces are naturally selected by producers because these plants seem to perform well in spite of the presence of a problem nematode. The International Agricultural Research Centre (IARC) that is mandated to conduct research on sweetpotato is Centro Internacional de la Papa (CIP). CIP has begun basic screening of germplasm for resistance to *M. incognita* and *M. javanica*. There is also collaboration with China for resistance to the storage rot nematode *Ditylenchus destructor* (Sharma et al., 1997). The release of a number of cultivars that have specific resistance against nematodes in the last several years indicates that plant breeders are attempting to resolve some of the problems with the use of resistance. Some recently released cultivars that have good resistance against *M. incognita* include 'Evangeline', 'Daichinoyume', 'Benimasari', 'Tamaotome', and 'Sree Bhadra' (LaBonte et al., 2008; Katayama et al., 2004; Ishiguro et al., 2003; Ishiguro et al., 2002; Vimala and Rajendran, 1998). Some fairly new cultivars that have been released with resistance against the stem nematode include 'Yushu 13', 'Lushu 78066', and 'Taishu 2' (Guohong et al., 1999; Sun and Chen, 1994; Sun and Dong, 1994).

Nematicides

Nematicides are widely used used to reduce nematode infestations and improve yield and quality of storage roots. Nielsen and Sasser (1959) found that the fumigants D-D and EDB increased both yield and quality of sweetpotato against root-knot nematodes. Birchfield and Martin (1965, 1968) also found that several nematicides including D-D, Telone, and Vorlex significantly increased number 1's and 2's of storage roots in fields infested with reniform. The most damaging populations of reniform nematode occurred at the time of planting. Other researchers in the U.S. have found fumigants or nematicides to be effective against nematodes. Several fumigant including D-D, Dowfume W-85, Telone, or Nellite and granular nematicides, Fuaradan, Temik, and Mocap, were found to increase marketable size of sweetpotato 'Centennial' against *M. incognita* (Johnson and Cairns, 1971). The systemic nematicide oxamyl was found to be very effective in reducing infection by root-knot nematode *M. incognita* in a greenhouse study (Rodríguez-Kábana et al., 1978). In Louisiana, Clark et al. (1983) found that Chloro-Pic and Mocap gave the best control of *R. reniformis* when applied to ground that was to be planted in sweetpotato (Plate 9.2d). Averre et al. (1993) also found the fumigants 1,3-D, Vorlex, and Metam significantly reduced root damage by *M. incognita*. Hall et al. (1988) showed a greater response with aldicarb and fenamiphos than the fumigant D-D against *M. incognita* and that fall applications were not considered as good as spring applications of nematicides, particularly the granular nematicides. The reverse of this was reported by Chalfant et al. (1992). The fumigant 1,3-D gave better control of *M. incognita* than aldicarb or fenamiphos. The granular nematicide, fenamiphos, greatly increased the yield of the USDA No. 1 storage root for two years but was not effective for the next two years (Johnson et al., 1992). Microbial degradation was suspected as being the cause of reduced efficacy of this nematicide. Johnson (1998) later found that there really was a more rapid degradation of fenamiphos where there was a history of its repeated application. Other nematicides evaluated against reniform in the U.S. include fosthiazate which increased yield dramatically with increasing rated (McLean and Lawrence, 1996). Aldicarb was found in Mississippi to significantly reduce reniform nematode numbers and produce more USDA Number 1 storage roots and jumbos than the untreated (Henn et al., 2006). Since the overall weights between treated and untreated did not differ, reniform may simply reduce marketable storage roots. Abel et al. (2007) showed a significant yield response of the cultivar 'Beauregard' against reniform nematode using either metam sodium, 1,3-D (Telone), or aldicarb. The lack of cracking or other visible symptoms suggested that even with high populations of this nematode, damage may not be recognized. Nemacur and Furadan are also effective against *Ditylenchulus dipsaci* (Zhang et al., 1990).

Cultural Management

A number of cultural practices can be followed to reduce the level of damage caused by plant-parasitic nematodes. In the U.S., sweetpotatoes are bedded out in the spring to produce sprouts that will be used for transplanting into production fields. Beds should be free of nematodes if possible, the storage roots used for propagation should be nematode-free, and slips cut above the soil line rather than pulled to reduce the possibility of spread. Although reniform nematode apparently is not found within storage roots, their presence in soil on the storage roots could provide a means of further spreading this pest to plant beds. Root-knot nematode survives very well in storage roots and could certainly be involved in spreading from one area to another. Crop rotation is one of the most widely used cultural practices with both traditional and subsistence agriculture (Bridge, 1996). It is considered to be extremely effective in low-input agriculture systems although probably rarely developed specifically for nematode management. Continuous production of the same crop within a field maintains high levels of nematodes that are specific for that crop. Crop rotation for nematode management should include nonhosts or poor hosts, resistant crops, and only occasionally a susceptible crop. Bridge (1996) suggested that a susceptible crop should not be planted in nematode infested soil more than once every fourth growing season. In Papua New Guinea, damage to the sweetpotato crop depended on crop rotation and susceptibility of cultivars (Bridge and Page, 1984; Bridge, 1996). Damage from *M. incognita* was most severe where sweetpotato was planted continuously and absent or greatly reduced when rotated with maize, goundnuts, and cassava. Hartemink et al. (2000) found similar trends with continuous planting of sweetpotato. Yields declined over time with the damage most likely attributed to *M. incognita*. Cropping sequences such as Irish potato and sweetpotato and potato/sweetpotato/peanut/sorghum were examined in the U.S. to determine effectiveness against several root-knot nematode species (Johnson et al., 1996). These two cropping systems promoted a shift in favor of *M. incognita* rather than *M. arenaria* or *M. hapla*. A cropping sequence of sweet corn/sweetpotato/vetch built up large numbers of *M. incognita* over time (Johnson et al., 1992). The buildup was attributed to occur primarily on the sweetpotato 'Jewel' although it is normally fairly resistant to *M. incognita*. Other cultural controls may include amendments of some type. Gapasin (1981) obtained a range of responses with either nematicides or chicken manure against *M. incognita* or *R. reniformis*. The chicken manure may have been more of a fertilizer effect rather than actually controlling nematodes. Another amendment that was evaluated was that of a mixture of soybean meal, urea, and a chitinous material derived from blue crab (*Callinectes sapidus*) for activity against *M. incognita* (Rodríguez-Kábana et al., 1990). Although these amendments did reduce populations of *M. incognita* on the sweetpotato cultivar 'Jewel', there was not a corresponding increase in yield. Piecemeal harvesting of storage roots over time rather than a single harvest was helpful against sweetpotato weevils (*Cylas* spp.) but not against nematodes (Ebregt et al., 2007). A sweetpotato/peanut rotation was compared to continuous sweetpotato culture in Papua New Guinea (D'Souza et al., 1986). However, no benefits were observed with this rotation on root-knot nematode.

Kassab (1990) evaluated the response of mycorrhizal associations of sweetpotato against a *Criconemella* sp., *R. reniformis*, and *Tylenchorhynchus* sp. Final populations of *Criconemella* and *R. reniformis* were higher on mycorrhizal plants and *Tylenchorhynchus* was higher on non-mycorrhizal plants. Sweetpotato plants that were associated with mycorrhiza were more tolerant of nematode infection.

Although resistance is usually considered to be desirable for the sweetpotato crop, resistance has also been proposed to help the succeeding crop. The cultivar 'J-red' is resistant to *M. incognita* and was evaluated as a nematode reducing crop to use with carrot as the main crop (Mochida et al., 2007). Population levels were much lower after the resistant sweetpotato and less galling was evident on carrots. The cultivar 'Sree Bradra' is highly resistant to *M. incognita* and was recommended as a trap crop to help manage nematodes (Mohandas and Ramakrishnan, 1996). The bud nematode can be eliminated from sweetpotato germplasm by the use of meristem culture (Tenente et al., 1993). *Pasteruia penetrans* is a bacterium currently being evaluated as a biological control organism against root-knot nematode. This bacterium was found to give good control of *M. incognita* race 1 against the moderately resistant 'Beniazuma' and highly susceptible 'Kokei 14, being almost as effective as the fumigant 1,3-dichloropropene particularly for the moderately resistant cultivar (Tateishi, 1998). Lower populations of the *Meloidogyne* spp. were reported by the third year in this eight year study on the susceptible cultivar 'Kukei 14' after treatment with *Pasteuria penetrans* (Tateishi et al., 2007).

Soil solarization is another cultural practice used to reduce damage by nematodes. The ground is covered with a thin layer of clear plastic sheeting, sealed around the edges, and allowed to remain in place for several weeks. The hottest months are usually best for heating the soil sufficiently to reduce nematodes. Soil solarization was found to reduce *M. incognita* by 90% down to a depth of 30 cm (Stevens et al., 1990). The cultivar 'Georgia Jet' had increased plant growth, higher yield, and reduced galls and egg production after solarization. Hot water treatment of sweetpotato can also eliminate root-knot nematode. Burk and Tennyson (1941) recommended 65 minutes for 46.7°C for sweetpotato roots. Martin (1962) showed that most root-knot nematodes (*M. incognita*) could be eliminated from sweetpotato roots by hot air treatment at 50°C for 4–8 hours without significantly affecting the keeping quality of the storage roots. Many of the crops grown or even weeds occurring in production fields may be hosts to more that one type of nematode. Several weed hosts of either *M. incognita* and *R. reniformis* are *Ipomoea hederacea, I. lacunosa*, and *I. trichocarya*. Several other species tested were susceptible to *M. incognita* but not *R. reniformis* (Clark and Watson, 1983).

Conclusions

Although nematode problems are likely to continue around the world on sweetpotatoes, there is great potential for the future to reduce the current losses. Both the use of resistant cultivars and tolerance to the various nematodes appears to be one of the best methods of managing nematodes. There are extensive breeding programs underway in a number of countries to pursue this type of management. However, since races of certain nematodes that can break resistance in cultivars have been found, other management options such as crop rotation or even nematicides must be utilized. In Louisiana, even the site-specific application of nematicides to fields is being investigated to manage nematode problems. Our understanding of nematodes has changed a lot in the past 100 years. Unfortunately, new pest problems such as *Meloidogyne mayaguensis* or resistance-breaking races continue to be found that will continue to keep nematodes as an important pest problem in sweetpotato.

References

Abel, C.A., Adams, L.C., and Stetina, S.R. 2007. Sweet potato yield reduction caused by reniform nematode in the Mississippi delta. Online: *Plant Management Network* doi:10.1094/PHP-2007-1115-01-RS

Agu, C.M. 2004a. Effect of *Meloidogyne incognita* and *Pratylenchus brachyurus* on leaf growth of sweetpotato. *Trop. Sci.* 44:48–50

Agu, C.M. 2004b. Growth and yield of sweetpotato as affected by *Meloidogyne incognita*. *Trop. Sci.* 44:89–91.

Agudelo, P., Robbins, R.T., Stewart, J.M., Szalanski, A.L. 2005. Intraspecific variability of Rotylenchulus reniformis from cotton-growing regions in the United States. *J. Nematol.* 37:105–114.

Almeida, A.M.S.F., Santos, M.S.N., and Ryan, M.F. 1997. Host status of selected plant species for *Meloidogyne megadora*. *Nematropica* 27:1–6.

Anguiz, R. and Canto-Sáenz, M. 1991. Reacción de cultivares de camote (*Ipomoea batatas* (L.) Lam.) a *Pratylenchus flakkensis* Sienhorst. *Nematrópica* 21:197–201.

Averre, C.W., Barker, K.R., Sorensen, K.A. and Westerbeek, P. 1993. Sweet potato root-knot management with soil fumigants. *F. & N. Tests* 48:213.

Babatola, J.O. 1979. Varietal reaction of rice and other food crops to the rice-root nematodes, *Hirschmanniella oryzae*, *H. imamuri* and *H. spinicaudata*. *Nematropica* 9:123–128.

Bessey, E.A. 1911. Root-knot and its control. U.S. Dept. Agri. Bur. Plant Industry Bull. 217, 89 pp.

Birchfield, W. and Martin, W.J. 1965. Effects of reniform nematode populations on sweetpotato yields. *Phytopathology* 55:497.

Birchfield, W. and Martin, W.J. 1968. Evaluation of nematocides for controlling reniform nematodes on sweetpotato. *Plant Dis. Rep.* 52:127–131.

Birchfield, W., Hollis, J.P., and Martin, W.J. 1978. A list of nematodes associated with some Louisiana plants. Louisiana State University Agricultural Experiment Station Technical Bulletin No. 101, 22 pp.

Bridge, J. 1988. Plant-parasitic nematode problems in the pacific islands. *J. Nematol.* 20:173–183.

Bridge, J. 1996. Nematode management in sustainable and subsistence agriculture. *Annu. Rev. Phytopathol.* 34:201–225.

Bridge, J. and Page, S.L.J. 1984. Plant nematode pests of crops in Papua New Guinea. *J. Plant Prot. Trop.* 1:99–109.

Brito, J, Powers, T.O., Mullin, P.G., Inserra, R.N., and Dickson, D.W. 2004. Morphological and molecular characterization of *Meloidogyne mayaguensis* isolates from Florida. *J. Nematol.* 36:232–240.

Burk, E.F. and Tennyson, G. 1941. Hot water treatment for control of nematodes in sweet potato seed roots. *Proc. Am. Soc. Hort. Sci.* 39:299–302.

Cervantes-Flores, J.C. and Yencho, G.C. 2002a. Efficient evaluation of resistance to three root-knot nematode species in selected sweetpotato cultivars. *HortScience* 37:390–392.
Cervantes-Flores, J.C. and Yencho, G.C. 2002b. Host reactions of sweetpotato genotypes to root-knot nematodes and variation in virulence of *Meloidogyne incognita* populations. *HortScience* 37:1112–1116.
Chalfant, R.B., Hall, M.R., Johnson, A.W., Seal, D.R., and Bondari, K. 1992. Effects of application methods, timing, and rates of insecticides and nematicides on yield and control of wireworms (Coleoptera: Elateridae) and nematodes (Tylenchida: Heteroderidae) that affect sweet potato. *J. Econ. Entomol.* 85:878–887.
Chitwood, B.G. 1949. Root-knot nematodes-Part I. A revision of the genus *Meloidogyne* Goeldi, 1887. *Proc. Helminthol. Soc.* 16:90–104.
Clark, C.A., Wright, V.L., and Miller, R.L 1980. Reaction of some sweet potato selections to the reniform nematode, *Rotylenchulus reniformis. J. Nematol.* 12:218.
Clark, C.A. and Watson, B. 1983. Susceptibility of weed species of Convolvulaceae to root-infecting pathogens of sweet potato. *Plant Dis.* 67:907–909.
Clark, C.A. and Moyer, J.W. 1988. Compendium of sweet potato diseases. APS, St. Paul, Minnosota. 74 pp.
Clark, C.A. and Wright, V.L. 1983. Effect and reproduction of Rotylenchulus reniformis on sweet potato selections. *J. Nematol.* 15:198–203.
Clark, C.A., Strand, C., and Watson, B. 1983. Comparision of soil fumigant with ethoprop for reniform nematode control in sweet potato, 1982. *F. & N. Tests* 38:12–13.
Clark, C.A., Hoy, M.W., and Valverde, R.A. 1997. The evolving disease situation for sweet potato. *La. Agric.* 40:15.
Coyne, D.L., Talwana, H.A.L., and Maslen, N.R. 2003. Plant-parasitic nematodes associated with root and tuber crops in Uganda. *African P. Protect.* 9:87–98.
Crozzoli, R., Cati, F., and Vovlas, N. 1994. Respuesta de diez selecciones de batata al nematodo agallador *Meloidogyne incognita. Fit. Venezolana* 7:50–54.
Dasgupta, D.R., and Seshadri, A.R. 1971a. Races of the reniform nematode, *Rotylenchulus reniformis* Linford and Oliveira, 1940. *Indian J. Nematol.* 1:21–24.
Dasgupta, D.R., and Seshadri, A.R. 1971b. Reproduction, hybridization and host adaptation in physiological races of the reniform nematode, *Rotylenchulus reniformis. Indian J. Nematol.* 1:128–144.
D'Souza, E., Bourke, R.M., and Akus, W.L. 1986. Intensification of subsistence agriculture on the Nembi planteau Papua New Guinea 3. Sweet potato cultivar trials crop rotation trials and crop introductions. *Pap. New Guinea J. Agric. For. Fish.* 34:1–4.
De Waele, D. and Elsen, A. 2007. Challenges in tropical plant nematology. *Ann. Rev. Phytopathol.* 45:457–485.
Dongro, C., Jaekook, L., Byeongyong, P., and Minam, C. 2006. Occurrence of root-knot nematodes in sweet potato fields and resistance screening of sweet potato cultivars. *Korean J. Appl. Entomol.* 45:211–216.
Dukes, P.D. and Bohac, J.R. 1994. Resistance in sweetpotato to root-knot nematode: its value and other benefits. *HortScience* 29:726.
Dukes, P.D., Jones, A., Cuthbert, F.P., and Hamilton, M.G. 1978. W-51 root-knot resistant sweet potato germplasms. *HortScience* 13:201–202.
Ebregt, E., Stuik, P.C., Odongo, B. and Abidin, P.E. 2007. Piecemeal versus one-time harvesting of sweet potato in north-eastern Uganda with special reference to pest damage. *Wageningen J. Life Sci.* 55:75–92.
Fargette, M. 1988. Présence de race B du nématode phytoparasite *Meloidogyne incognita* en Cote d'Ivoire et leur caractétisation par l'élecrophorése des estérases. *Comptes Rendus de l'Académie des Sciences, III (Sciences de la Vie)* 306:437–440.
Feng, G., Yifu, G., Pinbo, Z. 2000. Production and deployment of virus-free sweetpotato in China. *Crop Protection* 19:105–111.
Gapasin, R.M. 1979. Survey and identification of plant parasitic nematodes associated with sweet potato and cassava. *Ann. Trop. Res.* 1:120–134.

Gapasin, R.M. 1981. Control of *Meloidogyne incognita* and *Rotylenchulus reniformis* and its effect on the yield of sweet potato and cassava. *Ann. Trop. Res.* 3:92–100.
Gapasin, R.M. 1984. Resistance of fifty-two sweet potato (*Ipomoea batatas* (L.) Lam.) cultivars to *Meloidogyne incognita* and *M. javanica. Ann. Trop. Res.* 6:1–19.
Gapasin, R.M. and Valdez, R.B. 1979. Pathogenicity of *Meloidogyne* spp. and *Rotylenchulus reniformis* on sweet potato. *Ann. Trop. Res.* 1:20–25.
García-Martinez, R., Taylor, A.L., Smart, G.C. 1982. *Meloidogyne cruciani* n. sp., a root-knot nematode from St. Croix (U.S. Virgin Islands) with observations on morphology of this and two other species of the genus. *J. Nematol.* 14:292–303.
Gentile, A.G., Kimble, K.A., and Hanna, G.C. 1962. Reactions of sweetpotato breeding lines to *Meloidogyne* spp. when inoculated by an improved method. *Phytopathology* 52:1225–1226.
Giamalva, M.J., Martin, W.J., and Hernandez, T.P. 1963. Sweetpotato varietal reaction to species and races of root-knot nematode. *Phytopathology* 53:1187–1189.
Goodey, J.B., Franklin, M.T., and Hooper, D.J. 1965. *The Nematode Parasites of Plants Catalogued Under Their Host, (3rd ed.).* Famham Royal, Bucks, UK, CAB. 214 pp.
Guohong, Y., Zhihe, K., Lizhen, X., Lixia, H., Xiaoping, Y., and Jianhui, G. 1999. Breeding of a new starch-processing type sweet potato variety Yushu 13 with high yield and multi-resistance. *J. Henan Agric. Sci.* 3:3–5.
Hall, M.R., Johnson, A.W., and Smittle, D.A. 1988. Nematode population densities and yield of sweet potato and onion as affected by nematicides and time of application. *A. App. Nematol.* 2:15–21.
Hahn, S.K., Isoba, J.C.G., and Ikotun, T. 1989. Resistance breeding in root and tuber crops at the International Institute of Tropical Agriculture (IITA), Ibadan, Nigeria. *Crop Prot.* 8:147–168.
Hartemink, A.E., Poloma, S., Maino, M., Powell, K.S., Egenae, J., and O'Sullivan, J.N. 2000. Yield decline of sweet potato in the humid lowlands of Papua New Guinea. *Ag. Ecosys. Environ.* 79:259–269.
Henn, R.A., Burdine, B., Main, J.L., and Balbalian, C.J. 2006. Evaluation of nematicides for management of reniform nematode and sweet potato yield, 2004. *F. & N. Tests* 61:NO10.
Ishiguro, K., Yamakawa, O., Kumagi, T., and Kai, Y. 2002. Tamaotome: new sweetpotato cultivar for products dried after steaming. *Sweetpotato Res. Front* 14:3.
Ishiguro, K., Yamakawa, O., Kumagai, T, Yumi, K., Nakazawa, Y. 2003. Benimasari: new sweetpotato cultivar for table use. *Sweetpotato Res. Front* 15:3.
Iwahori, H. and Sano, Z. 2003. Distribution of main plant-parasitic nematodes in sweet potato and taro fields in Kyushu and Okinawa, Japan. 3. Survey in the northern part in Kyushu Island (Fukuoka, Saga, Nagasaki and Ohita Prefs.). *Kyushu Plant Prot. Res.* 49:83–87.
Iwahori, H. Sano, Z. and Ogawa, T. 2000. Distribution of main plant-parasitic nematodes in sweet potato and taro fields in Kyushu and Okinawa, Japan. 1. Survey in the central and southern parts in Kyushu Island (Kumamoto, Miyazaki and Kagoshima Prefs.) and development of an effective DNA analysis method for species identification. *Proceeding Assoc. Plant Protection of Kyushu* 46:112–117.
Iwahori, H., Sano, Z., and Torigoe, H. 2001. Distribution of the main plant-parasitic nematodes in sweet potato and taro fields in Fyushu and Okinawa, Japan. 2. Survey in several islands around Kyushu and the Nansei Islands. *Kyushu Plant Prot. Res.* 47: 112–117.
Jatala, P. 1991. Biology and management of plant-parasitic nematodes on sweet potato. In: Jansson, R.K. and Raman, K.V. (eds.), *Sweet Potato Pest Management, A Global Perspective.* Westview Press, Boulder, CO, pp. 359–378.
Jatala, P. and Russell, C. C. 1972. Nature of sweet potato resistance to *Meloidogyne incognita* and the effects of temperature on parasitism. *J. Nematol.* 4:1–7.
Jianhua, S., Deliang, P., Keli, Y., Ping, B., and Yongkang, P. 1998. SEM study on tissue pathology of stem nematode disease of sweet potato. *Acta Agric. Boreali-Sinica*13:101–105.
Johnson, A.W. 1998. Degradation of fenamiphos in agricultural production soil. *J. Nematol.* 30:40–44.
Johnson, W.A. and Cairns, E.J. 1971. Effects of different nematicides on yield and quality of Centennial sweetpotato and root-knot nematode damage. *J. Am. Soc. Hort. Sci.* 96:468–471.

Johnson, A.W., Dowler, C.C, Glaze, N.C., Chalfant, R.B., and Golden, A.M. 1992. Nematode numbers and crop yield in a fenamiphos-treated sweet corn-sweet potato- vetch cropping system. *J. Nematol.* 24:533–539.

Johnson, A.W., Dowler, C.C., Glaze, N.C., and Handoo, Z.A. 1996. Role of nematodes, nematicides, and crop rotation on the productivity and quality of potato, sweet potato, peanut, and grain sorghum. *J. Nematol.* 28:389–399.

Kassab, A.S. and Taha, A.H.Y. 1990. Aspects of the host-parasitic relationships of nematodes and sweet potato 1. Population dynamics and interaction of *Criconemella* spp. *Rotylenhulus reniformis*, *Tylenchorhynchus* spp. and endomycorrhiza. *Ann. Agric. Sci.*35:497–508.

Katayama, K., Kumagai, T., Kai, Y., Ishiguro, K., Yoshinaga, M., and Nakazawa, Y. 2004. Daichinoyume: new sweetpotato cultivar for starch production. *Sweetpotato Res. Front* 18:4.

Kistner, M.H., Daiber, K.C., and Bester, C. 1993. The effect of root-knot nematodes (*Meloidogyne* spp.) and dry land conditions on the production of sweet potato. *J. Southern African Soc. Hort. Sci.* 3:108–110.

Koenning, S.R., C. Overstreet, J.W. Noling, P.A. Donald, J.O. Becker and B.A. Fortnum. 1999. Survey of crop losses in response to phytoparasitic nematodes in the United States for 1994. *J. Nematol.* 31: 587–618.

Koenning, S.R., Kirkpatrick, T.L., Starr, J.L., Wrather, J.A., Walker, N.R., and Mueller, J.D. 2004. Plant-parasitic nematodes attacking cotton in the United States: Old and emerging production challenges. *Plant Dis.* 88:100–113.

Komiyama, A., Sano, Z., Murata, T., Matsuda, Y., Yoshida, M., Saito, A., and Okada, Y. 2006. Resistance to two races of *Meloidogyne incognita* and resistance mechanism in diploid *Ipomoea trifida. Breeding Sci. 56:81–83.*

Koshy, P.K. and Jasy, T. 1991a. Effect of the burrowing nematode, *Radopholus similis* on sweet potato. *J. Root Crops* 17:201–206.

Koshy, P.K. and Jasy, T. 1991b. Host preference of the burrowing nematode, Radopholus similis populations from India. *Indian J. Nematol.* 21:39–51.

Krusberg, L.R. and Nielsen, L.W. 1957. Pathogenesis of root-knot nematodes to the Porto Rico variety of sweetpotato. *Phytopathology* 48: 30–39.

Kukimura, H., Kimaki, K., Yoshinaga, M., Hidaka, M., Sakamoto, S., Yoshida, T., Tabuchi, S., and Ide, Y. 1992. The new sweet potato cultivar "Beniotome". *Bull.Kyushu Nat. Ag. Exp. Station* 27:249–267.

Kushman, L.J. and Machmer, J.H. 1947. The relative susceptibility of 41 sweet potato varieties, introductions and seedlings to the root-knot nematode, *Heterodera marioni* (Cornu) Goodey. *Proc. Helminthol. Soc.* 14:20–23.

LaBonte, D.R., Wilson, P.W., Villordon, A.Q., and Clark, C.A. 2008. 'Evangeline' sweetpotato. *HortScience* 43:258–259.

Lawrence, G.W. and Clark, C.A. 1986. Identification, race determination, and pathogenicity of root knot nematodes to resistant and susceptible sweet potato. *J. Nematol.* 18:617.

Lawrence, G.W., Clark, C.A. and Wright, V.L. 1986. Influence of *Meloidogyne incognita* on resistant and susceptible sweet potato cultivars. *J. Nematol.* 18:59–65.

Lewis and Smith, F.H. 1976. Host plants, distribution and ecological associations of *Hoplolaimus columbus*. *J. Nematol.* 8:264–270.

Lin, M.S., Fang, Z.D., and Xie, Y.P. 1993. Responses of sweet potato to exudate of potato rot nematodes (*Ditylenchus destructor*). *Acta Phytopathol. Sinica* 23:157–162.

Lin, M., Wen, L., and Fang, Z. 1999. Potato rot nematodes (*Ditylenchus destructor*) and stem nematode disease of sweet potato. *Jiangsu J. Agric. Sci.* 15:186–190.

Maluf, W.R., Azevedo, S.M., and Campos, V.P. 1996. Heritability of root knot nematode (Meloidogye spp.) resistance in sweet potato. *J. Gen. Breeding* 50:161–165.

Maosong, L., Limin, H., Ling, W. and Zhongda, F. 1996. Mechanism of morphological structure of sweet potato resistance to potato rot nematode (*Ditylenchus destructor*). *Scientia Ag. Sinica* 29:8–12.

Martin, W.J. 1960. The reniform nematode may be a serious pest of the sweetpotato. *Plant Dis. Rep.* 44:216.

Martin, W.J. 1962. Elimination of root-knot nematodes from infested sweetpotato roots and plants. *Plant Dis. Rep.* 46:21–23.
Martin, W.J. and Birchfield, W. 1973. Further observations of variability in *Meloidogyne incognita* on sweetpotatoes. *Plant Dis. Rep.* 57:199.
Massese, C.S.L. 1969. The principal plant nematodes of crops in the French West Indies. In: Peachy, J.E, (ed.), *Nematodes of Tropical Crops*, Commonwealth Bureau of Helminthology, St. Albans, Herts, England. pp 164–183.
Mcharo, M. LaBonte, D.R., Clark, C. Hoy, M. and Oard, J.H. 2005. Molecular marker variability for southern root-knot nematode resistance in sweetpotato. *Euphytica* 144:125–132.
McGawley, E.C. and Overstreet, C. 1995. Reproduction and pathological variation in populations of *Rotylenchulus reniformis*. *J. Nematol.* 27:508.
Milne, D.L. and Keetch, D.P. 1976. Some observations on the host plant relationships of *Radopholus similis* in Natal. *Nematropica* 6:13–17.
Mochida, H., Tateishi, Y. and Suzuki, T. 2007. Effect of nematode-resistant sweetpotato on root-knot nematode density and succeeding crop yield. *Sweetpotato Res. Front* 21:7.
McLean, K.S. and Lawrence, G.W. 1996. Chemical management of the reniform nematode on sweet potato, 1993. *F. & N. Tests* 51:190–191.
Mohandas, C., and Ramakrishnan, S. 1996. Sweet potato variety-Sree Bhadra-Atrap crop of root-knot nematode. *J. Root Crops* 22:135–136.
Montes, L., Crozzoli, R., and Vargas, G. 1998. Respuesta de selecciones de batata al nematodo agallador *Meloidogyne incognita* en Venezuela. *Nematropica* 27:113–117.
Morrison, L.S. 1970. Nematode diseases. In: Thirty years of cooperative Sweet Potato Research 1939–1969. Southern Cooperative Series Bulletin No. 159, Louisiana State University, Baton Rouge, LA. pp. 56–59.
Musarrat, A.R., Firoza, K., and Shahina, F. 2006. Study of root-knot nematodes (*Meloidogyne* spp.) in NWFP and Sindh, Pakistan. *Pak. J. Nematol.* 24:1–7.
Nielsen, L.W. and Sasser, J.N. 1959. Control of root-knot nematodes affecting Porto Rico sweetpotatoes. *Phytopathology* 49:135–140.
Nishizawa, T. 1974. A new pathotype of *Meloidogyne incognita* breaking resistance of sweet potato and trials to differentiate pathotypes. *Jpn. J. Nematol.* 4:37–42.
O'Bannon, J.H. 1976. Worldwide dissemination of Radopholus similes and its importance in crop production. *J. Nematol.* 9:16–25.
Okamoto, K. 1979. Pathogenicity and larval dimensions of resistance-breaking races of root-knot nematode, *Meloidogyne incognita*. *Jpn. J. Nematol.* 9:16–19.
Overstreet, C. and E.C. McGawley. 2000. Geographical distribution of reniform nematode in Louisiana. Proceeding Beltwide Cotton Conference, San Antonio, Texas, January 2000, pp.168–171.
Peixoto, J.R., Ferraz, F.M., Santos, L.C., Angelis, B. D., and Juliatti, F.C. 1998. Seleção de genótipos de batata-doce resitentes ao nematóide das galhas (*Meloidogyne* spp.). *Fitopatologia Brasileira* 23:51–53.
Ramakrishnam, S., Mohandas, C., and Amma, C.S.E. 1997. Reaction of sweet potato germplans against root-knot nematode, *Meloidogyne incognita*. *Indian J. Nematol.* 27:244–245.
Roberts, P.A. and Scheuerman, R.W. 1984. Field evaluations of sweet potato clones for reaction to root-knot and stubby root nematodes in California. *HortScience* 19:270–273.
Robinson, A.F., Inserra, R.N., Caswell-Chen, E.P., Vovlas, N., and Troccoli, A. 1997. *Rotylenchulus* species: Identification, distribution, host ranges, and crop plant resistance. *Nematropica* 27:127–180.
Rodríguez-Kábana, R., Turner, J. L., and Ingram, E.G. 1978. Tratamiento de raices de batata con el nematicida sistemico oxamyl para el control de nematodos fitoparasitos. *Nematrópica* 8:26–31.
Rodríguez-Kábana, R., Boubé, D., and Young, R.W. 1990.Chitinous materials from blue crab for control of root-knot nematode II. Effect of soybean meal. *Nematropica* 20:153–168.
Sano, Z. and Iwahori, H. 2005. Regional variation in pathogenicity of Meloidogyne incognita populations on sweetpotato in Kyushu Okinawa, Japan. *Jpn. J. Nematol.* 35:1–12.

Sasser, J.N. and Kirby, M.F. 1979. Crop Cultivars Resistant to Root-knot Nematodes, Meloidogyne Species. International Meloidogyne Project, Contract No. AID/ta-c-1234, North Carolina State University, Raleigh, North Carolina. 24 pp.

Sasser, J.N. and Carter, C.C. 1982. Root-knot nematodes (Meloidogyne spp.): Identification, morphological and physiological variation, host range, ecology, and control. In: Riggs, R.D. and Editorial Committee, S-76 and S-154 (eds.), Nematology in the Southern Region of the United States. Southern Cooperative Series Bulletin 276, pp 21–32.

Sasser, J.N. and Freckman, D.W. 1987. A world perspective on nematology: the role of society. In: Veech, J.A. and Dickson, D.W. (eds.), *Vistas on Nematology: A Commemoration of the Twenty-fifth Anniversary of the Society of Nematologists*. Society of Nematologist, Inc., Hyattsville, Maryland. pp 7–14.

Scurrah, M.I., Niere, B. and Bridge, J. 2005. Nematode parasites of Solanum and sweet potatoes. In: Luc, M., Sikora, R., and Bridge, J (eds.), *Plant Parasitic Nematodes in Subtropical and Tropical Agriculture, 2nd edition*. CABI Publishing, Wallingford, UK. pp 193–219.

Sharma, S.B., Price, N.S., and Bridge, J. 1997. The past, present and future of plant nematology in International Agricultural Research Centers. *Nematol. Abs.* 66:119–142

Shiga, T. and Takemata, T. 1981. Distribution of sweet potato clones with resistance to root-knot nematode in the Pacific Islands. *Proceedings of the 3rd Research Planning Conference on root-knot nematodes, Meloidogyne spp.*, Jakarata, Indonesia, 1981 pp.64–68.

Silveira, M.A. da and Maluf, W.R. 1993. Resistência de clones de batata-doce a *Meloidogyne* spp. *Hort. Brasileira* 11:131–133.

Stevens, C., Khan, V.A., Tang, A.Y. and Bonsi, C.K. 1990. Long-term effect of soil Solarization on controlling root-knot nematodes in sweetpotato. *HortScience* 25:855.

Struble, F.B., Moirrison, L.S., Cordner, H.B. 1966. Inheritance of resistance to stem rot and to root-knot nematode in sweetpotatoes. *Phytopathology* 56:1217–1219.

Sun, F.R. and Dong, Y.F. 1994 New high-yielding sweet potato variety Taishu 2. *Crop Genet. Res.* 4:57.

Sun, F.R. and Chen, H.C. 1994. New sweet potato variety with high resistance to stem nematode – Lushu 78066. *Crop Genet. Res.* 4:58.

Tarte, R., Pinochet, J., Gabrielli, C., and Ventura, O. 1981. Differences in population increase, host preferences and frequency of morphological variants among isolates of the banana race of *Radopholus similes. Nematropica* 11:43–52.

Tateishi, Y. 1998. *Pasteuria penetrans* for controlling Meloidogyne incognita race 1 on sweet potato cultivars with different susceptibility to the root-knot nematode. *Proceed. Assoc. Plant Protect. Kyushu* 44:72–75.

Tateishi, Y. 2003. Increase of virulent *Meloidogyne incognita* populations on a resistant sweet-potato cultivar, Benoitome. *Sweetpotato Res. Front* 16:4.

Tateishi, Y., Sano, Z., and Iwahori, H. 2007. Emergence of suppression of root knot nematodes in sweet potato cropping systems treated with *Pasteuria penetrans. Jpn. J. Nematol.* 37:1–9.

Taylor, A.L., and Sasser, J.N. 1978. *Biology, Identification and Control of Root-knot Nematodes (Meloidogyne species)*. North Carolina State University, Raleigh, North Carolina. 111 pp.

Tenente, R.C.V., Manso, E.S.B.G.C. and Filho, E.S.F. 1993. Tratamento de germoplasma vegetal visando à erradicacão de fitonematóides. *Nematol. Brasileira* 17:35–40.

Thies, J.A. 2000. Reactions of regional sweetpotato breeding clones and standard check cultivars to Southern root-knot nematode race 3 and Fusarium wilt, 1999. *Biol. Cult. Tests for Control of Plant Dis.* 15:194.

Thomas, R.J., and Clark, C.A. 1983a. Population dynamics of *Meloidogyne incognita* and *Rotylenchulus reniformis* alone and in combination, and their effects on sweet potato. *J. Nematol.* 15:204–211.

Thomas, R.J. and Clark, C.A. 1983b. Effects of concomitant development on reproduction of *Meloidogyne incognita* and *Rotylenchulus reniformis. J. Nematol.* 15:215–221.

Trudgill, D.L. 1991. Resistance to and tolerance of plant parasitic nematodes in plants. *Annu. Rev. Phytopathol.* 29:167–192.

Tyler, J. 1941. Plants reported resistant or tolerant to root knot nematode infestation. U.S. Dept. Agr. Misc. Pub. 406. 91 pp.

Vilsoni, F. and Hienlein. 1982. Influence of initial inoculum levels of the reniform nematode on the growth of mung, pawpaw, pigeon pea, and sweet potato. *Fuji Agric. J.* 44:67–70

Vimala, B. and Rajendran, P.G. 1998. "Sree Rethna" and "Sree Bhadra" – two promising short duration varieties of sweet potato. *J. Root Crops* 24:25–30.

Vovlas, N. and Inserra, R.N. 1982. Biological relationship of *Rotylenchulus borealis* on several plant cultivars. *J. Nematol.* 14:373–377.

Walters, S.A. and K.R. Barker. 1993. Reproductive and damage potentials of two populations of *Rotylenchulus reniformis* on sweetpotato and related comparisons with *Meloidogyne javanica* on tomato. *J. Nematol.* 25:830–835.

Wanderly, M.J.A. and Santos, J.M. 2004. Resistência de cultivares de batata-doce a *Meloidogyne incognita*. *Fitopat. Brasileira* 29:437–440.

Wang, Q.M., Chang, C.X., Wang, J.J., and Xi, G.H. 1995. Identification of resistance to stem nematode in sweet potato. *Crop Gen. Res.* 2:36–37.

Weimer, J.L. and Harter, L.L. 1925. Varietal resistance of sweet potatoes to nematodes, *Heterodera radicicola* (Greaf) Müller, in California. *Phytopathology* 15:423–426.

Xie, Y.Z, Qiu, R.L, Zhang, L.Y., Dai, Q.W., Xu, P.L., and Lin, C. 1992. Utilization of foreign sweet potato germplasm resources. *Crop Genet. Res.* 4:42–43.

Yik, C.P. and Birchfield, W. 1982. Reactions of sweet potato root tissue to the reniform nematode. *Plant Dis.* 66:707–709.

Yoshida, T. 1985. Correlation between successive yield tests for agronomic characters in sweet potato. *Jpn. J. Breeding* 35:204–208.

Zhang, F. and Schmitt, D.P. 1994. Host status of 32 plant species to *Meloidogyne* konaensis. *J. Nematol.* 26:744–748.

Zhang, S. and Zhang, S. 2007. Complex infection caused by *Ditylenchus destructor* and *Fusarium* spp. to sweet potato. *J. Fujian Agric. Foresty. Univ.* 36:361–364.

Zhang, S.B., Zhao, B.L., and Mi, L.Y. 1990. The chemical control of stem nematode on sweet potato. *Shangdong Agric. Sci.* 22:45–46.

Zhang, Y., Guo, X., Wu, J., Tang, J., Li, Y. 1998. Screening and evaluation of germplasm against *Ditylenchus dipsaci* in sweet potato. *Crop Genet. Res.* 2:37–39.

Zhang, S., Zhang, S., Wang, H., and Chen, Y. 2006. Characteristics of sweet potato stem nematode in China. *Acta Phytopathol. Sin.* 36:22–27.

Zhou, Z., Wang, X., Ma, D., Li, H., Xie, Y., Li, X. 2005a. Genetic diversity analysis based on RAPD for resistance and susceptibility of sweetpotato varieties to stem-nematode. *Jiangsu J. Agric. Sci.* 21:35–39.

Zhou, Z., Wang, X., Ma, D., Li, H., Xie, Y., Li, X. 2005b. Identification of RAPD markers linked to stem-nematode resistant gene in sweet potato. *J. Agric. Biotech.* 13:549–552.

Chapter 10
Sweetpotato Insects: Identification, Biology and Management

K.A. Sorensen

Introduction

Sweetpotato, *Ipomoea batatas* originated in northwestern South America around 8000–6000 Bc (Austin, 1988). Sweetpotato ranks seventh among all food crops worldwide, with an annual production of 115 million MT (FAO, 1980). Approximately 92% of the world's sweetpotato is produced in Asia and the Pacific islands; 89% is grown in China. The United States produces 1 million MT valued at $200 million. Sweetpotato is used as a staple food, vegetable (fleshy roots, tender leaves and petioles), snack food, animal feed, for industrial starch extraction and fermentation, and for processed products.

The major insect pests of sweetpotato in Java, Africa, and the Caribbean have been described. Over 100 arthropod pest species are listed (West, 1977) and most are leaf feeders, followed by stem, vine, root, and flower feeders. All but three species were insects. The biology and management of the major insect pests worldwide have been reviewed. Most recently, 270 insect and 17 mite species were listed as pests of sweetpotato in the field and in storage around the world, with weevils *Cylas formicarius formicarius, C.f. elegantulus*, and *C. puncticollis* being most damaging. Since most sweetpotatoes are produced in low-input agricultural systems, insect losses reach 60–100%.

The report entitled Ecology and Management of Sweetpotato Insects (Chalfant et al., 1990) provides an excellent review of the status of research on insect pests of sweetpotato, including the major insect pests in the Western Hemisphere, their biology and distribution, damage, sampling approaches, management programs, and includes the future of integrated pest management programs in developed and developing countries. It also contains a holistic list of references and is an invaluable resource used extensively in this chapter.

K.A. Sorensen (✉)
Department of Entomology, North Carolina State University, Raleigh, North Carolina 27695 USA
e-mail: kenneth_sorensen@ncsu.edu

G. Loebenstein, G. Thottappilly (eds.), *The Sweetpotato*,
DOI 10.1007/978-1-4020-9475-0_10, © Springer Science+Business Media B.V. 2009

Sweetpotato Weevils

Sweetpotato weevils (*C. formicarius* complex) are the most important insect pests attacking cultivated and stored sweetpotatoes worldwide (Southerland, 1986; Mullen and Sorensen, 1984). The taxonomy of the *C. formicarius* complex is composed of heterogeneous populations with three names, *C. f. formicarius, C. f. elegantulus*, and *C. f. turcipennis. C. formicarius* is considered the New World species and *C. f. formicarius* the Old World species while other important species *include* the *C. puncticollis* and *C. brunneus* species complexes, which occur in Africa and/or Madagascar. *C. f. elegantulus* has been revised and reverted to the following: *C. formicarius* (Fabricius) 1798-174. COLEOPTERA: Brentidae Cyladinae Charles O'Brien personal communication. *C. f. elegantulus* is currently found throughout the southern United States from southern Texas to coastal regions of North Carolina. In North Carolina the sweetpotato weevil is restricted to a wild *Ipomoea* host on the coastal outer banks (Sorensen, 1987).

CYLAS SPECIES All information on the biology of *Cylas* weevils comes from *C. formicarius* and is only included partially in this chapter (U.S.D.A., 1960). Weevil damage increases with the length of the growing season. Terpenoids and phenols are produced in response to weevil feeding, where they make even slightly damaged roots unpalatable. Hence, even low-level infestations can result in considerable economic loss and in the southern United States crop loss attributable to weevils exceeds $7 million annually.

Some 30 wild hosts of *C. formicarius* weevils are known and most are species of *Ipomoea* (Hill, 1983). Weevils have also been found on species of *Jacquemontia, Thunbergia, Cuscuta, Merremia, Dichondra, Calystegia*, and *Stictocardia*. Removal of weed hosts is standard practice for weevil control. Weevils are more abundant and injurious during the dry growing season, because of higher temperatures and soil cracks that expose fleshy roots. Weevil abundance patterns vary among cultivars and cultivars differ in their susceptibility to weevil damage.

WEST INDIAN WEEVIL OR SCARABEE Few comprehensive studies have been conducted on the biology of *E. postfasciatus*. Females lay grayish-yellow to yellow eggs singly, in a cavity excavated in either roots or stems, and seal the cavity with a fecal plug. Larval damage is similar to that of *Cylas* spp. and causes roots to produce terpenoids, making them unpalatable. Adults are smaller than *C. formicarius* adults and are reddish-brown to gray-black with short, stiff, erect bristles and scales covering the body. Mated females may lay an average of 106 eggs per month for 4–6 months and prefer to oviposit on roots within 2 cm of the soil surface. Adults may live for up to 6 months in the laboratory and do not fly. This weevil also utilizes wild *Ipomoea* hosts. Like *C. formicarius*, denser populations have been reported during the dry season than during the wet season.

Sex Pheromone

A synthetic sex pheromone that attracts only males has potential as a monitoring tool (Sorensen, 1988). Several workers studied the sex pheromone of *C. formicarius*;

which has been isolated, identified, synthesized, and bioassayed. The active component of the pheromone is (Z)-3-dodecen-1-ol (E)-2-butenoate and studies on trap type and height, time of day, pheromone dosage, lure age, and season have been conducted. Work continues to refine a pheromone-trap monitoring system for sweetpotato weevil in the southern United States, the Caribbean Basin and throughout the world.

Management

The following management approaches apply to both *Cylas* and *Euscepes* weevils. Strict quarantines have been placed at ports of entry and within 14 states in the United States on the shipment of sweetpotato vines and fleshy roots. Weevils can be controlled in storage with fumigants. However, because of problems associated with pesticide residues in food, environmental contamination, and phytotoxicity, alternative approaches to weevil management in storage such as the use of gamma irradiation have been explored. Recent reports show that chemical control in the field can reduce *C. formicarius* populations, although chemical insecticides may not adequately control weevils due to the cryptic nature of immatures developing within vines and roots. Chemical control also is limited by the availability of labelled products, and in developing countries, chemical control is too expensive and impractical. Hence alternative approaches to weevil management are needed.

Synthetic juvenile hormone and an insect growth regulator have produced adverse effects on various developmental stages. Certain organic materials (*Callophyllum* cake, Mahwah cake, and lemon-grass leaves) have been effective in reducing weevil infestations, whereas Neem cake and cashew shell powder have not.

The sex pheromone has potential for suppressing *C. formicarius* populations by mass trapping of males and perhaps by disrupting mating. It is also used extensively throughout the southern United States in monitoring and trapping every sweetpotato plant bed, greenhouse, field, storage facility and processing plant for first detection of adult sweetpotato weevils. Unfortunately, the pheromone attracts only males and work continues to identify plant-produced phytochemicals that attract females. Recent studies show that *C. formicarius* is attracted to *Ipomoea* plant extracts and that oviposition stimulants have been isolated. Thus, a female attractant could be isolated and integrated into a trapping system with the sex pheromone.

Parasitoids of *C. formicarius* include *Microbracon cylasovorus, Bassus cylasovorus*, *Microbracon* sp. (probably *M. punctatus), Metapelma spectabilis*, *Bracon* sp., and *Rhaconotus* sp. Other parasitoids have been found attacking *C. puncticollis* One predatory maggot, *Drapetis* sp., has also been found. An ant, *Pheidole megacephala* (F.), was more effective than chemical insecticides for managing weevil populations in Cuba and was compatible with *Bacillus thuringiensis* and *Beauveria bassiana*. Another parasitoid, *Eupelmus cushmani*, of *E. postfasciatus* has been reported in Hawaii while three parasitoids, *Heteroschema* sp, *Cerocephala* sp, and one unidentified species, have been found in Peru.

Various entomopathogenic fungi attack sweetpotato weevils and include *Beauveria bassiana*, *Isaria* sp, and *Metarhizium anisoplieae*. Both *B. bassiana* and *M. anisoplieae* show potential for *C. formicarius* management.

Entomopathogenic nematodes that attack *C. formicarius* include *Steinernema* sp. *(Neoaplectana)* and *Heterorhabditis heliothidis*. Unidentified *Heterorhabditid* sp. and other rhabditid nematodes have been isolated from weevils in Southern Florida. A HP88 strain is more effective than Steinennematid nematode ("All" strain) and chemical insecticides in suppressing weevil populations. Studies continue with these entomopath-ogenic nematodes as biological control agents of sweetpotato weevil.

Historically, weevil management has relied on cultural controls. These include the use of noninfested planting material, crop rotation, removal of volunteer plants and crop debris, prompt harvesting, removal of alternate wild hosts, planting away from weevil. infested fields, banking (filling in of soil cracks) or planting in light soils that do not crack, and sufficient irrigation to reduce soil cracking. Hilling of plants reduces weevil damage. Irrigation at 10-day intervals in combination with high levels of manuring also reduces weevil infestations. Intercropping also helps reduce weevil damage. Flooding weevil-infested fields for two or more weeks after harvest may reduce volunteer plant and weevil populations. Considerable research has also focused on breeding and evaluating sweetpotato germplasm for resistance to *Cylas* and *Euscepes* weevils (Hahn and Leuschner, 1982). Resistance mechanisms include antibiosis, antixenosis, and escape. Plant traits important in weevil resistance include fleshy root density, dry matter and starch content, root depth, vine thickness, and fleshy root chemistry.

Sweetpotato Vine Borer

Megastes grandalis is a serious pest in Trinidad and occurs in Brazil, Guyana, Venezuela, and Tobago. Although it feeds on other convolvulaceous plants in the laboratory, no wild hosts have been reported. The life history has been described. Eggs are laid singly or in clusters on the underside of leaf bases. Females may lay as many as 180 eggs which hatch in ca. one week. Larvae feed externally before tunnelling into the stem and roots. There are seven to eight larval instars, which may last as long as 5–7 weeks (West, 1977). Full grown larvae spin cocoons and pupate within the stem above the ground. Yield loss may be minimal if larvae remain in vines or attack plants late in the season.

Various insecticides and parasitoids have been used to control the vine borer. Cultivar selection helps reduce stem borer problems since certain cultivars seem to be preferred by vine borers.

Wireworms

In the southeastern United States several wireworm species cause serious economic losses. Adult click beetles do not feed on the crop but oviposit on the soil near the crop, weeds, or other vegetation. Larvae produce small, round shallow feeding holes on the root surface from root enlargement until harvest. Few crops are produced entirely free from damage (Cuthbert, 1967).

Manuals and keys to identification are available (Sorensen and Baker, 1994). Wireworm species complexes vary with location. The southern potato wireworm, *Conoderus falli* was reported to be the major species infesting sweetpotatoes in the southeastern United States. The tobacco wireworm, *C. vespertinus* and the gulf coast wireworm, *C. amplicollis*, are important in some areas. More recently *C. scissus* and *C. rudis* were found to be the predominant species in Georgia. *Melanotus communis*, the corn wireworm, is another destructive wireworm along the east coast of the United States.

Biology, Distribution, and Sampling

C. scissus has a life cycle of 2–3 years, with seven to ten larval instars. Larval movement is very sensitive to variations in soil moisture. The beetle also infests peanut, cowpea, and corn. *C. rudis* usually infests sweetpotatoes as an immigrant from nearby fields. The life cycle is short, with three to four stadia in 3–4 months. They are found near the soil surface and are prevalent in weedy fields. The gulf coast wireworm, *C. amplicollis*, is one of the most abundant species in Georgia, especially in corn fields. The life cycle lasts for 2–3 years. Larvae remain at a depth below 10 cm inside smooth, walled cavities. They move to the surface only for feeding and are less sensitive than *C. scissus* to moisture. The biology of *C. falli* has been studied extensively. There are multiple generations. *M. communis*, is a very large wireworm that attacks corn, potatoes and many other hosts where its damage is large cavities. It has a two or more year development period in the larval stage. Other minor species of wireworms found in Georgia sweetpotatoes include *C. lividus*. *C. vespertinus*, and *Glyphonyx* spp.

Wireworms change their vertical distribution seasonally. Highest populations occur during July in the upper 10–20 cm soil profile within 10 cm lateral to the plant, followed by a downward migration thereafter to a depth of 20–30 cm in November. Wireworm damage holes at harvest showed a clumped distribution pattern.

Several baiting techniques have been evaluated in sweetpotato fields for attracting wireworms. Bait performance varies with wireworm species, condition of the baits, and time of placement in the fields. A corn-wheat seed mixture attracts more wireworms then other baits.

White-Fringed Beetles

The white-fringed beetles *(Graphognathus* sp.) feed on numerous plant species. Adults are not voracious feeders, but they have been reported on 170 different host plants, while larvae are more destructive, with a host range of 385 species including peanuts, corn, sugarcane, cotton, cowpea, beans, cole crops, chufa, lucerne, Mexican clover, and sweetpotato. Larval injury to roots resembles damage caused by wireworms, *Diabrotica*, flea beetles and white grubs. However, the predominant injury is usually grub-like (Averre et al., 1991). Larvae gouge shallow channels, about 2–7 mm wide by several cm in length on the enlarging roots and can destroy the root system or reduce marketability.

White-fringed beetle is native to South America (Argentina, Chile, and Uruguay) and is widely distributed over the southern part of the United States (Roberts, 1952). They were first found in Florida in 1936 and now are scattered over 500,000 acres of Arkansas, Louisiana, Alabama, Georgia, Florida, and North and South Carolina. Adults are incapable of flight. They overwinter mainly as large larvae in the soil at a depth of about 23–30 cm but can also overwinter in the egg stage. Pupae are white and are found in the soil at a depth of 8–16 cm, from late May to the end of July. Adult emergence begins in May and continues until the middle of August. Newly emerged adults feed on the underside of leaves or at the base of the stems near the soil surface. Adults prefer to feed on the older plant foliage, but consume little. Females are parthenogenetic and initiate oviposition on the lower plant parts 10–12 days after emergence from the pupal stage. Females can live from 2 to 5 months. Eggs are deposited near the soil surface on objects such as plant stems and debris lying on or protruding from the ground, but they can also be deposited in the soil at depths of 3–8 mm. The whitish oval eggs are deposited in gelatinous masses, as many as 60 per cluster. A female may lay up to 3000 eggs. Eggs hatch in 2 weeks to up to 3 months, depending on season and the availability of water. Larvae are subterranean and feed on the root system of plants. They are yellow-white, C-shaped, legless, and sparsely covered with hairs.

Since the insect does not fly, its only means of dissemination is by walking and through commercial channels. Therefore, federal and state quarantine programs have been the initial method of suppression. To restrain movement special trenches, 30 cm wide by 30 cm deep, are constructed. The migrating insect is trapped in the trench and then destroyed by crushing or by applying kerosene. The development of the chlorinated hydrocarbons insecticides once provided excellent control. Currently, no effective insecticides for white-fringed beetle larvae are legally available in the United States. Long-term crop rotation can be used effectively to reduce infestation levels. The planting of oats or other small grains on heavily infested portions of a farm will reduce the population, as these crops are poor food for adults, and beetles feeding on them produce fewer eggs. In sweetpotato, alternative methods include avoidance of fields with white-fringed beetles, the use of nematode parasites and resistant cultivars.

White Grubs

Two species of white grubs, *Plectris aliena* and *Phyllophaga ephilida*. are serious pests of sweetpotatoes. Other white grubs in North Carolina include the Japanese beetle, *Popillia japonica* and the false Japanese beetle (spring rose beetle),

Strigoderma arboricola (Sorensen and Baker, 1994). Preferred hosts of grubs are mainly pasture grasses. Adults of *P. ephilida* are nocturnal and voracious feeders on deciduous trees. Adults of *P. aliena* live mainly in the soil. *P. aliena* is found in Argentina, Brazil, and Uruguay and was introduced into the United States and Australia about 1900. *P. ephilida* is a native of the United States. Adult Japanese beetles feed on many trees, vegetables and fruit, while spring rose beetle adults feed on many ornamental and vegetable flowers. Larvae of these white grub species feed mainly on pasture sod. However, severe damage by larval *P. aliena* to sweetpotato may result when it is planted in fields previously grown to pasture. *P. ephilida* is the only *Phyllophaga* species that attacks sweetpotato. Larvae of grub species gouge out broad shallow areas on the root, thereby reducing the marketability of the crop.

Adults of *P. aliena* are light tan with typical scarab-like characters. Adults emerge in late spring; do not feed and have poorly developed mouth parts. A mated female may lay 33 eggs and can survive 21 days without food. Eggs are laid in the soil during the spring and summer and hatch into C-shaped larvae. Larvae have three instars and vary in color from white to cream with light-tan heads and grayish areas on the tip of the abdomen. They may live 1–2 years, depending on the environment. This species appears to be restricted to light sandy soils, which may limit distribution.

P. ephilida adults emerge in late May, and peak flight occurs about the end of June. Adults are nocturnal and feed on the foliage of pecan and willow oak. The adult oviposition period probably lasts throughout the growing season; however, there is no information on mating. Adults oviposit on bare soil or on soil planted to crops. There are no known natural enemies of the immature stages of this pest. The larval stage is similar in appearance to *P. aliena*. The insect overwinters as a larva in the soil. There is only one generation per year. Crop rotation will not ensure protection.

Adults of spring rose beetles are active in May and June, while Japanese beetles are active from June through July. Both species over winter as grubs in the soil. There is one generation each year. These adults are attracted to yellow or white water buckets and the Japanese beetle has a feeding and sex attractant that is effective in mass trapping adults. *Bacillus popillae*, a bacterial disease known as milky spore disease is specific to Japanese beetle grubs and has been effective if a viable active strain can be maintained in the market.

Maladera matrida common name khumeini beetle is a beetle species in the family Scarabaeidae in the Middle East. Adults are active in the summer and spring, flying mostly in evening and are attracted to light. Adults range in length from 7 to 9 mm and they are brownish-red. Hence the name khumeini is probably due to the brown color. Adults feed on leaves, buds and flowers of roses, sweetpotato and citrus trees. The female lays between 60 to 100 eggs in groups. White grubs feed on the roots of sweetpotato and peanuts. The species name, matrida, in Hebrew means annoying. There are 2 generations a year with overwintering grubs emerging in April. Beetle movement has been correlated with wheat harvest. Hence placement of food baited traps around peanut and sweetpotato plantings before wheat harvest is useful in monitoring population size, time of movement, and when to apply insecticide treatments to field margins.

Cucumber Beetles (Diabrotica)

Cucumber beetles, *Diabrotica* spp., are also serious pests of sweetpotato. The principal species are *D. balteata*, the banded cucumber beetle, and *D. undecimpunctata howardi*, the spotted cucumber beetle. The distribution of *D. balteata* is from Columbia, South America to the southern United States. *D. undecimpunctata* is found from Central Mexico to east of the Rocky Mountains in the United States. Larvae eat small holes through the root periderm of sweetpotatoes and form irregular cavities under the skin. Feeding holes are in groups and enlarge as roots develop. The roots are often attacked during early development, which results in many unsightly healed holes at harvest. Adult feeding on sweetpotato foliage produces irregular holes. The elytra of *D. balteata* are marked with alternating green and yellow bands, while those of *D. undecimpunctata* have II spots on a yellowish-green background. Adults, which feed on the plant family Con-volvulaceae, lay eggs in the soil where they hatch in 1–2 weeks, depending on temperature. Larvae of these species are nearly indistinguishable. The larval stage may last 8–30 days, depending on the food supply. Pupae are found in cells just below the soil surface and emerge as adults in one week. In warm climates of the United States the insect can overwinter as adults.

The presence of a *D. balteata* sex pheromone has been identified and synthesized. Field tests show it to be a useful tool in integrated pest management (IPM). Nematodes, *Filipjevimermis leipsandra* and *Heterorhabditis* sp. also show potential for control of *D. balteata*. Foliar insecticides have given excellent control when applied on a regular basis.

Flea Beetles and the Sweetpotato Leaf Beetles

The elongate flea beetle, *Systena elongata* the pale-striped flea beetle, S. *blanda*, and *S. frontalis* feed on sweetpotato. The habits and life histories of the three species are similar, and in the immature stages they look alike. All have a wide host range, including many weeds. Larvae eat small holes through the periderm of the sweetpotato and make enlarged cavities and short tunnels just under the periderm surface. Although most of their damage is very similar to *Diabrotica*, the latter seldom tunnel into the roots. At harvest, early season damage appears as shallow healed scars which are elongate or irregularly shaped at harvest, in contrast to those of the cucumber beetle, which are usually round. Late season damage by these larvae is referred to as pinhole injury. The larvae are about 4 mm long with tiny legs; they are whitish, slender, delicate, and cylindrical, and they have a tubercle on the posterior of the abdomen. Larvae mature in 20–30 days, then curl up in an earthen cell and pupate. Adults of S. *elongata* are black with longitudinal white stripes on the elytra. Feeding is mainly restricted to one surface of the leaf with no penetration through the leaf. Adults of S. *blanda* are similar to S. *elongata* except that the legs, head, and thorax are not as dark. Feeding produces irregular holes through the leaf resembles adult cucumber beetle feeding. S. *frontalis* adults are larger than S. *elongata* and S. *blanda* and are a uniform black color. S. *frontalis* adults eat small holes through the leaf. Adults migrate into sweetpotato fields during the spring and summer and lay cream to yellow eggs in the soil. At least two generations per year occur in the southern United States.

The sweetpotato leaf beetles *(Typophorus nigritus viridicyaneus* and *T. nigritus nitidulus)* feed on sweetpotato, wild *Ipomoea* sp. and *Convolvulus arvensis*. They occur in the southern United States and Brazil. They are widely distributed in the United States but usually cause little crop loss. Adults begin feeding on the margins of the foliage and work inward. Larvae tunnel into the vines and roots. Root damage is easily recognized by large excrement-filled tunnels that penetrate into the roots. Larvae can be found in the tunnels sometimes a month after harvest. Adults are metallic blue-green. Eggs are laid during the spring and summer in clusters on the underside of leaves or just beneath the soil surface under sweetpotato plants. Larvae are pale yellow or yellow with brown heads. The insect overwinters as larvae and pupates as the weather becomes warmer. Adults emerge in the spring.

Management of Soil Insects

Wireworms, *Diabrotica, Systena* Complex, and the sweetpotato flea beetle feed on the surface of the sweetpotato fleshy root, causing holes, scars, surface tunnels, and other blemishes (Averre et al., 1991). Total root biomass is not reduced, but the quality is lowered, thereby reducing marketable yield. Damage tolerances are low, and sweetpotatoes must be free from insect damage to qualify for US No. 1 grades. Accordingly, growers in the United States depend heavily upon insecticides incorporated in the soil to produce damage-free crops. The various insect pest species vary considerably with respect to life cycle, distribution within the field, and response to insecticides. Most crops are planted in fields containing existing wireworm populations. Infestations of wireworms, *Diabrotica spp.*, and other soil insects from midsummer immigrants will cause additional injury. Long residual, broad-spectrum insecticides have given the most dependable control in the past but are no longer labelled for use. Current labelled insecticides have relatively short residuals, and control is erratic, influenced by formulation, method of incorporation, and soil conditions including composition, moisture, temperature, and presence of microorganisms.

Insecticidal activity is improved by incorporation and providing adequate soil moisture. Granular and liquid formulations of chlorpyrifos and fonofos provided adequate mortality of *Conoderus* spp. although granules are con-sistently more effective. Applications can be broadcast over the entire field or banded over the newly planted bed. Banded applications are more con-centrated than broadcast applications and generally provide better control.

In the southeastern United States, chlorpyrifos incorporated at transplanting has provided effective control of soil insects. Since the half-life of chlorpyrifos does not exceed 6 weeks, an insecticidal application of bifenthrin at root enlargement has been effective in the control of late infestations. Mechanical incorporation of the insecticide is difficult at this time due to the presence of the crop canopy, and prompt irrigation is recommended for activation. Some insecticides can also be applied in the irrigation water by chemigation. Pesticide registrations vary and the label should be followed for best results and applicator and food safety.

The species found in sweetpotato fields are dependent to some degree on the previous crop. *C. scissus* is more abundant following peanuts; C. *amplicollis* follows com; and *C. rudis* is attracted to weeds. Irish potatoes are infested more with wireworms following sorghum than following fallow. Thus, selection of fields to avoid adverse rotations and use of weed management could avoid planting sweetpotatoes in infested areas. Deep plowing may also have an adverse effect on wireworms by exposing them to predators and other natural stresses.

An alternative to insecticides is the development of insect resistant sweetpotato plants. Most sweetpotatoes produced in the United States are genetically similar and susceptible to most insect problems; therefore, in 1966 a mass selection breeding program to develop horticulturally acceptable breeding clones and cultivars with resistance to insects and diseases was initiated. The program emphasizes the development of sweetpotatoes with high levels of resistance to insects and diseases.

Sweetpotatoes are different from most other vegetables in that classical pedigree breeding procedures are difficult to follow. The sweetpotato, a hexaploid, has 90 chromosomes with a complex quantitative inheritance. Plants are clonally propagated, and pedigree records are of limited value, since it is not necessary to reproduce a particular genotype. Mass selection procedures with rapid generation turnover and high selection pressure provide a sound basis for crop improvement. A breeding program should involve both long- and short-term goals, with one or more mass selection populations to provide new parental types and to assure the development of a wide gene base. The mass selection program moves from one sexual generation to the next by advancing from true seed to true seed each year. In this process selected plants are not used in the next cycle; only the seed is used. Thus, seeds from selected plants can be bulked to start the next cycle.

The exact number of cultivars needed to start a mass selection program for insect resistance may vary from as little as 6 to as many as 350 but should include as wide a gene base as possible. The protocol used for mass selection for resistance to sweetpotato insects has been previously described and has resulted in the development of 8 breeding clones, and 6 cultivars with multiple resistance to the WDS complex, sweetpotato flea beetle, and grubs, and some resistance to the sweetpotato weevil.

Micropropagation techniques have been developed in sweetpotato propagation to stabilize plant mutations and minimize plant disease and viruses and have applications for insects in the future. Likewise is the use of *Bacillus* sp (BT) and other insect diseases in development and application of genetically modified organisms in sweetpotatoes.

Foliar Pests

Numerous species of insects feed on the above-ground parts of sweetpotatoes. Defoliators in the United States include *Agrius cingulatus*, the sweetpotato hornworm; *Typophorous nigritus viridicyanaeus*, and *T. nitidulus*, sweetpotato leaf beetles, and various sporadic Noctuidae. In areas where sweetpotato leaves are grown for human consumption or animal forage, defoliators impose a serious problem, while in the United States consumption is of storage roots only.

The relationship between insect defoliation and yield loss of sweetpotatoes has not been established. There is considerable variation depending upon cultivar, production practices, and growing conditions. "Jewel," and other major US cultivars, loose ca. 50% of their leaves naturally. Furthermore, energy conversion into the sweetpotato root can be interrupted for short periods of time without loss of yield. Thus, severe defoliation can have a temporary effect on yield for which the plant compensates by a longer period of growth. For growers of fresh market potatoes, delayed production of roots of suitable size can result in loss of profit early in the season when prices are highest. Increasing time of the root in the field causes longer exposure to soil pests. Late in the season, at the onset of cool weather, there will be a cessation of growth and loss of potential yield. However, under most conditions, it is unlikely that insecticidal application is justified.

Other foliage insects reported to be pests in the United States include *Bedellia* spp., a sweetpotato leaf miner which is not serious in the United States but is injurious to newly transplanted American varieties in Australia. A sweetpotato sawfly, *Sterictophora cellularis*, was reported to cause blister-like swellings on the leaves in eastern Virginia. However no recent reports of this insect have been found.

Challenges

Integrated pest management (IPM) programs for sweetpotato insect pests have a promising future. The future of *C. formicarius* management is exciting and points towards a greater emphasis on the use of biological and parabiological management approaches. These approaches are ideally suited for low-input agricultural systems. Efforts should continue with synthetic sex pheromone and entomopathogenic nematodes they should be integrated into weevil management programs worldwide.

In addition to entomopathogenic nematodes, bacterial and fungal pathogens also have potential. The recent development of *Bacillus thuringiensis* strains toxic to Coleoptera and improved strains of entomopathogenic fungi (primarily *B. bassiana)* may ultimately be integrated into weevil management programs. Also, the use of agrobacterium transfer genes for insect control on sweetpotato has been initiated in the USA and South Africa.

Some new pests and potential invasive insects include: sugarcane beetle, *Euetheola rugiceps*, and spider mites in the USA, symphylans in New Zealand, and various insect vector species (aphids, whiteflies, thrips) of plant diseases. Pest surveys, pest alerts, inspections and certifications of host plants/insects movement, enforcement of quarantines, training of regulatory officials, preparation and use of identification leaflets and Lucid diagnostics and continued cooperation of international agencies must be maintained and improved. Increased and rapid movement of plants and animals, global commerce and trade; and increased risks of bioterrorism demand international attention and action in order to minimize risks and keep traditional and invasive pests in abeyance.

With continued improvement in insect management programs of sweetpotato, particularly in developing countries, more information is needed on the biology of other *Cylas* weevils and *E. postfasciatus*. In addition to research, efforts should be directed to educate and train the trainers and growers in IPM. The active involvement of the International Centers in sweetpotato research and education worldwide should facilitate sweetpotato pest management in developing countries.

Demand for a high quality unblemished product in the United States places severe constraints but also increased dependence on an IPM approach to soil insects. Although strategies discussed above can and should be used, it is unlikely that use of pesticides can be eliminated. Existing insecticides are few and may be lost by attrition (insect resistance, microbial degradation, cancellation); and timely replacements are unlikely. Resistant cultivars have difficulty obtaining acceptance and may be site specific. Biological control, though slow and with some risks, should be continually examined.

Three additional challenges ahead will have added value for the sweetpotato. They represent new opportunities in sweetpotato production and IPM. First is the area of transgenics where studies in the United States on sweetpotato DNA finger printing are tracing the lineage and germplasm of cultivated and wild *Ipomoea*. Results will synergize sweetpotato breeding and increase the knowledge base on sweetpotato culture and pest resistance globally. The second is collaborative efforts in East Africa and the United States in engineering resistance to sweetpotato weevil and other pest insects using Bt gene technology. And the third is in the United States and Japan with studies on sweetpotato weevil using DNA finger printing for taxonomic classification and to trace the movement and introduction of this worldwide pest from its origin in the old world west into North America and into other continents and countries.

The future expansion and use of sweetpotato foliage and the storage root in human and animal consumption is exciting and promising. The sweetpotato will increase in importance and have added value as an organic and health food and coupled with its development as a biofuel alternative energy source will have social, economic and environmental impact and improve the quality of life throughout the world.

Key to Sweet Potato Pests (Sorensen and Baker, 1994)

A. Pests that feed on aboveground plant parts

1. Caterpillars with three pairs of legs and five pairs of prolegs

   a. **Southern armyworm** – Gray or nearly black larva up to 36 mm long with greenish or pinkish tint; lightly colored longitudinal stripes and paired triangular spots down back; pale yellow head capsule with bright reddish brown markings (Fig. 208A, B); feeds on leaves, tender stems, and tips of branches; congregates around bases of plants during hot portion of the day (see Fig. 221).

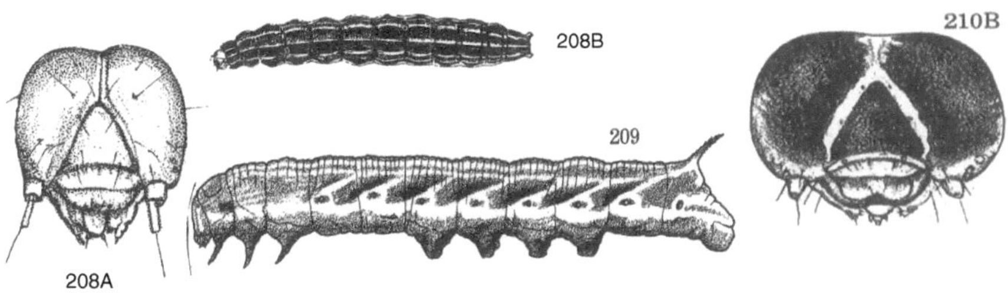

   b. **Sweetpotato hornworm** – First instar: white with black anal horn; later instars: green or brown with black angled marks plant p. 20 under large leaves down each side and black anal horn; body up to 90 mm long; head green or brown with black stripes (Fig. 209); defoliates plants; often hides near base of plant under large leaves (see Fig. 224).
   c. **Yellowstriped armyworm** – Pale gray to black caterpillar up to 45 mm long with yellow-orange stripes along each side and paired triangular spots on the back of most segments; head capsule brown with black markings and a white inverted V (Fig. 210B); feeds much like southern armyworm (see Fig. 227).

2. **Potato leafhopper** – Spindle-shaped pest up to 3 mm long; green body with yellowish to dark green spots (Fig. 211); usually jumps instead of flies; extracts sap from underside of leaf causing yellowing of leaf tips and margins; one of several leafhopper species attacking sweetpotato.

3. **Small fruit or vinegar fly** – Small yellowish fly about 3 mm long with red eyes (Fig. 212); hovers around overripe or decaying produce is often found with small creamy maggots in the cracks of sweetpotatoes (see Fig. 220).

211

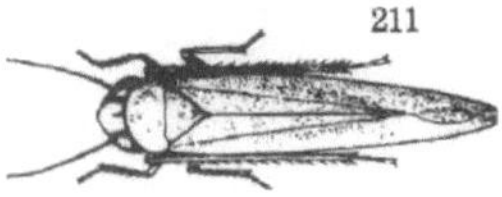

212

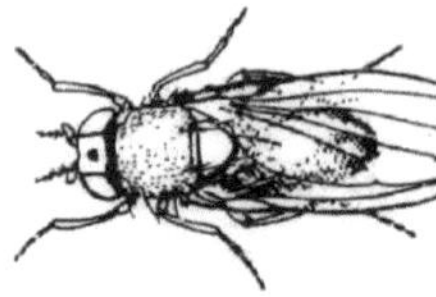

213

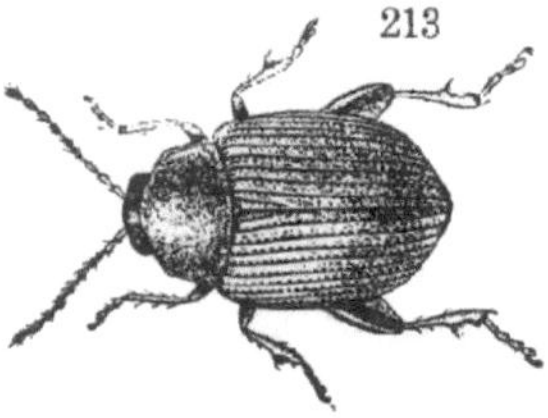

214A

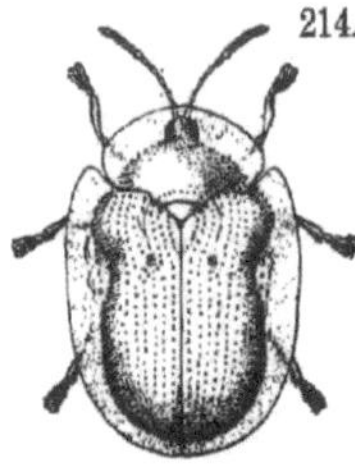

214B

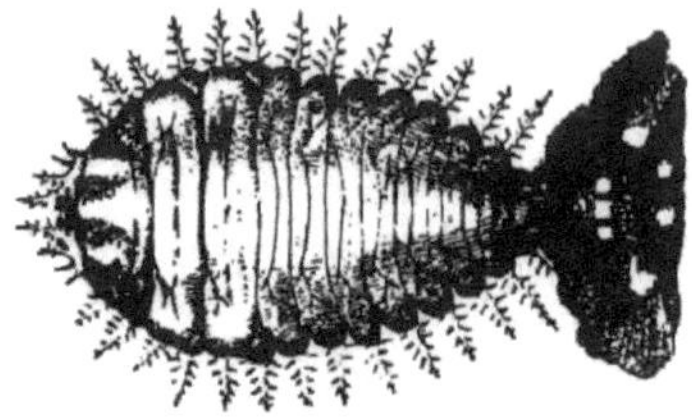

4. **Sweetpotato flea beetle** – Black oval beetle about 1.6 mm long with a bronze tinge, reddish yellow legs, and deeply ridged wing covers (Fig. 213); leaves narrow channels or grooves in upper surface of leaves; injured areas turn brown and die (see Fig. 223).
5. **Sweetpotato weevil adult and larva** – See B.2. for description. Feed on foliage, but primarily on underground plant parts (see Fig. 225).
6. **Tortoise beetle adult and larva** – Oblong-oval, basically gold-colored beetle, up to 8 mm long, with various black or red markings on its flattened, shell-like body (Fig. 214A); larva with dull yellow, brown, or green body up to 12 mm long with black head, legs, spots, and spines; long spines on larval abdomen hold excrement (Fig. 214B); adult and larva chew leaves leaving them riddled with holes (see Fig. 226).
7. **Spider mites** – Tiny pale or reddish spiderlike arthropods feed on the bottom of leaves (Fig. 215); heavily infested plants become yellowish, bronzed or burned in appearance.

215

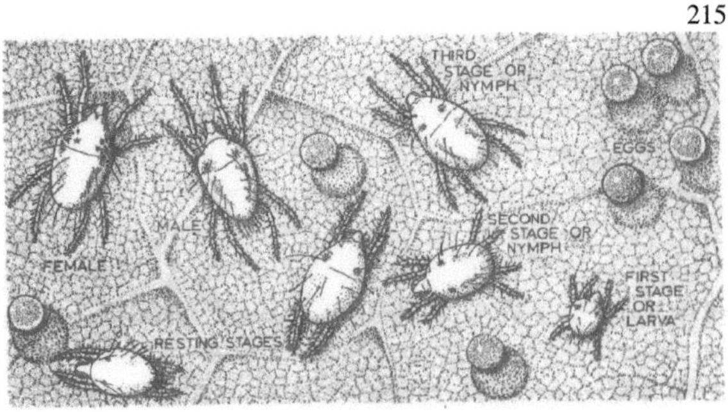

B. Pests that feed on belowground plant parts

1. **Sweetpotato flea beetle larva** – Slender, white, cylindrical larva, up to 5 mm long, with three pairs of legs near head (Fig. 216); etches shallow, winding tunnels on surface of sweetpotato roots and sweetpotatoes; tunnels darken, split, and leave scars (see Fig. 223).
2. **Sweetpotato weevil adult and larva** – Antlike snout beetles are about 6 mm long with dark blue wing covers and red-orange legs and thorax (Fig. 217A); fat, legless, dirty white larvae are about 9 mm long with pale brown head (Fig. 217B); beetle makes small holes over surface of sweetpotatoes particularly at stem end; larva tunnels inside filling tunnels with frass and causing sweetpotatoes to turn bitter (see Fig. 225).

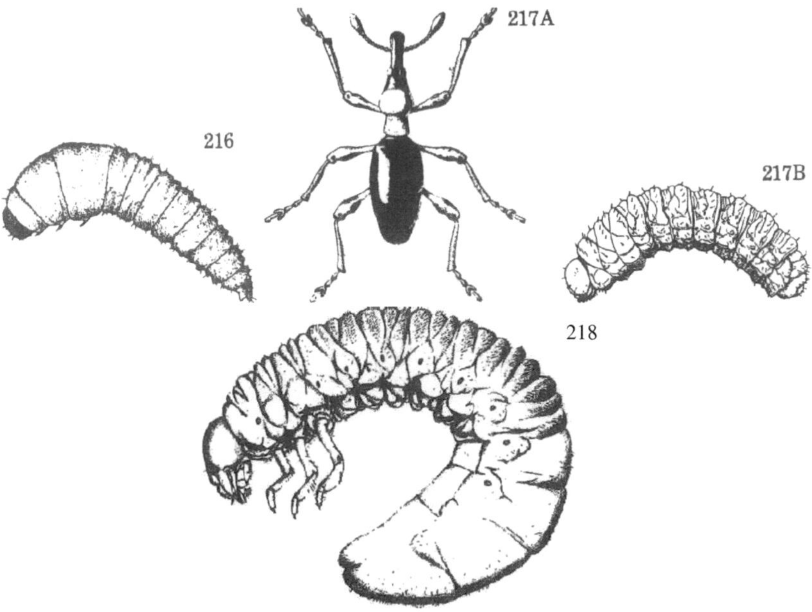

217A

216

217B

218

3. **White grub (spring rose beetle)** – Dirty white grub up to 25 mm long with brown head and three pairs of legs near head (Fig, 218); leaves large, shallow feeding scars on sweetpotatoes (see Fig. 222).
4. **Wireworms** – Several species of slender, Wirelike larvae with three pairs of short legs near the head and a pair of prolegs at the tip of the abdomen; large shallow cavities in sweetpotatoes evidence of early injury; deep ragged holes-later injury.

   a. ***Melanotus communis*** – Yellowish brown with darker head; body up to 25 mm long; last abdominal segment with scalloped edges (Fig. 219A).

b. **Southern potato wireworm** – Cream colored or yellowish gray with reddish orange head; body up to 17 mm long; closed oval notch in last abdominal segment (Fig. 219B).

c. **Tobacco wireworm** – White with brown head; body up to 19 mm long; V-shaped notch in last abdominal segment (Fig. 219C).

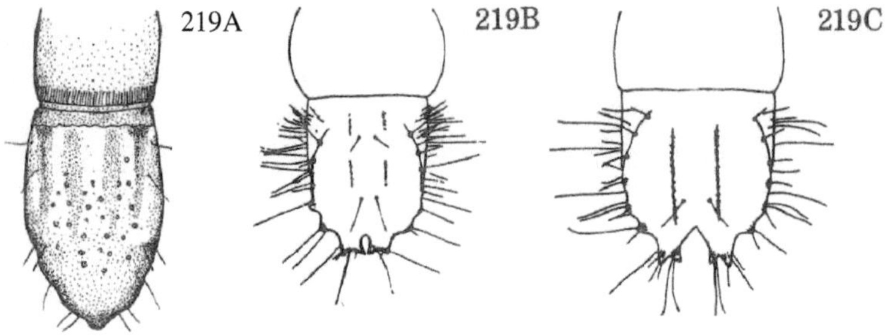

5. **Whitefringed beetle larvae** – these yellowish-white legless, 12-segmented grubs, up to 13 mm in length, have small, round, pale heads. They gouge on roots, reducing marketable sweetpotatoes

*Small Fruit Flies**

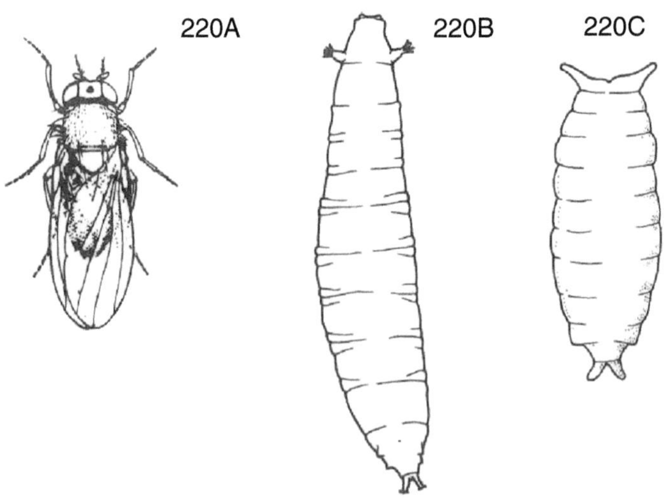

Fig. 220 Small fruit files. **A**, Adult. **B**, Larva. **C**, Pupa

**Drosophila* spp., particularly *Drosophila melanogaster* Meigan, Drosophilidae, DIPTERA

Description

Adult – About 3 mm long, these flies have red eyes and yellowish bodies with dark bands. Though commonly referred to as fruit flies, they are more correctly termed small fruit flies, vinegar flies, or pomace flies.

Egg – The tiny white elongate eggs are only about 0.5 mm long and have two slender filaments near the head end. Though individual eggs are too small to be easily noticed, clusters of eggs often resemble white mold on the surface of produce.

Larva – The cream colored maggots develop through three instars. They are about 5 mm long when fully grown.

Pupa – Yellowish white at first, the 3-mm-Iong pupae soon turn brown.

Biology

Distribution – Cosmopolitan in occurrence, small fruit flies are most likely to attain large populations around piles of overripe produce or in sweetpotato storage houses.

Feeding Habits – Drosophila flies consume yeast and bacteria associated with the initial decay of plant materials. Sap flows, mushrooms, and overripe produce are all very attractive to these flies.

Damage – Unlike real fruit flies, *Drosophila* flies do not break the skin of sound fruits and vegetables. They breed only in cracked or decaying overripe produce. As a result, these flies and their maggots are most likely to develop large populations in cull piles, storage houses, or processing plants.

Life History – Small fruit flies sometimes overwinter as larvae or pupae in sheltered locations with an abundance of dry fermented plant material. However, they have been known to breed throughout the winter in sweetpotato storage houses and in root cellars as far north as New Jersey. Egg laying, though, is much reduced at temperatures below 13 °C (55 °F) or above 38 °C (90 °F).

Eggs are deposited in cracked produce and incubate about 24 hours before hatching. When temperatures average 25 °C (77 °F), larvae feed and develop to maturity in about 4 days. Pupation then occurs within the shrunken skin of the last larval instar. About 5 days later, adult flies emerge. Within 2 days, females begin ovipositing at the rate of about 25 eggs per day. This process continues for several weeks, each female eventually depositing an average of 500 eggs.

The length of a complete life cycle (adult to adult) varies with temperature. At 20 °C (68 °F) about 15 days elapse, but. at 29 °G (85 °F) a life cycle is completed in only 8 days. Generations may be produced all year if temperature permits and fermenting produce is available.

Control

Small fruit flies are subject to many natural enemies. Adults are parasitized by protozoa, fungi, nematodes, and mites and preyed upon by spiders and certain species of flies. Maggots are parasitized by certain wasps and preyed upon by Staphylinid and Nitidulid beetle larvae.

Infestations can be prevented by the destruction of piles of culled produce. Storage houses should be well screened and sealed to minimize fly entrance as much as possible. Still, chemical control in storage areas may be necessary.

*Southern Armyworm**

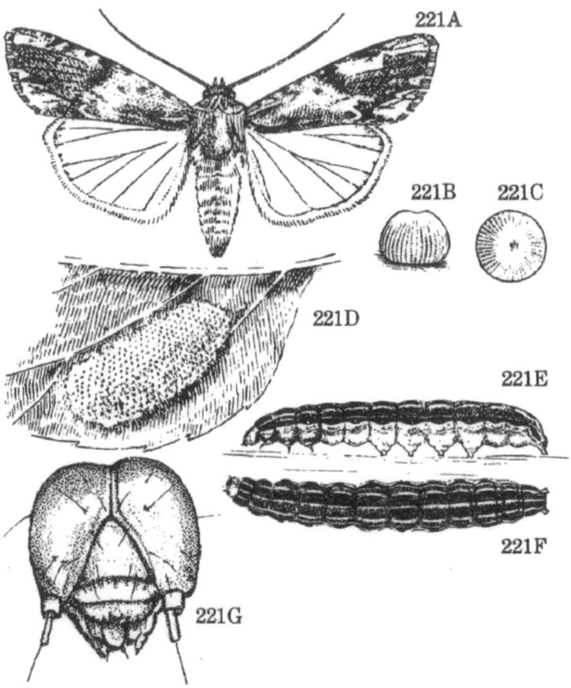

Fig. 221 Southern armyworm. **A**, Adult. **B–D**, Eggs (enlarged) and egg mass. **E–G**, Larvae with front view of head

Description

Adult – The southern armyworm moth has a wingspan of 30–36 mm. The forewings may vary in color from pale yellowish to dark brown. A darker streak extends from the center of each forewing almost out to the wing tip. The hind wings are white with brown veins and margins.

**Spodoptera eridania* (Cramer), Noctuidae, LEPIDOPTERA

Egg – The circular greenish egg is about 0.5 mm wide and 0.4 mm high. Viewed under magnification, the egg can be seen to have a ridged surface.

Larva – The fully grown caterpillar is gray or nearly black with whitish stripes tinged orange or pink and is about 36 mm long. The background body colour sometimes has a green or pink tint. Viewed from above, the larva has a pair of black triangular spots on each body segment, except the segment near the head bearing the first pair of legs. The larva has three pairs of true legs and five pairs of prolongs. The head capsule is pale yellow with bright reddish brown markings. The southern armyworm does not have a white inverted "V" on its head capsule.

Pupa – The darkly collared pupa is about 18 mm long and 5 mm wide.

Biology

Distribution – Florida, California, New Mexico, and central South America are year round homes for the. Southern armyworm. Each year moths migrate northward as far as Tennessee and Virginia. In North Carolina, this armyworm is only an occasional problem.

Host Plants – This armyworm is a general feeder with a wide host range. Weeds like spiny amaranth and pokeweed are preferred food plants. Some vegetable crop hosts include beet, cabbage, carrot, celery, collards, corn, cowpea, eggplant, okra, pepper, potato, rhubarb, sweetpotato, and tomato.

Damage – Though southern armyworms feed primarily on leaves, they have been known to consume tender stems and tips of branches. These caterpillars feed freely during the daytime but often are not observed because they tend to congregate around the bases of plants. Here they gnaw on stems or feed on potato tubers or sweetpotatoes near the soil surface. During the morning and evening, or on cloudy days, southern armyworms are likely to be found on foliage.

Life History – Southern armyworms overwinter either as larvae or pupae in Florida. Egg laying moths probably arrive in North Carolina in July. Each female deposits hundreds of eggs in masses on foliage. These masses are fuzzy in appearance since they are covered with scales from the bodies of moths. Eggs hatch in 4–6 days. For approximately 17 days, larvae feed and develop through six instars. At the end of this time, larvae drop by means of silken threads to the soil surface, enter the soil, and pupate. Nine to 13 days later a new generation of moths emerges. About 5 weeks elapse from egg stage to adult emergence during the summer. As many as five generations occur each year in Florida, but only two or 3 are likely to be completed in North Carolina.

Control

The variety NC Porto Rico 198 has been found to have some resistance to the southern armyworm. Populations of this armyworm species rarely reach high enough numbers to warrant chemical control.

*Spring Rose Beetle**

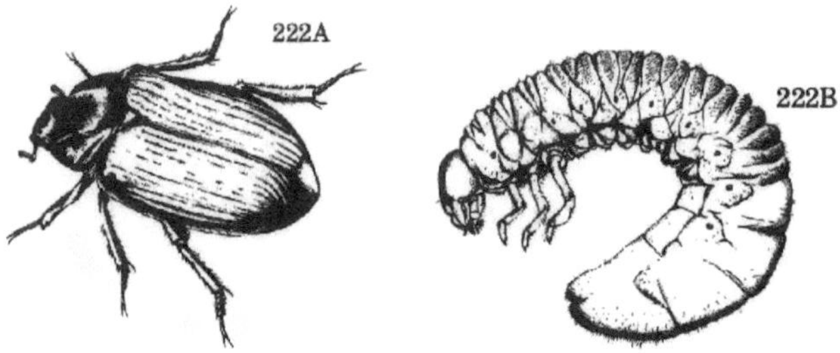

Fig. 222 Spring rose beetle. **A**, Adult. **B**, Grub

Description

Adult – This slightly hairy beetle, sometimes referred to as the spring rose beetle, is basically greenish black with a greenish purple iridescence. The wing covers, however, are dull brownish yellow in color. The beetle averages 10–12 mm long.

Egg – At first oval in shape, each white egg gradually enlarges, becoming more globose. From an initial size of 2 by 1.2 mm, the egg often increases in size to 2.5 by 2 mm.

Larva – The dirty white grub has a brown head and three pairs of forelegs. About 4 mm long when newly emerged; it reaches a maximum length of about 25 mm and resembles a common white grub in shape.

Pupa – The pupa is white when first formed but gradually darkens as it matures. It is approximately the same size and shape as the adult beetle.

Biology

Distribution – This beetle occurs from Canada southward through Kansas and North Carolina and is native to North America.

Host Plants – Adult beetles have been taken from many flowers, including those of clover, rose, blackberry, timothy, wild parsnip, dog fennel, and plantain. Larvae have been reported infesting roots of peanut, strawberry, sweetpotato, and certain pasture grasses.

Damage – These grubs feed on most underground plant parts. In certain cases they have been known to strip the taproot bare. Sweetpotatoes injured by these grubs have large but shallow feeding scars over their surface.

Life History – These insects overwinter as larvae in soil. In spring, grubs hollow out elongate, slightly curved earthen cells about 30 mm long. Within these cells, they spend approximately 6 days as inactive prepupae and 13 days as pupae. In Virginia, adult beetles usually emerge between May 15 and June 10. Further north, they often do not appear before the end of June. Several days after mating, females deposit

* *Strigoderma arboricola* Fabricius, Scarabaeidae, COLEOPTERA

eggs singly in soil (4–5 eggs/female based on lab studies). Eggs hatch an average of 17 days after deposition. By the time larvae begin feeding, at least one month has elapsed since adult emergence. Only one generation is completed each year.

Control

White grub infestations are typically associated with fields formerly in sod or pasture. Such a relationship has not been documented for *Strigoderma arboricola* grubs, but it may exist nonetheless. Chemically, these grubs are controlled by granular insecticides incorporated into the soil before planting.

*Sweetpotato Flea Beetle**

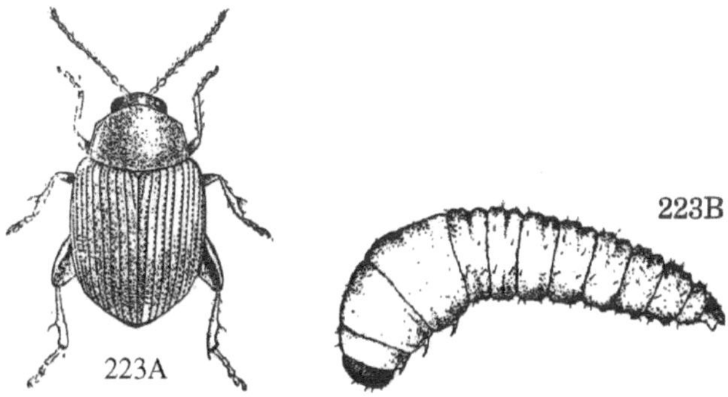

Fig. 223 Sweetpotato flea beetle. **A**, Adult. **B**, Larva

Description

Adult – The tiny oval beetle is black with a bronze tinge and about 1.6 mm long. It has reddish yellow legs and deeply ridged wing covers.

Egg – Each white, oblong-oval egg is about 0.2 mm long.

Larva – The slender, white, cylindrical larva has three pairs of legs near its head. It is about 4.8 mm long when fully grown. This larva has no dark spot or fleshy tubercle on its tail-end like cucumber beetle or palestriped flea beetle larvae.

Pupa – The pupa is white at first but gradually darkens and is approximately the same size and shape as the adult.

Biology

Distribution – The sweetpotato flea beetle occurs in practically all areas of this country where sweetpotatoes are grown.

* *Chaetocnema confinis* Crotch, Chrysomelidae, COLEOPTERA

Host Plant – Sweetpotato, corn, small grains, bindweed, raspberry, and sugar beet are the main food plants of this pest.

Damage – Adult flea beetles feed on foliage leaving narrow channels or grooves in the upper surfaces of leaves. These injured areas turn brown and die. Larvae live underground and feed on roots. Shallow winding tunnels etched into root surfaces indicate an infestation of flea beetle larvae. These tunnels eventually darken and split open leaving shallow scars. This type of damage usually is restricted to fibrous roots, but, during heavy infestations, larvae may injure the fleshy marketable portion of roots in the same manner as fibrous roots.

Life History – Sweetpotato flea beetles overwinter as adults under logs and leaves, along fence rows, and at the edges of wooded areas. They resume activity in spring and begin to deposit eggs in soil near host plants. A few days later eggs hatch. Newly emerged grubs feed for about 3 weeks before pupating in the soil. During summer, the entire life cycle is often completed in 30 days. Several generations per year are possible. From June onward, however, most eggs are deposited near bindweed, and flea beetle populations on sweetpotato decline.

Control

Cultural practices are instrumental in preventing flea beetle infestations. Controlling weeds along fence rows and plowing under crop debris destroy overwintering and egg laying sites. However, the use of resistant varieties such as Covington, Jewel or Centennial is the most effective means of preventing sweetpotato flea beetle injury.

In fields with a history of flea beetle infestation, chemical control may be justified. Preplant, soil-applied insecticides are available for this purpose.

*Sweetpotato Hornworm**

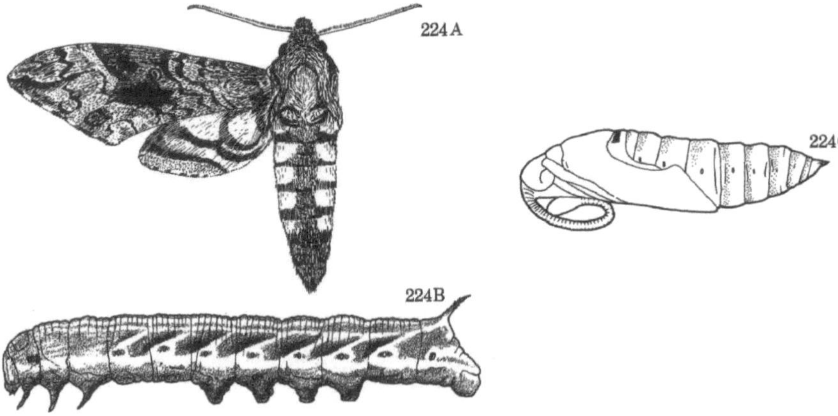

Fig. 224 Sweetpotato hornworm. **A**, Adult. **B**, Larva. **C**, Pupa

* *Agrius cingulatus* (Fabricius), Sphingidae, LEPIDOPTERA

Description

Adult – This grayish, heavy-bodied moth has a wingspan of 105–110 mm. The hind wings and abdomen bear bright pink bands.

Egg – Nearly spherical and about 1 mm in diameter, the translucent egg has a slightly greenish tint.

Larva – The first instar of the sweetpotato hornworm has a white body and a black anal horn. Later instars are basically green or brown with prominent, slanted black markings on each side of the body and a black anal horn. The head is also green or brown with three dark stripes on each side. A fifth instar hornworm may be 90 mm or more in length.

Pupa – The reddish brown pupa is about 15 mm wide and 64 mm long. The large tongue case has a pitcher-handle-like appearance.

Biology

Distribution – Sweetpotato hornworms are common in tropical America and the southern United States. Moths stray northward as far as Nova Scotia, but the larvae are too scarce to be pests that far north.

Host Plants – Sweetpotato and morning glory are the primary food plants of this horn worm although jimson-weed has also been reported as a host.

Damage – These large worms consume much foliage leaving only bare stems and petioles on plants. Sweetpotato hornworms have been reported to display armyworm like habits in Florida; however, their movement in large groups has not been observed here in North Carolina. Larvae often hide under large leaves at the base of plants.

Life History – The biology of this pest is not well documented. Its life history is probably very similar to that of tomato and tobacco horn worms. Moths appear in early June, again in August and September, and once more in early fall. There are probably 2 1/2 generations per year.

Control

In small gardens, hornworms can be controlled simply by picking them off plants. Chemical control, however, may be necessary in commercial production.

*Sweetpotato Weevil**

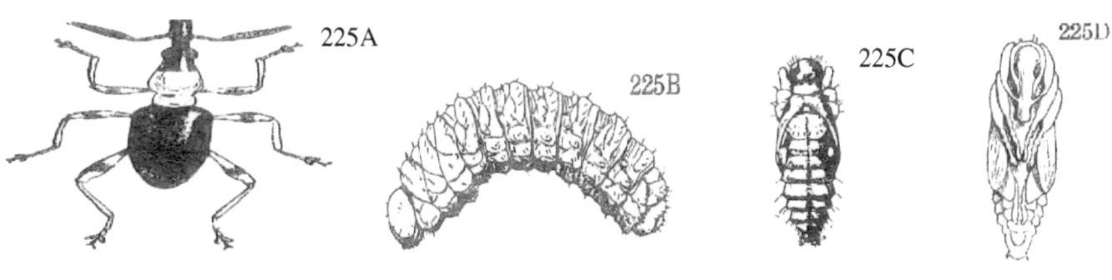

Fig. 225 Sweetpotato weevil. **A**, Adult. **B**, Larva. **C–D**, Pupae

* *Cylas formicarius* (Fabricius), Brentidae, COLEOPTERA

Description

Adult – This antlike snout beetle is about 6 mm long. The head and wing covers are metallic dark blue and the thorax and legs, bright red-orange.

Egg – Each white or pale yellow egg is inserted into a shallow hole in the vine. Broadly oval and about 0.6 mm long, the egg is slightly narrower at the attached end. The dark head of the larva becomes visible inside the egg just before hatching.

Larva – The fat, legless, slightly crescent-shaped larva has a dirty white to gray body and a pale brown head. When fully grown it is about 9 mm long.

Pupa – When newly formed, the pupa is the same color as the larva and about 5 mm long. Before transformation to the adult, the eyes, wing pads, and legs turn dark brown and the rest of the body is pale yellow. The last abdominal segment has two outward and backward curved tubercles.

Biology

Distribution – Sweetpotato weevils are a serious problem in some coastal areas from North Carolina to Texas. Discovered in eastern North Carolina in 1967, sweetpotato weevils are now largely under control in this area. Rarely, weevils are found as far north as New Jersey.

Host Plants – Sweetpotato and related wild plants such as morning glory are the only hosts of sweetpotato weevils.

Damage – These weevils and their larvae are the most destructive sweetpotato pests. Infestations may reduce plant growth during the first month after planting, but other damage often is not evident until harvest. Larvae and adults feed on foliage but they prefer to attack stems and sweetpotatoes underground. Small holes scattered over the surface of infested sweetpotatoes, particularly at the stem end, are the beetles' egg-laying and/or feeding punctures which cause sweetpotatoes to turn bitter. Such sweetpotatoes are unfit either for human consumption or stock feed.

Life History – Beetles become active in the field as soon as host plants are available. They first feed on leaves and stems (U.S. Department of Agriculture, 1960). As plant stalks enlarge and become woody, adult females prepare to deposit eggs. They make holes in stems and fleshy roots near the soil surface. Eggs are placed in these holes and covered with a jellylike secretion. Each female deposits an average of 120 eggs.

Larvae hatch less than a week after eggs are laid. They burrow deep into stems and fleshy roots for about 2–3 weeks. At the end of this period, third instar larvae return to the plant surface nearest the soil line to pupate. Pupae transform into adults in about a week, but another 4 days often elapse before the new beetles emerge from their pupal cells. Adults live about 2.5–3 months in summer and up to 8 months in winter.

Sweetpotato weevils continue to feed and breed throughout winter in stored sweetpotatoes. Development and activity, however, are much slower at temperatures below 15 °C (60 °F). As many as six to eight generations may be produced each year.

Control

Cultural practices such as crop rotation, use of weevil-free planting stock, and destruction of volunteer plants and crop residue are primary elements of weevil control. Planting sweetpotatoes in the same fields year after year leads to increased weevil populations. If slips for planting cannot be obtained from a weevil-free area, each sweetpotato chosen for seed should be examined carefully and destroyed if infested. Also, use of deep-rooted varieties such as Porto Rico over shallow-rooted varieties like Gold Rush is advisable.

Post-harvest insecticide treatments can be applied to prevent development of weevils in storage. Treated sweetpotatoes, however, will need to be washed thoroughly once they are removed from storage.

*Tortoise Beetles**

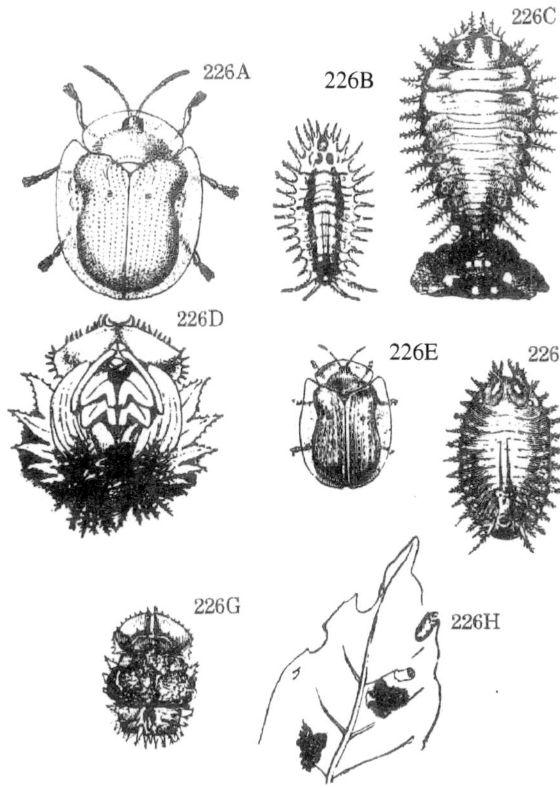

Fig. 226 Tortoise beetles. **A–C**, Blacklegged tortoise beetle. **A**, Adult. **B**, Larva without excrement cover. **C**, Larva with excrement cover pulled back. **D**, Pupa. **E–H**, Golden tortoise beetle. **A**, Adult. **B**, Larva without excrement cover **G**, Pupa. **H**, Damage to foliage

* *Argus tortoise beetle*, *Chelymorpha cassidea* (Fabricius); Blacklegged tortoise beetle, *Jonthonata nigripes* (Olivier); Golden tortoise beetle, *Metriona bicolor* (Fabricius); Mottled tortoise beetle, *Deloyala guttata* (Olivier); Striped tortoise beetle, *Agrioconota bivittata* (Say), Chrysomelidae, COLEOPTERA

Description (several species)

Adult – These oblong-oval beetles are basically gold in color with various black and/or red markings depending upon species. Slightly flattened and squared at the shoulders, tortoise beetles' bodies are somewhat shell-like in appearance. Body margins extend in a rooflike manner over much of the head and legs. Most species are 5–8 mm long.

Egg – Tortoise beetle eggs usually occur in masses. Each individual egg is stalked, the long stalk being attached on the plant surface by a gelatinous substance. The beige or white eggs of some species have a reddish tubercle on the, upper end. Eggs are about 1.6 mm long.

Larva – The spined larvae may be basically dull yellow, brown, or green depending upon the particular species. They all have black heads, prothoracic shields (area behind head), legs, spots, spinelike setae, and anal forks. The anal forks are long spines near the tip of the abdomen which hold large masses of excrement. Fully grown larvae are 10–12 mm long.

Pupa – Pupae are oblong-oval in shape like adult beetles but have spines along the abdomen like larvae. They are approximately the same size as adult beetles.

Biology

Distribution – Argus and mottled tortoise beetles occur in all arable sections of the United States and Canada. The golden and blacklegged species are most common from the Rocky Mountains eastward.

Host Plants – Most tortoise beetles feed on sweetpotato and closely related plants such as morning glory and bindweed. Argus tortoise beetles also infest cabbage, corn, raspberry, strawberry, milkweed, and plantain. Golden tortoise beetles have been found on eggplant.

Damage – Both larvae and adults feed on leaves causing them to be riddled with holes. This type of damage is most threatening to seedlings or newly set plants.

Life History – Tortoise beetles overwinter as adults under bark, in leaf litter, or in other dry, protected places. In spring, beetles emerge and feed on weed hosts until sweetpotato plants are available. Female adults deposit clusters of 15–30 eggs on the undersides of leaves. Larvae emerge 7–10 days later. After feeding for 2–3 weeks, larvae transform into pupae. About a week later, a new generation of beetles emerges. Several generations may occur each year in southern states.

Control

Tortoise beetles and other leaf-feeding insects do not affect sweetpotato production if growing conditions are satisfactory. Cultural practices, such as adequate fertilization, good weed control, and well-timed planting, effectively deter excessive tortoise beetle injury. Generally, chemical control is not necessary.

*Yellowstriped Armyworm**

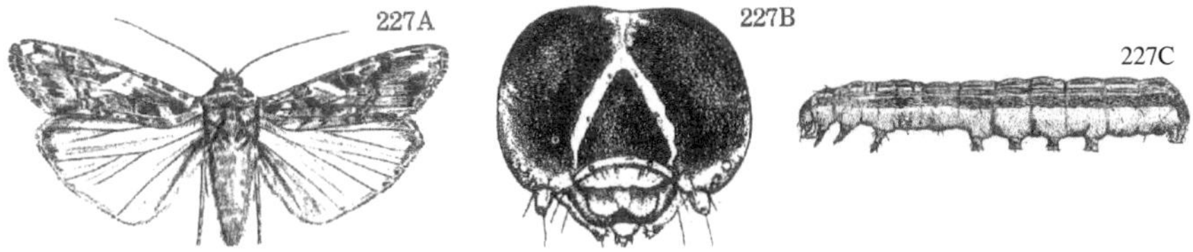

Fig. 227 Yellowstrlped armyworm. **A**, Adult. **B**, Front view of head of larva. **C**, Larva

Description

Adult – The yellowstriped armyworm moth has dark forewings with white and brown markings and white hind wings. The wingspan ranges from 32 to 38 mm.

Egg – Approximately 0.5 mm and 0.4 mm in diameter, the ribbed, greenish egg gradually becomes pale pink or brown before hatching. The egg mass is covered with scales from the moth's body.

Larva – This smooth-skinned, pale gray to jet black caterpillar has a yellowish orange stripe along each side and a pair of black triangular spots on the top of most segments. Its head capsule is brown with black markings and a white inverted "V." The sixth larval instar may be as long as 45 mm

Pupa – The brownish pupa is about 18 mm long and 5.5 mm wide.

Biology

Distribution – The yellowstriped armyworm occurs from New York southward into Mexico, westward to the Rocky Mountains, and in some areas of California and the West Indies. In this country, however, it is most common and most injurious in the southern states. In North Carolina, this caterpillar is observed annually in field and vegetable crops.

Host Plants – The yellowstriped armyworm is a general feeder. Some of its hosts include alfalfa, asparagus, bean, beet, cabbage, clover, corn, cotton, cucumber, grape, grass, jimsonweed, morning glory, onion, pea, peach, peanut, sweetpotato, tobacco, tomato, turnip, wheat, watermelon, and wild onion.

Damage – This foliage-feeding caterpillar is sporadically injurious to young crop stands. Defoliation at this stage can be a problem.

Life History – Yellowstriped armyworms overwinter as pupae in the soil. In southern states, moth emergence begins in early April and continues into May. After mating, females deposit egg masses on foliage, trees, or buildings. Approximately 6 days later, the eggs hatch and the larvae begin feeding. Although these caterpillars

---

* *Spodoptera ornithogalli* (Guenee}, Noctuidae, LEPIDOPTERA

emerge by late spring in the South, they may not appear before July in northern and midwestern states. Larvae feed during the day on tender foliage over a 3-week period. Mature sixth instars burrow into the soil and change into pupae. Two weeks later, the second generation of moths emerges. In southern states, including North Carolina, three to four generations occur each year.

Control

Yellowstriped armyworms seldom require control in North Carolina. Since large larvae are difficult to control with insecticides, early detection is important in maintaining populations below economic injury levels.

References

Austin, D. F. 1988. The taxonomy, evolution, and genetic diversity of sweetpotatoes and related wild species. *Exploration. Maintenance, and Utilization of Sweet Potato Genetic Resources*. International Potato Center, Lima, Peru, pp. 27–60.

Averre, C. W., Kenneth A. S. and George L. W. 1991. Know and manage sweetpotato pests. *North Carolina Cooperative Extension Service*, North Carolina State University. Poster 8 pp.

Chalfant, R. B., Richard K. J., Dakshina R. S., and James M. S. 1990. Ecology and management of sweetpotato insects. *Annu. Rev. Entomol.* 35:157–180.

Cuthbert, F. P. Jr. 1967. Insects affecting sweetpotatoes. *USDA Agric. Handb.* 329. Washington, DC. 28 pp.

FAO. 1984. *FAO Production Yearbook 1983*. Rome: FAO.

Hahn, S. K. and Leuschner, K. 1982. Breeding sweetpotato for weevil resistance. *Sweet Potato, Proc. 1st Int. Symp.* R. L. Villareal and T. D. Griggs (ed.) Asian Veg. Res. Dev. Cent., Shanhua, Tainan, Taiwan, pp. 331–36.

Harari, A. R., David B.-Y., Michael C. and David R. 1997, Population dynamics of *Maladera matrida* (Coleoptera: Scarabaeidae) in peanuts fields in Israel. *Environ. Entomol.* 26:1040–1048.

Hill, D. 1983. *Agricultural insect pests of the tropics & their control.* Cambridge University Press, Cambridge. 2nd edition.

Mullen, M. A. and Sorensen, K. A. 1984. Proc. of the sweetpotato weevil. *Proc. Sweetpotato Weevil Workshop*. North Carolina State University. pp. 1–73.

Roberts, R. A. 1952. Whitefringed beetle, pp. 608–11. *Insects. U.S. Dep. Agr. 1952 Yearbook of Agric.* 780 pp.

Sorensen, K. A. 1987. Cultural, regulational and educational programmes on the sweetpotato weevil in the United States. *Insect Sci. Appl.* 8(4/5/6): 825–830.

Sorensen, K. A. 1988. Sex pheromone trapping and management of the sweetpotato weevil, *Cylas formicarius elegantulus* (S.) (Coleoptera: Curculionidae) in North Carolina. *Proceeding XVIII International Congress of Entomology* Poster Abstract 99 p. 402.

Sorensen, K. A. and James R. B. 1994. Insect and related pests of vegetables: some important, common, and potential pests in the southeastern United States. *North Carolina Cooperative Extension Service* AG-295. 180 pp.

Sutherland, J. A. 1986. A review of the biology and control of the sweetpotato weevil *Cylas formicarius* (Fab). *Trop. Pest Manage.* 32:3–15.

U.S. Department of Agriculture. 1960. The sweetpotato weevil how to control it. *U.S. Dep. Agr. Leaflet* 431. 8 pp.

West, S. A. 1977. Studies on the biology and ecology of the sweetpotato stem borer *Megastes grandalis* Guen. Trinidad. MS. thesis. University of West Indies, Trinidad.

Chapter 11
Uses and Nutritional Data of Sweetpotato

G. Padmaja

Sweetpotato (*Ipomoea batatas* Lam.) is cultivated throughout the tropics and warm temperate regions of the world for its starch roots, which can provide nutrition, besides energy. The edible tuberous root is either long and tapered, ovoid or round with a skin colour ranging from white, brown, purple or red and the flesh colour ranging from white, pale cream, orange or purple. Besides, the plant is also much valued for its green tops, which are a concentrated source of many essential vitamins and minerals. Although China is the largest producer of sweetpotatoes, accounting for more than 80% of the world supply, only 40% of the production is used for human consumption and industrial uses, while, the rest goes as animal feed. Per capita production is the greatest in Solomon Islands (160 Kg/person/year), followed by Burundi (130 Kg/person/year), where sweetpotatoes are a staple food. The roots are also used to a great extent in countries like Japan, New Zealand, Papua New Guinea and parts of America.

Use as Food, Feed and Industrial Raw Material

Use of Sweetpotato as Food

The roots are most frequently used after boiling, baking or frying. They are also processed into starch, flour or puree to make secondary food products.

Primary Food Products

Sweetpotatoes are consumed at home level, mainly after cooking, baking or converting into fried chips. The roots are often converted to canned or pureed form, to enhance the shelf life. Sweetpotato based baby foods are preferred in many countries as the first solid food for infants.

G. Padmaja (✉)
Central Tuber Crops Research Institute, Trivandrum, Kerala, India
e-mail: g_padmaja1@yahoo.co.in

G. Loebenstein, G. Thottappilly (eds.), *The Sweetpotato*,
DOI 10.1007/978-1-4020-9475-0_11, © Springer Science+Business Media B.V. 2009

Sweetpotato is often boiled with its skin, till it becomes soft and mealy. The roots are classified into 'dry' and 'moist' types, depending on the texture of the cooked roots (Rao et al., 1974). Dry types are dry, firm and mealy after cooking while the moist types are soft, watery and sticky. The influence of cooking procedures on the structure and biochemical changes in sweetpotato was investigated by Valetudie et al. (1999). Many studies show that the sweet taste generated in cooked sweetpotato is due to the action of endogenous amylases (Walter et al., 1975). Starch, being the main ingredient, constituting up to 65–70% of dry weight, undergoes gelatinization during cooking. The presence of high levels of sugars, soluble organic and mineral components has been reported to affect the cooking behaviour of sweetpotatoes.

Ultrastructural studies on cooked sweetpotato showed that the intracellular spaces increased (Valetudie et al., 1999). Steaming is the most detrimental cooking method, as it leads to increase in the cell wall thickness and a sloggy texture of cooked roots.

The stability of amino acids during cooking of sweetpotatoes was studied by Purcell and Walter (1982), who found that the major nutritional change that occurred during baking, canning or flaking was lysine destruction. Both the canned and flaked sweetpotatoes ('Jewel' variety) was reported to contain 26% less lysine than the baked roots (Purcell and Walter, 1982). Low temperature blanching (LTB) treatment was reported to increase the firmness retention in cooked and canned sweetpotato (Truong et al., 1998).

Using moist, intermediate and dry varieties of sweetpotatoes, Truong et al. (1997) found that the dry matter content had high correlation with the alcohol insoluble solids (Table 11.1). While maltose production during cooking is pronounced in

Table 11.1 Chemical composition (g/100 g) fresh weight) of sweetpotato cultivars

	Raw samples	Cooked samples					
Cultivar	Dry matter	Dry matter	AIS	Glucose	Fructose	Sucrose	Maltose
Moist (soggy)							
Beauregard	18.59G	18.91G	9.21H	1.28B	0.73B	1.29G	4.61E
Hernandez	21.08F	22.05F	10.94G	1.91A	0.97A	1.90EF	5.54D
Jewel	20.57F	21.53F	10.33G	0.76C	0.44C	3.12A	5.49DE
Intermediate							
Oriental SP	27.98C	28.66C	15.73D	0.49EF	0.48C	1.96E	8.50A
Acc. 10–28	26.65D	27.20D	14.36E	0.55DEF	0.50C	2.06EF	8.37A
Acc. 12–5	24.62E	25.02E	11.90F	0.62DE	0.44C	2.72B	7.43B
Acc. 2–26	29.90E	29.59B	18.09C	0.30G	0.22D	2.42CD	6.53C
Acc. 15–13	27.65C	26.94D	18.80B	0.68CD	0.46C	1.88F	4.14E
Mealy (dry)							
Acc. 6–30	30.63A	32.01A	18.17BC	0.37FG	0.25D	2.18DE	8.81A
Acc. 8–22	30.03AB	30.03B	25.58A	0.46F	0.46C	2.51BC	ND[b]
Coefficient of variability (CV, %)	1.9	2.1	3.0	12.3	10.7	9.3	10.4

Source: Truong et al., 1997.

certain varieties, there is no production in a sweetpotato accession 8–22, due to lack of amylase activity (Truong et al., 1997; Takahata et al., 1994).

Baking sweetpotatoes in country ovens has been an old mode of processing of roots in countries *viz.*, China, India, Japan, Bangladesh, Indonesia, Kenya, Papua New Guinea etc. Traditional baking has now been replaced in many countries by microwave baking. A comparison of the carbohydrate components in sweetpotatoes, baked by convection heating and microwave heating was made by Purcell and Walter (1988). They found that the samples baked in microwave contained less total alcohol soluble, carbohydrates, reducing sugars and dextrins and more starch than the roots baked in convection oven.

Curing was reported to affect the sensory properties and carbohydrate composition of baked sweetpotatoes. Curing increased the rate of production of flavour notes in the initial days and the baked roots had better sensory qualities. However, as the days after harvest increased the difference between cured and uncured roots disappeared and both produced baked roots with same flavor characteristics. Nevertheless, several workers have reported increase in the level of natural amylolytic enzymes during curing of sweetpotatoes, leading to a sweeter flavour and moist mouthfeel in baked sweetpotatoes (Scott and Matthews, 1957; Sistrunk et al., 1954; Walter et al., 1975). Rheological properties of baked sweetpotatoes were affected by the amylolytic enzymes produced during curing and these were found to be correlated with the moist mouthfeel (Rao et al., 1975a) of baked sweetpotatoes.

Reddy and Sistrunk (1980) observed that baked or microwave cooked sweetpotato roots had more reducing sugars, total sugars and pectins than boiled or steamed roots. Microwave baked roots had higher hemicellulose and cellulose than the boiled roots, probably due to increased solubility of these components in the latter process. The percentage of starch conversion to maltose was reported to be around 54% in 'dry' cultivars, as compared to 63–69% in 'moist' cultivars (Martin, 1986). However subsequent studies have shown that many low or non-sweet clones with low/amylase activity respond differently with low or no production of maltose (Almazan, 1987; Kukimura et al., 1989; Martin, 1986).

Cooking has also been reported to increase the digestibility of starch. Cooked sweetpotato starch was more susceptible to enzymic hydrolysis than raw starch. Bradbury et al. (1985) reported significant increase in the dietary fibre in boiled and steamed sweetpotatoes, resulting probably from the conversion of part of the starch to 'resistant starch'.

Sweetpotato protein has been reported to be the least affected by the processes of steaming, boiling, baking etc (Bradbury et al., 1984). Loss of vitamins during the various cooking methods depends on the varieties as well as the conditions and exposure time. Generally, lower losses of carotenoids, ascorbic acid and vitamins like thiamine, riboflavin, niacin etc have been reported when the whole roots are boiled while the losses are the highest when sliced roots are boiled and water is discarded. Increased retention of carotenoids was reported in blanched or boiled sweetpotato slices from orange fleshed variety. However microwave baking retained almost totally the carotene while 75% retention was observed in baked roots (Lila Babu, 2006).

Minerals are lost during cooking mainly through leaching into the cooking water and loss of peels. The effect of boiling or steaming peeled roots on the loss of most minerals is reported to be negligible (Haytowitz and Mathews, 1984; Leung et al., 1972). Except potassium smallest changes in minerals have been reported to occur during baking of sweetpotato (Bradbury et al., 1984), which is the reason for its use in weaning foods.

Dehydrated Chips and Flour

Sweetpotato roots are dehydrated to enhance the shelf life of stored roots. The chips are further powdered to flour and used for making many snack foods. The roots are either peeled or unpeeled and sliced for drying. Discolouration of dried chips is a problem with certain cultivars having high activity of polyphenol oxidases and higher levels of phenols. Walter and Purcell (1980) found that the browning tendency of sweetpotato was correlated only with the phenolic content, while others found that PPO activity was also correlated with browning potential of sweetpotato (Scott and Kattan, 1957). Walter et al. (1979) found that the principal phenolics of sweetpotato *viz.*, chlorogenic acid and its isomers were effectively oxidized by sweetpotato PPO, resulting in browning in the processed product. Certain pre-treatments like soaking the sliced sweetpotato in 0.025% Potassium metabisulphate or 1.0% Citric acid for 1 h, followed by sun drying have been found to yield dried chips with a bright colour (Padmaja, unpublished data). Whilst small scale dehydration by sun drying is widely practiced in most countries, sophisticated devises like drum drying, spray drying or solar drying are adopted in developed countries. Extensive drying of sweetpotato is practiced in China to produce dried chips for its further use in starch, noodle and alcohol factories. Damp weather and prolonged drying periods can cause microbial contamination of the chips.

Dried sweetpotato cubes have been developed in the Philippines, using a fabricated sweetpotato slicer (Truong, 1990). The cubes can be cooked either alone or with other ingredients like coconut milk, rice, sugar and vanilla to make a traditional dish called 'guinata an'. Orange fleshed sweetpotatoes have been diced into long strips and soaked in 2% (w/v) metabisulphite prior to cooking in 60 °C Brix sugar syrup containing citric acid (0.8–1.0%). These are then dried and packed to make a product like sweetpotato candy. Data and Operario (1992) found that processing operations for dry chip production only slightly affected the chemical constituents of sweetpotato. Polyethylene sacks were found to be the best packing material, permitting 6 months storage without microbial or insect damage.

The pasting characteristics of drum dried and hot air-dried sweetpotato flour was investigated using a Rapid Visco Analyser. Water binding capacity, viscosity and total amylose decreased and starch digestibility increased in the hot air-dried flour due to native amylase action, compared to the drum dried flour. Scanning electron microscopic studies on the starch granules showed a tendency of clustering in the drum-dried flour.

Canning of Sweetpotato

Canning of sweetpotato is widely practised in the United States, to enhance the storage life and ensure round-the year availability of the product (Walter and Hoover, 1986). Canned sweetpotato is also widely available in countries like Australia, Taiwan and the Netherlands (Mason, 1982). The pre-processing steps in the production of canned sweetpotato are grading, cleaning, pre-heating, peeling and trimming (Bouwkamp, 1985). Grading helps to select undamaged and uniformly shaped roots (without crevices, grooves and disfigured shapes) for canning (Williams and Ammerman, 1968; Woodroof et al., 1955). Pre-heating is adopted by immersing the roots in heated water or live steam for small periods, which helps in driving off the intercellular gases, to facilitate good vacuum build up in cans (Bouwkamp, 1985). Scott (1952) found that 40 sec pre-heating in live steam can prevent the enzyme linked browning of sweetpotato. Peeling and trimming are essential stages in the preparation of canned sweetpotato. Several methods are available for peeling of sweetpotato roots, including manual peeling and the peeling losses vary from 15 to 60% (Bouwkamp, 1985). Exposing the roots to 7–10% (w/w) boiling lye for 6 min followed by washing with a high-pressure spray was found to be the best method, with a peeling loss of only 20–35%. Hand trimming is done to remove surface imperfections in the lye-peeled roots (Walter and Schadel, 1982).

Stages of Canning

1. Sizing and Cutting
   Sizing and cutting of trimmed sweetpotatoes are essential to improve the consumer acceptability. The roots are either halved or quartered or made into cubes, strips etc, which will also help for compact filling of the cans.
2. Blanching, can Filling and Syruping
   Blanching is an essential processing step to remove the trapped gases, so as to ensure proper vacuum build up in the cans. Usually 1–3 min blanching at 77 °C is practised. However, Truong et al. (1998) reported that 62 °C blanched samples gave more intact and firmer canned product than unblanched samples or samples blanched at higher temperatures. They also found that the sensory texture and acceptability was the greatest for samples blanched for 30–45 min at 62 °C. Hence, low temperature with longer time blanching is the best method to obtain a canned product having a firmer and intact texture. The blanched slices are manually filled or machine filled to cans, as tightly leaving only little air spaces for the syrup to trickle down. Sugar syrup of 20–40% concentration (w/w) is then added and the cans kept at 95–97 °C. Alternatively water or salt water is also used.
3. Exhausting Rretorting and Cooling
   Bigger size cans (>303) have to be exhausted for more time for the internal temperature to reach 77 °C. The cans are then vacuum closed and retorted, followed by rapid cooling to 35 °C.

4. Firmness of Canned Sweetpotato
   Firmness is decided to a large extent by the pre-processing factors. Delay in processing has been reported to lead to a less firm product when the roots were stored at 30 °C (McConnell and Gottschall, 1957; Kattan and Littrell, 1963; Baumgardner and Scott, 1963). However, low temperature storage (2–5 °C) increased the firmness. Cultivar differences, manurial practices etc have also been reported to lead to product with varying firmness (Kattan and Littrell, 1963). Curing of roots followed by storage for 1 month at 27 °C was reported to yield a firmer product than roots stored at 16 °C (Constantin and McDonald, 1968). Baumgardner and Scott (1963) found that decrease in the insoluble protopectin and increase in the oxalate soluble pectin in stored roots was responsible for low firmness of canned sweetpotato. Firmness could be increased by soaking the peeled roots in 1.0% or 2.0% (w/v) calcium chloride solution or citric acid. Soaking in calcium hydroxide or canning with 2% pectin also have been attempted by various-workers (Rao and Ammerman, 1974; Williams and Ammerman, 1968).
5. Color of Canned Sweetpotato
   The discoloration in canned sweetpotato has been correlated with the initial levels of phenol and phenol oxidizing enzymes in raw roots. While enzymic browning leads to discoloration during canning process itself, non-enzymic browning, resulting from the reaction of quinones with iron or tin of the metal cans, is responsible for the discoloration observed after the opening of cans.

   Pre-heating the roots prior to peeling has been reported to reduce the discoloration of the canned product (Scott, 1952). Pre-heating also facilitated rapid inactivation of the phenolases during the lye peeling stage (Walter and Schadel, 1982). The color of yellow/orange fleshed cultivars were reported to be improved during canning by using citric acid in the sugar syrup and holding for 24 h or by using a combination of citric acid and calcium chloride prior to canning. Sucrose syrup (40–50% w/w) was found to yield brighter product than corn syrup (Jenkins and Anderson, 1957; Woodroof et al., 1955).
6. Nutritional Changes in Canned Sweetpotato
   As high as 86% of the carotene content of the roots are reported to be retained after canning (Elkins, 1979). He also found that riboflavin was almost totally retained while thiamine was retained to the extent of 73% only. Isomerization of carotene during canning was demonstrated by several workers, who found that neo beta-carotene B was the main isomer in canned sweetpotato (Panalaks and Murray, 1970). Syrup concentrations or cooking time were not found to have effect on the carotene content of canned roots. Thermal processing has been reported to convert part of the biologically active all-trans form of β-carotene into less active cis-isomers.

   Leaching of potassium, copper and magnesium into the sugar syrup has been reported by Lopez et al. (1980). While Elkins (1979) reported complete retention of minerals and ascorbic acid in vacuum packed sweetpotatoes, Purcell and Walter (1982) reported that greater lysine destruction occurred in canned sweetpotato, as compared to baking.

Frozen Sweetpotatoes

Low temperature storage of sweetpotato is practiced in developed countries only, as the cost is prohibitive for adoption. Sweetpotatoes are frozen as whole roots or sliced cubes, pieces or as pastes. The roots are often blanched in water or with steam at 10 psi pressure (116 °C) to inactivate the enzymes associated with browning, off flavor development etc. Steam blanching was reported as the best method, as it does not lead to a soggy product (Woodroof and Atkinson, 1944). The slices/cubes are packed in plastic bags and blast frozen at 40 °C. The washed roots are sometimes steamed, crushed, mixed with 35% sugar (w/w) and filled to plastic bags under pressure before blast freezing at −40 °C. Frozen sweetpotato products are widely popular in Japan.

Fried Sweetpotato Products

Sweetpotato roots are transformed into more stable edible products like fried chips, crisps, French fries etc which are very popular in Japan, USA, China, Netherlands, Peru etc.

The roots are peeled, sliced into thin chips and deep fat-fried to obtain fried chips. Discoloration during frying at high temperature due to Maillard reaction is very common with cultivars having high amino acid and sugar contents. Pre-treatment of the slices with citric acid – sodium chloride solution (0.5% w/w each) for 1 h, followed by blanching for 2 min at 60 °C and deep-frying has been found to yield light golden brown fried chips (Padmaja, unpublished data). Sugar coated fried chips are popular in Japan, while salted or spicy chips are preferred in Papua New Guinea, Bangladesh and Peru. The quality is improved through treatments like blanching for 2 min at 93 °C in boiling water or a solution of sodium acid pyrophosphate (0.5–0.75% w/w) or diffusion extraction of sugars to eliminate the problem of browning etc. Picha (1986) reported that the glucose and fructose content of the sweetpotato slices determine the extent of browning of fried chips, rather than the sucrose content. Baba et al. (1981) found that the rapid production of maltose up to Ca 2% due to endogenous amylase action also led to considerable browning of the fried chips. A diffuser-extractor for removing the sugars almost totally was developed by Hannigan (1979).

The length of frying is influenced by the moisture content of the chips and the temperature of cooking oil. Fresh/blanched chips having around 50% moisture takes approximately 4.5 min at 138 °C (Hoover and Miller, 1973). The yield of chips is around 40% of the weight of precooked peeled root (Kelly et al., 1958). Higher temperature of frying has been reported to result in dark coloured chips (Martin, 1987). Optimum frying temperatures have been reported to be between 143 °C and 154 °C. Oil retention in the fried chips also affects consumer acceptability (Baba et al., 1981). High moisture and low dry matter content in the fresh slices leads to higher oil retention in the fried chips. Hoover and Miller (1973) reported reduction in the oil retention in fried chips made from partially dried blanched chips. Surface drying of blanched slices was also found to reduce blister formation in fried chips, which act as pockets for trapping oil (Padmaja, unpublished data). Blister formation has also been reported to lead to chip hardness and lack of crispness (Baba et al., 1981). Blanching and/or freezing and thawing of the sweetpotato slices could reduce the hardness of chips (Baba and Yamamura, 1981). Packaging of fried chips in moisture proof packs is essential to prevent leatheriness in the chips. Development of rancidity in fried chips was found to be delayed when the fresh slices are treated with an antioxidant like 1% w/v ascorbic acid solution before frying (Martin, 1987), which also increased the crispness.

Frozen French fry type products were developed from sweetpotato by deep-frying the slices, cooking and freezing the fries (Kelly et al., 1958). The frozen fries are cooked in an oven prior to eating. The process for frozen French fry making was subsequently modified by other workers (Walter and Hoover, 1986; Schwartz et al., 1987). In this process, the lye peeled roots are made into long strips (1.9 cm × 6.4 cm) blanched in boiling water containing 1% (w/v) sodium acid pyrophosphate, partial drying and freezing at 34 °C. Deep-frying is done immediately before eating. The surface crispness of French fries was found to be increased when blanched slices were dipped for 30 min in 1.0% citric acid, surface dried and deep fried (Padmaja, Tuber crops recipes). Reungmaneepaitoon et al. (2005) workers reported that good quality French fries with low oil content could be made by blanching sweetpotato strips for 3 min in boiling water at 100 °C, containing 0.25% sodium acid pyrophosphate and 0.25% calcium chloride and then partially frying at 180 °C for 20 sec.

Sweetpotato Puree

Sweetpotato puree, is a primary processed product from the roots, which is used directly as a baby food or used for mixing various food items like patties, flakes, reconstituted chips etc. High quality puree can be made from white, cream or orange-fleshed sweetpotatoes and also from tubers of any size or shape (Woolfe, 1992; Bouwkamp, 1985). Puree making also ensures round the year availability and better storage life. The initial process involved more cooking of the roots, peeling and than mashing. The process was subsequently modifies through a controlled alpha-amylase process, where commercial alpha-amylase was added to a portion of the puree for enabling partial hydrolysis of starch. The enzyme treated fraction was then treated with the remaining puree (Szyperski et al., 1986). This process resulted in puree with better rheological characteristics. Controlled heat processing of sweetpotatoes (Jewel variety) for puree making was investigated by Walter and Schwartz (1993). They prepared puree from 0.5 cm thick sweetpotato slices by pre-cooking at 100, 125 or 150 °C for varying periods; finish cooking for 15 min at 150 °C and pureeing. Comparison of this puree with the conventional puree made from baked sweetpotatoes (1.25 h at 195 °C) showed that the latter puree had higher maltose and lower starch content as well as lower viscosity as compared to the former. An endogenous enzyme activation process was developed for sweetpotato puree by heating the puree to 74–85 °C using steam, and this is usually used in flake manufacture (Hoover, 1967).

Maltose was found to be only sugar formed during puree making and more than 90% of its formation is completed during the initial 10 min (Walter and Purcell, 1976). The textural quality of puree has been studied by several workers and was found to be influenced by cultivars, storage period as well as length of time given for puree hydrolysis. Ice et al. (1980) found that pH of the puree affected the rheology of the stored puree and lower and higher pHs were found to lower the puree viscosity. Fasina et al. (2003) studied the thermal and dielectric properties of sweetpotato puree within a temperature range of 5–80 °C. Increase in temperature was reported to increase the specific heat, thermal conductivity and diffusivity of the puree while the density decreased. Sweetpotato puree, prepared under aseptic conditions has been reported to have a storage life of up to 9 months and flash heating at 123 °C was suggested as the best method for aseptic can filling of puree (Smith et al., 1982). The flow behavior of sweetpotato puree, coefficient of shear rate and apparent viscosity were reported to have good correlation with sensory scores (Rao et al., 1975b). The viscoelastic properties of sweetpotato puree based infant food were studied by Ahmed and Ramaswamy (2006). The puree behaved like an elastic solid with storage modulus (G') dominating over loss modulus (G''). Increase in temperature decreased both elastic and viscous moduli. Deviation in the rheological behavior at and above 65 °C was caused by gelatinization and possible amylase-lipid complex formation of sweetpotato starch

Sweetpotato Flakes

A procedure for the production of sweetpotato flakes was first described by Taubenhaus (1923), consisting of washing, cooking, mashing and drying on steam-heated drum dryers. The process underwent several steps for refining the technology and a much-advanced process is now available. The dehydrated flakes can be reconstituted to mashed sweetpotato or incorporated into various food products like pastries, cakes, bread, biscuits etc. The initial steps in flakes manufacture are similar to those in puree preparation, described under the section of sweetpotato puree. Hoover (1966) reported that the ratio of soluble to the insoluble solids in sweetpotato puree decided the final quality of the flakes produced. The high content of soluble solids in the sweetpotato flakes resulted in low water requirement to rehydrate the flakes. Further, the flavour also improved with increase in the proportion of enzyme treated puree in the mash. The earlier process was further modified for puree preparation using added alpha-amylases to partially hydrolyse the starch and increase the soluble solids content of puree. The hydrolyzed puree was then added to the control puree and subjected to drum drying. Spadaro and Patton (1961) observed that the quality of dehydrated sweetpotato flakes depended on the variety of sweetpotato and for each variety, the process parameters have to be optimized. The procedure was subsequently modified by adding amylase and/or sucrose after cooking and pureeing, leading to a more acceptable product. Curing of tubers and storage were reported to affect the amount of amylase required to produce acceptable flakes (Bertoniere et al., 1966). The extraneous addition of alpha-amylase was further eliminated by activating the endogenous amylases during processing (Hoover, 1967). The process consists of heating the puree almost instantaneously by steam injection to 160–185F. The pre-heating helped to activate the endogenous amylases in sweetpotato and a holding period of 2–6 min was sufficient to achieve the required conversion of starch to sugars. The partially cooked puree is then passed through a second steam injector, where high temperature in activation of the amylases is achieved at 200F. The treated puree was then drum dried to form dehydrated flakes. The modified process enables the use of freshly harvested and/or high starch cultivars also for the production of flakes with acceptable quality.

The factors affecting film thickness and uniformity on drum dryer during flake manufacture were studied by Wadsworth et al. (1967). They found that the thickness of the puree film is affected significantly by changes in drum spacing; drum velocity puree viscosity and the level of puree in the dryer. High drum surface temperature high puree viscosity and low puree level in the dryer adversely affected the film uniformity. The bulk density of the dehydrated sweetpotato flakes could be controlled within a range of 27–37 $1\text{b}/\text{ft}^3$ by varying the contact angle of the doctor blade against the drum through an arc of 60 °C.

Differences in the native amylases activity in sweetpotato cultivars could lead to inconsistency in the quality of flakes (Deobald et al., 1968). They reported that enhancement in amylase activity could be achieved through addition of calcium ions and increasing the holding time. Similar modulations in amylases could be done, when cured and stored roots having higher amylase activity are used, by reducing the conversion time.

Discoloration of sweetpotato flakes was controlled by the addition of sodium acid pyrophosphate (SAPP) and tetrasodium pyrophosphate (TSPP) to the puree or dried flakes. Flavor retention was maximum, when 0.2% of 3:1 mixture of SAPP and TSPP was used (Hoover, 1963). Citric acid, added to the puree @ 0.2% was also reported to improve the colour of the dehydrated flakes (Woodroof and Atkinson, 1944).

Fortified sweetpotato flakes were made through the incorporation of soy flour, cottonseed flour and wheat gluten flour into the puree. Although the protein levels could be elevated through fortification, the water holding capacity of the reconstituted flakes was reduced (Walter et al., 1978a). Development of off flavours in sweetpotato flakes has been reported during storage, resulting mainly from the autoxidation of carotene and lipid fractions (Purcell and Walter, 1968a; Walter et al., 1978b). It was found that the surface carotene was lost at about 100 times faster than bound carotene and oxygen-consuming reactions were activated during storage (Walter and Purcell, 1974). The fragmentation of β-carotene during autoxidation was studied using radioactive β-carotene and it was found that the major fractions of β-carotene remained unaffected. The 'hay-like' flavor of stored sweetpotato flakes has been reported to be due to the formation of monocarbonyls especially aldehydes or due to the decrease in the unsaturation ratio of fatty acid (Lopez et al., 1976). The shelf life of sweetpotato flakes could be extended through the addition of antioxidants to sweetpotato puree, prior to dehydration (Deobald and McLemore, 1964).

Amino acid changes in lipid autoxidised protein-enriched sweetpotato flakes were studied by Walter et al. (1978b), who reported that glutamic acid was lost from protein fortified as well as control sweetpotato flakes. Lysine content decreased with storage of protein-fortified sweetpotato flakes.

Other Puree-Based Products

Restructuring of sweetpotato puree into various shapes and sizes to improve the product texture, has been attempted by many workers. The physical and sensory characteristics of restructured sweetpotato puree texturized with methyl cellulose or methyl hydroxy propyl cellulose was studied by Truong and Walter (1994). Cooked puree were texturized using an alginate-calcium system, by first mixing the puree with trisodium phosphate and sucrose followed by alginate-sucrose and finally calcium chloride. The blended mixture was then extruded through 5.5 cm i.d sausage casings, cured overnight at 4 °C and held at –20 °C till use. Texture Profile Analysis showed that the texturized product made from sweetpotato cooked prior to puree preparation was the most fracturable and least springy.

The physical and sensory characteristics of restructured sweetpotato puree texturized with methyl cellulose or methyl hydroxyl propyl cellulose was studied by Truong and Walter (1994). Product equal in quality to baked roots could be produced using 0.25–0.50% methyl cellulose or MHPC at 60 °C.

Sweetpotato puree has been used to produce patties, which are suited to the frozen convenience food market. The process requires incorporation of sufficient starch into the puree, to develop it into a thin patty. The outer surface of the patty could be coated with various batters. Hoover et al. (1983) produced sweetpotato patties from cooked sweetpotatoes, mashing them into a paste, mixing with unmodified corn starch and sucrose. Other minor ingredients like mono and diglycerides, sodium chloride, sodium acid pyrophosphate were also added to the paste, mixed and pushed through a screen to separate fibers. This was then steam cooked at 104–116 °C, vacuum cooled and molded into patties of 2″ diameter and $^1/_2$″ thickness. Frozen patties are cooked in peanut oil before use. Walter and Hoover (1984) reported that the starch content after cooking of the patty mix was the most important factor deciding the consistent quality of sweetpotato patties. Scanning electron microscopic studies on cooked and pureed sweetpotatoes demonstrated almost complete rupture of the cells and indicated that the patty was held together by an amorphous matrix containing added ingredients and spilled cellular contents (Walter and Hoover, 1984).

Johnson et al. (1992) reported the development of sweetpotato yogurts from puree. Sweetpotato puree was made from baked sweetpotatoes prepared using a convectional oven at 176 °C. Sweetpotato puree was then mixed with fruit pulp, lemon juice, sugar etc. The formulated puree was then combined with yogurt base in proportion of 1:4 and the mixture was refrigerated and evaluated. The yogurt made with sweetpotato and apricot was rated as good. Sweetpotato roots (anthocyanin rich accessions ST13 and 'Gouri') were used for the production of curd, using traditional starter cultures (Ray et al., 2005).

The anthocyanin content in sweetpotato curd made with ST13 was found to be around 8–11 mg per 100 g curd, when the boiled sweetpotato puree was added at the rate of 8–16% (Ray et al., 2005). Higher consumer acceptability was obtained for curd with 8% anthocyanin rich sweetpotato, while higher level of incorporation (Up to 16%) was acceptable in the case of β-carotene rich varieties (Table 11.2).

Lactic acid fermentation of β-carotene rich sweetpotato for the production of lacto-juice was investigated by Panda and Ray (2007). Sweetpotato roots (boiled or non-boiled) were fermented with *Lactobacillus plantarum* MTCC 1407 for 48 h to make lacto-juice, having a pH of 2.2–3.3 and lactic acid content of 1.19–1.27 g/Kg root. Use of boiled roots was found to affect only the β-carotene retention in lacto-juice, with a lower value 130 mg/Kg roots compared to fresh roots 165 mg/Kg roots.

Collins and Hutsell (1987) prepared 'vegetable leather' from sweetpotato, possessing a supple leathery texture. The ingredients included dextrinized sweetpotato puree, apple puree, honey, evaporated milk, pineapple puree and margarine, besides spices. The leather mix was spread on waxed paper to a thickness of 1.5 mm and dried. The dry sheets were rolled into a scroll and covered with wax paper and foil. These workers found that puree made from baked sweetpotatoes was the best for the preparation of leather. The leather had a crude protein content of Ca 6.0% and dietary fibre content of 4.1%.

Spray drying of amylase hydrolyzed sweetpotato puree was attempted by Grabowski et al. (2006). The effect of viscosity reduction of sweetpotato puree with alpha-amylase, maltodextrin addition and inlet air temperature on the physico-chemical characteristics of spray-dried powder was investigated. A steam-jacketed mixer was used to elevate the temperature of the puree as well as for blending it with alpha-amylase. After a reaction period of 30 min, the temperature was raised to 90 °C and maltodextrin was added as a drying aid and the mixture was spray-dried. Maltodextrin significantly increased the powder solubility, altered the hue value and raised the glass transition temperature of the powder. Alpha-amylase

Table 11.2 Biochemical characteristics of anthocyanin rich sweetpotato curd*

Sweetpotato (%) in curd	pH	Lactic acid (g/100 g)	Anthocyanin (mg/100 g)	Protein (g/100 g)
8	3.5	0.61	4.8	16.5
12	3.5	0.62	7.8	15.5
16	3.49	0.57	11.0	17.0

* Reproduced from Ray et al. (2005).

treatment could reduce the glass transition temperature and decrease the particle size of the powder. The spray-dried powder was further characterized for nutritional value and rheological properties by Grabowski et al. (2007). Spray drying significantly reduced the β-carotene and ascorbic acid contents. The all-trans form of β-carotene was found to be transformed into cis-isomers during the spray drying. The viscosity of the reconstituted solution was found to be much lower than that of puree and was rheologically similar to pre-gelatinized starch solution.

Intermediary Food Products

Sweetpotato roots can be termed as a '3-in-1' product, as it integrates the qualities of cereals (high starch), fruits (high content of vitamins, pectins etc.) and vegetables (high content of vitamins, minerals etc.). The beneficial effects of these ingredients have been appropriately put to use by molding the roots into a number of intermediary food products like jam, jelly, soft drinks, pickles, sauce, candies etc. However, these are produced to a commercial level only in a few countries like Philippines and Bangladesh.

Sweetpotato is processed into jams, in the Philippines, making use of the available water-soluble pectin in the roots. The process consists of cooking a mixture of 20.7% sweetpotato, 45% sugar, 34% water and 0.3% citric acid until a solids content of 68° Brix was reached. Sensory evaluation of fruit-flavoured sweetpotato jam scored high for taste, but gelling consistency was slightly softer than fruit jam due to the high content of starch in the roots. Sweetpotato jam is also prepared on a small scale in parts of China (Sheng and Wang, 1987). The high carotene content of orange-fleshed variety '*Kamala Sundari*' of sweetpotato is utilized to develop naturally coloured jam in Bangladesh (Shah Chaudhury, 1992). Jam is prepared from sweetpotato pulp (1.0 Kg), sugar (0.90 Kg), citric acid 1% and pectin (0.5%) or alternatively using sugar cane or date molasses instead of sugar. The shelf life of the jam was found to be more than one year. Sweetpotato pulp is mixed with fruit pulp like mango, banana or apple for making jams in India, which will also help to mask the typical sweetpotato flavour (Padmaja and Premkumar, 2002). However, the consistency was slightly softer due to the high starch in the roots, permitting the incorporation of only up to 60% levels.

Sweetpotato is processed into candies in parts of Japan and China (Woolfe, 1992). The pale cream or purple-fleshed cultivars are used for this purpose. The sweetpotato mash is mixed with barley malt for 1.5 h at 55 °C to enable the hydrolysis of starch to maltose and dextrin. The expressed juice is then concentrated under fire and pulled by hand/machine to introduce air and harden it. The sheeted candy is cut into fixed shapes. Sweetpotato candy was prepared in Malaysia using coconut milk, skimmed milk, refined sugar and glucose syrup (Samsiah et al., 2005). The best formation based on sensory scores was found to have 10% steamed sweetpotato mash, cooked at 120 °C.In China, sweetpotato pieces are candied in sugar syrup, containing flavouring and then dried in an oven for 8 h (Wiersema et al.,

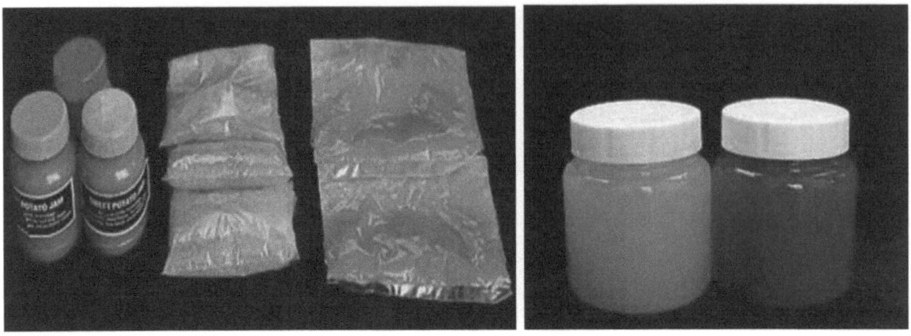

Plate 11.1 Sweetpotato jam (Courtesy: Padmaja and Premkumar, 2002) (See also Plate 7 on page xx)

1989). Jelly is prepared from orange-fleshed sweetpotato (variety: *Kamala Sundari*) by extracting the juice and mixing with sugar (50:50), citric acid (1.5%), pectin (2%) and flavourings. Jelly has a lighter consistency than the sweetpotato jam (Shah Chaudhury, 1992) (Plate 11.1).

Pickling of sweetpotato has been attempted in India, Philippines and Bangladesh. Tan et al. (2005) used orange-fleshed varieties of sweetpotato *viz.*, VSP-1 and RC-2000 for making sweetpotato pickles. Both low salt and high salt curing methods were adopted. It was found that low salt concentrations promoted the growth of undesirable yeasts and molds. Acceptability was more for the pickles made from blanched sweetpotato shreds than that from raw shreds.

Sweetpotato (variety: *Kamala Sundari*) has been pickled in Bangladesh using the traditional ingredients. The sweetpotato pieces were slightly fried, to enhance the taste (Shah Chaudhury, 1992). In India, the pale cream varieties of sweetpotatoes were found to give the most acceptable pickles. The diced cubes are first treated with 1.0% acetic acid solution for 1 h to prevent browning and impart acid taste to the slices. The cubes are then made into pickles using the standard ingredients. The shelf life of tightly bottled pickles was found to be more than six months (Padmaja and Premkumar, 2002).

The soft texture of the sweetpotato pulp is suited for making soft drinks, by mixing with thick or thin fruit pulps/juices. The cooked/mashed and sieved pulp is mixed with ripe mango pulp or orange/lemon/pineapple juice and made into soft drinks. Appropriate flavouring has been found to enhance the acceptability (Padmaja et al., 2005a) (Plate 11.2).

Sweetpotato Leaves as Human Food

Sweetpotato leaves, though a rich source of vitamins, minerals and protein have been much less used as a human food. Sweetpotato green tips are used as a vegetable in parts of the world (Villareal et al., 1979).The nutritive value of sweetpotato leaves has been attributed to the high content of antioxidants especially phenolic compounds in them (Islam et al., 2002; Yoshimoto et al., 2005). The various phenolic

Plate 11.2 Sweetpotato soft drink (Courtesy: Padmaja and Premkumar, 2002) (See also Plate 8 on page xxi)

fractions have been characterised from sweetpotato leaves and the caffeoylquinic acid derivatives have been associated with the anti-mutagenic effect of the leaves (Yoshimoto et al., 2005). Ishiguro and Yoshimoto (2005) also reported the high content of a α-xanthophylls, lutein in sweetpotato leaves, which has got eye-protectant effect. Lutein, present to the extent of 29.5 mg/100 g fresh weight was more than the levels present in around 120 fruits and vegetables.

The cooking of young shoots of sweetpotato has been reported to decrease its total protein from 3.7% to 2.5% (fwb) during a period of 4 min (Onate et al., 1970). Similar decreases have been reported in the case of American and Asian sweetpotato varieties also (Haytowitz and Mathews, 1984; Leung et al., 1972). Leaching losses of ascorbic acid, carotene and minerals have also been reporting during cooking in water for 4 min. Maeda and Salunkhe (1981) found that the fresh leaves of sweetpotato contained around 49.6 mg/100 g (fwb) of carotene and 1374 mg/100 g Vitamin C (fwb) and drying the leaves in open sunlight led to 96% loss in carotene and 98% loss in Vitamin C. Blanching sweetpotato leaves in boiling water for 50 sec followed by drying in an enclosed solar drier retained 34% of carotene.

Secondary Food Products

Sweetpotato flour or starch is the base material for making a number of secondary food products.

Noodles and Other Extruded Foods

Sweetpotato is processed into noodles in many countries of the Far east *viz.*, China, Japan, Taiwan and Korea. A major part of sweetpotato starch produced in China and Korea is utilized for the production of noodles (Wiersema et al., 1989; Hong, 1982). There are many home scale and cottage level processing units in China, which make traditional noodles from sweetpotato starch. The process consists in gelatinizing sweetpotato starch slurry in a big vessel at 80 °C, treating with sulphate to prevent discoloration and mixing with native dry sweetpotato starch (5%) to form a dough. The dough is then filled to long cylindrical column (30 cm × 40 cm) and pressed to extrude the dough into strings into hot water. This is then separated manually to prevent adhesion. The strings are then suddenly put to cold water, when the outside hardens and stickiness is reduced. The noodles are then dried slowly so that both inside and outside get dry (Wiersema et al., 1989). In the Korean method, sweetpotato starch is used to replace part of the wheat flour. The noodles are made from sweetpotato starch, wheat flour and salt. Steam cooking of wet sweetpotato starch, over boiling water is adopted in Vietnam. The extruded starch is then dried in bamboo racks and semi-dried material is cut into strips. Noodles are often dark coloured due to phenol oxidation, which limits its marketability.

The physicochemical properties of sweetpotato starches and their applications in noodle making were studied by Chen (2003). The process consisted of gelatinizing 5% sweetpotato starch with water (1:9 w/v) and then mixing with 95% native starch. The mixture was made into dough at 40 °C (moisture content 55%) and then extruded through a lab-scale extruder to strips of 1.5 cm diameter, directly into hot water (95–98 °C). The noodles were kept for 50–70s at this temperature and then transferred to cold water. Pre-cooling was done at 40 °C for 6 h and then frozen at 5 °C for 8 h followed by drying. Chen (2003) reported varietal differences in the cohesiveness of sweetpotato starch noodles. Freezing was found to be an important step in noodle manufacture as this could considerably reduce the cohesiveness.

Sweetpotato flour (white-fleshed varieties) was used for extrusion into vermicelli by Thirumaran and Ravindran (1992). The process consisted of mixing sweetpotato flour with refined wheat flour (50:50) or with added legume flour also (10–30%), steaming the dough, extruding, drying and packaging. Although the energy values of sweetpotato flour: wheat flour vermicelli was high (234 Kcal/100 g), protein was low (6.1%), and this could be enhanced through legume fortification. In Japan, sweetpotato flour is mixed in varying proportions with other cereal flours and extruded into noodles. Orange-coloured noodles prepared from OFSP and green noodles with sweetpotato leaf incorporation are sold in Taiwanese markets. Chang and Lee (1974) reported that addition of sweetpotato flour beyond 20% to wheat flour affected the organoleptic quality of the noodle. Sweetpotato flour enriched with soy flour utilized for replacing part of the wheat flour in noodle making (Collins and Pangloli, 1997). Combinations of sweetpotato and defatted soy flour (DSF) increased the protein and dietary fibre content of the noodles. Orange-fleshed sweetpotato increased the β-carotene (5.0 mg/100 g) as compared to 0.2 mg/100 g in wheat flour based noodles. Although cooking loss was increased by 12%, the product had acceptable flavour and reduced stickness. Lee (2005) developed extruded ready-to-eat breakfast food and snack foods from sweetpotato. The sweetpotato flour was either used alone or blended with cassava flour and extruder through the Brabender extruder. A screw compression ratio of 3:1 and round die with i.d of 2 mm were used. Increasing the proportion of sweetpotato flour was reported to reduce the diameter of the extruded snack.

Restructured sweetpotato sticks were made from cooked and mashed sweetpotato using extrusion technology (Utomo et al., 2005). White, yellow and orange varieties of sweetpotato were blanched, mashed and mixed with 0.3% Carboxymethyl cellulose as binder. The mash was then mixed with 5% sweetpotato flour and extruded through a tube to produce sticks of 1.0 cm diameter and 5.0 cm length. The sticks were deep fried in oil at 163 °C for 1 min and frozen at – 20 °C till use. Further frying was done at 175 °C for 2 min, prior to eating. The product from yellow varieties had the lowest values for firmness, hardness and shearing force, while that from the white variety was more cohesive and chewy. Response surface analysis of the color of single screw extruded blends of soy-sweetpotato flour- was done by Iwe et al. (2000), who found that the whiteness decreased with increase in the level of incorporation of sweetpotato in the blends. While redness increased as the content of sweetpotatoes increased, yellowness increased especially as a result of die diameter. The effect of extrusion cooking of soy-sweetpotato mixtures on available lysine content and browning index were studied by Iwe et al. (2004). Increase in screw speed and a reduction in die diameter enhanced lysine retention in the product and increased the browning index.

Sugar Syrups

Both glucose syrup and HFS are commercially made from sweetpotato starch in China (Wiersema et al., 1989). Sweetpotato starch is converted to glucose syrup or high fructose syrup for use in confectionery industries, pharmaceutical applications etc. Microbial enzymes with high conversion efficiency are available to effect the liquefaction and saccharification reactions, which have advantages of the earlier acid-linked hydrolysis.

The optimum temperature for liquefaction using liquezyme X was standardized as 90 °C for 1 h. Viscosity profile analysis during liquefaction of sweetpotato starch indicated that small amount of liquezyme X (6.0 mg/100 ml slurry) could bring down the viscosity of a suspension (1: 10 w/v) from 3653 cP to 972 cP. Approximately 96% conversion of sweetpotato starch to glucose could be obtained with 48 h of action by the saccharifying enzyme, Dextrozyme X on liquefied sweetpotato starch slurry. Glucose isomerase (Sweetzyme T) could effect the conversion to HFS at 80 °C and pH 7.0 (Regy Johnson et al. unpublished data).

Commodity Chemicals from Starch

Sweetpotato starch is commercially utilized for the production of a number of commodity chemicals like citric acid, mono sodium glutamate, microbial enzymes etc. which are used in the food industry. Most of these are produced on small scale in China and Japan where sweetpotato starch is industrially produced. The starch is first converted to sugars and fermented to citric acid by *Aspergillus niger* (Wiersema et al., 1989).

Monosodium glutamate, a flavour enhancer for various foods is manufactured from sweetpotato starch in China. Sweetpotato starch is first hydrolysed using enzymes to glucose, which is then converted by *Brevibacterium glutamicus* to glutamic acid. Monosodium glutamate is produced using alkali treatment of glutamic acid (Wiersema et al., 1989).

Non-alcoholic and Alcoholic Beverages

Non-alcoholic beverages have been prepared in the Philippines and India from sweetpotatoes. A fruity sweetpotato beverage developed from sweetpotato contains in addition to mashed sweetpotato (variety VSP-1), sugar, citric acid and ascorbic acid (Truong and Fementira, 1990). Orange-fleshed sweetpotatoes yielded beverages with pleasant color and aroma. Truong (1992) observed that the aroma could be further enhanced by adding pulp of fruits like guava, pineapple, lemon etc @ 0.6–2.4% w/v. The sweetpotato beverage was found to have an acidic pH of 3.2 with a solids concentration of 13 Brix and insoluble solids content of 9.4 mg/100 ml. An intake of 8 ounces of the fruity-sweetpotato beverage (VSP-1) could provide the daily requirement of vitamin A for adults (Truong, 1992).

Non-alcoholic beverage has been prepared from cream or orange-fleshed variety of sweetpotato by mixing the cooked and mashed pulp of sweetpotato with pulp of ripe mango or fruit juices from orange, lemon, pineapple etc in India (Padmaja and Premkumar, 2002).

Alcoholic Beverage

'Shochu' is traditional distilled liquor made from sweetpotato or other sources like rice, barley, buckwheat etc. The process consists in first preparing a fermentation broth from rice by crushing white rice, steeping in water for 3–4 h, steaming, cooling and then adding seed 'Koji' to the steamed rice, as a starter. The starter 'Koji' contains *Aspergillus niger* or *A. Kewachii* and the mould growth is facilitated at 38–40 °C for 24 h followed by 18 h fermentation at 34–36 °C. The 'Koji' is then mixed with traditional yeast, *Saccharomyces cerevisiae* and adequate water. The seed mash after 5–7 days of incubation at 25–30 °C is added to steamed sweetpotato slurry. Further incubation at 30 °C for 10–12 days yields a broth having 13–15% alcohol, which is distilled and blended to form 20 to 40% alcohol.

The 'Koji' process for 'Shochu' making has been hitherto improved using microbial enzymes, which helped to improve the quality and reduce the reaction time (Ogawa et al., 1982). Kudo et al. (1979) reported vacuum distillation as a better method, compared to steam distillation, as the former reduces the content of furfural, acetaldehyde etc in the alcohol, contributing to off flavours. The presence of the furanoterpenoid compound, ipomeamarone has been detected in Shochu, made using weevil damaged sweetpotato roots. This being a lung toxin has to be removed by treatment of Shochu with 0.03% activated carbon for 5 h (Kudo and Hidaka, 1984). Ohtai et al. (1990) characterized the flavour components of 'Shochu' and found that it contains several monoterpene alcohols, like linalool, α-terpineol, citronellor, nerol and geraniol.

Flour Based food Products

The possibility of replacing part of the wheat flour with sweetpotato flour for making baked foods was investigated by many workers (Seralathan and Thirumaran, 1990; Montemayor and Notario, 1982). Most of these studies showed that acceptability is reduced due to dominant sweetpotato flavor, when higher substitution was attempted. Replacement of wheat flour up to 30% was acceptable for cakes, biscuits, muffins etc. Collins and Aziz (1982) reported that up to 21% sweetpotato flour addition to wheat flour did not affect the quality of doughnuts.

Sweetpotato flour has been used for the production of sauce, a product similar to soy sauce in the Philippines (Datta et al., 1986). The process consists in mixing steamed soybeans with roasted sweetpotato flour (50:50) and spreading the mix on trays, mixing with starter culture of *A. oryze* or *A. sojae* and incubating for 4–5 days at room temperature. The culture broth was stirred to permit aeration and the 'Koji' was then transferred to plastic containers and mixed with salt and water. Further incubation for 3 months followed by straining to extract the sauce yielded slightly thin sauce. Color and viscosity were improved through the addition of molasses and pasteurized at 80 °C for 30 min prior to bottling.

Sweetpotato flour (white, orange and purple fleshed varieties) was used to develop an instant gulab jamun mix in India (Padmaja et al., 2005b). Gulab jamun is a popular sweet dessert of India, made traditionally from refined wheat flour and milk powder or 'khoa' (concentrated milk). The possibility of replacing part of the milk powder and refined wheat flour (RWF) with sweetpotato flour was investigated. It was found that 20% sweetpotato flour addition to milk powder: RWF mix (31: 29) gave an instant mix, which produced highly acceptable gulab jamuns. White-fleshed sweetpotato varieties, having a soft texture on cooking are suited for making the instant mix (Padmaja et al., 2005b) (Plate 11.3).

Industrial Utilization of Sweetpotato

Sweetpotato roots with 20–30% starch are one of the major sources for the commercial extraction of starch. The roots are extensively used for starch extraction in

Plate 11.3 Sweetpotato based gulab jamuns and instant mix (Padmaja et al., 2005b) (See also Plate 9 on page xxi)

China, Japan, Korea and Taiwan. The basic process for manufacture of starch is similar to the process for other sources. The crushed roots are separated from fibre and allowed to settle in long columns in the sun. Saturated limewater is used during grinding and sieving, to brighten the sweetpotato starch, by preventing the action of phenol oxidizing enzymes. In modern starch plants, jet nozzles are used to separate the coloring matter, protein, slimy starch etc and the clean starch is separated using centrifugal separators. Flash drier is used to reduce the moisture content to less than 18%.

Sweetpotato starch finds commercial application in the production of noodles, sugar syrups, thickeners etc. China uses a major fraction for ethanol production.

Alcohol

Sweetpotato starch is similar to any other starch source for use as a raw material for the production of alcohol. As high as 98% conversion to glucose could be obtained using improved enzymes for the liquefaction and saccharification of starch, *viz.*, liquezyme X and Dextrozyme GA (Regy Johnson, unpublished data). However, use of raw tubers for ethanol production may not be economical, unless otherwise the pectin and cellulose matrix is broken using enzymes, to release the starch and make it available for the alpha-amylase action. Cooked sweetpotatoes, fermented with *Rhizopus niveus* was reported to yield up to 37.5 ml ethanol per 100 g starch (Sreekantiah and Rao, 1980). The native amylases in sweetpotato were activated by heating the root slurry to 70–80 °C and up to 90% conversion of the total solids could be achieved. Glucoamylase, pectin depolymerase and yeast were added together to minced sweetpotato roots and incubated for 5 days to yield 100 g per Kg of roots (Svendsby et al., 1981). Use of membrane reactors in the enzymic hydrolysis of sweetpotato for alcohol production was studied by Azhar and Hamdy (2004). They found that the immobilized enzymes in the thin-channel system showed a much better performance compared to the stirred cell system.

Other Commercial Value Added Products

Sweetpotato starch as well as the solid and liquid wastes from starch factory has been used for the production of single cell protein. The wastewater from starch factories has been utilized for culturing yeasts, yielding single cell protein in Japan (Sugimoto et al., 1967; Takakuwa et al., 1972). The whole sweetpotato root based media have also been attempted for the production of single cell protein (El-Ashwah et al., 1980).

Orange and purple-fleshed varieties of sweetpotato have been commercially used in Japan, for the production of pigments for use in food products, soft drinks etc. the important pigments are β-carotene in OFSP and anthocyanins in purple-fleshed varieties.

Roots and Vines as Livestock Feed

Sweetpotato roots and vines have been used as animal feed for many years in China, Japan and Taiwan. Besides, the wastes from starch and alcohol factories are also used as animal feed. Many research studies are available dealing with the use of sweetpotato roots as cattle, pig poultry and fish feed.

Pig Feed

The effect of feeding sweetpotato roots and vine on the growth performance of growing and finishing pigs was investigated by Koh et al. (1976). They found that a crude protein intake of 334 g and energy intake of 8.5 MCal could lead to satisfactory growth performance of pigs. Feeding raw sweetpotatoes to pigs requires soybean meal or other protein sources in adequate amounts to compensate for the trypsin inhibitors. Many feed experiments show that the growth performance of pigs is affected when dried sweetpotato chips was fed to pigs, compared to those fed corn. However the daily weight gain and feed-to-gain ratio were superior, when sweetpotato was substituted in corn rations @ 25% (Yeh et al., 1977). Tai and Lei (1970) reported that the back fat thickness was reduced in pigs, as the proportion of sweetpotato chips was increased. The porcine fat melting point and unsaturated fatty acid content are significantly lowered when sweetpotato was fed to pigs. Formula feeds incorporating sweetpotatoes should have 16% crude protein and 3.2–3.3 M Cal DE/Kg at the starter (growing) stage and 14% crude protein with the same energy value in the finishing feed.

Rose and White (1980) reported high digestibility of 93–94% for energy and 42–57% protein when fresh sweetpotato was fed to pigs. However, high protein digestibility of 77% was reported for dry sweetpotato meal (Wu, 1980). The net energy of sweetpotato was about 79% that of Corn. Furuya and Nagano (1986) reported higher percentage of digestible energy in boiled and oven-dried roots than freeze-dried roots. Sweetpotato meal prepared by mixing the roots and tops in the ratio 3:1 followed by pressing and sun-drying was found to have significantly lower digestible energy than corn (Han et al., 1976), probably due to the high fibre content of the tops.

Growth performance and nutrient digestibility were studied in male piglets fed on boiled sweetpotato root based ration (Gupta et al., 2006). The rations contained, besides boiled sweetpotato, maize, deoiled rice polish and soymeal and were enriched with 2% minerals and 0.1% DL-methionine. These workers reported high dry mater and protein digestibility in piglets fed with boiled sweetpotato (40% of the total DM intake). High protein sweetpotato roots (7.6% dwb) were used as pig feed in Korea and the performance of pigs was better than those fed corn meal (Han et al., 1976)

Yeh and Bouwkamp (1985) found that pigs fed 100% sweetpotato together with vines and soymeal took 201 days to reach 90 Kg, as pared to only 118 days for pigs fed 100% corn-soymeal. This was imparted to the effect of the trypsin inhibitors in sweetpotatoes. Raw sweetpotato based diets also have low starch digestibility in pigs. Yeh et al. (1977) found that microwave cooking of dried chips at 80–90 °C up to 4.5 min could not change the starch availability. However, popping the dried chips with 9–10% moisture at 6–8 Kg/cm^2 pressure at 164–175 °C was found to improve the starch availability by 194% and the daily gain and feed-to-gain ration were significantly higher.

Sweetpotato root and vine silage also finds use as pig feed. Castillo et al. (1964) described the construction silos for the preparation of such silage. They found that corn can be completely replaced with sweetpotato silage without affecting the performance of the pigs. Sweetpotato roots ensiled with barley bran have been used for pigfeeding in Japan (Kurihara and Imamura, 1956). However, initial growth rate was reported to be affected by feeding 60% roots, 30% vines and 10% barley bran silage (Jung and Lee, 1968). Tomita et al. (1985) found that the apparent digestibilities of dry matter, organic matter, total protein and energy in sweetpotato root silage were 91%, 91%, 32% and 89% respectively.

Cattle Feed

Although both roots and tops can be used as cattle feed, most of the studies have been conducted on the feeding value of foliage (Chen and Chen, 1979). Sweetpotato vines has gained popularity as a commercial cattle feed in Peru, China, Japan and Taiwan (Achata et al., 1988). There are several reports on the feeding value of sweetpotato roots to dairy coves. Replacement of corn with dehydrated sweetpotato roots led to 88–90% of the milk yield compared to those fed corn based diet (Mather et al., 1948). Increase in the vitamin A and carotene content of milk by 22% and 30% respectively has been reported when orange-fleshed sweetpotato is fed to cattle (Mather et al., 1948). Replacement of corn-soybean meal silage with dehydrated sweetpotatoes and sorghum silage with fresh, chopped sweetpotato roots has also been reported not to affect the milk yield. Replacement of corn by 50% and 100%

with dehydrated sweetpotato roots has been reported to affect the daily weight gain of beef cattle by only an increase of 100 g/day (50% replacement) and decrease of 90 g/day (100% replacement) (Southwell and Black, 1948). Sweetpotato trimmings, a byproduct of sweetpotato canning industry was found to be inferior as a beef cattle ration, resulting in poor carcass quality (Bond and Putnam, 1967). Backer et al. (1980) reported lowest weight gains from beef cattle fed all vines and best weight gains when 50% vines and 50% fresh roots were fed. Supplementation of sweetpotato vines with soybean meal has been reported to increase the growth rate of cattle. However, dehydrated vines were less palatable and required more concentrate to reach the same weight.

The production of sweetpotato vine silage for cattle/ruminant feeding has been studied by many workers (Konaka, 1962; Sutoh et al., 1973). Addition of in organic acids like hydrochloric acid and sulphuric acid in the ratio 19:1 and molasses was found necessary to control the fermentation in vine-root silage. Carpio (1984) observed better feed conversion efficiency, when 1:1 ratio of sweetpotato leaves and grass was fed to fattening goats when compared to 100% grass forage. Aregheore (2003) investigated the nutritive value of sweetpotato forage and its mixtures with batiki grass as goat feed. Better live weight gains were obtained when mixed diets were fed to goats and based on the data; he suggested that sweetpotato forage was a cheap nitrogen source in the diets of growing goats.

Fish Feed

There are also a few reports on the use of sweetpotato leaves and roots as fish feed. Mwanga and Wanyera (1988) reported that the leaves could be fed to *Tilapia* spp to supplement its growth. Highest weights and body length have been reported in crayfish fed *ad libitum* with dehydrated sweetpotato vines and leaves (Goyert and Avault, 1977). Use of sweetpotato distillery byproducts as a feed for red carp (*Cyprinus carpio*) was studied by Mokolensang et al., 2003. They found that fish attained greater weight coupled with lower feed conversion ratio when the sweetpotato distillery by product was fed.

Poultry Feed

Limited numbers of studies have been conducted on the feeding value of sweetpotatoes for poultry. Lee and Yang (1979) reported that chicks fed with diets containing 20% dried sweetpotato chips grew well and their performance was similar to those fed 100% corn diet in mash or pellet form. They also found that in broiler rations, 20% incorporation of sun-dried sweetpotato chips was possible, while in the layer chicks only 10% was preferred. Job et al. (1979) reported that upto 60% replacement of corn meal with dried sweetpotato meal was possible in broiler chick diets. Though protein content in the major tissues was not affected, fat content was considerably low. Contrary to this Tillman and Davis (1943) observed that only up to 20% substitution was possible in broiler diets and higher levels led to decreased weight gain. Increased digestibility of nutrients has been reported when cooked roots were dehydrated (65%) and fed to young chick than uncooked roots (Yoshida et al., 1962).

A diet containing 25% orange-fleshed sweetpotato meal (130 mg carotene/Kg) fed to layers was found to yield eggs rich in vitamin A in the egg yolks (Lease and Mitchell, 1940). Weber (1969) also reported higher yolk pigmentation in eggs from layers fed dried orange-fleshed 'centennial' variety of sweetpotato. Garlich et al. (1974) also found that broiler skin and egg yolk pigmentation was better, when carotene rich sweetpotato roots and leaves were fed to poultry.

Nutritional Profile of Roots and Leaves

The starchy roots and the green tops are the two nutritionally valuable parts of the sweetpotato plant. Sweetpotatoes are considered as one of the world's healthiest foods, due to the high content of nutraceuticals like phenolic acids, carotenes, xanthophylls and anthocyanins.

Major Nutrients in Roots

Starch and Sugars

Carbohydrates are the highest nutrient available in sweetpotato roots. It accounts for around 80–90% of the total dry mater. The dry matter contents varies depending on the cultivars, cultural practices, climate etc and is reported to range from 13 to 48% (Bradbury and Holloway, 1988; Cereda et al., 1982). The carbohydrate in sweetpotato root is made up of starch, cellulose, hemicellulose, pectins and sugars. Starch accounts for 60–70% of the DM, although varieties having much less starch content, are also reported (Cereda et al., 1982; Truong et al., 1986). Wide variations in starch content, grown under identical conditions have also been reported, indicating that cultivar characteristics decide the starch content (Shanmugan and Venugopal, 1975; Prabhuddham et al., 1987). The content of various carbohydrate fractions in raw and baked sweetpotatoes (variety 'Garnet') has been worked out by Shen and Sterling (1981), who found that baking led to sharp increase in sugars from 22.4% (dwb) to 37.6%. Sweetpotato starch is composed of 30–40% amylose and 60–70% amylopectin (Madamba et al., 1973; Bertoniere et al., 1966).

The total sugars in sweetpotato cultivars also show wide variations, with a usual range of 0.38% to 5.64% (Bradbury and Holloway, 1988; Picha, 1985b). Truong et al. (1986) reported a range of 5.6%–38.3 % on dry weight basis for varieties grown in the Philippines. Sucrose is the most abundant sugar, followed by glucose and fructose in raw roots. Maltose could not be detected by several workers in fresh roots; however, there is tremendous production as a consequence of boiling or baking of roots. Picha (1985b) reported that in the 'Centennial' variety of sweetpotato, maltose was formed up to 9.33% during baking, while in another variety 'White star', as high as 14.12% maltose was present in baked roots. Maltose production during baking was also reported by Lila Babu (2001) for the Indian Cultivars. Varietal variations have also been reported in the glucose and fructose levels of sweetpotato roots, with glucose dominating over fructose in some cultivars and vice versa in others (Bradbury et al., 1985; Picha, 1985b).

Non-starch Polysaccharides

The non-starch polysaccharides comprising cellulose, hemicellulose and pectin contribute towards the 'dietary fibre' fraction of sweetpotato roots. Shen and Sterling (1981) reported a total dietary fibre content of 7.0% (dwb) in raw sweetpotato roots, which decreased to 3–8% during baking, resulting possibly from the degradation at the high temperature. Lund and Smoot (1982) reported that the raw sweetpotatoes contained around 0.44% lignin, which also contributed to the dietary fibre fraction. Owing to the prophylactic significance of dietary fibre in diseases like colon cancer, diabetes, heart ailments and conditions like obesity, sweetpotato could be considered as a health promoting food. Though many studies have not been conducted on the relative content of soluble and insoluble fibre in the total dietary fibre, one report indicates that out of a total content of 3.14 g in baked roots, 1.01 g is contributed by soluble fibre and the rest by insoluble fraction. The pectic constituents of sweetpotato have been studied in more detail by various workers, as they contribute to the rheological properties of cooked roots (Heinze and Appleman, 1943; Ahmed and Scott, 1958). Very high pectin content of 5.1% (fwb) has been reported for certain varieties.

Protein and Amino Acids

Sweetpotato protein is the well-studied nutrient of the roots. The crude protein content in most varieties ranges from 1.3 to 10% (dwb) (Purcell et al. 1972; Splittstoesser, 1977). Many factors besides cultivar differences, like cultural practices and environmental influence have been found to affect the protein content in sweetpotato (Constantin et al., 1974). Research on sweetpotato protein has been reviewed by Walter et al. (1984), who report that the possibility of enhancing the protein nutritional quality through cultural and genetic manipulations is high.

The scope of biofortification of protein in sweetpotato through breeding was investigated by many workers (Dickey et al., 1984). Distribution of protein within the sweetpotato roots was investigated by Purcell et al. (1976). The protein content was reported to be slightly more at the proximal end (near to stem) than at the distal end. Also, outer layers were found to have more concentration of protein (Purcell et al., 1976; Bradbury et al., 1984). Various workers have reported widely different values for the amino acids in 100 g crude protein. Whilst threonine and valine occur in high amounts, the sulfur containing amino acids are present in small amounts only, having an amino acid score of 37–63. However, a high score of 93 has been reported for the total sulfur containing amino acids by Purcell and Walter (1982).

Purcell and Walter (1980) found that the main amino acids of the NPN fraction were asparagin (61%), aspartic acid (11%), glutamic acid and serine (4% each) and threonine (3%) in 'Jewel' sweetpotatoes, stored for 107 days. Sweetpotato crude protein value includes the non-protein nitrogen also, occurring in high proportions (15–35% of cP) and hence the values for protein reported for sweetpotato do not represent the true protein.

Most of the storage protein in sweetpotato is sporamin, accounting for 75–80% of the total protein in roots (Maeshima et al., 1985). Two related proteins having similar molecular weights, but different amino acid profiles *viz.*, sporamin A and sporamin B constitute the total sporamin in the roots and lysine was the limiting amino acid for sporamin A.

The protein nutritional quality of sweetpotato has been studied by several workers (Walter and Catignani, 1981; Walter et al., 1983). Most of these reports indicated the total sulfur containing amino acids as the first limiting factor. Lysine was reported as the 2nd limiting amino acid. Though variations among the reports are evident with regard to the limiting amino acids, sulfur-containing amino acids have been reported as limiting by more workers.

Based on animal feeding studies, the protein efficiency ratio (PER) of sweetpotato protein was reported as 1.9 (Horigome et al., 1972). Addition of lysine and methionine was reported to elevate the protein to 2.5.

Minor Nutrients in Roots

The total lipid content in sweetpotato roots is in the range of 0.29–2.7% (dwb). However, the type of lipids present is important, as they decide the keeping quality and off-flavour development during storage of processed products like flakes and chips (Boggess et al., 1970). Linoleic acid is the major fatty acid and around 42% of the total lipids are neutral lipids followed by 27% by triglycerides in 'Centennial' variety. The lipid content of the peel was reported to be much higher than the flesh and hence peeling leads to loss of lipids (Faboya, 1981). The lipid fraction contained 44.7% of linoleic acid and 29.3% of palmitic acid (Boggess et al., 1970; Opute and Osagie, 1978). Most of the workers have found the sweetpotato lipids to be highly unsaturated.

Lot of variability has been reported by the different workers in the mineral content of sweetpotato roots. The ash content ranges from 3 to 4% (dwb). Makki et al. (1986) observed that whilst the peel contained 14.1% ash, the flesh contained only 4.6% ash (dwb). Most workers have reported high content of K, followed by P, Ca or Mg (Lopez et al., 1980; Picha, 1985a). Sodium content is low 13–30 mg/100 g (fwb), and other minerals like iron, zinc etc are very low.

Sweetpotato Root Vitamins

Sweetpotato roots are rich in carotenoids and vitamin C and contain reasonably good amounts of thiamin (B1), riboflavin (B2) and pantothenic acid. Beta-carotene is the most abundant pigment (provitamin A) in sweetpotato and orange-fleshed varieties contain as high as 16 mg/100 g (fwb). It is now recognized as one of the best sources of vitamin A and lot of global efforts are in to popularize the consumption of orange-fleshed sweetpotato especially in countries and among population, where vitamin A deficiency is a major problem. The content of carotenoids decide the flesh color of sweetpotato roots and ranges from white, cream, light orange and dark orange. The yellow and orange-fleshed cultivars of sweetpotato contain a major fraction of carotenoids as β-carotene (Purcell and Walter, 1968b). The carotenoids of 'centennial' variety of sweetpotato have been characterized by Purcell and Walter (1968b). They found that 86.35% of total carotenoids existed in the β-carotene form. Other major carotenoids were phytoene, phytofluene, α and γ-carotenes etc. Martin (1983) reported that beta-zea-carotene was the major carotenoid in many white-fleshed and light cream-fleshed sweetpotato roots. Wang and Lin (1969) reported that the carotenoid content of Taiwanese roots fell in the range of 0.40 mg/100 g to 24.8 mg/100 g (fwb). Garcia et al. (1970) reported values in the range of trace to 11.45 mg/100 g for 26 varieties grown in the Philippines. Gruneberg et al. (2005) reported that the orange-fleshed sweetpotato accessions in the CIP germplasm were very high in carotene and most of them had 8–14 mg/100 g (fwb) pro-vitamin A. Consumption of 100 g of OFSP per day by pre-school children ($<$ 5 years) was sufficient to provide the recommended dietary allowance (RDA). The carotenoids in sweetpotato were extracted using the enzymes, pectinase and cellulase by Cinar (2005). Stability studies conducted in the carotenoids at different temperatures and light regimes showed highest retention in pigments extracted from blanched roots and stored at 4 °C.

Walter and Giesbrecht (1982) observed no carotenoid destruction in lye peeled sweetpotato roots. However 15 min lye peeling (10% NaOH solution) was necessary to prevent discolouration. The carotenoid retention in boiled and mashed orange-fleshed sweetpotato (OFSP) was investigated by Jaarsveld et al. (2006). They found that as compared to chloroform: methanol (2:1), acetone, tetrahydrofuran etc, tetrahydrofuran: methanol (1:1) was more efficient in extracting carotenoids from both raw and cooked OFSP. As high as 92% retention was observed, when the OFSP roots covered with water was boiled for 20 min. Loss increased with cooking time. However, an increase of β-carotene in OFSP on blanching or boiling in water was reported by Lila Babu (2006), resulting probably from the better extractability from boiled roots. Microwave cooking retained the carotenoids almost totally, while baking retained only 75% of the original content.

Bradbury and Singh (1986) reported that the South Pacific sweetpotato cultivars contained around 14.3 mg/100 g of fresh roots. Schmandke and Gurra (1969) obtained only 9.5 mg/100 g in Cuban sweetpotato varieties. Holland et al. (1991) reported high content of 23 mg vitamin C/100 g in orange-fleshed sweetpotato varieties. They also observed very high content of 4.56 mg vitamin E/100 g (fwb), a potent anti-oxidant in sweetpotato roots, compared to 0.56 mg/100 g in carrots and 1.22 mg/100 g in tomatoes. Different cooking methods were reported to retain the vitamin contents to varying extents. (Lanier and Sistrunk, 1979). Baking for 75–90 min at 190 °C was found to retain more vitamins (ascorbic acid, niacin, riboflavin and total carotenoids) than microwave cooking (18–35 min). Maximum loss of vitamin C occurred in canned sweetpotatoes.

Nutritional Profile of Sweetpotato Leaves and Tops

Sweetpotato leaves and tops are consumed in many parts of the world as a nutritious vegetable. The green tips were reported to contain around 86% moisture, 2.7% crude protein, 2.0% fibre and 1.7% ash. The tips are also a rich source of vitamin A (5580 IU/100 g) and calcium (74 mg/100 g) (Anon, 1968). The dietary fiber content in the tips has been reported to increase with age, although the aged roots are less preferred for human consumption. Candlish et al. (1987) reported that the sweetpotato tips contained around 2.4 g/100 g dietary fibre (fwb). About 80% of the edible fibre was constituted by pectin, cellulose and hemi cellulose.

The leaves are also a very good source of vitamins especially β-carotene and vitamin C. Sutoh et al. (1973) observed that the carotenoids of the leaves decreased as the leaf matured. Maeda and Salunkhe (1981) reported that fresh sweetpotato leaves contained around 50 mg/100 g (dwb) of carotene and 1374 mg/100 g (dwb) vitamin C. However, Leung et al. (1972) reported β-carotene values of only 2.7 mg/100 g (fwb) in fresh sweetpotato leaves. Villareal et al. (1979) observed that the ascorbic acid content of 10 cultivars from Taiwan ranged from 32 to 73 mg/100 g (fwb) while the Malaysian varieties were reported to contain 71–136 mg (fwb) vitamin C in the leaves.

Sweetpotato as a Health-Protectant Food

Sweetpotato is increasingly recognized as a health food, due to several of its nutraceutical components *viz.*, dietary fibre, vitamin A, vitamin C, anthocyanins, xanthophylls etc. The roots are considered as a highly functional, low calorie food, with anti-diabetic effects (Kusano and Abe, 2000). It has been reported to stabilize blood sugar levels and lower insulin resistance (Ylonen et al., 2003). Most of the yellow and orange-fleshed varieties of sweetpotato contain Beta-carotene, the precursor of vitamin A, which also possess anti-oxidant activity. Another antioxidant vitamin, ascorbic acid, also occurs in high amounts in the roots. High content of vitamin B_6 (pyridoxine) in the roots help in reducing the blood levels of homocysteine, which is associated with the increased risk of cardio vascular diseases.

The most studied nutraceuticals in sweetpotatoes are carotenoids and anthocyanins. Purple-fleshed sweetpotatoes are a rich source of anthocyanins, which have medicinal value as anti-oxidant and cancer preventing agent. Besides, in Japan, the colored roots are used for extracting the pigment, which is further used in various food products. Suda et al. (2003) reported that the sweetpotato anthocyanins have multiple physiological functions such as radical-scavenging, anti-mutagenic, hepato-protective, anti-hypertensive and hypoglycaemic activities.

Sweetpotato cultivar 'Ayamurasaki' (purple fleshed) was reported to have high ACE-inhibitory activity than those from white or orange-fleshed sweetpotatoes (Yamakawa et al., 1998). Working with five sweetpotato cultivars, Oki et al. (2002) found that anthocyanins possessed major radical-scavenging activity in 2 cultivars *viz*., 'Ayamurasaki' and 'Kyushu-132'. However, in other two varieties *viz*., 'Miyanou 36' and 'Bise', phenolic compounds like chlorogenic acid was responsible for the major radical-scavenging activity. These workers also found that anthocyanins with peonidin aglycon had more radical scavenging effect than those with cyanidin aglycon.

The chemical structures of two major anthocyanins from purple-fleshed sweetpotato were established by Odake et al. (1992). The anthocyanin fractions have been characterized from purple-fleshed sweetpotato as YGM 1a, YGM 1b, YGM2, YGM3, YGM4b, YGM5a, YGM5b, YGM6 etc of which the first 4 belong to the cyanidin type and the latter 4 belong to the Peonidin type. Suda et al. (2003) reported 9 anthocyanin pigment peaks by HPLC in the sweetpotato variety, Yamagawa-murasaki and 13 peaks in the Tanegashima-murasaki variety. Further most of the pigment fractions; being acylated had high heat and UV ray stability. Several types of foods and beverages are available in the Japanese food market. Terahara et al. (1999) isolated 8 acylated anthocyanins from the purple-fleshed sweetpotato variety 'Yamagawamurasaki'.

The purple-fleshed sweetpotato juice with high anthocyanin content has been reported to have ameliorative effect on the CCl_4 induced liver injury (Suda et al., 2003). The anti-hyperglycaemic effect of diacylated anthocyanin from 'Ayamurasaki' was found to be due to the α-glucosidase inhibitory action.

Yoshimoto et al. (1999) found that the anti-mutagenic effect of purple colored sweetpotatoes was due to the cyanidin type of anthocyanins. However, in the yellow, white and orange-fleshed varieties, phenolic compounds contributed to the anti-mutagenic effect of the outer portions. Huang et al. (2004) reported that the sweetpotato leaves contained high amounts of total phenolic and flavonoid compounds, which were responsible for its DPPH radical scavenging activity.

Sweetpotato leaves are also rich source of nutraceuticals. Ishiguro and Yoshimoto (2005) reported high content of polyphenols and xanthophylls in sweetpotato leaves. They reported the presence of high concentration of lutein, a member of the xanthophylls family of carotenoids, in sweetpotato leaves. (Variety 'Sujoh'). Lutein was present to the extent of 29.5 mg/100 g (fwb) and hence the consumption of leaves of this variety were capable of mitigating eye diseases like age-related macular degeneration and cataract.

In contrast to the earlier concepts of phenolics as anti-nutrients, their presence in foods and crop plants is now regarded as having high nutraceutical value. The phenolic flavonoids have been reported to have high anti-oxidant activity. Thompson (1981) isolated and separated four isomers of caffeoylquinic acid from fourteen sweetpotato cultivars. The distribution of phenols in various parts of sweetpotato was studied by Walter and Schadel (1981). They found that more than 78% of the phenolics were located in the skin. Histochemical localization of phenolics in sweetpotato showed that the phenolics were present in phellem, phellogen and phelloderm and also in the latex of laticifers, phloem, cambium etc. Truong et al. (2007) characterized the phenolic acids of sweetpotato roots and leaves and found that chlorogenic acid was the major phenolic acid in the roots, while 3, 5-di-o-caffeoylquinic acid and 4,5 –di-o-caffeoylquinic acid were the major compounds in sweetpotato leaves. The sweetpotato leaf phenolics were also characterised by Islam et al. (2002), who found that caffeoylquinic acid (CQA) derivatives and caffeic acid were the major polyphenolics. Of these, caffeic acid and 3, 4, 5, tri-caffeoylquinic acid exhibited very high depression of melanin production and the action was through the potent inhibition of tyrosinase activity.

Flavor Components

The flavor components of sweetpotato have been studied by several workers and the strong flavor associated with cooked or baked sweetpotato is identified as a factor favouring its consumption. Several volatile compounds have been isolated from cooked sweetpotato (Kays and Horvat, 1983; Nagahama et al., 1977; Purcell et al., 1980) (Table 11.3).

Varietal differences have been reported among the volatile components (Kays and Horvat, 1984). Apart from the 30 volatile constituents identified in 'Jewel' variety having strong sweet caramel aroma, Kays and Horvat (1984) identified 7–8 more compounds. 'Morado', a white-fleshed variety having little aroma contained 13–14 new volatile compounds not present in 'Jewel'. Tiu et al. (1985) studied the volatiles from a number of baked sweetpotato roots having different olfactory responses. They reported that out of the 27 volatiles, 5 had association with cultivars having good flavor and 8 volatiles were present in roots with poor flavor.

The effect of cooking method on the aroma constituents of sweetpotatoes was investigated by Wang and Kays (2001). GC-olfactory analysis led to the identification of 37 compounds that were odor active. Aroma active compounds were found to be retained to a higher extent in conventionally baked compounds.

Anti-nutritional Factors in Sweetpotato

Enzyme Inhibitors in Sweetpotato

Proteinaceous inhibitors of digestive enzymes are present in many food crops. Two types of enzyme inhibitors *viz.*, proteinase inhibitors and amylase inhibitors are ubiquitous in occurrence in several cereal and legumes and many have been characterized.

Table 11.3 Components of sweetpotato volatiles identified by mass spectroscopy

Peak no.[a]	Compd	Retention time, min
1	Methanol	5.6[b]
2	Ethanol	5.8[b]
	Acetone	5.9[b]
	Diethyl ether	6.1[b]
3	Dichloromethane	6.6
4	2, 3- butanedione (diacetyl)	7.2[b]
5	3- methyl pentane	7.7
6	Hexane	8.3[b]
7	Tetrahydrofuran	8.6
	Methylcyclopentane	8.9
8	2,3- pentanedione	10.0
9	Methylbenzene (toluene)	13.3[b]
10	2- methyltetrahydrofuran- 3- one	14.7
11	Furfuraldehyde	15.6[b]
12	Dimethylbenzene (xylene)	18.1[b]
	Isobutyronitrile	18.3
	2- pyrone	18.6
13	Heptanal	19.7[b]
14	2- furyl methyl ketone	20.0
15	Benzaldehyde	22.4[b]
16	5- methyl-2 – furaldehyde	22.8
17	Trimethylbenzene (mesitylene)	26.3
18	Octanal	26.6
19	2- pentylfuran	27.6
20	Phenylacetaldehyde	30.0[b]
21	Nonanal	33.8
22	Linalool	34.7
23	Decanal	41.0
24	β- ionone	44.7[b]
25	4- (2,2,3,3- tetramethylbutyl) phenol	55.5

Source: Purcell et al., 1980

a – Refers to the peak number in the chromatogram shown as Figure 1 in the original article;

b – Retention time in comparison to standard reference compounds

Proteinase Inhibitors

Sohonie and Bhandarkar (1954) were the first to report a trypsin inhibitor from sweetpotato. They isolated a highly thermolabile protein from sweetpotato roots. Subsequently, after almost 2 decades, three trypsin inhibitors were isolated from sweetpotato roots by Sugiura et al. (1973). They reported molecular weights of 23,000 and 24,000 for two isoinhibitors II and III respectively. Obidairo and Akpochafo (1984) isolated several smaller molecular weight inhibitors from local Nigerian cultivars. Seven trypsin isoinhibitors were characterized from American sweetpotato cultivars by Dickey and Collins (1984). Sugiura et al. (1973) reported loss in trypsin inhibitor activity, when the arginyl groups were modified and hence suggested that arginine was essential for the activity. Traces of chymotrypsin inhibitor activity were first reported in sweetpotato roots by Bradbury et al. (1985).

Table 11.4 Purification of Trypsin inhibitors from sweetpotato (Cultivar S1139)

	Total TIU	Total protein (mg)	STIU	Yield %	Fold purification
Crude extract	342	478	0.72	100	1
Heat treated fraction	403	331	1.22	117	1.7
$(NH_4)_2SO_4$ ppt. fraction	239	134	1.78	69	2.5
Isoinhibitor I (SPI_1)	12	11.5	1.04	4	1.5
Isoinhibitor II (SPI_2)	84	20.8	4.04	24	5.6
Isoinhibitor III (SPI_3)	54	15.1	3.57	16	4.9
Isoinhibitor IV (SPI_4)	64	18.1	3.53	18	4.9

Source: Sasikiran (2000)

Ogiso et al. (1974) also found that inhibition of both the proteotylic and esterolytic activities were reduced, as the arginyl groups of sweetpotato trypsin inhibitor were modified.

Sasikiran (2000) isolated four trypsin isoinhibitors from five accessions of sweetpotato. Of the four isoinhibitors, SP I_2, SP I_3 and SP I_4 had high T1 activity, while SP I_1 had prominent chymotrypsin inhibitor activity (Table 11.4). He also reported non-competitive inhibition of trypsin by sweetpotato TI, while chymotrypsin was inhibited un-competitively by SP I_1. Sweetpotato TI were rapidly inactivated within 2 h at 90 °C. The SP I_1 and SP I_2 fractions had high molecular weight of around 67 KDa while SP I_3 and SP I_4 were proteins with molecular weights of 25 and 17 KDa respectively.

Heating sweetpotato slices at 100 °C led to rapid inactivation of the inhibitors. Microwave baking was found to be the best method to eliminate the TI and CI from sweetpotato roots (Sasikiran and Padmaja, 2003).

A gradual increase in trypsin inhibitor concentrations was reported during growth phase, with a maximum at 90 days after planting. Decrease in both TI and CI was observed during sprouting of roots and high concentration of both the inhibitors was detected in the emerging sprouts, indicating a dual role as N-reserve proteins and as defence chemicals (Sasikiran et al., 2002). Inhibition of insect proteases by sweetpotato trypsin inhibitors was also noticed, confirming their defensive role in sweetpotato (Rekha et al., 2004).

Amylase Inhibitors

Sweetpotato roots also have been reported to contain α-amylase inhibitors (Rekha et al., 1999). Four isoinhibitors could be purified from sweetpotato by ammonium sulphate precipitation and a two-step ion-exchange chromatography on DEAE-cellulose (Fig. 11.1).

The isoinhibitors were completely inactivated after 4 h at 90 °C (Rekha, 2000). Among the four isoinhibitors, SA I_1 was found to be a high molecular protein (63 KDa), comprised of three subunits of 21 KDa each. The other three isoinhibitors had molecular weight in the range of 14–20 KDa. Lysine and disulphide linkages were essential for the AI activity. SAI_1 and SAI_2 were resistant to pepsin, trypsin and chymotrypsin. Cooking for 30 min did not completely inactivate the AI of

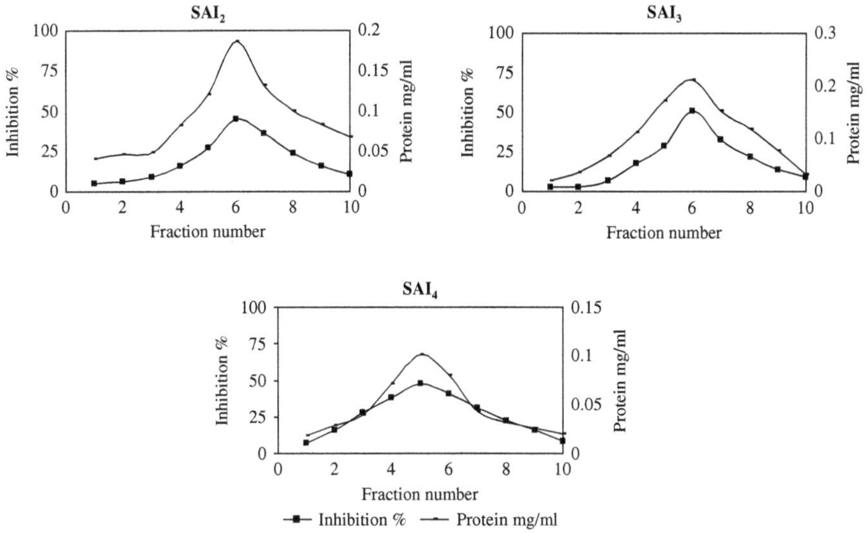

Fig. 11.1 Rechromatography of the major amylase isoinhibitors of Sweetpotato (Cultivar RS-III-2) (Source: Rekha, 2000)

sweetpotato and around 19–30% retention of AI activity was reported in roots after microwave baking (Rekha and Padmaja, 2002). Flour preparation after grating and drying was the best method to almost completely eliminate the AI activity. High AI activity was observed during the root bulking phase in sweetpotato (Sasikiran et al., 2002). Decrease in AI activity was evident during sprouting, while pronounced increase was noticed in wounded roots during 4–8 h of wounding (Table 11.5). These studies indicated that as in the case of the trypsin inhibitors, the amylase inhibitors of sweetpotato have a role as N-reserve protein as well as a regulatory role in amylase.

Table 11.5 Changes in α-amylase inhibitor activity of sweetpotato tuber on wounding

Hours after wounding	AIU/g fresh weight					
	Proximal end wounded			Distal end wounded		
	Proximal	Middle	Distal	Proximal	Middle	Distal
0	416 ± 18	435 ± 16	445 ± 21	416 ± 9*	435 ± 12	445 ± 11
4	465 ± 22^{NS}	505 ± 19	590 ± 18	470 ± 11*	515 ± 8	658 ± 13
	(11.8)**	(16.1)	(32.6)	(13)	(18.4)	(47.9)
8	685 ± 36	645 ± 29	710 ± 26	467 ± 14	521 ± 11	766 ± 12
	(64.7)	(48.3)	(59.6)	(12.3)	(19.8)	(72.1)
16	615 ± 28	445 ± 20^{NS}	450 ± 17^{NS}	445 ± 10^{NS}	466 ± 15^{NS}	480 ± 9^{NS}
	(47.8)	(2.3)	(1.1)	(7.0)	(7.3)	(7.9)
24	405 ± 13^{NS}	425 ± 11^{NS}	460 ± 12^{NS}	395 ± 9^{NS}	(–)396 ± 10^{NS}	486 ± 11^{NS}
	(–)	(–)	(3.4)		(–)	(9.2)

Source: Rekha, 2000

* Significant at $p < 0.05$; NS-not significant. All other values significant at $p < 0.01$

** Figures in parentheses indicate the percentage changes in α-amylase from the initial value

Enhanced amylase activity during sprouting is ensured through a reduced level of the AI activity in sprouting roots. Besides, they also have defensive role, as evidenced from their inhibitory potential on insect amylases (Rekha et al., 2004) and wounding associated synthesis (Sasikiran et al., (2002).

Toxic Stress Metabolites

Sweetpotato roots synthesise many stress metabolites, as a consequence of injury, infestation by insects, fungal infection etc. The compounds of great significance are furano-terpenoids (Uritani, 1967), which have been shown to produce lung toxicity (Boyd et al., 1974). Damage caused in the roots due to infection by the black rot fungus, *Ceratocystis fimbriata* and *Fusarium solani* or due to weevil infestation has been reported to result in accelerated synthesis of sesquiterpenes (Uritani et al., 1975; Schneider et al., 1984). Suzuki et al. (1975) found that HMG-CoA (Hydroxy methyl glutaryl CoA) reductase was a major enzyme in the biosynthetic pathway of Ipomeamarone, the major furanoterpene. Although the compound was isolated from fungal infected sweetpotato roots, its structural elucidation could be achieved only a decade later (Kubota and Matsuura, 1953). Yang and Yu (1976) isolated the hydroxylated derivative of ipomeamarone from mould-damaged sweetpotato, *viz*., ipomeamaronol that was found to be equally toxic to liver as ipomeamarone. Besides these two, Wilson et al. (1971) isolated and characterized a lung toxic compound 4-ipomeanol from moldy sweetpotatoes. Studies using artificially inoculated (*Fusarium solani*) sweetpotato roots showed the synthesis of four closely related lung toxins *viz*., 4-ipomeanol, 1-ipomeanol, 1, 4-ipomeadiol and ipomeanine (Boyd et al. (1974).

Toxicity studies on the sesquiterpenses were done in mice by intraperitoneal injection and LD 50 was 200 mg Kg^{-1} body weight for ipomeamarone ipomeamaronol, 4-hydroxymyporone, 7-hydroxymyoporone (Seawright and Mattocks, 1973; Wilson and Burka, 1979). The most lethal compound was 6-myoporol, (LD 50 $=$ 84 mg Kg^{-1} b.w), which produced considerable necrosis of hepatocytes and led to enlarged liver. In roots, artificially inoculated with *Fusarium solani*, the level of ipomeamarone was much higher (30–35 times) than those of 4-ipomeanol and 1, 4-ipomeadiol (Clark et al., 1981). However, the latter two lung toxins were reported to have a much lower LD 50 than ipomeamarone (Burka et al., 1974).

Flatulence Factors

The undigested carbohydrates in sweetpotato have been associated with flatulence induction after sweetpotato consumption. Although sweetpotato roots are known to cause flatulence, the phenomenon as such has not been totally understood. Tsou and Yang (1984) observed that gas formation induced by sweetpotato was partially reduced by cooking. Varietal differences were also observed in the ability to cause flatulence. These workers have also found that sweetpotato starch has high correlation with the hydrogen production. Tsou and Yang (1984) found that the volume of hydrogen production, induced by cooked sweetpotato ranged from 3.2 to 58 ml g^{-1} dry matter intake although cooking reduced gas production by 3–10 times. Fermentation of carbohydrates that escape to the large intestine or colon, by microorganisms has been often cited as responsible for gas production. The galactoside oligosaccharides, verbascose, stachyose and raffinose present in legumes have been correlated with their gas formation. However, Truong et al. (1986) found that these sugars were either not present or present in only trace amounts in sweetpotato and hence these were not responsible for flatulence induction in sweetpotato. Wagner et al. (1976) reported that dietary fiber could be fermented in the large intestine, to produce gas. Rekha et al. (1999) reported varying concentrations of α-amylase inhibitors in sweetpotato accessions. Further, the α-amylase inhibitors varied in their heat stability and hence the retention of α-amylase inhibitors in processed sweetpotatoes can lead to undigested roots entering the large intestine and the residual starch can produce gases through fermentation. However, this hypothesis needs to be further investigated for conclusive results.

References

Achata, A., Fano, H., Goyas, H., Chiango, O. and Andrade, M. 1988. Cited by: *Sweetpotato: an untapped food resource* (J.A. Woolfe, Ed.).Cambridge University Press, Cambridge, 643pp.

Ahmed, E.M. and Scott, L.E. 1958. Pectic constituents of the fresh roots of the sweetpotato. *Proc. Am. Scoc. Hort. Sci.* 71: 376–387.

Ahmed, J. and Ramaswamy, H.S. 2006. Viscoelastic properties of sweetpotato puree infant food. *J. Food Engg.* 74 (3): 376–382.

Almazan, A.M. 1987. Selecting non-sweet clones of sweetpotato from the IITA germplasm collection. In: Tropical root crops: root crops and the African food crisis: *Proc. Third Triennial Symposium of the International Society for Tropical Root Crops* (E.R. Terry, M.O. Akoroda and O.B. Arene, Eds.), IDRC, Ottawa, pp. 76–78.

Anon, 1968. *Food Composition Table, Handbook 1, (4th rev.).* Food and Nutrition Research Center, Manila, Philippnes.

Aregheore, E. M. 2003. Nutritive value of sweetpotato (Ipomoea batatas (L) Lam.) forage as goat feed: voluntary intake, growth and digestibility of mixed rations of sweetpotato and batiki grass (Ischaemum aristatum var. indicum), *Small Ruminant Res.* 51: 235–241.

Arthur, J.C. Jr. and McLemore, T.A. 1957. Effect of processing conditions on the chemical properties of canned sweetpotatoes. *Agric. Food Chem.* 5: 863–867.

Azhar, A. and Hamdy, M.K. 2004. Alcohol fermentation of sweetpotato: Membrane reactor in enzymatic hydrolysis. *Biotechnol. Bioeng.* 23 : 1297–1307.

Baba, T. and Yamamura, E. 1981. Development of snack foods produced from sweetpotatoes. II. Optimization of processing conditions in blanching and freezing for sweetpotato chips. (Japanese) *J. Jap. Soc. Food Sci. Technol.* 28 (7): 355–359.

Baba, T., Kouno, T. and Yamamura, E. 1981. Development of snack foods produced from sweetpotatoes. I. Factors affecting hardness and discoloration of sweetpotato chips. (Japanese) *J. Jap. Soc. Food Sci. Technol.* 28 (6): 318–332.

Backer, J., Ruiz, M.E., Munoz, H. and Pinchinat, A.M. 1980. The use of sweetpotato (Ipomoea batatas (L) Lam) in animal feeding. II. Beef production. *Trop. Anim. Prod.* 5: 152–160.

Baumgardner, R.A. and Scott, L.E. 1963. The relation of pectin substances to firmness of processed sweetpotatoes. *Proc. Am. Hort. Sci.* 83: 629–632.

Bertoniere, N.R., McLemore, T.A. and Hasling, V.C. 1966. The effect of environmental variables on the processing of sweetpotatoes into flakes and on some properties of their isolated starches. *J. Food Sci.* 31: 574–582.

Boggess, T.S., Marion, J.E. and Dempsey, A.H. 1970. Lipid and other compositional changes in 9 varieties of sweetpotatoes during storage. J. Food Sci. 35 (3): 306–309.

Bond, J. and Putnam, P.A. 1967. Nutritive value of dehydrated sweetpotato trimmings fed to beef steers. *J. Agric. Food Chem.* 15 (4): 726–728.

Bouwkamp, J.C. 1985. *Sweetpotato products: A natural resource for the tropics*. CRC Press, 271pp.

Boyd, M.R., Burka, L.T., Harris, T.N. and Wilson, B.J. 1974. Lung toxic furanoterpenoids produced by sweetpotatoes (Ipomoea batatas) following microbial infection. *Biochim. Biophys Acta. 337* : 184–195.

Bradbury, J.H. and Holloway, W.D. 1988. Chemistry of tropical root crops: significance for nutrition and agriculture in the Pacific. *ACIAR Monograph Ser. No.* 6. Canberra.

Bradbury, J.H. and Singh, U. 1986. Ascorbic acid and dehydroascorbic acid content of tropical root crops from the South Pacific. *J. Food Sci.* 51 (4): 975–978, 987.

Bradbury, J.H., Baines, J., Hammer, B., Anders, M. and Millar, J.S. 1984. Analysis of sweetpotato from the highlands of Papua New Guinea: relevance to the incidence of Enteritis necroticans. *J. agric. Food Chem.* 32: 469–473.

Bradbury, J.H., Beatty, R.E., Bradshaw, K., Hammer, B., Holloway, W.D., Jealous, W., Lau, J., Nguyen, T. and Singh, U. 1985. Chemistry and nutritive value of tropical root crops in the South Pacific. *Proc. Nutr. Soc. Austr.* 10: 185–188.

Burka, L.T., Kuhnert, L., Wilson, B.J. and Harris, T.M. 1974. 4-Hydroxymyoporone, a key intermediate in the biosynthesis of pulmonary toxins produced by Fusarium solani infected sweetpotatoes. *Tetrahedron Lett.* 46: 4017–4020.

Candlish, J.K., Gourley, L. and Lee, H.P. 1987. Dietary fibre and starch contents of some Southeast Asian vegetables. *J. Agric. Food Chem* 35: 319–321.

Carpio, W.N. 1984. A comparative study of feedlot fattening performance of native goats fed with sweetpotato leaves and combination with kikuyu stargrass forage. *B Sc Thesis, Mountain State Agricultural College*, La Trinidad.

Castillo, L.S., Aglibut, F.B., Javier, T.A., Gerpacio, A.L., Garcia, G.V., Puyaoan, R.B. and Ramin, B.B. 1964. Camote and cassava tuber silage in swine rations. *Agric. Los Banos.* pp. 11–13.

Cereda, M.P., Conceicao, F.A.D., Cagliari, A.M., Heezen, A.M. and Fioretto, R.B. 1982. Comparative study of sweetpotato (Ipomoea batatas) varieties to estimate their utilization in the food industry. (Portuguese). *Turrialba* 32 (4): 365–370.

Chang, K.J. and Lee, S.R. 1974. Development of composite flours and their products utilizing domestic raw materials. IV. Textural characteristics of noodles made of composite flours based on bariey and sweetpotato (Korean). *Korean J. Food Sci. Technol.* 6 (2): 65–69.

Chen, Z. 2003. Physicochemical properties of sweetpotato starches and their application in noodle products. *PhD thesis, Wageningen University*, the Netherlands.

Chen, M.C. and Chen, C.P. 1979. Studies on the utilization of sweetpotato chips and cassava pomace for cattle. *J. Agric. Assoc. China NS* 107, 45–48.

Cinar, I. 2005. Stability studies on the enzyme extracted sweetpotato carotenoproteins. *Food Chemistry* 89 (3): 397–401.

Clark, C.A., Lawrence, A. and Martin, F.A. 1981. Accumulation of furanoterpenoids in sweetpotato tissue following inoculation with different pathogens. *Phytopathology* 71 (7): 708–711.

Collins, J.L. and Pangloli, P. 1997. Chemical, physical and sensory attributes of noodles with added sweetpotato and soy flour. *J. Food Sci.* 62(3): 622–625.

Collins, J.L., and Aziz, A. 1982. Sweetpotato as an ingredient in yeast-raised doughnuts. *J. Food Sci.* 47 (4): 1133–1139.

Collins, J.L. and Hutsell, W.L. 1987. Physical, chemical, sensory and microbiological attributes of sweetpotato leather. *J. Food Sci.* 52 (3): 646–648.
Constantin, R.J. and McDonald, R.E. 1968. Effects of curing, storage and processing temperatures on firmness of canned sweetpotatoes. *Proc. Assoc. South Agric. Workers*. p. 166.
Constantin, R.J., Hernandez, T.P. and Jones, L.G. 1974. Effects of irrigation and nitrogen fertilization on the quality of sweetpotatoes. *J. Amer. Soc. Hort. Sci.* 99: 308–310.
Data, E. and Operario, J.J.A.1992. Processing and storage characteristics of sweetpotato chips. In: *Sweetpotato Technology for the 21st Century* (W.A. Hill, C.K. Bonsi and P.A. Loretan, Eds.) Tuskegee University, Al.
Datta, E.S., Diamante, J.C. and Forio, E.E. 1986. Soy sauce production utilizing root crop flour as substitute for wheat flour (100% substitution). *Ann. Trop. Res.* 8 (1): 42–50.
Deobald, H.J. and McLemore, T.A. 1964. The effect of temperature, antioxidant and oxygen on the stability of pre-cooked dehydrated sweetpotato flakes. *Food Technol.* (Chicago) 18: 145–148.
Deobald, H.J., McLemore, T.A., Hassling, V.C. and Catalano, E.A. 1968. Control of sweetpotato alpha-amylase for producing optimum quality precooked dehydrated flakes. *Food Technol.* (Chicago) 22: 627–630.
Dickey, L.F. and Collins, W.W. 1984. Cultivar differences in trypsin inhibitors of sweetpotatoes. *Bot. Bull. Acad. Sinica* 21: 7–13.
Dickey, L.F., Collins, W.W., Young, C.T. and Walter, W.M. Jr. 1984. Root protein quantity and quality in a seeding population of sweetpotatoes. *Hort. Sci.* 19 (5): 689–692.
El-Ashwah, E.T., Musmar, I.T., Ismail, F.A. and Alian, A. 1980. *Fungi imperfecti* as a source of food protein. II. Effect of sweetpotato based media on yield and protein concentration. *Egypt J. Microbiol.* 15($^1/_2$): 63–71.
Elkins, E.R. 1979. Nutrient content of raw and canned green beans, peaches and sweetpotatoes. *Food Technol.* (Chicago) 33 (8): 66–71.
Faboya, O.O.P. 1981. The fatty acid composition of some tubers grown in Nigeria. *Food Chem.* 7 (2): 151–154.
Fasina, O.O., Farkas, B.E. and Fleming, H.P. 2003. Thermal and dielectric properties of sweetpotato puree. *Inter. J. Food Prop.* 6 (3): 461–472.
Furuya, S. and Nagano. R. 1986. Differences in nutritive values among the variety of sweetpotato and among the processing methods including drying and boiling of sweetpotato, potato and cassava. (Japanese). *Jpn. J. Swine Sci.* 23 (2): 62–67.
Garcia, E.H., Querido, I.B. and Cahanap, A.C. 1970. The relation of carotene and starch content of some twenty six varieties of sweetpotatoes (Ipomoea batatas) *Phil. J. Plant Food.* 35: 203–213.
Garlich, J.D., Bryant, D.M., Covington, H.M., Chamble, D.S. and Purcell, A.E. 1974. Egg yolk and broiler skin pigmentation with sweetpotato vine meal. *Poult. Sci.* 53 (2): 692–699.
Goyert, J.C. and Avault, J.W. 1977. Agricultural by-products as supplemental feed for crayfish, Procambarus clarkia. *Trans. Am. Fish. Soc.* 106 (6): 629–633.
Grabowski, J.A., Troung, V.D. and Daubert, C.R. 2006. Spray-drying of amylase hydrolysed sweetpotato puree and physicochemical properties of powder. *J. Food Sci.* 71(5): 209–217.
Grabowski, J.A., Troung, V.D. and Daubert, C.R. 2007. Nutritional and rheological characterization of spray-dried sweetpotato powder. *LWT Food Sci. Technol.* 41: 206–216.
Gruneberg, W.J., Bonierbak, M., Roca, W., Reynoso, D., Espinoza, J. and Burgos, G. 2005. Biofortification of sweetpotato for pro-vitamin A (β-carotene), iron and zinc: What has been achieved and what is within reach? In: *14th Triennial Symposium of the International Society for Tropical Root Crops*, 20–26 November, 2006, Kerala, India.
Gupta, J.J., Bardoloi, R.K. and Bujarbaruah, K.M. 2006. Growth performance and nutrient digestibility in different age groups of pigs fed raw sweetpotato tuber at various levels In: *14th Triennial Symposium of the International Society for Tropical Root Crops*, 20–26 November, 2006, Kerala, India.
Han, I.K., Yoon, D.J. and Kim, C.S. 1976. Studies on the nutritive values of dried sweetpotato meal. 1. Partial substitution of dried sweetpotato meal for corn in the ration of growing swine. (Korean). *Seoul. Natt. Univ. Coll. Agric. Bull.* 1 (1): 67–74.
Hannigan, K. 1979. *Sweetpotato chips. Food Eng. Int.* 4 (5): 28–29.

Haytowitz, D.B. and Mathews, R.H. 1984. Composition of foods: vegetables and vegetable products. Human Nutrition Information Ser., *USDA Agric. Handbook No. 8–11*, pp. 428–439.

Heinze, P.H. and Appleman, C.O. 1943. A biochemical study of curing processes in sweetpotatoes. *Plant Physiol.* 18: 548–550.

Holland, B., Unwin, I.D. and Buss, D.H. 1991. Vegetables, herbs and spices. *The fifth supplement to McCane and Widdowson's the Composition of Foods (4th edition)*, Royal Society of Chemistry, Cambridge.

Hong, E.H. 1982. The storage, marketing and utilization of sweetpotatoes in Korea. In: *sweetpotato: Proc. First International Symposium* (R.L. Villareal and T.D. Griggs, Eds.), AVRDC, Tainan, pp. 405–411.

Hoover, M.W. 1966. An enzyme process for producing sweetpotato flakes from starchy and uncured roots. *Food Technol.* (Chicago) 20: 84–87.

Hoover, M.W. 1963. Preservation of the natural coor in processed sweetpotato products. I. Flakes. *Food Technol.* (Chicago) 17: 128–130.

Hoover, M.W. 1967. An enzyme activation process for producing sweetpotato flakes. Food Technol. (Chicago), 21: 322–325.

Hoover, M.W. and Miller, N.C. 1973. Process for producing sweetpotato chips. *Food Technol.* 27 (5): 74, 76, 80.

Hoover, M.W., Walter, W.M. and Giesbrecht, F.G. 1983. Method of preparation and sensory evaluation of sweetpotato patties. J. Food Sci. 48 (5): 1568–1569.

Horigome, T., Nakayama, N. and Ikeda, M. 1972. Nutritive value of sweetpotato protein produced from the residual products of the sweetpotato industry. *Nippan Chikusan Gakkahi*-Ho 43: 432–435.

Huang, D.J., Lin, C.D., Chen, H.J. and Lin, Y.H. 2004. Antioxidant and anti-proliferative activities of sweetpotato (Ipomoea batatas L. Lam Tainong 57) constituents. *Bot. Bull. Acad. Sin.* 45: 179–186.

Ice, J.R., Hamann, D.D., and Purcell, A.E. 1980. Effects of pH, enzymes and storage time on the rheology of sweetpotato puree. *J. Food Sci.* 45: 1614–1618.

Ishiguro, K. and Yoshimoto, M. 2005. Content of eye-protective nutrient lutein in sweetpotato leaves. In: *2nd International Symposium on sweetpotato and cassava*, 14–17 June, 2005, Kuala Lumpur. Malaysia pp. 213–214.

Islam, M.S., Yoshimoto, M., Yahara, S. and Yamakawa, O. 2002. Identification and characterization of foliar polyphenolic composition in sweetpotato (Ipomoea batatas Lo) genotypes. *J. Agric. Food Chem.* 50: 3718–3722.

Iwe, M.O., Zuilichem, D.J.V., and Ngoddy, P.O. 2000. Color of single-screw extruded blends of soy-sweetpotato flour-A response surface analysis. *Plant Foods for Hum. Nutr.* 55 (2): 159–168.

Iwe, M.O., Zuilichem, D.J.V., Stolp, W. and Ngoddy, P.O. 2004. Effect of extrusion cooking of soy-sweetpotato mixtures on available lysine content and browning index of extrudates. *J. Food Engg.* 62 (2): 143–150.

Jaarsveld, P.V., Maraisa, D.W., Harmsea, E., Nestelb, P. and Rodriguez-Amayac, D.B. 2006. Retention of β-carotene in boiled, mashed orange-fleshed sweetpotato. *J. Food Comp. Anal.* 19 (4): 321–329.

Jenkins, W.F. and Anderson, W.S. 1957. Quality improvement in canned sweetpotatoes. *Miss. Farm. Res.* 20 (6): 2–6.

Job, T.A., Oluyemi, J.A. and Entonu, S. 1979. Replacing maize with sweetpotato in diets for chicks. *Brit. Poult. Sci.* 20 (6): 515–519.

Johnson, L.D., Hunt, S.K. Colvin, R.L. and Moorman, V.T. 1992. Consumer acceptability of sweetpotato yogurts. In: *sweetpotato technology for the 21st century* (W.A. Hill, C.K. Bonsi and P.A.Loretan, Eds.) Tuskegee University, pp. 415–419.

Jung, C.H. and Lee, J.H. 1968. A study on substitution of sweetpotato ensilage for concentrates in pork production. (Korean). *Res. Rep.* 11 (4): 35–43.

Kattan, A.A. and Littrell, D.l. 1963. Pre- and post-harvest factors affecting firmness of canned sweetpotatoes. *Proc. Amer. Soc. Hort. Sci.* 83, 641–643.

Kays, S.J. and Horvat, R.J. 1984. A comparison of the volatile constituents and sugars of representative Asian, Central American and North American sweetpotatoes. *Proc. 6th Symp. ISRTC* pp. 577–586.
Kays, S.J. and Horvat, R.J., 1983. Insect resistance and flavor chemistry: Integration into future breeding programs. In: Breeding new sweetpotatoes for the Tropics. (F.M. Martin, Ed.), Proc. *Amer. Soc. Hort. Sci. Trop. Region* 27 (B) 97–101.
Kelly, E.G., Baum, R.R. and Woodward, C.F. 1958. Preparation of new and improved products from eastern (dry type) sweetpotatoes: chips, dice, julienne strips and frozen French fries. *Food Technol.* (Chicago), 12: 510–514.
Koh, F.K., Yeh, T.P. and Yen, H.T. 1976. Effects of feeding sweetpotatoes, dried sweetpotato chips and sweetpotato vine on the growth performance of growing finishing pigs. (Chinese). *J. Chin. Soc. Anim. Sci.* 5 (3–4): 55–67.
Konaka, N-1962. Use of sweetpotato stolon for silage. (Japanese). *J. Japan. Grassland Sci.* 8 (1): 64–67.
Kubota, T. and Matsuura, T. 1953. Chemical studies on black rot disease of sweetpotato. (Japanese). *J. Chem. Soc. Jap*an 74: 101–109, 197–199, 248–251 and 668–670.
Kudo, T., and Hidaka, T. 1984. Removal of bitter components from shochu produced from diseased sweetpotatoes by carbon treatment. (Japanese). *J. Soc. Brewing Japan* 79 (12): 900–901.
Kudo, T., Hamakawa, S., Nakayama, K. and Hidaka, T. 1979.Production of shochu by vacuum distillation. (Japanese). *J. Soc. Brewing Japan* 74 (7): 484–486.
Kukimura, H., Yoshida, T., Komaki, K., Sakamoto, S., Tabuchi, S., Ide, Y and Yamakawa, O. 1989. 'Satsumahikari': a new sweetpotato *cultivar. Bull. Kyushu Nat. Agric. Expt. Stn* 25(3): 250–251.
Kurihara, T. and Imamura, T. 1956. Experiments on feeding of sweetpotatoes, barley bran silage, sweetpotatoes-barley bran meal and dry sweetpotatoes for growing pigs. (Japanese). *Kyushu Agric. Res.* 17: 140–141.
Kusano, S. and Abe, H. 2000. Anti-diabetic activity of white skinned sweetpotato (Ipomoea batatas L.) in obese zucker fatty rats. *Biol. Pharm. Bull.* 23 (1): 23–26.
Lanier, J.J. and Sistrunk, W.A. 1979. Influence of cooking method on quality attributes and vitamin content of sweetpotato. *J. Food Sci.* 44 (2): 374–380.
Lease, E.J. and Mitchell, J.H. 1940. Biochemical and nutritional studies on dehydrated sweetpotato. *S.C. Agric. Exp. Stn. Bull.* 329–331.
Lee, P.K. and Yang, Y.F. 1979. Nutritive value of high protein sweetpotato chips as feed ingredients for broilers. (Chinese). *J. Agric. Assoc. China* 106: 71–78.
Lee, S.Y. Extruded products from sweetpotato from breakfast food to snacks. 2005. In: *2nd Inter. Symp. on sweetpotato and cassava Kuala Lumpur*, Malaysia, 14–17 June, 2005.
Leung, W.T.W., Butrum, R.R. and Chang, F.H. 1972. Food Composition Table for use in East Asia. Part I. Proximate composition, mineral and vitamin contents of East Asian foods. *U.S. Department of Health, Education and Welfare, Bethesda*, M.D, FAO, Rome.
Lila Babu. 2001. Starch to sugar conversions during processing in sweetpotato-role of β-amylase. In: *Sweet Potato: Food and Health for Future. Int. Symp.* 26–29 Nov. 2001, Lima, Peru.
Lila Babu. 2006. Carotene content in sweetpotato varieties and its retention after processing. In: *14th Triennial Symposium of the International Society for Tropical Root Crops*, 20–26 November, 2006, Kerala, India. pp. 30–31.
Lopez, A., Williams, H.L. and Cooler, F.W. 1980. Essential elements in fresh and canned sweetpotatoes. *J. Food Sci.* 45 (3): 675–678, 681.
Lopez, A., Wood, C.B. and Boyd. E.N. 1976. Changes in some monocarbonyl classes during processing and storage of sweetpotato flakes. *J. Food Sci.* 41: 524–526.
Lund, E.D. and Smoot, J.M. 1982. Dietary fibre content of some tropical fruit and vegetables. *J. Agric. Food. Chem.* 30: 1123–1125.
Madamba, L.S.P., Bustrillos, A.R. and San Pedro, E.L. 1973. Sweetpotato starch: physicochemical properties of the whole starch. *Philipp. Agric.* 54: 350–353.
Maeda, E.E. and Salunkhe, D.K. 1981. Retention of ascorbic caid and total carotene in solar dried vegetables. *J. Food Sci.* 46: 1288–1290.
Maeshima, M., Sasaki, T. and Asahi, T. 1985. Characterization of major proteins in sweetpotato tuberous roots. *Phytochemistry* 24 (9): 1899–1902.

Makki, H.M., Abdel-Rahmann, A.Y., Khalil, M.K.M. and Mohamed, M.S. 1986. Chemical composition of Egyptian sweetpotatoes. *Food Chem.* 20 (1): 39–44.

Martin, F.W. 1987. Fried chips from staple type sweetpotatoes. *J. Agric. Univ. Puerto Rico.* 71 (4): 373–378.

Martin, F.W. 1983. The carotenoid pigments of white fleshed sweetpotatoes-reference to their potential value as source of vitamin A activity. *J. Agric. Univ. Puerto Rico* 67 (4): 494–500.

Martin, F.W. 1986. Sugars in staple type sweetpotatoes as affected by cooking and storage. *J. Agric. Univ. Puerto Rico.* 70 (2): 121–126.

Mason, R.L. 1982. Sweetpotato canning investigations. *Food Technol. Austr.* 34 (12): 574–576.

Mather, R.E., Linkous, W.N. and Eheart, J.F. 1948. Dehydrated sweetpotatoes as a concentrate feed for dairy cattle. *J. Dairy Sci.* 31: 569–573.

McConnell, E.R. and Gottschall, P.B. 1957. Effects of canning on new and familiar sweetpotato varieties with particular emphesis on breakdown and firmness. *Food Technol.* (Chicago), 11, 209–212.

Mokolensang, J.F., Yamasaki,S. And Onoue, Y.2003. Utilization of sweetpotato distillery byproducts as a feedstuff for red carp. *J. World Aquacultural Soc.* 34(4): 512–517.

Montemayor, N.R. and Notario, J.N. 1982. Development of food products using soybean and root crop flours. *CMU J. Agric. Food Nutr.* 4 (3): 274–288.

Mwanga, R.O.M. and Wanyera, N.W. 1988. Sweetpotato growing and research in Uganda. In: *Improvement of sweetpotato (Ipomoea batatas) in East Africa, with some references of other tuber and root crops*, pp. 187–196.

Nagahama, I., Inoue, K., Nobori, Y., Fujimoto, S. And Kanie, M. 1977. On some components in steam distillate of sweetpotato. *J. Agric. Chem. Soc. Jpn.* 51: 597–600.

Obidairo, T.K. and Akpochafo. M. 1984. Isolation and characterization of some proteolytic enzyme inhibitors in sweetpotatoes (*Ipomoea batatas*). *Enz. Microb. Technol.* 6: 132–134.

Odake, K., Terahara, N., Saito, N., Toki, K. and Honda, T. 1992. Chemical structures of two anthocyanins from purple sweetpotato, *Ipomoea batatas. Phytochemistry* 31 (6): 2127–2130.

Ogawa, K., Toyama, H. and Toyama, N. 1982. Utilization of enzyme preparations in shochu making. (Japanese). *Bull. Fac. Agric. Univ. Miyazaki* 29 (1): 203–211.

Ogiso, T., Tamura, S., Kato, Y. And Sugiura, M. 1974. Studies on trypsin inhibitors of sweetpotatoes. II. Modification of amino acid residues in inhibitor III. *J. Biochem.* 76: 147–156.

Ohtai, T., Ikutai, R., Nakashima, M., Morimitsu, Y., Samutai, T. and Saikii, T. 1990. Characteristic flavor of Kamho-shochu (sweetpotato spirit). *Agric. Biol. Chem.* 54 (6): 1353–1357.

Oki, T., Masuda, M., Furuta, S., Nishiba, Y., Terahara, N. and Suda, I. 2002. involvement of anthocyanins and other phenolic compounds in radical. Scavenging activity of purple-fleshed sweetpotato cultivars. *J. Food Sci.* 67 (5). 1752–1756.

Onate, L.U., Arago, L.L., Garcia, P.C. and Abdon, I.C. 1970. Nutrient composition of some raw and cooked Philippine vegetables. *Phil. J. Nutr.* 23 (3): 33–44.

Opute, F.I and Osagie, A.U. 1978. Fatty acid composition of total lipids from some tropical storage organs. *J. Sci. Food Agric.* 29 (11): 959–962.

Padmaja, G. and Premkumar, T. 2002. Tuber Crops Recipes, *Technical Bulletin Series 36, CTCRI, Kerala*, India, pp. 26.

Padmaja, G., Moorhty, S.N., Nanda, S.K. and Sajeev, M.S. 2005a. Innovative approaches in the value addition of sweetpotato and cassava for mini-agri-business in developing countries. In: *Second International Symposium on sweetpotato and cassava, Kuala Lumpur*, Malaysia, 14–17 June, 2005.

Padmaja, G., Sajeev, M..S.., Moorthy, S..N. and Naskar, S.K. 2005b. Nutritional and textural evaluation of gulab jamuns from sweetpotato based instant mixes. In: *Root and Tuber Crops: Post harvest Management and value addition* (G. Padmaja, T. Premkumar, S. Edison and B. Nambisan, Eds.) CTCRI, Kerala, India, 400pp.

Panalaks, T. and Murray, T.K. 1970. The effect of processing on the content of carotene isomers in vegetables and peaches. *Can. Inst. Food Technol.* J. 3: 145–148.

Panda, S. and Ray, R.C. 2007. Lactic acid fermentation of β-carotene rich sweetpotato (*Ipomoea batatas* L.) into lacto-juice. *Plant Foods Hum. Nutr.* 62 (2). 65–70.

Picha, D.F. 1986. Influence of storage duration and temperature on sweetpotato sugar content and chip color. *J. Food Sci.* 51 (1): 239–240.
Picha, D.H. 1985 a. Crude protein, minerals and total carotenoids in sweetpotatoes. *J. Food Sci.* 50 (6): 1768–1769.
Picha, D.H. 1985 b. HPLC determination of sugars in raw and baked sweetpotatoes. J. *Food Sci.* 50 (4): 1189–1190.
Prabhuddham, S. Tantidham, K., Poonperm, N., Lertbawornwongsa, C. and Tongglad, C. 1987. A study of sweetpotato quality and processing methods. Cited by: *Sweetpotato: an untapped food resource* (J.A. Woolfe Ed.).Cambridge Uuv. Press, Cambridge, 643pp.
Purcell, A.E. and Walter, W.M. 1968 b. Carotenoids of centennial variety of sweetpotato, Ipomoea batatas (Lam.) during storage. *J. Agric. Food Chem.* 16(5): 769–770.
Purcell, A.E. and Walter, W.M. Jr. 1968 a. Autoxidation of carotenes in dehydrated sweetpotato flakes using 14 C-B-carotene. *J. Agric. Food Chem.* 16:650–653.
Purcell, A.E. and Walter, W.M. Jr. 1980. Changes in composition of the non-protein nitrogen fraction of 'Jewel' sweetpotatoes Ipomoea batatas (Lam.) during storage. *J. Agric. Food Chem.* 28 (4): 842–844.
Purcell, A.E. and Walter, W.M. Jr. 1982. Stability of amino acids during cooking and processing of sweetpotatoes. *J. Agric. Food Chem.* 30: 443–444.
Purcell, A.E. and Walter, W.M. Jr. 1988. Comparison of carbohydrate components in sweetpotatoes baked by convection heating and microwave heating. *J. Agric. Food Chem.* 36 (2): 360–362.
Purcell, A.E. Walter, W.M. Jr. and Giesbrecht, F.G. 1976. Distribution of protein within sweetpotato roots (Ipomoea batatas L.). *J. Agric. Food Chem.* 24 (1): 64–66.
Purcell, A.E., Swaisgood, H.E. and Pope, D.H. 1972. Protein and amino acid content of sweetpotato cultivars. *J. Am. Soc. Hort. Sci.* 97: 30–33.
Purcell, A.E., Walter, D.W. and Lee, M.L. 1980. Analysis of the volatile constituents of baked, 'Jewel' sweetpotatoes. *J. Agric. Food Chem.* 28: 939–941.
Rao, C.S. and Ammerman, G.R. 1974. Canning studies on sweetpotatoes. *J. Food Sci. Technol.* 11 (3): 105–109.
Rao, V.N.M., Hamann, D.D. and Humphries, E.G. 1975 b. Flow behaviour of sweetpotato puree and its relation to mouthfeel quality. *J. Texture Studies*. 6: 197–209.
Rao, V.N.M., Hamann, D.D and Humphries, E.G. 1974. Mechanical testing as a measure of kinesthetic quality of raw baked sweetpotatoes. *Trans. Am. Soc. Agric. Eng.* 17, 1187–1190.
Rao, V.N.N., Hamann, D.D. and Humphries, E.G. 1975a. Apparent viscosity as a measure of moist mouthfeel of sweetpotatoes. *J. Food Sci.* 40: 97–100.
Ray, R.C., Naskar, S.K. and Sivakumar, P.S. 2005. Sweetpotato curd. *CTCRI Technical Bull.* Ser. No. 39, pp. 24.
Reddy, N.N. and Sistrunk, W.A. 1980. Effect of cultivar, size, storage and cooking method on carbohydrates and some nutrients of sweetpotatoes. *J. Food Sci.* 45 (3): 682–684.
Rekha, M.R. 2000. Biological variations, metabolic significance and kinetic properties of the amylase inhibitor of tuber crops. *PhD thesis, University of Kerala*, India, pp. 206.
Rekha, M.R. and Padmaja, G. 2002. Alpha amylase inhibitor changes during processing of sweetpotato and taro tubers. *Plant Foods Hum. Nutr.*, 57(3–4), 285–294.
Rekha, M.R., Padmaja, G., Easwari Amma, C.S. and Sheela, M.N. 1999. Cultivar differences in the alpha-amylase inhibitor activity of sweetpotato and yam tubers. *J. Root Crops* 25 (2): 185–191.
Rekha, M.R., Sasikiran, K. and Padmaja, G. 2004. Inhibitor potential of protease and alpha-amylase inhibitors of sweetpotato and taro on the digestive enzymes of storage pests. *J. Stored Products Research* 40: 461–470.
Reungmaneepaitoon, S. Poolperm, N., Jariyavattanavijit, C. and Teangpook, C. 2005. Production of sweetpotato frozen French-fry-type products from sweetpotato grown in Thailand. In: *2nd International symposium on sweetpotato and cassava, Kuala Lumpur*, Malaysia, 14–17 June, 2005.
Rose, C.J. and White, G.A. 1980. Apparent digestibilities of dry matter, organic matter, crude protein, energy and acid detergent fibre of chopped, raw sweetpotato (Ipomoea batatas L.)

by village pigs (Susscrofa papuensis) in Papua New Guinea. *Papua New Guinea Agric. J.* 31: 69–72.

Samsiah, M.S., Latifah, M.S. and Ong, C.Y. 2005. Sweetpotato candy. In: *2nd International symposium on sweetpotato and cassava, Kuala Lumpur*, Malaysia, 14–17 June, 2005, pp. 223–224.

Sasikiran, K. 2000. Isolation, characterization and metabolic significance of trypsin and chymotrypsin inhibitors in tuber crops. *PhD thesis, University of Kerala*, India, 247 pp.

Sasikiran, K. and Padmaja, G. 2003. Inactivation of trypsin inhibitors in sweetpotato and taro during processing. *Plant Foods Hum. Nutr.* 58: 153–163.

Sasikiran, K., Rekha, M.R. and Padmaja, G. 2002. Proteinase and alpha-amylase inhibitors of sweetpotato: changes during growth phase, sprouting and wound induced alterations. *Bot.Bull. Acad. Sin.*, China, 43: 291–298.

Schmandke, H. and Gurra, O.O. 1969. The carotene, L-ascorbic acid, dehydrpascorbic acid and tocopherol content of Cuban vegetable products. (German). *Nahrung* 13 (6): 523–30.

Schneider, J.A., Lee, J., Naya, Y., Nakanishi, K., Oba, K. and Uritani, I. 1984. The fate of the phytoalexin, Ipomeamarone: furanoterpenes and butenolides from Ceratocytis fimbriata infected sweetpotatoes. *Phytochemistry* 23 (4): 759–764.

Schwartz, S.J., Walter, W.M., Carroll, D.E. and Giesbrecht, F.G. 1987. Chemical, physical and sensory properties of a sweetpotato French-fry type product during frozen storage. *J. Food Sci.* 52 (3): 617–619, 633.

Scott, L.E. 1952. Sweetpotato discoloration in relation to HP steam peeling. *Food Packer* 33 (3): 38–39.

Scott, L.E. and Kattan, A.A. 1957. Varietal differences in the catechol oxidase content of the sweetpotato root. *Proc. Am. Soc. Hort. Sci.* 69: 436–442.

Scott, L.E. and Matthews, W.A. 1957. Carbohydrate changes in sweetpotatoes during curing and storage. *Proc. Amer. Soc. Hort. Sci.* 70: 407–418.

Seawright, A.A. and Mattocks, A.R. 1973. The toxicity of two synthetic 3-substituted furan carbamates. *Experientia* 29 (10): 1197–1200.

Seralathan, M.A. and Thirumaran, A.S. 1990. Utilization of sweetpotato (Ipomoea batatas) flour in South Indian dishes. In: Tropical root and Tuber crops: Changing role in a modern world. *Proc. Eighth Symposium of the International Society for Tropical Root Crops*. (R.H. Howeler, Ed.), October 30-November 5, 1988, Bangkok, Thailand, 1990, 724p.

Shah Chaudhury, J.C. 1992. Processing of sweetpotato. *Bangladesh Agricultural Research Institute Bulletin*, 40pp.

Shanmugan, A. and Venugopal, K. 1975. Starch content of sweetpotato (Ipomoea batatas Lamb.) varieties. *Sci. Cult.* 41 (10): 504–505.

Shen, M.C. and Sterling, C. 1981. Changes in starch and other carbohydrates in baking *Ipomoea batatas Starch/Starke*. 33: 261–268.

Sheng, J. and Wang, S. 1987. The status of sweetpotato in China from 1949-to the present. Cited in: *Sweetpotato: an untapped food resource* (J.A. Woolfe, Ed.). Cambridge University Press, 1992, 643pp.

Sistrunk, W.A., Miller, J.C. and Jones, L.G. 1954. Carbohydrate changes during storage and cooking of sweetpotatoes. *Food Technol.* 8: 223–225.

Smith, D.A., McCaskey, T.A., Harris, H. and Rymal, K.S. 1982. Improved aseptically filled sweetpotato purees. *J. Food Sci.* 47: 1130–1132, 1147.

Sohonie, K. and Bhandarkar, A.P. 1954. Trypsin inhibitors in Indian foodstuffs. I. Inhibitors in vegetables. *J. Sci. Ind. Res.* (India), 131: 500–503.

Southwell, B.L. and Black, W.H. 1948. Dehydrated sweetpotatoes for fattening steers *Ga. Agric. Exp. Stn. Bull.* 45–47.

Spadaro, J.J. and Patton, E.L. 1961. Pre-cooked dehydrated sweetpotato flakes. *Food Eng.* 33: 46–49.

Splittstoesser, W.E. 1977. Protein quality and quantity of tropical roots and tubers. *Hort. Sci.* 12: 294–298.

Sreekantiah, K.R. and Rao, B.A.S. 1980. Production of ethanol from tubers. *J. Food Sci. Technol.* 17: 194–199.

Suda, I., Oki, T., Masuda, M., Kobayashi, M., Nishiba, Y. and Furuta, S. 2003. Physiological functionality of purple-fleshed sweetpotatoes containing anthocyanins and their utilization in foods. *JARQ* 37 (3): 167–173.

Sugimoto, K., Shiga, I. and Goto, F. 1967. Treatment of waste water from the sweetpotato starch factory. III. Effects of the addition of nutrients. (Japanese). *J. Jap. Soc. Starch Sci.* 14 (4): 120–127.

Sugiura, M., Ogiso, T., Takeuti, K., Tamura, S. and Ito, A. 1973. Studies on trypsin inhibitors of sweetpotatoes. *Biochim. Biophys. Acta* 328: 407–417.

Sutoh, H., Uchida, S. and Kaneda, K. 1973. Studies on silage making. XXII. The nutrient content of sweetpotato (Ipomoea batatas L. var. edulis) at the different stages and the quality of sweetpotato vine silage. (Japanese). *Sci. Rep. Fac. Agric. Okayama Univ.* 41: 61–68.

Suzuki, H., Oba, K. and Uritani, I. 1975. The occurrence and properties of 3-hydroxyl-3-methylglutaryl Coenzyme. A reductase in sweetpotato roots infected by Ceratocytis fimbriata. *Physiol. Plant Pathol.* 7 (3): 265–276.

Svendsby, O., Kakutani, K., Matsumura, Y., Izuka, M. and Yamamoto, T. 1981. Ethanol fermentation of uncooked sweetpotato with application of enzymes. *J. Ferment. Technol.* 59: 485–489.

Szyperski, R.J., Hamann, D.D. and Walter, W.M. 1986. Controlled alpha-amylase process for improved sweetpotato puree. *J. Food Sci.* 51 (2): 360–363, 377.

Tai, N.L. and Lei, T.S. 1970. Determination of proper amount of yellow corn and dried sweetpotatoes in swine feed. *J. Agric. Assoc. China* NS 70, 71.

Takahata, Y., Noda, T. and Nagata, T. 1994. Effect of beta-amylase stability and starch gelatinization during heating on varietal differences in maltose content in sweetpotatoes. *J. Agri. Food Chem.* 42: 2564–2569.

Takakuwa, M., Ikeda, K. and Murakami, T. 1972. Utilization of by-products in manufacture of sweetpotato starch. Part 4: Utilization of Saccharomyces cerevisiae cultivated in sweetpotato juice as food yeast or materials of nucleic acids. (Japanese). *J. Jap. Soc. Starch Sci.* 19 (3): 117–121.

Tan, J.D., Forio, E.E. and Estoy, L.S. 2005. Production and market potential of sweetpotato flakes. In: *2nd International Symposium on sweetpotato and cassava*, Malaysia, 14–17 June, 2005, pp. 215–216.

Taubenhaus, J.J. 1923. *The culture and diseases of the sweetpotato*. E. P. Dutton, New York, 179pp.

Terahara, n., Shimizu, T., Kato, Y., Nakamura, M., Maitani, T., Yamaguchi, M. and Goda, Y. 1999. Six acylated anthocyanins from the storage roots of purple sweetpotato, Ipomoea batatas. *Biosci. Biotechnol. Biochem.* 638: 1420–1424.

Thirumaran, A.S. and Ravindran, D.M. 1992. Processing of sweetpotato vermicelli in India. *In: Sweetpotato Technology for the 21st Century* (W.A. Hill, C.K. Bonsi and P.A. Loretan, Eds.), Tuskegee university, Tuskegee, Al. p 389–399.

Thompson, D.P. 1981. Chlorogenic acid other phenolic compounds in fourteen sweetpotato cultivars. *J. Food Sci* 46: 738–740.

Tillman, A.D. and Davis, H.J. 1943. Studies on the use of dehydrated sweetpotato meal in chick rations. *Louisiana Expt. Stn. Bull.* 358.

Tiu, C.S., Purcell, A.E. and Collins, W.W. 1985. Contribution of some volatile compounds to sweetpotato aroma. *J. Agric. Food Chem.* 33: 223–226.

Tomita, Y., Hayashi, K. and Hashizume, T. 1985. Palatability to pigs of sweetpotato-silage and digestion trical by them. (Japanese). *Bull. Fac. Agric. Kagoshima Univ.* No. 35, pp. 75–80.

Truong, V.D. 1990. Development of small scale technology for dehydrated sweetpotato cubes for traditional food preparations In: Tropical root and tuber crops: Changing role in a modern world. *Proc. Eighth Symposium of the International Society for Tropical Root Crops.* (R.H. Howeler. Ed.), October 30-November 5, 1988, Bangkok, Thailand, 1990, 724p.

Truong, V.D. 1992. Sweetpotato beverages: Product development and technology transfer. In: *Sweetpotato Technology for the 21st Century* (W.A. Hill, C.K. Bonsi and P.A. Loretan, Eds.), Tuskegee university, Tuskegee, Al. p 389–399.

Truong, V.D. and Fementira, G.B. 1990. Formulation, consumer acceptability and nutrient content of non-alcoholic beverages from sweetpotato. In: Tropical root and tuber crops: Changing role in a modern world. *Proc.8th Symposium. International Society for Tropical Root Crops.* (R.H. Howeler, Ed.), October 30-November 5, 1988, Bangkok, Thailand, 1990, 724p.

Truong, V.D. and Walter, W.M. Jr. 1994. Physical and sensory properties of sweetpotato puree texturized with cellulose derivatives. *J. Food Sci.* 59: 1175–1180.

Truong, V.D., Bierman, C.J. and MarLelt, J.A. 1986. Simple sugars, oligosaccharides, and starch determination in raw and cooked sweetpotato. *J. Agric. Food Chem.* 34 (3): 421–425.

Truong, V.D., Mcfeeters, R.E., Thompson, R.T., Dean, L.L. and Shofran, B. 2007. Phenolic acid content and composition in leaves and roots of common commercial sweetpotato (Ipomoea batatas L. Lam) cultivars in the United States. *J. Food Sci.* 72 (6): 343–349.

Truong, V.D., Walter, W.M. Jr. and Belt, K.L. 1998. Textural properties and sensory quality of processed sweetpotatoes as affected by low temperature blanching. *J. Food Sci.* 63 (4): 739–743.

Truong, V.D., Walter, W.M. Jr. and Hamann, D.D. 1997. Relationship between instrumental and sensory parameters of cooked sweetpotato texture. *J. Text. Stud.* 28: 163–185.

Tsou, S.C.S. and Yang, M.H. 1984. Flatulence factors in sweetpotato. *Acta Horticulturae* 163: 179–186.

Uritani, I. 1967. Abnormal Substance produced in fungus-contaminated foodstuffs. *J. Assoc. Offic. Anal. Chem.* 50: 105–108.

Uritani, I., Saito, T., Hondo, H. and Kim, W.K. 1975. Induction of furanoterpenoids in sweetpotato roots by the larval components of the sweetpotato weevils. *Agric. Biol. Chem.* 39 (9): 1857–1862.

Utomo, J.S., Che Man, Y.B., Rahman, R.A., Saad, M.S. 2005. Physical and chemical characteristics of restructed sweetpotato sticks made from three sweetpotato cultivars. In: *2nd International Symposium on sweetpotato and cassava, Kuala Lumpur*, Malaysia, 14–17 June, 2005, pp. 221–222.

Valetudie, J.C., Gallant, D.J., Bouchet, B., Colonna, P and Champ, M. 1999. Influence of cooking procedures on structure and biochemical changes in sweetpotato. *Starch/Starke* 51 (11 & 12), 389–397.

Villareal, R.L., Tsou, S.C.S., Lin, S.K. and Chiu, S.C. 1979. Use of sweetpotato (Ipomoea batatas) leaf tips as vegetables. II. Evaluation of yield and nutritive quality. *Exp. Agric.* 15: 117–122.

Wadsworth, J.I., Gallo, A.S., Ziegler, G.M. and Spadaro, J.J. 1967. Controlling bulk density of high sugar products during drum drying,. *Food Technol.* (Chicago), 21: 1038–1041.

Wagner, J.R., Becker, M.R., Gumbmann, M.R. and Olson, A.C.1976. Hydrogen production in the rat following ingestion of raffinose, stachyose and oligosaccharide free bean residue. *J. Nutr.* 106 (4): 466–470.

Walter, W.M. and Hoover, M.W. 1984. Effect of pre-processing storage conditions on the composition, microstructure, and acceptance of sweetpotato patties. *J. Food Sci* 49: 1258–1261.

Walter, W.M. and Hoover, M.W. 1986. Preparation, evaluation and analysis of a French-fry type product from sweetpotatoes. *J. Food Sci.* 51: 967–970.

Walter, W.M. and Schadel, W.E. 1981. Distribution of phenols in 'Jewel' sweetpotato (Ipomoea batatas L. Lam) roots. *J. Agric. Food Chem.* 29 (5): 904–906.

Walter, W.M. Jr. and Catignani, G.L. 1981. Biological quantity and composition of sweetpotato fractions. *J. Agric. Food Chem.* 29: 797–799.

Walter, W.M. Jr., Catignani, G.L., Yow, L.L. and Porter, D.H. 1983. Protein nutritional value of sweetpotato flour. *J. Agric. Food Chem.* 31 (5): 947–949.

Walter, W.M. Jr. and Purcell, A.E. 1974. Lipid autoxidation in pre-cooked dehydrated sweetpotato flakes stored in air. *J. Agric. Food Chem.* 22: 298–302.

Walter, W.M. Jr. and Purcell, A.E. 1976. Changes in amyloid carbohydrates during preparation of sweetpotato flakes. *J. Food Sci.* 41: 1374–1377.

Walter, W.M. Jr. and Purcell, A.E. 1980. Effect of substrate levels and polyphenol oxidase activity on darkening in sweetpotato cultivars. *J. Agric. Food Chem.* 28: 941–944.

Walter, W.M. Jr. and Schadel, W.E. 1982. Effect of lye peeling conditions on sweetpotato tissue. *J. Food Sci.* 47: 813–817.

Walter, W.M. Jr. and Schwartz, S.J. 1993. Controlled heat processing of Jewel sweetpotatoes for puree production. *J. Food Qual.* 16: 71–80.

Walter, W.M. Jr. Purcell, A.E., Hoover, M.W. and White, A.G. 1978 a. Preparation and storage of sweetpotato flakes fortified with plant protein concentrates and isolates. *J. Food Sci.* 43: 407–410, 419.

Walter, W.M. Jr., and Giesbrecht, F.G. 1982. Effect of lye peeling conditions on phenolic destruction, starch hydrolysis and carotene loss in sweetpotatoes. *J. Food Sci.* 47:810–812.

Walter, W.M. Jr., Collins, W.W. and Purcell, A.E. 1984. Sweetpotato protein: a review. *J. Agric. Food Chem.* 32: 695–699.

Walter, W.M. Jr., Purcell, A.E and Nelson, A.M. 1975. Effects of amylolytic enzymes on 'moistness' and carbohydrate changes of baked sweetpotato cultivars. *J. Food Sci.* 40: 793–796.

Walter, W.M. Jr., Purcell, A.E. and McCollum, G.K. 1979. Use of high pressure liquid chromatography for analysis of sweetpotato phenolics. *J. Agric. Food Chem.*. 27(5): 938–941.

Walter, W.M. Jr., Purcell, A.E., Hoover, M.W. and White, A.G. 1978 b. Lipid autoxidation and amino acid changes in protein-enriched sweetpotato flakes. *J. Food Sci.* 43: 1242–1244.

Wang, H. and Lin, C.T. 1969. The determination of the carotene content of sweetpotato parental varieties and their offspring (Chinese). *J. Agric. Assoc. China* 65: 1–5.

Wang, Y. and Kays, S.J. 2001. Effect of cooking method on the aroma constituents of sweetpotatoes (Ipomoea batatas (L.) Lam. *J. Food Qual.* 24 (1): 67–78.

Weber, C.W. 1969. Trends and forecasts. In: *Proc. 13th Annu. Poult. Ind. Day. Univ. Arizona*, pp. 55, 58.

Wiersema, S.G., Hesen, J.C. and Song, B.F. 1989. Report on a sweetpotato post harvest advisory visit to the People's Republic of China, 12–27 January, *International Potato Center, Lima.*

Williams. J.L. and Ammerman, G.R. 1968. Canning of stored sweetpotatoes. *Miss. Farm Res.* 31: 6–7.

Wilson, B.J. and Burka, L.T. 1979. Toxicity of novel sesquiterpenoids from the stressed sweetpotato (Ipomoea batatas). *Food Cosmet. Toxicol.* 17 (4): 353–355.

Wilson, B.J., Boyd, M.R., Harris, T.M. and yang, D.T.C. 1971. A lung oedema factor from mouldy sweetpotatoes (Ipomoea batatas). *Nature* (Lond.) 231: 52–53.

Woodroof, J.G. and Atkinson, I.S. 1944. Preserving sweetpotatoes by freezing. *Ga. Agric. Exp. Stn. Bull.* 232.

Woodroof, J.G., Dupree, W.E. and Cecil, S.R. 1955. Canning sweetpotatoes. *Ga Agric. Exp. Stn. Bull.* No. 12.

Woolfe, J.A. 1992. *Sweetpotato: an untapped food resource, University Press, Cambridge*, 643pp.

Wu, J.F. 1980. Energy value of sweetpotato chips for young swine. *J. Anim. Sci.* 51 (6): 1261–1265.

Yamakawa, O., Suda, I. and Yoshimoto, M. 1998. Development and utilization of sweetpotato cultivars with high anthocyanin content. *Food & Food Ingre. J. Jpn.* 178: 69–77.

Yang, Y.S. and Yu, P.H. 1976. Studies on the contamination of furanoterpenoids in sweetpotato in Taiwan. (Chinese). *Plant Protect. Bull.* 18 (1): 1–12.

Yeh, T.P. and Bouwkamp, J.C. 1985. Roots and vines as animal feeds. In: *Sweetpotato products: A natural resource for the tropics.* (J.C. Bouwkamp, Ed) CRC Press, Boca Raton, Florida, 271pp.

Yeh, T.P., Wung, S.C., Koh, F.K., Lee, S.Y. and Wu, J.F. 1977. Improvement in. the nutritive value of sweetpotato chips by different methods of processing. *Animal Industry Research Institute, Taiwan. Ann. Res. Rep.* 89–95.

Ylonen, K., Alfthan, G., Groop, L. 2003. Dietary intakes and plasma concentrations of carotenoids and tocopherols in relation to glucose metabolism in subjects at high risk of type 2 diabetes: the Botnia Dietary Study. *Am. J. Clin. Nut.* 77 (6): 1434–1441.

Yoshida, M., Hoshii, H. And Morimoto, H. 1962. Nutritive value of sweetpotato as carbohydrate source in poultry feed. IV. Biological estimation of available energy of sweetpotato by starting chicks. *Agric. Biol. Chem.* 26 (10): 679–682.

Yoshimoto, M., Kurata, R., Okuno, S., Ishiguro, K., Yamakawa, O., Tsubata, M., Mori, S. and Takagaki. 2005. Nutritional value of and product development from sweetpotato leaves. In: *2nd International symposium on sweetpotato and cassava, Kuala Lumpur*, Malaysia, 14–17 June, 2005, pp. 183–184.

Yoshimoto, M., Okuno, S., Yoshinaga, M., Yamakawa, O., Yamaguchi, M. and Yamada, J. 1999. Antimutagenicity of sweetpotato (Ipomoea batatas) roots. *Biosci. Biotechnol. Biochem.* 63 (3): 537–541.

Chapter 12
Economics of Sweetpotato Production and Marketing

T. Srinivas

Introduction

The major Roots and Tubers, Cassava, Potato, Sweetpotato and Yam occupied about 53.93 million ha world wide, producing 736.747 million tonnes annually (FAO, 2008). Individually Cassava, Potato, Sweetpotato and Yam rank among the most important food crops worldwide and in terms of annual volume of production. Cassava, Potato and Sweetpotato rank among the top ten food crops produced in developing countries. Many of the developing world's poorest producers and the most undernourished house holds depend on roots and tubers as an important source of food and nutrition (Scott et al., 2000). Roots and tubers produce large quantities of energy per day, in comparison of cereals. Sweetpotato is an important tropical root and tuber crop as it ranks second after cassava among the tropical tuber crops. The crop can be considered promoting nutritional security particularly in agriculturally backward areas. Besides carbohydrates, it is a rich source of protein, lipid, calcium and carotene. It becomes an ideal crop for popularisation in areas with poor soils and poor agricultural infrastructure facilities.

In annual production, sweetpotato ranks as the fifth most important food crop on a fresh-weight basis in developing countries after rice, wheat, maize, and cassava. Sweetpotato is cultivated in 114 countries and ranks among the five most important food crops in over 50 countries. Asia has the world's major production area for sweetpotato. In Asia, the greatest share of production is in China that accounts for about 85% of global production.

Feed, processed food and other non-food uses for sweetpotato expanded considerably in Asia over the last three decades (Pham et al., 1996; Scott, 1992; Titapiwatanakun, 1998). Rapid growth in demand for meat has created growth opportunities in more remote areas to use Sweetpotato as animal feed like in Sichuan province, China (Huang, 1999). Small farmers in China who have long cultivated

T. Srinivas (✉)
Senior Scientist, Section of Social Sciences, Central Tuber Crops Research Institute, Thiruvananthapuram, Kerala, India
e-mail: srinictcri@yahoo.com

G. Loebenstein, G. Thottappilly (eds.), *The Sweetpotato*,
DOI 10.1007/978-1-4020-9475-0_12, © Springer Science+Business Media B.V. 2009

sweetpotato as food security crop now process roughly half of their annual harvest into animal feed, 20–30% annual sweetpotato output into starch for noodles and other processed products (Huang, 1999; Timmins et al., 1992).

Sweetpotato also helps alleviate poverty by providing employment opportunities in production, processing and marketing. Surveys on sweetpotato production found that labor is the most important cost of production (Srinivas and Ramanathan, 2005; Pham et al., 1996; Cabanilla, 1996). Sweetpotato also holds a promise of helping to alleviate environment problem as it can serve as a quick cover crop to reduce soil erosion (Orno, 1991).

Before discussing sweetpotato marketing, an idea regarding the spatio-temporal variations at global level is essential.

Spatio-Temporal Growth Trends of Sweetpotato

Sweetpotato area, production and productivity data for the major growing countries in Asia, Africa, Latin America and USA for the period from 1961 to 2006 were collected from www.faostat.org. The data were classified into six 10-year periods starting from 1961.

Major sweetpotato growing countries in each continent were considered in understanding the trends in area, production and productivity. The following major sweetpotato growing countries were considered in the estimation of compound growth rates (CGR).

Asia – China, Indonesia, Japan, Vietnam, India, Philippines, Korea, Bangladesh
Africa – Angola, Burundi, Congo, Ethiopia, Ghana, Guinea, Kenya, Madagascar, Nigeria, Rwanda, Tanzania, Uganda
Americas – Brazil, Cuba, Haiti, USA
Polynesia – Papua New Guinea

Area Trends

Spatio-temporal variations in sweetpotato area were presented in Table 12.1. Sweetpotato area globally has been showing a declining trend. It has come down from 13.36 million ha in 1961 to 8.85 million ha in 2006. It declined significantly at the rate of one percent per annum and during 1961–1970, two percent per annum. This decline was more predominantly seen in Asia followed by in Latin America. Global decline was mainly due to area decrease in the Asian continent from 12.24 m ha in 1961 to 5.44 m ha in 2006. However, it is heartening to note that the area in Africa has seen five fold increase during the same period. The effect of this significant decline in Asia and Latin America was offset by more than three percent growth in sweetpotato area in Africa.

Table 12.1 Compound growth rates of sweetpotato area (Country wise and Period wise from 1961 to 2006)

Area	1961–70	1971–80	1981–90	1991–2000	2001–2006	1961–2006
China	−0.023**	−0.01NS	−0.16**	−0.007**	−0.034**	−0.017**
Indonesia	−0.033NS	−0.028**	−0.018NS	−0.022*	−0.006NS	−0.023**
Japan	−0.103**	−0.052**	−0.008**	−0.032**	−0.004NS	−0.043**
Vietnam	−0.009**	0.081**	−0.028**	−0.053**	−0.065*	0.002NS
India	0.064**	0.013NS	−0.043**	−0.023**	0.015NS	−0.014**
Philippines	−0.014NS	0.074**	−0.026**	−0.011*	−0.008**	−0.005**
Korea	0.007**	−0.07**	−0.100**	0.012NS	0.052*	−0.060**
Bangladesh	0.081**	0.010NS	−0.037**	−0.021**	−0.02*	−0.009**
Asia	−0.022**	−0.003NS	−0.018**	−0.01**	−0.033**	−0.017**
Brazil	0.034**	−0.09**	−0.029**	−0.041**	0.015**	−0.038**
Cuba	0.053**	0.00NS	0.015NS	−0.020NS	−0.023NS	0.003NS
Haiti	0.043**	0.026**	0.0002NS	−0.013*	−0.009NS	0.008**
Latin America	0.026**	−0.031**	−0.008NS	−0.019**	−0.011NS	−0.014**
Angola	0.014NS	0	0.007**	0.073*	0.112**	0.034**
Burundi	0.004NS	0.039**	0.038**	0.003NS	0.007NS	0.02**
Congo	0.009NS	0.023**	0.014**	−0.065*	0.005NS	0.004NS
Ethiopia	0.036**	0.016**	0.008**	0.044**	0.098**	0.027**
Ghana	–	–	–	–	0.004NS	—
Guinea	0	0.002NS	0.047**	0.005NS	0.284NS	0.026**
Kenya	0.02**	0.029**	−0.018NS	0.057*	−0.064NS	0.015**
Madagascar	0.002NS	0.035**	0.003NS	–	0.067**	0.016**
Nigeria	0.010NS	0.008NS	0.064**	0.0293**	0.197**	0.098**
Rwanda	−0.028**	0.056**	0.035**	−0.022NS	−0.071*	0.028**
Tanzania	0.29**	0.116**	0.101**	0.069**	0.005NS	0.064**
Uganda	0.064NS	−0.067*	0.011NS	0.03**	0.003NS	0.027**
Africa	0.025NS	0.001NS	0.28**	0.053**	0.020NS	0.031**
USA	−0.046**	−0.007NS	−0.032**	0.012**	−0.008NS	−0.016**
Papua New Guinea	0.011**	0.012**	0.007**	0.002*	0.001NS	0.008**
World	*−0.018***	*−0.003NS*	*−0.012***	*0.002NS*	*−0.005NS*	*−0.010***

Note: 1. ** indicates significant at 1% and * indicates significant at 5%.
2. Estimated by Author

China, Indonesia, Japan and India in Asia showed significant declining trends in area under sweetpotato. Though China continued to have the largest area under the crop globally, significant area reduction was observed between 1961 and 2006. 81% of the global sweetpotato (10.85 m ha) in 1961 in China has become 52% of global sweetpotato (4.71 m ha) in 2006. With increasing economic growth and rapid urbanization in many parts of Asia, consumers decreased their demand for traditional starchy stables, such as fresh sweetpotato in favour of meat, bread, potato and other preferred foods. Decline in crop area in India is attributed mainly to lack of diversified uses to produce value added products, availability of food grains in sufficient quantities, changed food habits due to increased standard of living *etc.* Hybrid maize or imported feed rations displaced sweetpotato as feed source in Korea (Chin, 1989) and Taiwan (Chiang, 1992). Bulkiness, perishability and erratic year-to-year, season-to-season movements in supply and prices made it difficult to establish local sweetpotato based agro industries in the Philippines (Cabanilla, 1996).

Area growth was negative in Latin America. Only during the period 1961–1970, the area recorded a 2.6% growth in Latin America. Thereafter a declining trend was observed with varying percentage of decline in different periods. Brazil, which was in first position from 1961 to 1990, recorded nearly 4% reduction per annum and went down to third position in 2006, with only 0.046 m ha. in 2006. Haiti and Cuba attained first and second position by 2006.

All African countries recorded significant growth in sweetpotato area during 1991–2000 and 2001–2006. As a whole 3.1% annual growth was recorded in the African continent. Currently 34.5% of global sweetpotato area is in Africa (3.14 m ha) from a meager 4.6% crop area in 1961 (0.62 m ha). Nigeria, Tanzania, Angola, Rwanda, Uganda and Ethiopia recorded 9.8, 6.4, 3.4, 2.8, 2.7 and 2.7% annual area growth rates respectively since 1961.

USA recorded negative growth rate in all the periods except during 1991–2000 when a 1.2% positive annual growth was recorded. An overall 1.6% decline in area was observed during 1961–2006.

Production Trends

Spatio-temporal variations in sweetpotato global production are presented in Table 12.2. Globally sweetpotato production has showed an increasing trend with 0.4% growth per annum during 1991–2000. Positive growth was observed in all the periods except in 1981–1990 and 2001–2006 with a declining trend in production. The positive and significant growth in global production was due to significant growth in Africa and Asia. This compensated the declining trend in Latin America. Currently 122.79 million tonnes of sweetpotato is produced compared to 98.19 million tonnes in 1961.

Among the major sweetpotato growing Asian countries, Vietnam and China recorded a significant growth of 1.2 and 0.4% per annum respectively between 1961 and 2006 while Japan, Indonesia, Bangladesh, India and Philippines recorded a significant decline in sweetpotato production. Production growth for sweetpotato in China is now positive although it had contracted. The rebound is largely due to the explosive demand for meat and animal feed in the inland sweetpotato production centers. Growth in demand both at home and abroad for processed food products made from sweetpotato has also contributed to the upsurge in sweetpotato output (Fugile et al., 1999; Zhang, 1999). Improved, small scale processing of sweetpotato roots has also boosted production by making household or village-level processing less onerous and more profitable (Wheatley et al., 1997). In addition new varieties have been adopted more widely in part because of the rebound in off farm demand.

In Latin American countries a declining trend was observed from 1971 onwards. Only in the period 1961–1970 a growth of 4.3% per annum in sweetpotato production was observed. Brazil and Haiti recorded significant negative growth from 1961 to 2006 at 3.7 and 0.9% per annum respectively. Only in Cuba, it showed 0.8% growth per annum during the same period. Production of sweetpotato stagnated or

Table 12.2 Compound growth rates of sweetpotato Production (Country wise and Period wise from 1961 to 2006)

Production	1961–70	1971–80	1981–90	1991–2000	2001–2006	1961–2006
China	0.046**	0.012^{NS}	−0.007^{NS}	0.014^{NS}	−0.029**	0.004**
Indonesia	−0.047*	−0.005^{NS}	0.007^{NS}	−0.02*	0.01^{NS}	−0.01**
Japan	−0.105**	−0.005^{NS}	−0.002^{NS}	−0.016^{NS}	−0.006^{NS}	−0.037**
Vietnam	−0.007^{NS}	0.09**	−0.019^{NS}	−0.048**	−0.033**	0.012**
India	0.078**	−0.009^{NS}	−0.031**	−0.013**	0.027**	−0.007**
Philippines	−0.006^{NS}	0.06**	−0.03**	−0.027**	0.009^{NS}	−0.007**
Korea	0.073^{NS}	−0.043**	−0.094**	0.018^{NS}	−0.003^{NS}	−0.007**
Bangladesh	0.0121**	0.007^{NS}	−0.041**	−0.026**	−0.023*	−0.012**
Asia	0.037**	0.011^{NS}	−0.008^{NS}	0.012^{NS}	−0.027*	0.003**
Brazil	0.058**	−0.132**	−0.014*	−0.04**	0.012**	−0.037**
Cuba	0.031^{NS}	0.016^{NS}	−0.002^{NS}	0.033^{NS}	−0.014^{NS}	0.008**
Haiti	0.042**	0.012^{NS}	−0.038**	−0.017**	−0.02**	−0.009**
Latin America	0.043**	−0.07**	−0.006^{NS}	−0.004^{NS}	0.02^{NS}	−0.018**
Angola	0.045**	0.008**	0.001^{NS}	0.017**	0.132**	0.025**
Burundi	0.004^{NS}	0.041**	0.039**	0.001^{NS}	0.01^{NS}	0.02**
Congo	−0.016^{NS}	0.02**	0.008**	−0.063**	0.007^{NS}	0.001^{NS}
Ethiopia	0.036**	0.016**	0.008**	0.061**	0.06**	0.03**
Ghana	–	–	–	–	0.004^{NS}	–
Guinea	−0.002^{NS}	−0.009^{NS}	0.002^{NS}	0.027*	0.134^{NS}	0.012**
Kenya	0.055**	0.039**	−0.019^{NS}	0.059*	−0.002^{NS}	0.024**
Madagascar	0.022**	0.032^{NS}	0.024**	0.008^{NS}	0.009^{NS}	0.014**
Nigeria	0.001^{NS}	−0.005^{NS}	0.029^{NS}	0.319**	0.069*	0.073**
Rwanda	−0.03^{NS}	0.095**	−0.022**	−0.014^{NS}	−0.088*	0.024**
Tanzania	0.022*	0.001^{NS}	0.018^{NS}	0.024^{NS}	0.008^{NS}	0.032**
Uganda	0.082**	0.001^{NS}	0.018^{NS}	0.024^{NS}	0.008^{NS}	0.027**
Africa	0.024*	0.032**	0.010*	0.055**	0.001^{NS}	0.027**
USA	−0.016^{NS}	0.002^{NS}	−0.018^{NS}	0.016*	0.033^{NS}	0.0001^{NS}
Papua New Guinea	0.018**	0.015**	0.01*	0.0001^{NS}	0.0007^{NS}	0.012**
World	*0.036***	*0.01^{NS}*	*−0.007^{NS}*	*0.014**	*−0.015***	*0.004***

Note: 1. **indicates significant at 1% and * indicates significant at 5%.
2. Estimated by Author

contracted due to urbanization and its associated shifts in eating habits. Imports of wheat flour for food and maize or concentrates for feed provided shifting competition for sweetpotato in Peru (Blondet and Espinola, 1998). Sweetpotato use in Argentina was confined to niches for processed products in the domestic market or to exports of fresh roots.

All the major sweetpotato growing African countries showed positive and significant growth in production of 2.7% from 1961 to 2006. In the African continent, Nigeria showed the highest growth (7.3%) followed by Tanzania (3.2%), Uganda, Kenya, Rwanda (2.7% each). Most of the growth in production occurred in Eastern, Central and Southern Africa in response to steadily increasing pressure on local food system due to the population growth, civil war and economic hardship (Bashaasha et al., 1995; Tardif-Douglin, 1991). In the kivu region of Democratic Republic of the Congo, sweetpotato has been used as a staple food for disaster relief (Tanganik et al., 1999). Decline or stagnation in output of other staples has also contributed to the interest by farmers and consumers in sweetpotato in Malawi (Phiri, 1998). Cash sales of tubers and nascent processing sector have added to the momentum in production in Uganda (Scott et al., 1999).

Sweetpotato production in USA was stable and recorded positive and non-significant growth in all the periods, except during 1999–2000 where it showed 1.6% growth.

Productivity Trends

Spatio-temporal variations in sweetpotato productivity were presented in Table 12.3. Sweetpotato yields globally showed increasing trend in all the periods, except

Table 12.3 Compound growth rates of sweetpotato productivity (Country wise and Period wise from 1961 to 2006)

Productivity	1961–70	1971–80	1981–90	1991–2000	2001–2006	1961–2006
China	0.07**	0.018*	0.01NS	0.022**	0.005NS	0.021**
Indonesia	−0.014**	0.023*	0.024**	0.003NS	0.016**	0.013**
Japan	−0.002NS	0.002NS	0.006NS	0.016NS	−0.002NS	0.006**
Vietnam	0.002NS	0.011NS	0.009NS	0.005NS	0.032**	0.01**
India	0.014NS	−0.022**	0.012*	0.01NS	0.012NS	0.007**
Philippines	0.009**	−0.01NS	−0.004NS	−0.016**	0.018**	−0.002**
Korea	−0.003NS	0.028**	0.006NS	0.006NS	−0.055**	0.005**
Bangladesh	0.039**	−0.003NS	−0.003NS	−0.004**	−0.002NS	−0.003**
Asia	0.59**	0.014*	0.01NS	0.022**	0.006NS	0.02**
Brazil	0.024**	−0.043**	0.016**	0.001NS	−0.003NS	0.0001NS
Cuba	−0.022**	0.015**	−0.017NS	0.053*	0.009NS	0.005*
Haiti	0.00001NS	−0.014NS	−0.04**	−0.004**	−0.011**	−0.017**
Latin America	0.016**	−0.04**	0.003NS	0.015**	0.003NS	−0.004**
Angola	0.032**	0.008**	−0.006NS	−0.056NS	0.02*	−0.01**
Burundi	0.001NS	0.002NS	0.001NS	−0.003NS	0.003NS	0.001*
Congo	−0.025**	−0.001NS	−0.006**	0.002NS	0.002NS	−0.002**
Ethiopia	–	−0.00004NS	–	0.016**	−0.038NS	0.003**
Ghana	–	–	–	–	0.001NS	–
Guinea	−0.002NS	−0.011**	−0.045**	0.022**	−0.15*	−0.014**
Kenya	0.035**	0.009NS	−0.001NS	0.002NS	0.062NS	0.009**
Madagascar	0.02NS	−0.003NS	0.022**	0.011*	−0.058**	−0.003*
Nigeria	−0.01NS	−0.013**	−0.035**	0.026NS	−0.128**	−0.0026**
Rwanda	−0.002NS	0.39**	−0.57**	0.008NS	−0.017NS	−0.004NS
Tanzania	−0.007NS	−0.002NS	−0.082*	0.059**	0.019**	−0.036**
Uganda	0.018NS	0.067**	0.007NS	−0.006NS	0.005NS	0.005**
Africa	0.0001NS	0.03**	−0.018**	0.002NS	−0.018NS	−0.003**
USA	0.03**	0.008NS	0.014*	0.003NS	0.040**	0.016**
Papua New Guinea	0.007**	0.003NS	0.003NS	−0.001NS	0.004NS	0.004**
World	***0.054****	***0.013****	***0.005NS***	***0.012****	***−0.01****	***0.013*****

Note: 1. ** indicates significant at 1% and * indicates significant at 5%.
2. Estimated by Author

during 2001–2006 where it recorded significant decline at one percent per annum. Average sweetpotato yield in the world over has almost doubled from 7.35 t ha^{-1} to 13.87 t ha^{-1}. This to some extent compensated the affect of decline in sweetpotato area on production. Only in Asia, sweetpotato yields have showed increasing trend at 2% per annum. Productivity of sweetpotato in Latin America and African continents declined at 0.4 and 0.3% respectively.

In all the major sweetpotato growing Asian countries, sweetpotato yields have recorded positive and significant growth except in Bangladesh and Philippines. It has increased three times in China from 7.12 t ha^{-1} in 1961 to 21.28 t ha^{-1} in 2006 recording 2.1% annual growth. But China's earlier isolation from western science and sweetpotato's much lower priority than cereals or industrial crops such as cotton have handicapped more rapid improvement in productivity. It has increased at 0.7% per annum in India, which helped in compensating the affect of area reduction on its production.

Among the Latin American countries, productivity growth was positive and significant only in Cuba (0.5% per annum). The same has declined at 1.7% per annum in Haiti. For Latin America as a whole sweetpotato productivity has been affected by weak demand. Most farmers have had little incentives to use yield-increasing technologies because potential commercial opportunities have yet to be exploited. Existing market out lets are limited and relatively thin. Remaining growers are generally resource poor farmers who choose to cultivate these commodities in part to avoid the financial risks associated with more input and cash incentive crops. As area planted with sweetpotato has declined or at best stagnated, cultivation has been pushed into or confined to more marginal soils.

Tanzania, Guinea, Angola in Africa showed declining trends in yield of sweetpotato during 1961–2006. Kenya, Uganda, Ethiopia and Burundi showed positive growth trend in sweetpotato yield. Increase in yield especially in Sub-Saharan Africa is difficult to achieve in the region because of nutrient poor soils, lack of irrigation and weak infrastructure (Spencer and Badiane, 1995).

USA and Papua New Guinea recorded a growth of 1.6 and 0.4% per annum in the productivity of sweetpotato during 1961–2006.

Sweetpotato Marketing – Asia

Among the major sweetpotato growing countries in Asia, marketing in China, Japan, Bangladesh and India are covered in this chapter. Sweetpotato production in Asia is characterized by four trends:

1. Continued overwhelming dominance of China.
2. Shrinking area planted to sweetpotatoes, a trend that has accelerated in much of the region during the last ten years.
3. Levelling off of yields as the rate of growth has slowed in many countries, including China. As sweetpotato cultivation has been pushed into more marginal

land and average yields have improved to 17 t ha^{-1}, it has become more difficult to maintain the rate of growth of improvement in yields.

4. The possible shift in future prospects for regional sweetpotato production due to recent changes in relative prices for sweetpotato versus traditional substitutes such as imported wheat flour, as a consequence of the economic crises in Southeast Asia.

Sweetpotato – China[1]

The most common names for sweetpotato in China are *ganshu*, "sweetpotato" and *hongshu* "red potato". Sometimes nomenclature overlaps for different root and tuber crops in different places. For example, a variety of yam *Dioscorea esculenta* is usually called *shanyao*, mountain medicine, but is also occasionally referred to as *ganshu*, sweetpotato. This crop has been cultivated in South China since at least the first century B.C. Sweetpotato, a comparatively modern crop, has effectively stolen the name. Chinese farmers produce more sweetpotato than anywhere else in the world.

Utilization Pattern

Food security aspects of sweetpotato cultivation are probably the reason for its original acceptance. More recently, as production of fine grains has developed rapidly in most regions, the role of sweetpotato has changed. A sizeable amount of production is still consumed by humans; although an increasingly greater proportion now is utilized by the processing or livestock feed industries. Thus fresh root is used for human consumption, industrial processing and feed. It is eaten fresh after cooking and dried as a staple food in poorer areas. In more affluent urban areas it is consumed as a vegetable, the resurgence in popularity driven by an increasing health awareness trend. Roasted sweetpotato is a popular snack throughout China. These can take the form of very basic products in rural areas or sophisticated niche marketed products (i.e. health benefits) in wealthy urban areas (Table 12.4). Industrially, fresh root is processed into starch and alcohol products, and both fresh root and vines are used as livestock feed.

Utilization trends vary enormously between provinces and depend on extent of development and relative geographic isolation. Richer provinces such as Jiangsu and Guangdong use less sweetpotato for food, and more for processing and feed; these provinces tend to have lower post-harvest losses than poorer provinces such as Sichuan and Guangxi, In terms of developing sweetpotato for industrial use Shandong is rapidly becoming a major player. This province has an aggressive

[1] Information on marketing of sweetpotato in China, Japan and Australia was drawn from the Report of Rural Industries Research and Development Corporation for the selected markets for taro, sweetpotato and yam compiled by Grant Vinning, Asian Markets Research, 2003.

Table 12.4 Principal sweetpotato products produced in China

Type of product	Name of the product
Food products	1. Unprocessed fresh, dried and boiled tubers as food 2. Canned and candied tubers, cakes, frozen fried tubers, ice cream, sherbet, pancakes, paste 3. Refined dried tubers including Red Heart brand dried sweetpotato and dried strips
Refined starch products	1. Starch 2. Vermicelli, sheet jelly, noodles
Saccharification products	1. Dextrose, fructose, glucose, maltose, amylose, amylose paste 2. Sugar residues

policy focused on developing sweetpotato as an industrial raw material. Farmers in Shandong generally grow two different major types of sweetpotato varieties for table consumption and those for industrial use. The later are normally harvested the earliest.

Processing

Processing of sweetpotato occurs at scales ranging from artisan farmer household to industrial scale. Simple processing of sweetpotato by farmers includes slicing and field-drying chips after harvest. Dried chips are sold to factories or sent to market by farmers themselves if they have the means to transport the product to the factory gate or through middlemen. Fresh root is also sent to factories, again either by farmers or middlemen. Dried chips are used as a snack food or occasionally as a raw material for starch production.

There are many staple and non-staple food uses for sweetpotato in addition to fresh root, *viz.*, starch, noodles, alcoholic beverages, vinegar and monosodium glutamate (Table 12.4). In Sichuan, sweetpotato chip snacks made from starch have been developed for farm-level operations. Boiled, sun dried flavoured chips and strips are also produced and have become quite popular.

Marketing

Depending on the cultivar, roots develop to marketable size in 90 to 150 days after transplanting. Normally, harvest begins when most of the roots are at a size suitable for the market, which may vary with intended end use.

The marketing of sweetpotato has changed dramatically since the reforms of 1978. The structure of demand has changed as incomes have increased, with sweetpotato food use declining and industrial and feed use increasing. Utilization is now directed towards higher value uses, such as industrial processing, animal feed and processed food products in more developed regions of the country, rather than staple consumption. The marketing structure has changed as a result of government withdrawal from former procurement and marketing functions. Prices also increased during the initial reform period but over the last few years have remained remarkably stable.

Many enterprises and traders have joined in vying for farm output with the government purchasing organizations that formerly were the only procurement agent for most rural products. These new marketing channels allow multiple entries into the market. In the case of suburban production for the urban fresh consumption market, farmers may market sweetpotato themselves, individually or collectively, if they have access to transportation, for example a bicycle, tricycle, or small tractor-drawn cart. Otherwise they may sell their crop to traders. Fresh root is not transported long distances. But because dried whole or chipped roots may be transported long distances with little damage, they are attractive industrial raw material sources. This attractiveness to users far away has caused some conflicts with local authorities over price instability.

However, the actual extent of trade is unknown, but regional and international trade of sweetpotato and its products has existed for some time and is growing. This is especially so for sweetpotato products for industrial processing such as chips or semi-processed products; factories often procure raw material within a radius of 150–200 km often buying from a number of provinces.

Dried chips are transported far greater distances, for example, from northern Anhui to processors in Beijing, Tianjin, and Shanghai, distances of more than 1,000 km. Also chip products produced in Sichuan can be found in Jilin and Heilongjiang. Long-haul transport of fresh root is not often seen. However, the opening of the Beijing-Shanghai expressway, which traverses the major sweetpotato growing areas of Shandong, may have an impact on supply of fresh root to the urban markets and for export. Some dry strips and chips are exported for starch processing or feed use. Sweetpotato-based snacks and candies, however, are popular in Southeast Asia, Hong Kong, and Japan. High-quality snack foods, including candied sweetpotato chunks and strips, and fried starch chips similar to shrimp chips, have been developed to accommodate this export market. These products have sold well abroad. In addition, canned sweetpotato is exported.

Prices

Gaining a reasonable estimate of prices in China is difficult. Prices increased during the 1980s by as much as 40–60%. Though during the late 1990s farm gate prices in processing regions, for sweetpotato, across provinces and at the beginning and end of the season were remarkably stable.

The data in Table 12.5 are based on sweetpotato procured in processing areas; these trends will not reflect the situation for the table variety market. Processors only operate for a limited period, which is governed by supply of fresh roots at an economical price.

There is contradictory evidence about seasonal price movements of fresh root for the table market. In Beijing, many retailers claim that price movements are fairly stable. In Chengdu (Sichuan) and Jinan (Shandong), retailers often comment on seasonal price fluctuations. Retail prices are usually the highest just before harvest,

Table 12.5 China – Farm gate prices for sweetpotato in selected provinces, 1998–2000

	Root price (RMB/kg)	
Province	Season start	Season end
Shandong	0.33–0.42	0.32–0.40
Sichuan	0.32–0.34	0.36–0.40

Source: Report of Rural Industries Research and Development Corporation for the selected markets for taro, sweetpotato and yam compiled by Grant Vinning, Asian Markets Research, 2003.

dropping sometimes by as much as half with the appearance of a new crop on the market. At times of the year when sweetpotato is in great demand, especially around certain holidays, the price increases greatly, sometimes by as much as 100%. The retail prices farmers charge for fresh root is usually about 20% less than that charged by traders and the wholesale cost to traders is also usually about 20% less than the retail price charged by farmers (a price of 0.32 RMB/kg wholesale to traders vs. a price of 0.40 RMB/kg retail).

Sweetpotato Marketing–Japan

Sweetpotato is called both *kansho* and *Satsuma-imo*. In general, farmers and older people use the name kansho whilst younger consumers use the name Satsuma-imo. Traders use both terms.

Utilization Pattern

Kansho is used directly as a food and indirectly in other forms. Between 10–15% of production is consumed in the processed form. The raw product is often steamed and then eaten (*mushiimo*-steamed potato). *Mushiimo* is also commonly referred to as *muneyake* -heartburn. In the warmer southwestern regions it is often eaten together with fish and shellfish. Three basic forms are utilized:

1. The tuber is first sliced cross-wise into round slices and then further cut into smaller squares. This is then boiled along with rice. It is popularly referred to as *imo meshi* -potato rice or *imo gayu*-potato rice gruel- a more liquid soupy version of *imo meshi*.
2. *Namakirihoshi* strips of dried potato and *namakirihoshikona* dried potato powder. The tuber is cut into strips and dried and/or then further crushes into a powder. This is stored as a food source. In rural districts and farms it is often stored in straw bags. After soaking the dried potato in water it can be used to make *imo meshi* and *imo gayu*. The powder is used to make *dango*, dumplings and other products such as *mushipan* steamed bread.
3. Using a quality powder potato, the skin is slightly peeled and together with special *mochi* sticky and glutinous rice, is steamed and then pounded to make rice cakes. *Kansho* is consumed widely in a variety of between-meal forms. Two of the most common forms are:

a. *Mushi imo* steamed potato: Also popular is cutting the tuber into rings, battering it in wheat flour, and then steaming. The latter form when it is made with bean paste has a similar texture and taste to *manju* bean paste buns. Because it is prepared quickly and simply, in the Kumamoto region of southern Japan it is referred to as *ikinari*.

b. *Yaki imo* roasted sweetpotato: The tuber is usually roasted whole but can be sliced into rings. In former times, during the night from autumn to the end of winter it was a common sight throughout Japan to see small trucks selling *ishi yaki imo* sweetpotato roasted on hot stones. Sadly, this sight, sound (as the salesman sings the particular noise peculiar to the craft announcing their presence), and smell that is disappearing in the larger cities.

Kansho supplied early in the season command the highest prices for these uses. *Kansho* is commonly served as a *tempura* quickly deep-fried side dish.

Sweetpotato has a large variety of processed foods made from it. Most of these are traditional in form but some are adaptations of western style chips and snacks. The important ones are:

1. *Imo an s*weetpotato bean paste: The tuber is peeled, steamed and then kneaded with sugar. Some genuine red bean paste made from Azuki beans is added to the original mix.
2. *Imo youkan* sweet jelly: This made in a manner similar to *imo an*. To prevent breakdown sugar is added along with gelatin and agar to harden the product.
3. *Imo senbei* sweetpotato rice crackers: The tuber is cut very thinly and covered in sugar.
4. *Imo karintou* sweetpotato fried cookies: Rather fat slices of potato are dried, cooked in oil, and then covered in sugar. Recently, imo karintou is being cut more finely or else made from sweetpotato powder.
5. *Mushikirihoshi* steamed, cut, dried sweetpotato: The potato is peeled, steamed, then cut into 7–8 mm slices and then dried. Also commonly referred to as *mushi imo* and *imo surume* dried potato cuttlefish. Originally, it was a product from the Hamaoaka region in Shizuoka Prefecture but now it is more generally associated with the central eastern section of Ibaraki Prefecture. A similar product is produced in the Tenkusa region of Kumamoto Prefecture.
6. Here, a small round potato is steamed, and becomes just like the popular dried persimmons.
7. *Daigakku imo* university potato: The tuber is cut diagonally into irregular chunks then fried in oil. Sugar syrup or honey is added, and finally sprinkled with sesame seeds.
8. Sweetpotato Chips: One of the types of *karintou* fried dough cookies. Cutting the potato into extremely thin strips and then adding table salt. It comes in various shapes and forms.
9. Sweetpotato Mash: The potato is peeled, steamed, put through a roller to flatten, and dried to finally produce a powder. Used in vegetable salads, croquettes, and also as a raw material for a variety of confectionaries. Products made from

the carotene rich potatoes, which have a reddish colour, are said to be rich nutritionally and also referred to as "Golden Mash".
10. Frozen Products: Steamed potatoes are sealed into packs and refrigerated.
11. Others: Canned and Square Cuts: These two variations have been introduced from other countries, primarily the USA mainland and Hawaii. The raw materials are essentially carotene potatoes. The potatoes are cut into large slices and steamed at high pressure then canned. Square cuts are steamed at high pressure, cut into squares then dried. Often added to soups.

As Japan moves increasingly towards the presentation of foods in a more ready-to-eat form through the convenience stores, it is most likely that greater emphasis will be placed on *kansho* in the processed form.

A major outlet for processed kansho is as *imojochu*, a vodka-like distilled spirit. *Imojochu* is a particular form of *shochu*, a low-valued indigenous spirit. It is considered that *imojochu* has a particular market niche compared with barley-based *shochu* because the *kansho* imparts a distinctive aroma and flavour. About 80 000 tonnes of *kansho* annually go into *imojochu*.

Another major non-direct food use involves the utilization of *kansho's* starch and flour properties, and as a food pigment. It is noted that nearly all of these uses are as low-value products. An emerging potential use is in biotechnology with the utilization route being through the production of lactic acid to make a biodegradable plastic

Annual Wholesale Performance

The data indicate that overall output is declining but that there can be increases in any given year. Over the 14 years, annual average wholesale prices can be considered to have risen slightly. In 2000, the price was 167 yen/kg. (RIRDC Report, 2003).

Monthly Wholesale Performance

Kansho has distinct seasonal patterns of monthly throughput and wholesale prices.

This allows for confident predictions.

The indications are that prices will rise in May or the beginning of summer, peak in June-July and decline in autumn. This period coincides with the traditional low usage of *kansho* in summer. Indications are that retail prices have a one-month lag after wholesale prices (RIRDC Report, 2003).

Annual household expenditure patterns show that the volume being purchased is declining at around the same rate as retail prices are increasing.

Given that the rate of decline in household purchases is less than the rate of decline in total production, the implication is that households are consuming more *kansho* in the processed form. Household expenditure is lowest in late spring to early summer, before prices peak in mid-summer. As to be expected, the quantity purchased by households is lowest when prices are highest and vice versa.

Kansho is imported into Japan in the frozen form both as a stand-alone product but increasingly in combination with other products such as burdock, carrot, bamboo, and lotus root. In 2000, 713 t were imported at an annual average price of 89 yen/kg. Data indicate that "fresh" sweetpotato is imported into Japan. Japan does not allow the importation of sweetpotato for quarantine reasons. Importations of fresh produce must follow the post-entry quarantine requirements, that is, it cannot be propagated. It is most likely that the data cited as "fresh" refer to dried product. The White Paper on sweetpotato in Japan estimates that 15,000 t of sweetpotato product in the form of bakery items, flour, dough and cereals are imported annually.

Sweetpotato in Bangladesh[2]

Sweetpotato is ideally suited for Bangladesh as it provides generous quantities of Vitamin A and C and iron, all of which are grossly deficient in the diets of a high percentage of Bangladeshis.

Jamalpur, Mymensingh, Noakhali, Comilla, Kurigram, Natore, Fraidpur and Madaripur are the major sweetpotato growing districts in Bangladesh. It is grown mainly in the Rabi season (October-November) and harvested from March to May. The harvested produce are consumed within two months due to poor storage facilities, poor storability of cultivated varieties and limited capacity in existing (or in absence of) cold storage facilities. The perishability of sweetpotato affects its marketability and prices received by farmers. Sweetpotatoes are allowed to dry in the shade of a tin roof. Dowl, dali, bamboo baskets, sandy soil, cane baskets, gunny bags, bamboo fences and open floors are used in storage.

Utilization Pattern

Lower income groups are reported to consume more than middle-income consumers. Tubers are eaten either roasted or boiled or cut into pieces and fried. Roasted or boiled sweetpotato is eaten with curry. Vines and roots are also used as animal feed. In some areas, it is eaten as staple as a substitute to rice during peak season when it is cheapest. Sweetpotato mash is mixed with wheat flour for preparing bread or mixed with vegetables. It is also used to produce starch, fructose and liquid glucose during the harvest season for three months in limited quantities.

---

[2] Information on Sweetpotato marketing in Bangladesh is drawn from Anonymous 1997 Sweetpotato marketing in Bangladesh: Results of a Rapid Market Appraisal. Summary Report Working Paper. Social Science Dept. Working paper No. 1997–3. International Potato Center (CIP), Lima, Peru.

Marketing

Marketing channels and processes vary from district to district. The major players in sweetpotato marketing are growers, baparies, wholesalers, retailers and consumers. The flow of sweetpotato from grower to consumer as well as the methods, in which this flow takes place, varies depending on the constraints and infrastructure under which the market developed. Product moves from market to market to reach the consumer. Perishable sweetpotato loses its quality through transport damage, which results in deterioration during storage. Most growers take their produce to the nearest village for sale, while keeping part for home consumption and animal feed. Large farmers sell directly to primary markets through farias/baparies and wholesalers. Middlemen also buy directly from farmer's fields. From primary markets, middlemen go to rural assembly markets and from here to semi-urban and urban markets after which the produce is taken to retailers and to the consumers. Some farmers store the produce in anticipation of price increase. Marketing channel for sweetpotato in Bangladesh is given in Fig. 12.1.

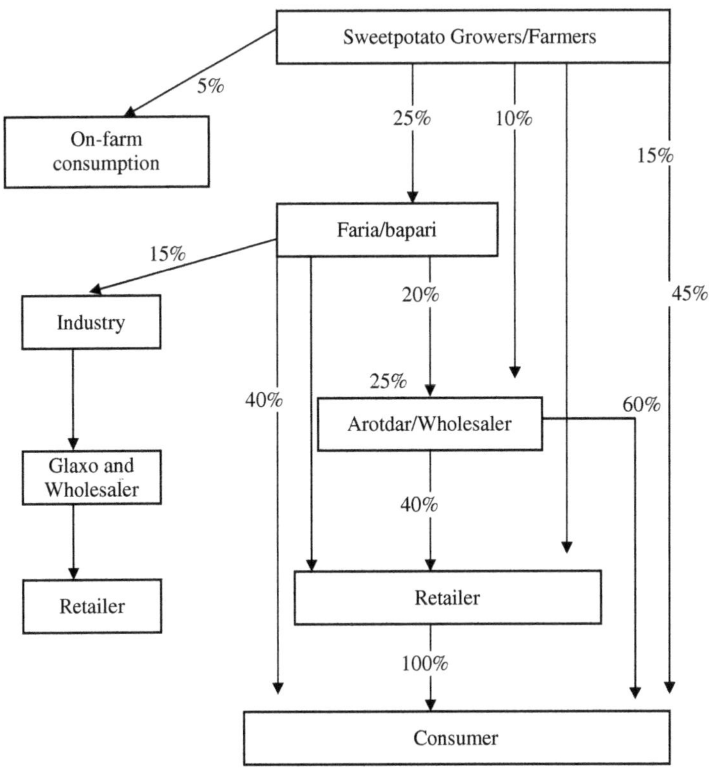

Fig. 12.1 Sweetpotato marketing channel in Bangladesh by percentage

Some of the marketing constraints/problems identified in marketing sweetpotato in Bangladesh were:

1. Lack of market information.
2. Limited demand for sweetpotato.
3. Limited availability and demand created for value added products from sweetpotato and a lack of knowledge of diversified products.
4. High cost of transport and lack of efficient transport facilities from remote marginal production areas. The majority of consumers are in the rural areas, which involve higher transport costs to move produce to these areas. Low sales in urban areas give lower margins to traders, which does not compensate for transport and handling costs.
5. Lack of storage facilities.
6. Lack of physical facilities for marketing sweetpotato in the rural, assembly and secondary markets as well as poor infrastructure and proper communication facilities.
7. Excess market charges.
8. Lack of consumer interest in using sweetpotato as a substitute for a staple food, except during times when a preferred staple is high priced.

Sweetpotato – India

Sweetpotato is predominantly cultivated as a rainfed crop in eastern India especially in Orissa, Bihar, Uttar Pradesh, West Bengal and Jharkhand states. Area under this crop is declining year after year due to less importance given to this crop in the programmes of State Department of Horticulture/Agriculture in many states. Industrially this crop is exploited very little. So far in India sweetpotato is not utilized for industrial purposes like extraction of sweetpotato starch, alcohol *etc*. Even though technologies to produce many value added products from sweetpotato are available, diversification and value addition has not taken any momentum as understood from the surveys undertaken in the production centres of the crop.

India occupies twelfth, eighth and fifth rank globally in terms of sweetpotato area, production and productivity respectively while fifth rank in Asia in area, production and productivity respectively (Table 12.6). Sweetpotato is predominantly cultivated as a rainfed crop in eastern India accounting for 77% of area and 82% of production of India (Table 12.7). Even in eastern India, the crop is concentrated in few districts only. Koraput, Sundargarh, Keonjhar, Ganjam and Bolangir are the major sweetpotato growing districts covering fifty per cent of the area under crop in Orissa. Ranchi, Hazaribagh districts in Jharkhand and Muzaffarpur and Samastipur in Bihar are the major sweetpotato growing districts covering 50% of the area under the crop in these states. So far very little work has been carried out on the economic aspects of sweetpotato cultivation and marketing in India.

Table 12.6 Rank of India in global sweetpotato scenario

Crop	Asia	World
Area	5	12
Production	5	8
Productivity	5	5

Source: www.fao.org

Table 12.7 Sweetpotato in India

State	Area ('000 ha)	Production ('000 t)	Productivity (t ha^{-1})
Orissa	47.1	394.3	8.37
Uttar Pradesh	20.0	240.8	12.04
Bihar	4.6	63.9	13.89
Jharkhand	7.6	103.0	13.55
West Bengal	26.1	193.7	7.42
India	136.5	1211.0	8.87

Source: Agriculture, Centre for Monitoring Indian Economy, 2006

Important Features of Sweetpotato Cultivation

Important features of sweetpotato cultivation in major growing states were presented in Table 12.8. Average holding size of sweetpotato farmers was same in Orissa (1.12 acres) and Bihar (1.12 acres) states while it was low in Jharkhand (0.48 acres). High yielding varieties developed by Central Tuber Crops Research Institute (CTCRI) like Sree Bhadra, RS III-3, X4, Kalmegh *etc.*, were very popular in Jharkhand and Bihar while local varieties of sweetpotato were predominantly grown in Orissa. Sweetpotato is grown round the year in Bihar and during kharif in Jharkhand and Orissa. The main growing seasons are kharif (rainy season), September to January/February and summer (January/February to May/June) in Bihar. It is also grown as a short duration crop after devastating floods. It is grown in bheet (upland), diara (sandy land between a large river and a tributary which remains under water during the rains) and flood prone areas.

Table 12.8 Important features of sweetpotato cultivation

Features	Jharkhand	Bihar	Orissa
Ava. Sweetpotato holding size (acres)	0.48	1.12	1.12
Varieties	Sree Bhadra RSIII-3, Kalmegh	RS47, X4, Sree	Local
Planting Time	June-July	February, June-July, September	June-July
Planting method	Mound	Ridge & Furrow	Ridge & Furrow
NPK ratio	60:25.4:6.41	35:67:46	19:8:5
Ava. Yield (t per ha)	7.67	24.7	8.22
Price per kg	2.18	2.22	3.34

Kharif is the major sweetpotato-growing season in the bheet upland areas. This crop is grown with the major objective of selling in the market as vegetable. Summer crop is generally grown on chaur (low land) areas after harvesting the paddy. In some areas along the rivers, farmers start planting sweetpotato mainly for commercial purpose as and when water recedes from their fields. Primary and secondary nurseries were raised before transplanting in the main field. Both ridge and furrow method and mound method of planting were practiced for planting sweetpotato. NPK ratio of 60:25.4:6.41, 35:67:46 and 19:8:5 was applied by farmers in Jharkhand, Bihar and Orissa states respectively. Significant difference in yields was observed among these three states. In Bihar owing to very fertile soils of gangetic belt, an average yield of about 25 tones per ha were obtained. While in Orissa and Jharkhand average yields recorded were to the tune of eight tones per ha only. Poor lateritic soils and low management practices adopted by farmers in these two states are the reasons for low yields. In Orissa, barter system was predominant in the study area. Farmers exchange sweetpotato for Ragi, Paddy etc.

Cost Concepts and Farm Income Measures for Sweetpotato

Cost concepts and farm income measures for sweetpotato cultivated in Jharkhand, Bihar, Uttar Pradesh and Orissa states were estimated. Based on the primary data collected from Hazaribagh and Ranchi districts in Jharkhand, Muzaffarpur and Samstipur districts in Bihar, Faizabad and Sultanpur districts in Uttar Pradesh and Koraput and Keonjhar districts in Orissa, Cost concepts and farm income measures estimated for sweetpotato cultivation in Uttar Pradesh state were presented in Table 12.9. Labor days in sweetpotato cultivation in Uttar Pradesh were presented in Table 12.10.

Jharkhand: Sweetpotato is cultivated as rainfed kharif crop in Jharkhand. Crop is planted during June-July following mound method. Sree Bhadra, RS III-3 and Kalmegh were the important sweetpotato varieties popularly cultivated in the study area. Average sweetpotato holding size was estimated to be 0.19 ha and most of the farmers are marginal to small with limited resources. NPK ratio of fertilizers applied by farmers in the study area was 60:25.4:6.41. Estimated average yield was 7.67 tonnes per ha. which was less than the national average yield of sweetpotato. On an average farmers were selling at Rs. 2.18 per kg of tubers.

It was estimated that gross cost (Cost C) of Rs. 16,429.28 was incurred in the cultivation of one ha. of sweetpotato. Family labour was mostly involved in farm operations in sweetpotato cultivation. Therefore gross labour cost of only Rs. 3,839.60 was incurred. Gross material costs of Rs. 3,810.28 was incurred in the sweetpotato farming. Operational cost (Cost A_1) of Rs. 7,930.35 was incurred in the cultivation of sweetpotato. It was estimated that the cost of production of sweetpotato was Rs. 2.14.

It was estimated that gross income of Rs. 16,911.73 was obtained from sweetpotato cultivation. Even though net income (Rs. 482.45) was low, farm business income (Rs. 8,981.38) was higher indicating the profitability of sweetpotato

Table 12.9 Cost concepts and farm income measures for Sweetpotato cultivation in Uttar Pradesh in India (in rupees per ha)

Particulars	Cost (Rs.)	Percentage to the total cost
Land preparation	3605.56	10.41
Nursery	84.22	0.24
Transplanting	2554.94	7.37
Application of manures	312.50	0.90
Application of fertilizers	8.93	0.03
Interculture	2823.01	8.15
Irrigation	887.65	2.56
Harvesting	3455.75	9.97
Gross labour costs	13732.56	39.63
Planting material	2685.34	7.75
Manures	574.40	1.66
Fertilizers	861.00	2.48
Gross material costs	4120.74	11.89
Land revenue	190.14	0.55
Interest on working capital	796.21	2.30
Depreciation	1861.80	5.37
Cost A1	20701.81	59.74
Lease amount	0	0
Cost A2	20701.45	59.74
Imputed value of owned land	10608.10	30.61
Interest on owned fixed capital	2087.83	6.03
Cost B	33397.39	96.38
Imputed value of family labour	1253.69	3.62
Cost C	34651.07	100.00
Farm income measures:		
Yield (t)	12.51	
Price/kg tubers	4.20	
Gross income	52490.81	
Net income	17439.74	
Farm business income	35195.06	
Owned farm business income	35195.06	
Family labour income	22499.13	
Farm investment income	30535.67	

cultivation. Family labour income and farm investment income were estimated to be Rs. 5,534.11 and Rs. 3,929.72 respectively. Benefit cost ratio was found to be very low (1.03:1) in Jharkhand in the cultivation of sweetpotato.

It was estimated that 154 man labour days were required in sweetpotato cultivation in Jharkhand. Since sweetpotato growers were resource poor, family labour involvement was more (90.15 man labour days) than hired labour (63.85 man labour days). Family labour was largely engaged for transplanting, weeding and harvesting operations while hired labour was engaged for land preparation.

Bihar: Important feature of sweetpotato cultivation in Bihar is that it is cultivated in three seasons viz., February planting, June–July planting and September planting. Ridge and furrow method was adopted by farmers during planting the crop. Sree Bhadra, RS 47 and X4 were the important sweetpotato varieties popular in the study

Table 12.10 Labour days involved in Sweetpotato cultivation in Uttar Pradesh states in India (per ha)

State	Particulars	Hired labour days	Family labour days	Total labour days
Uttar Pradesh	Land preparation	0.71	0	0.71
	Nursery	0.09	0.71	0.80
	Transplanting	64.15	3.12	67.27
	Application of manures	6.25	2.20	8.45
	Application of fertilizers	0.18	2.10	2.28
	Interculture	110.94	5.46	116.40
	Irrigation	0	3.52	3.52
	Harvesting	81.21	10.42	91.63
	Gross labour days	263.54	27.53	291.07
	Tractor labour days			1.63
	Bullock labour days			10.12

area. Average sweetpotato holding size was estimated to be 0.45 ha and most of the farmers were medium to large farmers. NPK ratio of fertilizers applied by farmers in the study area was estimated to be 35:67:46. Average yield was very high and was 24.7 tonnes per ha which was more than double of national average yield of sweetpotato. Farmers were selling tubers at an average price of Rs. 2.22 per kg.

It required Rs. 33,428.91 for cultivating one ha. of sweetpotato. Labour cost was estimated to be Rs. 9,321.02 while material cost was worked out to be Rs. 6,555.51. Operational cost of cultivation of sweetpotato was estimated to be Rs. 16,485.91. Even though cost of cultivation was the highest in this state, cost of production was found to be lower in Bihar than in Jharkhand and Orissa states due to very high productivity of the crop in this area.

Gross income of Rs. 55,305.00 and net income of Rs. 21,876.09 were obtained from one ha of sweetpotato cultivation. Farm business income was estimated to be Rs. 38,819.09 indicating high profitability from this crop. Family labour income and farm investment income were estimated to be Rs. 23,696.09 and Rs. 36,999.09 respectively. Benefit-cost ratio was found to be very high (1.65:1) among the four states.

Hired labour days involved were more (122.54) than family labour days (25.46) in sweetpotato cultivation. It required 148 man labour days in sweetpotato cultivation. Family labour was engaged mainly for weeding and transplanting operations.

Orissa: Sweetpotato is cultivated as kharif rainfed crop mostly by tribals in Orissa. It is planted during June–July following ridge & furrow method. Local varieties were predominantly cultivated in the study area. Average sweetpotato holding was estimated to be 0.46 ha and most of the farmers are small and resource poor. Farmers did not practice recommended fertilizer schedule. NPK ratio of fertilizers applied by farmers in the study area was 28: 9: 2. Estimated average yield was 6.97 tonnes per ha which was less than the national average productivity of sweetpotato. Barter system was predominant in the study area. Farmers exchange sweetpotato for ragi, paddy etc. On an average farmers were selling at Rs. 2.73 per kg of tubers. Primary and secondary nurseries were raised before transplanting in the main field.

It was estimated that the gross cost (Cost C) of Rs. 12,816.00 was incurred in the cultivation of one ha of sweetpotato. Family labour involvement in the farm operations was on par with the hired labour engaged. Labour cost was estimated to be Rs. 3,050.39 while material cost was worked out to be Rs. 3,034.39. Operational cost of cultivation (cost A_1) was worked out to be Rs. 6,376.51 per ha of sweetpotato. Cost of production was estimated to be Rs. 1.84 per kg of tubers.

Gross income of Rs. 16,468.28 and a net income of Rs. 3,652.28 were obtained from one ha of sweetpotato cultivation. Farm business income was estimated to be Rs. 10,091.77 indicating the profitability from the crop cultivation. Family labour income and farm investment income were estimated to be Rs. 5,752.48 and Rs. 8,010.07 respectively. BCR was found to be 1.28:1.

Number of hired (81.27) and family labour days (78.35) involved in sweetpotato cultivation was observed to be on par with Orissa in the study area. It required 159.62 man labour days in sweetpotato cultivation. Family labour involvement was more in intercultural and harvesting operations while hired labour was engaged mainly for land preparation and transplanting operations.

Uttar Pradesh: In Uttar Pradesh, sweetpotato is cultivated under irrigated production system in Faizabad and Sultanpur districts. Sweetpotato nurseries are grown two months before the planting season using the vines collected from the previous season. They are transplanted in June–July months on flat beds. Sree Bhadra, Narendra Sovereign and local varieties are under cultivation. Average sweetpotato holding size in the study area was observed to be 0.38 ha. On an average farmers were getting a yield of 12.51 tonnes per ha. At the time of data collection, price of the sweetpotato tubers was Rs. 4.20 per kg.

It was estimated that the gross cost of cultivation was Rs. 34,651.07 per ha. Expenditure on planting material was maximum (Rs. 2,685.34) followed by fertilizers (Rs. 861.00) and manures (Rs. 574.40). Among the labour costs, operations like land preparation, harvesting, interculturing and transplanting and were found to be expensive. Cost of production of sweetpotato was estimated to be Rs. 2.77 per kg of tuber. On an average farmers were getting a yield of 12.51 tonnes per ha with gross income of Rs. 52,490.81. Benefit cost ratio was worked out to be 1.51:1. Farm business income, owned farm business income, farm investment income and family labour income were estimated to be Rs. 35,195.06, Rs. 35,195.06, Rs. 30,535.67 and Rs. 22,499.13 respectively. Sweetpotato cultivation required on an average 263.54 hired labour days comprising 27.53 family labour days, 10.12 bullock labour days and 1.63 tractor labor days.

Resource Use Efficiency of Sweetpotato Farmers

Production elasticities and their respective standard errors for irrigated sweetpotato farmers in Uttar Pradesh were given in Table 12.11.

It was observed that co-efficient of multiple determination (R^2) was 0.3889 (Uttar Pradesh), 0.9649 (Bihar), 0.4741 (Orissa) and 0.2248 (Jharkhand), indicating that 39, 96, 47 and 22 per cent variation in gross income has been influenced by the selected variables respectively.

The sum of elasticities (Σbi) of production is an indication of the returns to scale. It is observed from the table that the sum of elasticities was −0.1538, 1.0955, 0.0451

Table 12.11 Production elasticities of sweetpotato cultivation under irrigated production system in Uttar Pradesh

Production system and State	Particulars	Production elasticity	Standard error
Irrigated (Uttar Pradesh)	Constant	9.9048	2.5038
	Farm size	−0.2148	0.1743
	Hired human labour	0.1918	0.2519
	Family labour	−0.0293	0.0195
	Manures and fertilizers	0.0424**	0.0155
	Planting material	−0.1322	0.2691
	Irrigation	−0.0117	0.0138
	Returns to scale	−0.1538	
	R^2	0.3889	
	F	2.2282	

and 0.3043 for irrigated sweetpotato farmers in Uttar Pradesh and Bihar and rainfed farmers of Orissa and Jharkhand respectively indicating decreasing returns to scale in operation for irrigated sweetpotato farmers in Uttar Pradesh and rainfed farmers of Orissa and Jharkhand and increasing returns to scale in operation for irrigated sweetpotato farmers in Bihar.

Production Elasticities of Variables (b_i)

Irrigated Production System

Uttar Pradesh: Out of six independent variables selected, manures alone were found to influence gross production and in turn gross income positively. Human labor was showing positive but non-significant relationship with gross income while farm size, family labor and planting material and irrigation were showing negative and non-significant relationship with the gross income.

Bihar: Out of the six independent variables selected, planting materials, manures and fertilizers were found to influence gross production and in turn gross income positively. Farm size and family labor were showing positive but non-significant relationship with gross income while hired human labor was showing negative and non-significant relationship with gross income.

Rainfed Production System

Orissa: Out of five independent variables selected, planting materials alone was found to influence gross production and in turn gross income negatively. Hired human labour, farm size, family labour and manures & fertilisers were showing positive but non -significant relationship with gross income.

Jharkhand: Out of the six independent variables selected, hired human labour was found to influence gross production and in turn gross income positively. Planting materials, family labour, farm size and fertilizers were showing positive and non-significant relationship with gross income while manures was showing negative but non-significant relationship with gross income.

Resource Use Efficiency

The estimated marginal value products (MVP), opportunity costs (OC) and the ratio of MVP and OC values for farmers of sweetpotato in Uttar Pradesh were presented in Table 12.12.

Table 12.12 Marginal Value Product (MVP), Opportunity Cost (OC) and ratio of marginal value product and opportunity cost for the sweetpotato farmers under irrigated production system in Uttar Pradesh

Type of Production system	Factors	MVP	OC	MVP/OC
Irrigated (Uttar Pradesh)	Farm size	−15066.73	10608.10	−1.42
	Hired human labour	0.73	1.00	0.73
	Family labour	−179.23	1.00	−179.23
	Manures & fertilizers	3170.43	1.00	3170.43
	Planting material	−2.50	1.00	−2.50
	Irrigation	−15842.15	1.00	−15842.15

Irrigated Production System

Uttar Pradesh: The inputs – manures and fertilizers were being used optimally indicating that the utilization of these inputs could be increased sufficiently while utilization of all other variables were not at optimum and there were significant difference between MVP and OC as MVP were much less than acquisition costs. Hence these inputs are being used excessively at present than required by this group of sweetpotato farmers. There is a need to reduce these inputs to so as to reduce the cost of cultivation, to achieve higher productivity and thereby higher gross returns.

Bihar: The inputs – planting materials, family labour, manures and fertilizers were being used optimally indicating that the utilization of these inputs could be increased sufficiently while utilization of all other variables were not at optimum and there was significant difference between MVP and OC as MVP were much less than acquisition costs. Hence these inputs are being used excessively at present than required by this group of sweetpotato farmers. These inputs weren't being used efficiently. There is a need to reduce the use of these inputs to so as to reduce the cost of cultivation, to achieve higher productivity and thereby higher gross returns.

Rainfed Production System

Orissa: The inputs – manures and fertilizers were being used optimally indicating that the utilization of these inputs could be increased sufficiently while utilization of all other variables were not at optimum and there was significant difference between MVP and OC as MVP were much less than acquisition costs. Hence these inputs are being used excessively at present than required by this group of sweetpotato farmers. These inputs weren't being used efficiently. There is a need to reduce the use of these inputs to so as to reduce the cost of cultivation, to achieve higher productivity and there by higher gross returns.

Jharkhand: The inputs – planting materials and farm size were being used optimally indicating that the utilization of these inputs could be increased sufficiently. The ratio between MVP and OC for fertilizers was negative indicating the underutilization of this resource, while utilization of all other variables were not at optimum and there was significant difference between MVP and OC as MVP were much less than acquisition costs. Hence these inputs are being used excessively at present than required by this group of Sweetpotato farmers. These inputs aren't being used efficiently. There is a need to reduce the use of these inputs to so as to reduce the cost of cultivation, to achieve higher productivity and there by higher gross returns.

Utilization Pattern

Current utilization pattern of sweetpotato was presented in Fig. 12.2. Of the total sweetpotato produced, 80–85% is consumed as fresh and around 15–20% in the processed form. The common method of consumption is in the form of snack food after roasting/baking or boiling. Boiled or baked tubers are preferred with curd, milk and lassi. The poor prepare roti also from sweetpotato floor, but only a few families consume it. The majority does not prefer it because reportedly, the bread becomes too sweet. Boiled sweetpotato is also used as an adulterating material for Khoya (milk concentrate). Farmers felt that the most important factor, which limits sweetpotato consumption in Bihar is the gastric and acidity problem due to its consumption, followed by easy availability of alternate food grains and social prestige. It is believed by the local population that sweetpotato consumption along with milk and buttermilk is good for health but without these it is harmful for human health.

It is also occasionally used in vegetable preparations. In Bihar, fresh tubers are chipped, dried and ground into flour. This flour is mixed with wheat flour for making chapattis. On certain religious occasions when women folk are not expected to eat cereal food, sweetpotato flour is mixed with milk and sugar/jaggery and consumed. The vines are mostly fed to livestock as animal feed. 80–90% of vines are used as animal feed and 10–20% are used as planting materials for the next season crop.

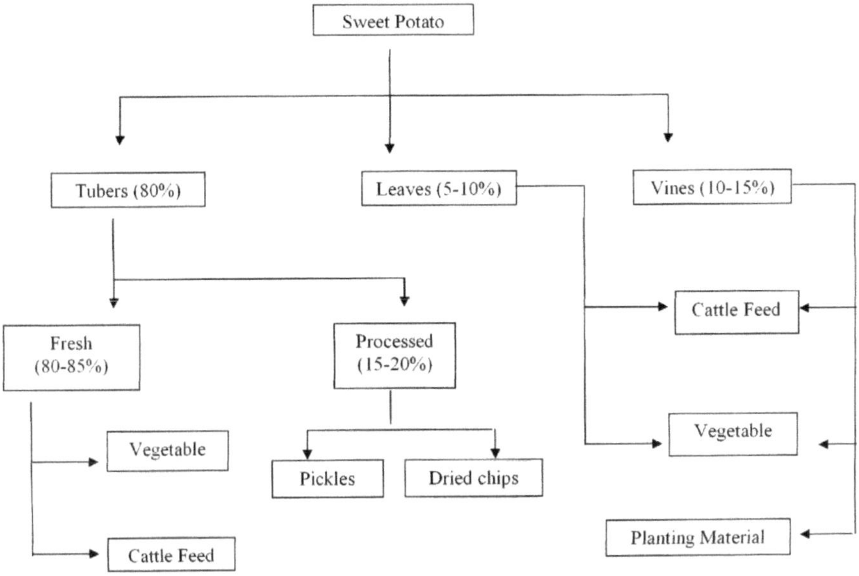

Fig. 12.2 Current Utilisation Pattern of Sweetpotato in India

In Bihar and Jharkhand, the crop is harvested one week before Diwali festival. During this festival, sweetpotato tubers are offered to the sun god on the occasion of "Chatt" festival after Diwali and consume as prasadam.

In Orissa, sweetpotato is predominantly cultivated by tribals. After harvesting tubers are stored for exchanging with cereals like ragi, paddy, jowar etc. under barter system during off-season. So far in India, sweetpotato is not utilized for industrial purposes like starch, alcohol and liquid glucose.

Marketing

Majority of farmers market sweetpotato tubers by themselves at the nearby rural markets. However to certain extent agents are also involved in the marketing of sweetpotato. They collect the tubers at the farm site and take them to the neighbouring towns and cities.

Farmers retain one to two quintals of tubers for daily consumption as breakfast food. Sweetpotato is harvested generally one week before Diwali festival in Jharkhand, which comes in the month of October or November. Traders from Uttar Pradesh and North Bihar purchase sweetpotato tubers from Jharkhand for selling in Uttar Pradesh and Bihar. Sweetpotato tubers are offered to the Sun god on the occasion of "Chatt" festival after Diwali and consume tubers as prasadam. More than 70% of the produce in Jharkhand is marketed for this purpose. Market price during this season ranges between Rs. 2/- to Rs. 2.50/-per kg. If the crop is harvested

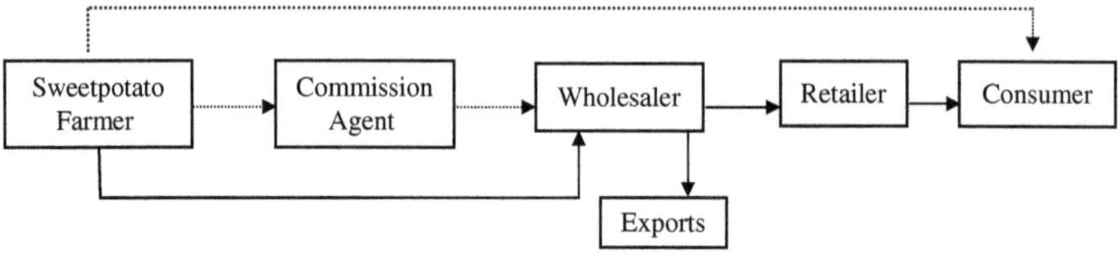

Fig. 12.3 Marketing Channel for Sweetpotato in India

after Diwali, it fetches very low price. Commission agents collect tubers from sweetpotato growing villages during harvesting period and supply to wholesalers.

Some wholesalers are exporting tubers in raw form to Sri Lanka, USA in small quantities. Retailers purchase tubers from wholesalers for further distribution to consumers. Some farmers sell tubers directly in the Vegetable market by themselves. Marketing channel identified for sweetpotato is presented in Fig. 12.3.

Farmers harvest the tubers and kept as heaps on the field. Village agents in the sweetpotato production centres at the time of harvest purchase from farmers by paying the existing market price and pack the tubers in jute/sugar gunny bags. Village agents collect the value of 10–15% of the produce from farmers to cover the weight loss due to drainage while transporting and also as commission. Village agents supply to wholesalers. Some commission agents in the vegetable markets also act as wholesalers. Sweetpotato from eastern India is also marketed in Azadpur market in New Delhi. The produce comes here from as far as south India. Some wholesalers export sweetpotato fresh tubers occasionally. Retailers purchase from wholesaler for further distribution to consumers. Some farmers directly sell the tubers to wholesalers. No grading of the produce at farmer level. Grading can be seen only at wholesaler level, who exports the fresh tubers.

Scope for Value Addition

Sweetpotato can be utilized in a variety of ways in home front, in animal feed and also as a raw material for industrial products. Central Tuber Crops Research Institute developed many technologies for the production of value added products like Sweetpotato chips, candies, Jams, Pickles, Squash, Sauce, starch etc. Pigments like carotenoids and anthocyanins present in sweetpotato are natural colours and technologies are available for commercial extraction. Enzymes like β – amylase, alcoholic beverages and fermented products like citric acid; Monosodium glutamate etc can be produced from sweetpotato. Both vines and roots can be used either in fresh or dried form as animal feed. Roots supply energy and vines supply protein to animals. Product development aimed at home level, cottage level, small scale and large-scale industries with the help of conventional and modern technologies will help to develop rural agro-processing industries.

Sweatpotato Marketing in India

Sweetpotato field

Harvesting sweetpotato crop

Harvested produce heaped in the field

Packed sweetpotato in jute bags in the market ready for sale

Sweetpotato in Azadpur market yard

Purchased sweetpotato transported from wholesaler to retailer in Azadpur market, New Delhi

Sweetpotato – Australia

Area and production of sweetpotato is not as high as in Asia. Interest in sweetpotato as a commercial crop can be traced to work undertaken by the Queensland Department of Agriculture and Stock in the 1920s. The introduction of the American dessert types has transformed the Australian sweetpotato industry. Consumers are now able to purchase smooth easy to peel varieties such as Beauregard. The introduction of Beauregard has improved the convenience and preparation of sweetpotato and consumer demand has improved. Queensland is the major producer, growing around 11,000 t in 2001. The Bundaberg area is that State's largest sweetpotato growing region with an estimated 7,500 t grown there. Other major producing centres are in the Rockhampton area with 1,500 t and on the Atherton Tablelands with 1,200 t.

The major supermarket chains have reported improvements in sales in the last four years and attribute much of this success to the introduction of Beauregard. Whilst sweetpotato only makes up a small percentage of the major retailers total fresh vegetable turnover, the rate of growth in sweetpotato sales has been very high. This growth has been more apparent in the higher value end of the market with many stores in affluent suburbs now showing the highest sales. This led the major retailers to target sweetpotato grading and hence tightening of specifications. The major retailers consistently pay higher than market floor price for product that meets their specifications. This improvement in uniformity and consistency is thought to fuelling improvements in consumption.

Marketing

The preferred varieties have gold flesh and include Beauregard, Beerwah Gold, and Hernandez. These constitute around 95% of the total with the red-skinned-white-fleshed types being the major component of the remaining 5%. In many parts of Australia white skinned purple-fleshed types are sought after particularly by ethnic groups such as Vietnamese and Cambodians, however this is only a very small niche market. Shape and size are major considerations as they improve convenience for consumers. The preferred size sweetpotato is 180–250 mm with a diameter of 60–75 mm and around 600 g in weight. The most sought after shape is almost submarine-like with minimal grooves and shallow eyes; this facilitates the convenient use of a potato peeler when preparing the product, thus meeting the all-important modern criteria of convenience. The majority of the market is for gold-fleshed types because of their consistency and convenience. Twelve-month supply lines exist for gold flesh varieties with many retailers stating that the only thing that stops them from running white skinned and purple- fleshed varieties is the lack of availability of smooth consistent tubers.

Whilst Queensland is the major supplying state, the major outlet is Sydney. The only distinguishable pattern is that volume declines in February and increases in March. In contrast, prices show a more definable pattern with prices being markedly higher in the second half of the year.

Sweetpotato – USA

Sweetpotatoes are as American as apple pie, and even more so. Native Americans were already growing sweetpotatoes in the Caribbean, Central and South America when Columbus came to these shores in 1492. Sweetpotatoes have been growing in the Southeastern United States from as early as 1648. Today, commercial production of sweetpotatoes is centered in eleven states in the country. They are North Carolina, Louisiana, Mississippi, California, Arkansas, Alabama, Georgia, New Jersey, South Carolina, Texas and Virginia. The sweetpotato is a native crop in North Carolina grown in the Coastal Plains and is the official vegetable of North Carolina. Mississippi now ranks third in the nation in the production of sweetpotatoes, just behind North Carolina and Louisiana.

North Carolina sweetpotatoes are available year-round, but are in abundance from September through June. North Carolina is the number one producer of sweetpotatoes in the United States. Today more than 40% of the national supply of sweetpotatoes comes from North Carolina. There are several different types of sweetpotatoes found in North Carolina. They are the Beauregard, Carolina Rose, Carolina Ruby, Cordner, Hernandez, Jewel and NC Porto Rico 198.

Utilization Pattern

Sweetpotatoes are used in many different dishes. They can be eaten raw, steamed, boiled, microwaved and fried. Earlier all sweetpotatoes were eaten fresh (baked, candied, etc). A fairly new way of eating sweetpotatoes is due to increased technology, which changed the way people could prepare them. Sweetpotatoes can be found in chips, frozen and microwave products on grocery store shelves.

The sweetpotato industry has changed especially in the area of storage and curing. Storage systems have improved availability so that sweetpotatoes are available all year. New varieties have increased the production per acre and curing systems have improved overall quality. The fresh crop grading capabilities are becoming more specific now too. There is an effort to get sweetpotatoes more involved in the foodservice or restaurant business such as in fries, chips and salads.

Though there have been advancements in the industry, the number of sweetpotatoes eaten has decreased over the years. This is because sweetpotatoes are considered a regional food of the South and as the population grows more quickly in other areas, the number of sweetpotatoes eaten across the country decreases.

Marketing

Sweetpotato councils are operating in major sweetpotato growing states like North Carolina, California etc. They provide all the information to the growers from production to marketing their produce. Provide amenities required by the producer during selling the produce like warehousing; educate growers on the latest practices to improve their product and their livelihood the need for an organized and professional campaign to promote sweetpotatoes etc.

The first sweetpotatoes are ready to be harvested in late August and the process continues until early November. Automatic harvesters, called diggers, are used to harvest the crop, however, they cause excessive skinning and a majority of the potatoes are plowed and then harvested by hand. Once sweetpotatoes are harvested, they go to a packinghouse. Here they must be graded. This is called the "green crop." Others are cured and stored in large bins until they are needed for market. "Cured" is the process of allowing the skin on the potatoes to tighten, the starches to turn to sugars, and the abrasions to heal. "Cured" potatoes are sweeter than "green stock" and are more resistant to skinning. When sweetpotatoes are cured they are kept at a constant temperature of about 85 degrees and at a relative humidity of 85 to 95% for 5 to 7 days. Once the curing process is over, sweetpotatoes are placed in storage at 55–65 degrees until needed for market.

Most packinghouses grade their product according to buyer specifications and use in-house quality control measures. State officials of the North Carolina Department of Agriculture and Consumer Services inspect on a random basis for food protection, but other state officials are used upon the request of the packinghouse to certify certain quality or grade. These inspections are most often required by the buyer. The guidelines to grade are set by the U.S. Dept. of Agriculture. These samples are graded on similar type, reasonable firmness, shape, color and if it's free from damage and disease. These samples are then used to determine the grade for the whole crop. Most North Carolina sweetpotatoes are sold fresh, although frozen and dehydrated product is available. Fresh product is packaged in 40 lb. cardboard boxes. It is done this way for two reasons. One, it is cost effective and two, it's an industry standard meaning all sweetpotato growers pack the same way especially to retail outlets. Though this is the standard sweetpotatoes can be bagged or individually pressure wrapped. Once the sweetpotatoes are packaged, they are ready to be sold and shipped. They are sold to a store warehouse and then they are shipped from here to individual stores. Sweetpotatoes are mostly sold domestically (within the United States) throughout the Eastern U.S. They are also exported internationally to Canada and Great Britain. When the sweetpotatoes leave the farm to go to the warehouses they are shipped on a truck in an enclosed trailer. This is so that the sweetpotatoes are not exposed to the environment. Upon the invention of the railroads during the Industrial Revolution, trains were used. As technology developed, trucks became the most common method of transportation. Sweetpotatoes in bags or in their large boxes must carry a label stating its net weight, distribution point and grade statement. All labels may look a bit different, but they all contain this information.

Summary

Sweetpotato area globally has been showing a declining trend. The effect of the significant decline in Asia and Latin America was offset by more than three percent growth in sweetpotato area in Africa. Sweetpotato yields globally showed increasing

Sweetpotato Marketing in USA

Bagging Machine

New Pressure Wrap Packaging

trend at one percent per annum. Average sweetpotato yield in the world over has almost doubled from 7.35 t ha^{-1} to 13.87 t ha^{-1}. This to some extent reduced the affect of decline in sweetpotato area on production.

Sweetpotato is grown for selling as vegetable and diversification is less resulting in low income to the farmer especially in India and Bangladesh, thereby area under the crop is declining year after year. The low income generated as compared to other horticultural crops, have placed sweetpotato in the category of "Orphan crops". In order to maintain the rhythm in the supply of food material and to keep pace with the geometrically increasing population, secondary and tertiary staple food crops like sweetpotato have to be linked with industry to retain sweetpotato within the cropping system of marginal farmers. Better post harvest management, which includes diversification for the production of value-added products, is one of the methods to retain the crop in the cropping system. China dominates the world figures for production of sweetpotato, with animal feed and industrial products the major uses of sweetpotato. Availability is year-round (but seasonal according to specific geographical location) and prices in production zones rise before, and fall after, harvest season(s). In Japan sweetpotato as fresh product is concentrated to their summer/autumn season. Between 10–15% of production is consumed as processed product. Wholesale market is declining steadily annually, and annual price has remained static, but seasonal price varies with maximum price/kg is achieved during June-August. Frozen imports offset this to some extent; but fresh sweetpotato cannot be imported to Japan. Australian production of sweetpotato is largely confined to Queensland and varieties in the main have golden flesh (whereas Japanese varieties are normally red-skinned with white flesh). Fresh product supply to the market is constant year round, but price tends to rise in the second half of the year. Sweetpotato councils operating in different states are playing an important role in production and marketing of sweetpotato in USA.

Expenditure on manures and fertilizers can be enhanced sufficiently without any adverse effect on the productivity of the crop. Expenditure on planting materials in Bihar can be increased sufficiently. There is a need to reduce the use of these inputs viz., hired labour and irrigation so as to reduce the cost of cultivation, to achieve higher productivity and there by higher gross returns. In the Indian scenario, 80–85% is consumed as fresh and around 15–20% in the processed form. Commission agents play an important role in marketing sweetpotato in India.

References

Agriculture. 2006. Centre for Monitoring Indian Economy, Mumbai.

Anonymous, 1997. *Sweetpotato Marketing in Bangladesh: Results of a Rapid Market Appraisal.* Summary Report Working paper. Social Science Department. Working paper No. 1997–3. International Potato Center (CIP), Lima, Peru.

Bashaasha, B., Mwanga, R., Ocitti, C., Oboyoa, P. and Ewell, P. 1995. Sweetpotato in the farming and food systems of Uganda: *A farm survey report.* Lima, Peru: International Potato Center (CIP).

Blondet, C., and Espinola, N. 1998. Capacitacion y evaluacion de la diffusion de tecnologia panificadora de camote en panaderias populares de Lima Metropolitina. *International Potato Center (CIP) Social Science Department Working paper* 1998–4. Lima, Peru, CIP.

Cabanilla, L.S. 1996. *Sweetpotato in the Philippines: Production, processing and future prospects.* Lima, Peru: International Potato Center (CIP) and University of the Philippines at Los Banos.

Chiang, W. 1992. Sweetpotato production and utilization in Taiwan. In *Product development for roots and tubers. Vol. 1-Asia*, ed. G. Scott, S. Wiersema, and P. Ferguson. Lima, Peru: International Potato Center (CIP)

Chin, M.S. 1989. The outline for sweetpotato in Korea. In *Improvement of sweetpotato (Ipomoea batatas) in Asia.* Report of the workshop on sweetpotato improvement in Asia. Held at ICAR, Trivandrum, India. October 24–28, 1988. Lima, Peru: International Potato Center (CIP).

Fugile, K., Zhang, L., Salazar, L.F. and Walker, T. 1999. Economic impact of virus free sweetpotato seed in Shandong Province, China. In *International Potato Center Program Report 1997–1998.* International Potato Center (CIP), Lima, Peru.

Huang, J. 1999. Sweetpotato economy in China. *Report submitted to the International Potato Center*, Lima, Peru, Mimeo.

Orno, J. 1991. Farmer's perception on economic and environmental importance of Sweetpotato. *Proceedings of the 2nd Annual UPWARD International Conference*, UPWARD, University of the Philippines at Los Banos, Laguna, Philippines.

Pham, T.B., Nguyen, M.H., Le, C.T. and Henry, G. 1996. Socio-economic aspects of cassava production, marketing and rural processing in Vietnam. *In A benchmark study on cassava production, processing and marketing in Vietnam.* Proceedings *of a workshop, Hanoi*, Vietnam, October 29–31, 1992, ed. R.H.Howeler, Bangkok, Thailand: Centro Internacional de Agricultura Tropical (CIAT).

Phiri, M.A.R. 1998. Grain Legumes: Issues and Options for research in Malawi, production, marketing and policy issues. *Paper presented at a training course on Methods for Analyzing Agricultural Markets, November 9–15. Nairobi*, Kenya.

Report of Rural Industries Research and Development Corporation for the selected markets for taro, sweetpotato and yam compiled by Grant Vinning, Asian Markets Research, 2003.

Scott, G. 1992. Sweetpotato for animal feed in developing countries: present patterns and future prospects. In *Roots, tubers, plantains and bananas in animal feeding*. Ed. D. Machin and S.Nyvold. Proceedings of the FAO expert consultation held at CIAT, Cali, Colombia. January 21–25, 1991. FAO, Rome.

Scott, G.J., Otieno, J., Ferris, S.B., Muganga, A.K. and Maldonado, L. 1999. Sweetpotato in Ugandan food systems: Enhancing food security and poverty alleviation. In. *International Potato Center Program Report 1997–1998.* International Potato Center (CIP), Lima, Peru.

Scott, G.J., Rosegrant, M.W. and Ringler, C. 2000. Roots and tubers for the 21st Century: Trends, Projections and policy Options. Food, Agriculture and the Environment Discussion Paper 31. *International Food Policy Research Institute, Washington, D.C.* USA and International Potato Center (CIP), Lima, Peru.

Spencer, D.S.C. and Badiane, O. 1995. Agriculture and economic recovery in African countries. In *Agricultural Competitiveness: Market forces and policy change*. Ed. G.H. Peters and D.D. Hedley. Proceedings of the Twenty Second International Conference of Agricultural Economists, 22–29 August 1994, Harare, Zimbabwe, Aldershot, England, Dartmouth.

Srinivas, T. and Ramanathan, S. 2005. Production and marketing of sweetpotato in India. *Indian J. Agric. Marketing*, 2:68–77.

Tanganik, M., Phezo, P., Ewell, P.T., Lutaladio, N.B. and Scott, G. 1999. The role of potato and sweetpotato in disaster relief: The case of Rwandan refugees in South Kivu, Democratic Republic of the Congo (ex-Zaire) 1994–1996. In *International Potato Center Program report 1995–1996*. International Potato Center (CIP), Lima, Peru.

Tardif-Douglin, D. 1991. The marketing of sweetpotatoes in Rwanda: Commercializing a perishable crop under adverse conditions. *Ph.D thesis, Department of Agricultural Economics, Cornell University, Ithaca*, New York, USA.

Timmins, W.H., Marter, A.D. Westby, A. and Rickard, J.E.1992. Aspects of sweetpotato processing in Sichuan Province, People's Republic of China. In *Product development for roots and tubers*, Vol. I-Asia, ed. G.Scott, S.Wiersema and P.Ferguson. Proc. Interna. workshop April 22–May 1, 1991, at Visayas State College of Agriculture (ViSCA), International Potato Center (CIP), Baybay, Leyte, Philippines, Lima, Peru.

Titapiwatanakun, B. 1998. Opportunities for the use of cassava in Thailand. In *Global cassava market study: Business opportunities for the use of cassava*. Assembled by dTp Studies Inc., Guelph, Ontario, Canada. Mimeo.

Wheatley, C., Liping, L. and Bofu, S. 1997. Enhancing the role of small-scale sweetpotato starch enterprises in Sichuan, China. *In International Potato Center Program Report 1995–1996*. International Potato Center (CIP), Lima, Peru.

WWW.agr.state.nc.us/agscool/commodities/swwetkid.htm.

WWW.faostat.org 2008. Food and Agriculture Organization of United Nations, Rome Italy.

Zhang, X. 1999. Agricultural marketing in a country in transition: case of sweetpotato products in Sichuan, P.R. China *Ph.D. Thesis, Wageningen Agricultural University, Wageningen*, The Netherlands.

Chapter 13
Marketing Sweetpotatoes in the United States: A Serious Challenge for Small-to-Moderate Volume Growers

E.A. Estes

Introduction

Historically, U.S. sweetpotato marketing efforts have focused primarily on the holiday seasons of Thanksgiving and Christmas and the intervening weeks. Typically, November and December sweetpotato sales are nearly 1/3 of total annual sales, a figure that far exceeds the period's normal expected sales rate of 17% (Lucier, 2008). While sweetpotatoes remain a popular holiday food for Americans, most consumers have grown accustomed to year-round availability of fruits and vegetables so recent marketing efforts have stressed the year-round availability of nutritious, healthy, high beta-carotene sweetpotatoes. Year-round availability of sweetpotatoes became common about 20 years ago when growers and shippers built environmentally controlled storage facilities, adopted improved curing and storage technologies, and minimized major pest and disease problems in stored sweetpotatoes. Industry leaders believe that advertising and promotional efforts, which stress nutrition, health, and processed uses, have helped to stabilize the decline in per capita sweetpotato consumption. Sweetpotatoes have become a more regularly consumed year-round vegetable in Southern U.S. households (Johnson-Langdon, 2008). For generations, Southern households have served sweetpotatoes in a variety of ways including baked, candied, and marshmallow-topped. With the notable exception of California, most sweetpotatoes are grown and consumed in the U.S. South. The top producers of sweetpotatoes are growers located in North Carolina (38% of U.S. annual production), in California (23%), in Mississippi (19%), and in Louisiana (16%) (Lucier, 2008). Thus, growers located in the remaining states collectively supply less than 4% of the average domestic crop (Table 13.1). Government survey data from 2002 indicated that 35% of the U.S. population lives in the South but Southern households consumed 42% of all fresh-market sweetpotatoes shipped and ate 54% of all processed sweetpotatoes sold nationally. According to US Census of Agriculture data, approximately 2,375 farms, located in nine states,

E.A. Estes (✉)
Department of Agricultural and Resource Economics, College of Agricultural and Life Sciences, North Carolina State University, Raleigh, North Carolina, 27695-8109 USA
e-mail: eaestes@ncsu.edu

G. Loebenstein, G. Thottappilly (eds.), *The Sweetpotato*,
DOI 10.1007/978-1-4020-9475-0_13,

Table 13.1 Top 5 states in US in Sweetpotato production, 2007

State	Area harvested (hectares)	Yield (kg)	Price (kg)	Production (million kg)
North Carolina	17,402	7,484	$45.98	3,218.2
California	5,382	14,515	$41.80	1,930.4
Mississippi	8,094	7,938	$42.46	1,587.6
Louisiana	6,070	8,845	$44.66	1,326.8
Alabama	971	5,443	$70.18	130.6
Other states	1,538	5,126	$43.45	175.9
Total U.S.	39,457	8,573	$44.66	8,369.5

grew and sold a commercial crop of sweetpotatoes in 2002 (Census of Agriculture 2007). Since 2002, the sweetpotato industry has expanded total output despite a decline in the number of U.S. farms earning income from sweetpotato production. In the U.S., 2007 United States Department of Agriculture (USDA) data suggested that the number of sweetpotato farm operations had declined slightly, U.S. planted acreage was up 18%, yield per hectare was up 21%, and total national output had increased 44% when compared with 2002 data (Lucier, 2008).

Cursory analysis of marketing and price trends over the past two decades suggests that U.S. demand for sweetpotatoes has increased modestly, despite little gains in per capita consumption levels. This is in contrast to most fruit and vegetable commodities that, on average, have increased about 15% since 1987 (Lucier and Jerardo, 2007). Recent USDA data indicated that U.S. sweetpotato output increased about 42% (2% annually) over the past two decades while per capita usage (consumption) has remained flat around 2 kilograms (kg) per person (Lucier, 2008; Lucier and Plummer, 2002). While stabilization of sweetpotato per capita consumption might signify industry underachievement and be a source of concern for grower-shippers, consumption stabilization has reversed a longer-term decline in consumption that began in the early 1920s when U.S. sweetpotato consumption peaked around 13.6 kg per person. In 1965, U.S. sweetpotato per capita consumption had declined to 2.7 kg per person, or roughly 20% of its 1920s level. For 2008, USDA anticipates that annual sweetpotato consumption will be about 2.36 kg per person, a figure that is still below the 1965 level but greater than the 2006 value of 2.09 kg per person (Lucier, 2008). Long-term U.S. marketing prospects for sweetpotatoes depend primarily on new and expanded uses for value added sweetpotatoes as well as modest improvements in per capita consumption of fresh market sweetpotatoes (Bliss, 2008).

While long-run consumption patterns suggest that fewer Americans eat sweetpotatoes, recent production and marketing uses offer a more positive view of the U.S. industry. National sweetpotato output has increased 38% over the past decade (1997–2007) and U.S. season average shipping-point prices also have increased, on average, about $1.10 per 100 kg over the past decade. For the July 1, 2006-June 30, 2007 marketing year, USDA estimated that the U.S. average free-on-board (FOB) shipping-point price was about $44.50 per 100 kg, or nearly 33% higher than the July 1996-June 1997 shipping-point price (Lucier, 2008). In general, for seven of the most recent ten marketing years, year-to-year comparisons of season average shipping-point prices revealed that price increased relative to the previous year's price (Lucier, 2008). Thus, significant increases in U.S. output and shipping point prices suggested that sweetpotato growers realized demand-expansion over the past decade despite very little growth in per capita consumption levels. It is likely that a number of factors have impacted demand such as industry-wide promotional efforts to stress the nutritional and health benefits of eating sweetpotatoes along with innovative firms stressing value-added concepts such as fries and chips with buyers. It is anticipated that plantings will continue to increase modestly in 2009 and the near-term future as value-added demand appears strong. Increased availability of sweetpotatoes in restaurants and in processed forms would result in small-to-moderate gains in per capita consumption and result in strengthened sweetpotato demand across all regions of the U.S, especially the South.

Declining demand and flat per capita consumption was an industry-wide concern in the 1950s. Consumption continued its downward trend during the 1960s and 1970s and industry reaction was the creation of state-wide Sweetpotato Commissions in the major producing states (North Carolina Sweetpotato Commission, California Sweetpotato Commission, Louisiana Sweetpotato Commission, and the Mississippi Sweetpotato Commission) as well as establishment of the U.S. Sweetpotato Commission currently located in Columbia, South Carolina (Johnson-Langdon, 2008; Estes, 2006). The main goals of the national and state commissions were to promote and advertise the benefits of sweetpotato consumption. In this way, Commission members (often current sweetpotato growers and shippers) hoped to increase sweetpotato sales, expand market outlets, increase in-store shelf space, educate consumers about the nutritional benefits of eating sweetpotatoes, and expand foreign marketing opportunities. Secondary goals of state Commissions also included collecting assessment fees that enabled them to fund sweetpotato research at land grant universities. Commissions often represented growers, shippers, handlers, and processors in dealing with a variety of state and federal regulators, legislative policy makers, and the media. Since 2000, state commission promotional efforts have focused on the overall health benefits and nutritive value obtained from including sweetpotatoes with at-home dinners beyond the holiday season.

Americans often have a choice of purchasing one or two basic types of sweetpotatoes: (1) moist, orange-flesh varieties that are often marketed incorrectly as "yams" (*Dioscorea spp.*) but are in fact sweetpotatoes (*Ipomoea batatas*); and (2) dry white-flesh or yellow-flesh varieties that have a firmer flesh and are often grown in more Northern U.S. climates (AgMRC, 2008). Dry-flesh types were grown first in the U.S. and then moist, orange flesh varieties became widely available in the 1960s. Many growers, shippers, and distributors switched to moist-flesh types and wanted to differentiate them from the drier, traditional varieties so new growers decided to market the new, moist-flesh types as "yams". Today, many Americans still believe that yams and sweetpotatoes are exactly the same. Of course, yams are a tropical crop not grown in the mainland U.S. and are unrelated botanically to sweetpotatoes. To eliminate consumer confusion, current USDA regulations require that any sweetpotatoes marketed as "yams" can be sold in the U.S. only if the seller also identifies and labels the shipping container with the "sweetpotato" name also displayed (Estes, 2006).

Basic Global Trends

Although the primary focus of this chapter is on U.S. sweetpotato marketing, it is useful to review how the U.S. sweetpotato industry fits into the global sweetpotato production and marketing system. Sweetpotatoes likely were first grown about 5,000 years ago near Central America, perhaps in the West Indies islands off the coast of Mexico (AgMRC, 2008). Today, sweetpotatoes are grown broadly throughout the world and often are raised as a cheap substitute for corn and rice in diets. In contrast, for American consumers, sweetpotatoes are a holiday starch consumed regularly only in the Southern region of the U.S. For developing economies, sweetpotatoes rank as the fifth most important food crop on a fresh-weight basis ranking only behind rice, wheat, corn (maize), and cassava (Scott, 2001). In 2006, the U.S. ranked 12th in world sweetpotato production at about 800,000 metric tons for all uses (Lucier, 2008). For Americans, sweetpotatoes rank 12th in vegetable consumption and are not among the broad spectrum of foods that consumers eat regularly. Asian and African nations often are among the largest sweetpotato producers, with China ranked first, Nigeria ranked second, and Uganda ranked third in world production (Scott, et al., 2000; Lucier, 2008). However, it should be noted that these statistics likely understate output since in many parts of Asia and Africa food is often in short supply so many small-volume farmers raise sweetpotatoes for family home consumption. Therefore, it is likely that in many poorer parts of Africa and Asia local production has expanded since 2000 despite overall reductions in world output. In contrast, since 2000, it is evident that fewer sweetpotatoes are grown and consumed as food in China and Indonesia. Sweetpotato plantings in China have declined, in part, as human consumption declined but animal feed use expanded. Worldwide, yield per hectare has leveled off after increasing steadily during the 1990s as improved varieties became available and pest pressures moderated. China is the world's dominant sweetpotato producer accounting for 81% of global production annually (Anonymous, 1997; Lucier, 2008). In China, nearly 20.6 billion kg of dried sweetpotatoes are used each year to feed swine and other livestock (AgMRC, 2008). U.S. growers supply approximately 3.8% of the world's output (800 million kg of sweetpotatoes). In the U.S. expanded uses for sweetpotato are being investigated as part of University research projects because the high-starch content of sweetpotatoes can be increased through breeding and high-starch sweetpotatoes would be useful as an industrial product (for example, flour and pectin) or they could be used as a biofuel (ethanol). At North Carolina State University, a major research effort is underway led by Dr. Craig Yencho to determine the feasibility and sustainability of producing a sweetpotato-based biofuel (Estes, 2006).

In the U.S., sweetpotatoes are primarily consumed as food although in cattle and pork regions of the Midwest some farmers crush sweetpotatoes and then feed them to their livestock. The bulk of the U.S. crop (80%) is sold for fresh market consumption (retail, foodservice, and exports) while processed sweetpotatoes account for the remaining 20% (canned, baby food, chips, and frozen) of sales (Gonzalez, 2008). U.S. distributors imported about 6,350,300 kg (about 2.9% of supplies) in 2007 but at the same time U.S. firms exported nearly 6% of available supplies. Most exports (90% of 2007 export volume) were sold to firms located in Canada or England (Lucier, 2008; Gonzalez, 2008).

U.S. sweetpotato yield per hectare can vary greatly. Among the four major producing states (North Carolina, Mississippi, California, and Louisiana), yield per hectare averages about 22,408 kg/ha except in California where growers expect to average 33,612 kg, or about 50% more/ha than the rest of the U.S. (Johnson-Langdon 2008; May and Scheuerman, 1998). It is unclear why California growers average much higher yields but nearly ideal growing conditions and superior crop management skills likely contribute to above average yields (Lucier, 2008). Sweetpotato yield is more variable among growers located in the non-major producing states. During the past decade, sweetpotato yield per hectare in other states that report sweetpotato yield estimates ranged from a low of 5,041 kg/ha in Texas (in 1998 and 2000) to 21,286 kg/ha in Alabama (2003) and Virginia (1999).

In North Carolina, the largest volume state in the U.S., the 2006–2007 sweetpotato output exceeded 204 million kg Approximately 81% of North Carolina's 2006–2007 crop was sold to fresh market outlets including food service operators, retail grocers, and or export markets, a proportion very similar to the national market breakdown (Fig. 13.1). In addition, nearly 39 million kg of North Carolina sweetpotatoes were processed, mostly by canneries. Combined frozen and chip utilized

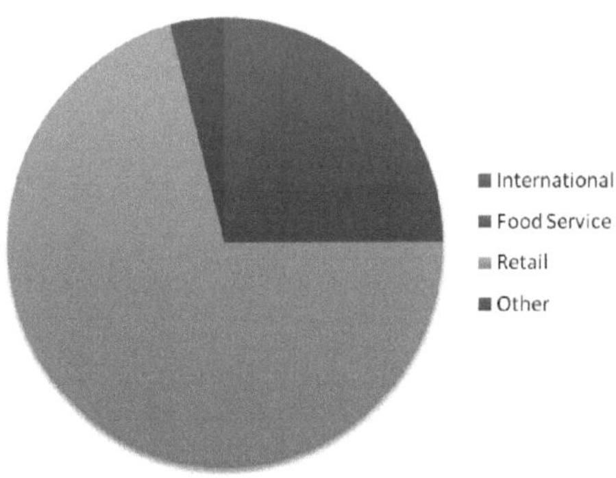

Total lbs: 364.7 million lbs

Retail: 258,937,000 lbs

Food Service: 72,940,000 lbs

International: 18,235,000 lbs

Other: 14,588,000 lbs

Fig. 13.1 2006 Fresh market breakdown for North Carolina SweetPotato Sales

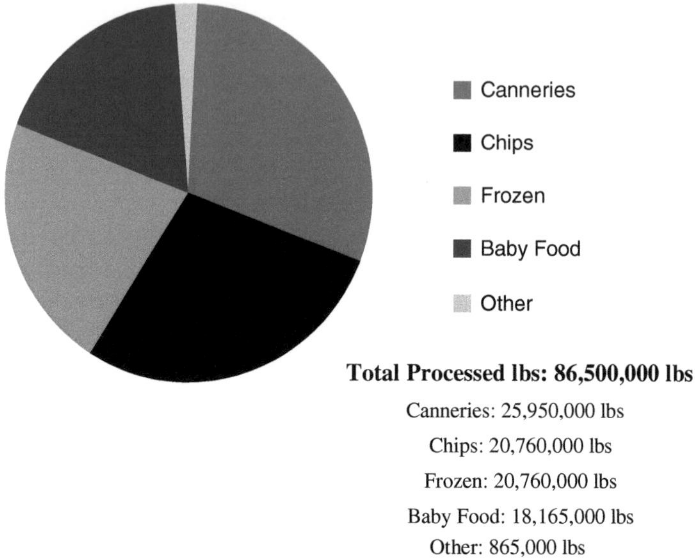

Fig. 13.2 2006 breakdown of total lbs processed for North Carolina Sweetpotatoes

approximately 48% (18.7 million kg) of the processing crop (Fig. 13.2). Finally, baby food processors used 8 million kg, or about 21% of the processed sweetpotatoes (Gonzalez, 2008).

U.S. Utilization Patterns

Since 2005, American growers have produced, on average, 805 million kg of sweetpotatoes per year (Lucier], 2008). Growers obtained this output from 36,826 ha harvested. Farm receipts between 2005 and 2007 were, on average, in excess of \$350 million each year. In the major U.S. production areas, sweetpotato plants are set in fields during April and May. The crop reaches maturity in 105–130 days, depending on the variety planted, the soil temperatures, and the environmental conditions (see Plates 13.1 and 13.2). Typically, harvesting occurs in major production areas between late August and mid-November. Some sweetpotatoes are marketed as "green" (that is, uncured) sweetpotatoes and are sold immediately after harvest. However, the vast majority (perhaps 90%) of the U.S. crop is 'cured', that is, heated to nearly 30 °C in an enclosed, high humidity (90%) room for 4–8 days, and then the temperature is lowered to 12.8 °C (May and Scheuerman, 1998). Curing sweetpotatoes allows the skin to harden thus preventing the entrance of decay organisms. After curing, sweetpotatoes are placed in an environmentally-controlled facility held at 15.6 °C and stored in bulk bins until they are sold. Bulk bins contain between 227 kg and 431 kg of sweetpotatoes, depending on the grower-shipper preference, the buyer preference, and the amount of storage space available. At packing time, sweetpota-

Plate 13.1 Sweetpotato grading and packing line located at grower-shipper packing shed in Columbus County, North Carolina (See also Plate 10 on page xxii)

Plate 13.2 Commercial grower-shipper bulk bin storage room equipped with environmental management controls, Nash County, North Carolina (See also Plate 11 on page xxii)

toes are dumped into a water tank to be washed, cleaned, disinfected, and then dried via a heater. Three market grades are recognized by packer-shippers: 1) U.S. #1's; 2) U.S. #2's; and 3) jumbos and canners grade. Grades are differentiated by differences in weight, maximum and minimum diameter, and shape. The highest prices paid by buyers for fresh market are for US #1's, the next highest price paid is for jumbos, and finally the lowest fresh market price paid is for US # 2's. For processing uses, buyers often pay the lowest price for canner grade sweetpotatoes. Contract and spot market prices range between \$6.60 and \$8.80 per 100 kg Fresh market sweetpotatoes are graded and placed in 18 kg shipping cartons for market distribution. In contrast, canners and processing uses are sold in bulk bins weighing between 227 kg and 450 kg When transporting sweetpotatoes, distributors must recognize that sweetpotatoes are sensitive to both ethylene and chilling injury. Therefore, sweetpotatoes should never be shipped together with ethylene-producing produce or ripening fruit such as apples or melons. In addition, transport temperature should remain between 12 °C and 13 °C rather than the common produce transport temperature of 2.2 °C.

Shipper-handlers try to minimize the number of times that sweetpotatoes are moved or handled in order to minimize skin damage. Sweetpotatoes lose about 2% of their weight per month in storage so proper storage conditions must be maintained in order to minimize shrinkage and losses. Under ideal conditions, shipper-dealers can hold cured sweetpotatoes for 6 to 9 months without any significant impact on marketability if high-quality sweetpotatoes were placed into storage bins. The risks associated with storing sweetpotatoes are numerous but common problems include pest infestations, internal rotting, and disease. In addition, there are also significant marketing and price risks associated with storage. For example, analysis of recent Spring (April, May, and June) FOB shipping prices for North Carolina distributors indicated that spring sales prices can be below the previous November's FOB shipping point price. In this circumstance, the shipper has incurred added storage costs (from November till they are sold in the following spring) plus sweetpotatoes lose moisture (i.e., lose weight). Unfortunately for the grower-shipper, however, the average April-May FOB price is occasionally less than the 'uncured', or green price. This is the marketing risk associated with storage. In North Carolina, since 2000 lower April-May prices were observed in 2001, 2003, and 2006. Higher storage (April-May) prices, relative to November, were observed in the remaining years since 2000 (Lucier, 2008; Gonzalez, 2008). This suggests significant price variability and uncertainty for grower-shippers. If they are unwilling to assume this marketing risk, then they simply do not store sweetpotatoes but instead sell their crop immediately to buyers or other shipper-growers who are willing to assume the price and marketing risks. Therefore, both production and price uncertainty contribute to larger shipping volumes and increased sales every November and December despite attempts by state Sweetpotato Commissions to increase and distribute sales more evenly throughout the marketing year. Finally, of course, holiday demand also contributes to increased fall sales remains as growers observe grocers willing to buy sweetpotatoes for customers as soon as the new crop is harvested. It is likely that U.S. shipment volume will remain high during the last quarter of the year simply because of supply, demand, and price uncertainty conditions.

Who Eats Sweetpotatoes in the U.S.?

Relatively little information is available about who eats sweetpotatoes in the U.S. because few marketing surveys are conducted concerning people who eat sweetpotatoes. Instead, surveys tend to focus on heavily consumed vegetables such as tomatoes, sweet corn, and white potatoes. In 2007, U.S. Department of Agriculture (USDA) reported that U.S. consumption exceeded 816.5 million kg, that is, about 2.27 kg per person per year (Lucier and Dettman, 2008) In contrast, Americans eat four times as much fresh market tomatoes, or roughly 9.25 kg per person per year. Relatively few Americans eat sweetpotatoes regularly, with about 1.5% of U.S. residents eating a fresh-market sweetpotato on any given day while fewer people (0.5%) eat processed sweetpotatoes (chips, fries, canned, or patties) daily. For consumption information, data were gleaned from USDA reports; particularly the 1994–1996 -food consumption survey entitled "Continuing Survey of Food Intakes by Individuals" (CSFII 1994–96, 1998). Much of the CSFII information has been summarized in a variety of USDA food publications but sweetpotato consumption was featured in the USDA-ERS publication entitled "Vegetable and Melon Outlook Reports – Sweetpotato Highlights" (Lucier and Jerardo, 2007).

More than any other vegetable purchased, Americans tend to eat sweetpotatoes primarily at home. Sweetpotatoes are consumed at home (89% of total consumption) because relatively few U.S. restaurants offer sweetpotatoes as an option to customers and very few processed forms are available to institutional suppliers. Of course, the popularity of French fries as a starch alternative (1 in 7 U.S. residents eat French fries daily) also hurts away-from-home sweetpotato consumption since people often view potatoes as a substitutable product (Powers, 1994). U.S. sweetpotato consumption also has been hurt by three recent trends in American food consumption: (1) more Americans are eating at away-from-home establishments but many food service suppliers do not offer sweetpotatoes as a vegetable choice; (2) Americans are broadening diets to include spicier foods but this trend does not benefit sweetpotatoes; and (3) U.S. population ethnic diversity has expanded produce choices as many Americans substitute newer or alternative vegetables for more traditional mainstays such as sweetpotatoes (Estes, 2006).

It appears that older men and women are more likely to eat sweetpotatoes than any other age population. Men 60 years and older consume about 16% of all sweetpotatoes but they represent only 7% of the U.S. population, that is, they consume double the U.S. average rate (CSFII 1994–1996, 1998). Older women (over 60 years) also eat a lot of sweetpotatoes, roughly 3.17 kgs per woman, or nearly 50% more than the average American. In general, male and females of all ages tend to eat sweetpotatoes in similar amounts, that is, between 1.9 and 2.0 kg annually. As noted earlier, Southern U.S. residents consumed the most sweetpotatoes at 2.59 kg per person while Western U.S. residents ate the least amount of sweetpotatoes at 1.18 kg per person. Midwest (1.95 kgs) and Northeast (1.77 kg) residents eat moderate levels of sweetpotatoes, with people in both regions consuming only slightly below the U.S. average rate of 2.04 kg per person. African-American consumers tend to eat a lot of sweetpotatoes since they eat 21% of all domestic sweetpotato supplies but represent less than 13% of the U.S. population. Slightly more than 50% of all African-American U.S. citizens live in the South while less than 10% live in Western states. In the South, native populations are accustomed to eating sweetpotatoes with at-home meals so sweetpotatoes remain popular with native Southerners (CSFII, 1994–1996, 1998; US Census of Agriculture, 2007). In general, middle and upper income consumers tend to eat more fresh market sweetpotatoes than their population share so this would imply that lower income residents tend to eat less-than-their proportionate share of fresh market sweetpotatoes. CSFII surveys found that lower income consumers tended to eat more processed forms of sweetpotatoes while upper income consumers favored fresh market sweetpotatoes (bakers).

Marketing Options & Sales Channels for U.S. Sweetpotatoes

The marketing of fruits and vegetables is big business in the U.S. although sweetpotatoes remain a small contributor to overall retail produce sales. Produce analysts (Cook, 2004; Kaufman et al., 2000; Estes, 2006) estimate that combined U.S. retail and foodservice sales for fruits and vegetables will be nearly $100 billion by the end of 2008 (Kaufman, 2008-personal communication; Estes, 2006). Distribution of sweetpotatoes and other vegetables are changing as the U.S. market matures and disposable income increases. Some consumers may view sweetpotatoes as a low-income vegetable similar to cabbage. As household income increases, shoppers purchase higher value or more exotic speciality items. Consumer choices have increased dramatically in recent years as American supermarkets typically offer an array of more than 400 separate stock keeping units (SKUs) in their produce departments (Estes, 2006)This expansion reflects changing consumer demand for more convenience, changing tastes, and greater ethnic diversity in the population.

Primarily, producers, wholesalers, integrated chains, independent distributors, and supermarket retailers have tended to consolidate through mergers, acquisitions, and takeovers as sales per firm have increased. Despite consolidation, competition has increased at all market levels from farm to retail and exerted short-term downward pressure on prices. Industry concentration and competition have intensified, in part, because of the influx of companies that were historically not involved in retail food sales such as Wal-Mart and Super Target stores. For example, Wal-Mart has improved sweetpotato marketing efficiency by requiring all produce suppliers to deliver product at their distribution centers in standard, plastic bulk bins equipped with radio frequency identification (RFID) technology and a digitized electronic bar code (EBC). RFID facilitates food safety traceback problems and offers customers an improved measure of food safety protection in that if a recall notice is issued then Wal-Mart knows which inventory to remove from shelves. Traditional food retailers have been much slower to adopt new RFID and trace back technologies but chain stores were forced to adjust when competitors offered greater safety assurances and lower prices. Also, global markets have evolved and expanded for sweetpotato shippers because of U.S. participation in free trade agreements, which reduce barriers to trade and standardize phytosanitary handling and treatment options. U.S. export expansion efforts remained focused on European Union (EU) countries, especially Great Britain. Since 2000, U.S. export volume has increased 67%, with most exported sweetpotatoes going to Canada and Great Britain (Lucier, 2008).

Ten years ago U.S. consumers most often bought their fruit and vegetables from a traditional grocery store (Powers, 1994). Today, USDA studies show that U.S. consumers obtain their fruits and vegetables from a variety of sources, including specialty grocers and direct farmer-to-consumer sales. Cook (2004) reports that U.S. consumers obtain a majority of their fruit and vegetable daily servings from food service outlets (restaurants, institutions, etc). Roughly 55% of fruit and vegetable purchases are made from food service suppliers. Unfortunately, sweetpotatoes have a limited presence in food service outlets because very few sweetpotato products are popular and available to food service buyers. The two main marketing channels for sweetpotatoes are retail grocery stores including chain supermarkets and direct sales to consumers via community farmers' markets. As additional forms (chips, patties, and fries) of sweetpotatoes become available to the food service sector, then sales expectations will increase rapidly because of the nutritional and health benefits associated with eating sweetpotatoes.

The U.S. fruit and vegetable marketing system, including sweetpotatoes, operates primarily on free market principles. The marketing system coordinates delivery of sweetpotatoes in the form, place, and time that is preferred by consumers. Unlike the grain and oilseed sectors, U.S. fruit and vegetable producers do not receive government subsidies as part of national legislation. Until recently, few sweetpotato producers were eligible for income insurance or crop insurance. Instead, the government's role (federal and state) was to facilitate commerce, enhance international trade, improve market and price information and transparency, and minimize market distortions concerning sweetpotato transactions. In addition, state and federal agencies established grades and standards to ensure buyer and seller understand how quality is defined. In addition, if grades and conditions are well established then transactions can occur via electronic computer, fax, or telephone without the need to visually inspect each load of sweetpotatoes.

The land grant University system, state departments of agriculture and USDA analysts also assist sweetpotato producers and marketers by conducting market research, enhance demand through advertising and promotional programs, finance and minimize risk exposure for sweetpotato growers, and disseminate timely market information concerning volume and price. As noted earlier, decades earlier the major sweetpotato producing states had established state Sweetpotato Commissions as well as the U.S. Sweetpotato Council. Commissions do not sell product or control sales volume but instead use assessments (voluntarily-contributed funds) to assist the industry by financing research and using funds to promote and advertise sweetpotatoes throughout the U.S. and in individual states. It is important to note that interstate shipment of fresh sweetpotatoes (includes all fresh fruits and vegetables) is regulated by the federal government through the Perishable Agricultural Commodity Act (PACA) of 1930 (modified and updated several times since 1930). PACA is a federal law that is administered by the regulatory branch of the Fruit and Vegetable Division of USDA-Agricultural Marketing Service. Buyers and sellers of fresh produce (including sweetpotatoes) must apply for and receive a PACA license in order to buy or sell product if it is shipped across state borders. Thus, PACA also governs trading practices. Failure to follow fair trade practices as defined by PACA can result in USDA suspending and/or revoking a firm's PACA license. Without a PACA license, buyers or sellers cannot legally buy and sell commodities that travel across states. PACA regulations define "fair" trade as one in which both buyers and sellers are assured that they know what to expect in a business transaction. In essence, fair trade terms focus on when and how buyers must pay sellers and when shippers must deliver volume and quality at the agreed on price to buyers. Well-defined grades, publicly posted prices, PACA licenses, breeding and market research, and promotion programs all contribute to a marketing process that minimizes risk, decreases losses, and increases transaction efficiency for sweetpotato growers and shippers which contributes to reduce prices paid by the consumer.

In the U.S., the most common way to market sweetpotatoes is for the grower-shipper, who has assembled shipping loads of sweetpotatoes, to sell directly to a grocery chain store. Regional or area chain stores operate central buying offices where they buy sufficient quantities of fresh sweetpotatoes (and other perishables) that they can buy directly from high volume shipper-distributors. Chain store buyers typically send loads of sweetpotatoes to one or more of their wholesale distribution centers where each load is repackaged and stored in its proper environment and temperature. Inventory is rotated to assure that first-in sweetpotatoes are first-out as store produce managers order quantities needed. Specific quantities ordered by produce managers are then loaded onto company-owned distribution trucks for delivery to individual stores.

In addition to direct shipper-to-chain retailer sales, independently owned or small-chain companies (9 stores or less) purchase sweetpotatoes from the spot or open market and this option is the second most-frequently used method of marketing sweetpotatoes in the U.S. In this situation, sales agents, specialized produce selling brokers, specialized produce buying brokers, truck brokers, and specialized produce wholesalers interact with each other in order to move sweetpotatoes from farm to a sales outlet irrespective of buyer location. This marketing option results in grower's sales agent, shippers, selling brokers, or commission merchants contacting buying brokers (located in or near large U.S. city) or merchant wholesalers who are interested in buying and/or reselling sweetpotatoes at the market price (ownership transfers at time of sale). Sweetpotato growers must stay abreast of current market conditions and price, even if they hire an agent or specialized broker to handle sales for them, because the marketing system can take advantage of uninformed growers.

Next, general-line foodservice wholesalers purchase sweetpotatoes from grower-shippers since some restaurants and institutional customers (hospitals, schools, and baby foods) utilize sweetpotatoes. For most food service wholesalers, produce sales are a small percentage of total sales and for sweetpotato growers, foodservice products are limited so foodservice sales are a small but important source of income for some specialty produce suppliers. Finally, direct farmer-to-consumer sales such as those at community farmers' markets represent a small amount of income for geographically dispersed growers who live in rural areas of the South. Southerners prefer to purchase locally grown sweetpotatoes directly from their neighbor-farmers, especially if the direct price is less than the local grocery store price.

Sweetpotato Distribution Patterns

In general, U.S. sweetpotato shipping patterns flow predominantly from south-to-north, that is, North Carolina-grown sweetpotatoes are marketed primarily along the East Coast, including eastern Canada. Similarly, Mississippi and Louisiana grown sweetpotatoes are sold to Mid-Western customers while California growers market their sweetpotatoes primarily to buyers located along the West Coast and western Canada (Gonzalez, 2008). While this selling pattern holds for most grower-shippers, it is certainly true that the largest volume distributors, irrespective of location, sell to any and all customers. Indeed, large-volume grower-shippers have arrangements with other large volume competitors to cooperate with one another to supply sweet-potatoes to customers. Sometimes this is cited as an example of cross-marketing sweetpotatoes. For example, a North Carolina grower-shipper may have a Dallas, Texas customer so the dealer asks that an allied Louisiana grower-shipper firm supply sweetpotatoes to the Dallas customer using Louisiana-grown sweetpotatoes but packed in the North Carolina firm's shipping cartons. Of course, the North Carolina dealer would reciprocate for the allied Louisiana-based dealer if the customer wanted sweetpotatoes delivered to Maryland. Sweetpotato grower-shippers, dealers, and distributors compete and cooperate with each other, depending on specific circumstances. Finally, some growers hire a broker who negotiates specific details of a sales contract between buyer and seller. Brokers can work for either the buyer or the seller, depending on who pays the brokerage fee (between 8% and 10% of the price). Typically, brokers can arrange transportation for the buyer or the seller. Brokers sell information about buyers and sellers and do not take ownership of sweetpotatoes; they simply arrange deals and collect a fee for their knowledge about market opportunities.

In addition to the south-to-north market flow, shipper-growers recognize that Southerners eat proportionately more sweetpotatoes per capita than the rest of the U.S. so specific marketing emphasis is directed toward Southern markets, especially to consumers located in nearby and adjacent states. In general, consumers eat more sweetpotatoes, as temperatures get cooler. In recent years, innovative marketers have cello-wrapped individual sweetpotatoes so they can be prepared quickly by cooking in a microwave oven (see Plate 13.4). In 2008 new market opportunities

Plate 13.3 Yams and sweetpotatoes for sale side-by-side at Lowe's Grocery Stores, Wake County, North Carolina (See also Plate 12 on page xxii) (photo by E. Estes)

Plate 13.4 Cello-wrapping of sweetpotatoes on grading line at grower-shipper packing shed in Columbus County, North Carolina (See also Plate 13 on page xxiii)

seem to be focused in two areas: (1) value-added uses such as sweetpotato puree; and (2) ethanol biofuel. Shelf-stable puree was developed by Dr. Van-Den Truong, a USDA-Agricultural Research Service Food Scientist stationed at North Carolina State University in Raleigh, N.C. (Bliss, 2008). The puree can be used as a nutritious ingredient for use in soups, baby food, beverages, gluten-free pancakes, and nutraceuticals. Also at North Carolina State University, sweetpotato breeder Dr. Craig Yencho is developing a high dry-matter, industrial sweetpotato variety that can be used in biofuel production. Preliminary results suggest that high-dry matter varieties have potential for fuel use but the economic feasibility of using sweetpotatoes remains uncertain. Feasibility will be investigated in 2010 after several new varieties are field-tested to determine flesh dry matter content.

References

Agricultural Marketing Resource Center (AgMRC). 2008. Sweet potato profile, http://agmrc.org/agmrc/commodity/vegetables/sweetpotatoes/sweetpotatoes.htm. Accessed on 3 june 2008.

Anonymous. Overview of root and tuber crops. 1997. International Potato Center (CIP) report, www.fao.org/Wairdocs/TAC/TACX5791E/x5791e0b.htm. Accessed on 22 April 2008.

Bliss, R.M. 2008. Sweet potato puree adds to bottom line, USDA-AMS-ARS, www.are.usda.gov/is/pr/2008/080326.htm. Accessed on 7 may 2008.

Continuing survey of food intakes by individuals (CSFII) in 1994–96, 1998. 2000. USDA-ARS-Beltsville, MD, Human Nutrition Research Center, Food Surveys Group, http://www.ars.usda.gov/SP2UserFiles/Place/12355000/pdf/Csfii98.pdf. Accessed on 6 June 2008.

Cook, R. 2004. Supermarket challenges and opportunities for fresh fruit and vegetable producers and shippers; lessons from the US experience, paper presented at the Conference on *Supermarkets and Agricultural Development in China-Opportunities and Challenges*, May 24.

Conversion metric. 2008. http://www.conversion-metric.org/weight_conversion/pound_to_kilogram_conversion.php.Accessed on 3 june 2008.

Estes, E. 2006. Economics and trends in sweet potato industry: Sweet Potato Speaks, News for NC Processors and Growers, *NC Sweet Potato Commission*, Selma, N.C., March.

Gonzalez, K. 2008. *Marketing North Carolina Sweet Potatoes including Louisiana and Mississippi, 2006–2007 crop*. North Carolina Department of Agriculture and Consumer Services, Market News Section, Division of Marketing, Raleigh N.C., May 2008.

Johnson-Langdon, S. 2008. North Carolina Sweet Potato Commission website, www.ncsweetpotatoes.com.Accessed on 6 june 2008.

Kaufman, P. et al. 2000. Understanding the dynamics of produce markets, *USDA-ERS-AIB* Number 758, August.

Lucier, G. 2008. US Sweet Potato Statistics, *USDA-ERS, Economics, Statistics, and Market Information System*, Stock number 03001, table01.xls, table42.xls, http://usda.mannlib.cornell.edu/MannUsda/viewDocumentInfo.do?documentID=1492, May. Accessed 3 June 2008.

Lucier, G. and Dettman, L. 2008. Vegetables and melons outlook, VGS-326, *USDA-ERS-Market and Trade Economics Division*, April 17.

Lucier, G. and Jerardo, A. 2007. Vegetables and melons situation and yearbook, *USDA-ERS-Market and Trade Economics Division*, VGS-2007, July 26.

Lucier, G. and Plummer, C. 2002. Commodity highlights: sweet potatoes, vegetables and melon outlook, *USDA-ERS*, VGS-293, pp. 10–12, October 24.

May, D. and Scheuerman, R. 1998. Sweet Potato Production in California. *University of California Coop Ext Serv*. Pub.7237, Division of Agriculture and Natural Resources, Davis, C.A.

Powers, N. 1994. Marketing practices for vegetables, *USDA-ARS, AIB* 702, August.

Scott, G.J. 2001. Sweetpotato/Sweetpotato Facts, *International Potato Center*, La Molina, Peru, http://www.cipotato.org/sweetpotato/facts/facts.asp. Accessed on 25 April 2008.

Scott, G.J., Rosegrant, M.W. and Ringler, C. 2000. *Roots and Tubers for the 21st Century: Trends, Projections, and Policy*, 2020 Brief No.66, IFPRI, www.ifpri.org/2020/BRIEFS/number66.htm, Accessed on 30 April 2008.

US Census of Agriculture. 2007. *USDA-NASS*, http://www.agcensus.usda.gov/. Accessed on 23 May 2008.

Part II
Sweetpotatoes in Different Regions

Chapter 14
Sweetpotato Production in the United States

T.P. Smith, S. Stoddard, M. Shankle and J. Schultheis

Historical Background

The sweetpotato, *Ipomea batatas* (L.) Lam., is tropical in origin and was considered an important food plant for ancient civilizations in tropical America; however, the plant was not grown extensively by North American Indians. The sweetpotato became widely established in North America by the end of the 18th century (Edmond, 1971a). It was disseminated by explorers from Mexico and the West Indies after the discovery of America. Records indicate that the sweetpotato was grown by colonists in Virginia as early as 1648, and in Carolina and New England by 1723 and 1764, respectively. According to Gray (1933), a noted historian of agriculture in the United States from colonial times to a time predating the United States Civil War, "Sweetpotatoes (during Colonial times) were raised for home use by a large proportion of farmers and planters of the lower south. They were a principal food product in the eastern part of the Carolinas and Georgia and in the sandy lands of the Gulf coastal plain. They frequently occupied a paramount position in the diet of the poorer classes, being used for many purposes ranging from bread to beer. They were an important contribution to the diet of the slaves and they were frequently employed to supplement corn in the fattening of stock. Crops of 400–600 bu per acre *(1 bushel is equal to 50 lbs or 22.70 kg)* were occasionally reported, and in one instance, 1000 bu per acre, but ordinarily yields ranged from 100–200 bu per acre." (Edmond, 1971a).

In the United States, agricultural specialization began to occur in many areas at the beginning of the 20th century. At this time, climates and soil types suitable for the production of sweetpotatoes were identified. Areas in Delaware, Maryland, Virginia and southern New Jersey emerged as the first commercial sweetpotato districts in the United States around 1909 (Edmond, 1971a). As the industry grew, consumer preferences became distinctly different between the northeast and the southern United States. Consumers in the northeast wanted a dry flesh type

T.P. Smith (✉)
Sweetpotato Research Station, Louisiana State University Agricultural Center,
Chase, LA 71324, USA
e-mail: TSmith@agcenter.lsu.edu

G. Loebenstein, G. Thottappilly (eds.), *The Sweetpotato*,
DOI 10.1007/978-1-4020-9475-0_14,

of sweetpotato, while those in the southeast preferred a moist, sweeter product. Understandably, producers of the crop in these areas grew what consumers demanded and virtually monopolized markets along geographic lines (Edmond, 1971a).

In the 1960's new commercial districts in Vardaman-Batesville, Mississippi, Oak Ridge in Louisiana and the San Joaquin Valley of California were established and became competitive in the market place (Edmond, 1971a). The release of the Unit 1 Porto Rico by the Louisiana State University Agricultural Experiment Station in 1953 caused a change in preference among consumers in the southeast. The Louisiana Sweetpotato Commission began an active advertising campaign promoting the Porto Rico variety, which had a light orange flesh color and ample beta-carotene. The end result was that consumers chose more on the basis of flesh color as opposed to moistness of the potato when baked (Edmond, 1971a). Baker 1947 indicated that consumer preference for moist, orange-flesh sweetpotatoes was overwhelming that they eventually dominated the U.S. market (Edmond, 1971a).

After orange-flesh sweetpotatoes were introduced, southern United States producers and shippers desired to distinguish them from the more traditional white-flesh types. The African work "nyami" referring to the starchy, edible root of the *Dioscorea* genus of plants was adopted in its English form, "yam" (Schultheis and Wilson, 1993). Now, the term "yam" is market terminology to denote a sweetpotato with moist texture and orange after it is cooked or baked. Most yams produced in the United States are sweetpotatoes, and the United States Department of Agriculture requires that they be labelled as such.

The development and commercialization of the sweetpotato industry in the United States in the latter part of the 20th century was in large part due to the breeding and variety development programs in sweetpotato producing states across the country. The Sweetpotato Collaborators Group originated in 1938, when Julian C. Miller (Louisiana State University Agricultural Experiment Station), George B. Hoffman (USDA Horticultural Field Station, Meridian, MS), and W. S. Anderson (Mississippi Agricultural Experiment Station) convened to discuss the need for research and development in the sweetpotato industry. In 1939 the United States Congress appropriated $32,000 for research that would be cooperative and beneficial to the various states involved (Bowers et al., 1970). The first formal meeting of the Sweetpotato Collaborator's group was held in the fall of 1939 in Meridian, Mississippi, and included representatives from the states of Louisiana, Maryland, Virginia, Georgia, Mississippi, Texas and South Carolina.

Many successes were achieved in sweetpotato breeding in the years following World War II (Bowers et al., 1970). As a result, several breeding programs were established to enhance and compliment the work that originated with the first sweetpotato breeding program in the United States under the direction of Dr. Julian Miller at Louisiana State University Agricultural Experiment Station. Breeding programs were soon established in North Carolina, Maryland, South Carolina, Georgia, Mississippi, Virginia and Oklahoma.

The primary focus of the sweetpotato collaborators group then and now is the breeding and development of new varieties; however, scientists in the group work in many facets of sweetpotato production. Examples include research in cultural practices, soil and fertilizer management, integrated pest management, and food science. The formation of sweetpotato breeding programs across the country resulted in a rapid increase of seedling materials and seed stock available for testing. The exchange of material was streamlined and facilitated through individuals in the sweetpotato Collaborators Group who worked together and evaluated potential new varieties across state boundaries on different soil types and production systems (Bowers et al., 1970). Researchers have evaluated sweetpotato breeding lines for potential use as table stock (fresh consumption) and for industrial use such as starch (Bowers et al., 1970). The implications of past and present research and of the sweetpotato varieties that have been released since the release of the Unit I Porto Rico are far-reaching and evident in the sweetpotato industry of the United States, as it exists today.

Certain varieties released between the years of 1950–1970 were instrumental to the industry and strengthened the growth and development of the commercial districts described previously. While over 40 varieties were released from various institutions during this time period, varieties such as Centennial (Louisiana, 1960) and Jewel (North Carolina, 1970) became the leading varieties of the 1960s and 70s and were grown extensively in all production regions across the United States.

Centennial was released by the Louisiana Agricultural Experiment Station in 1960. At the time of its release it consistently yielded higher and had improved quality compared to other varieties being grown at the time, such as the Unit I Porto Rico, Goldrush, and Acadian varieties.

Jewel was released from the North Carolina State University Breeding Program in 1970. Jewel has a copper skin color, deep orange flesh, and excellent taste and baking qualities. In addition, the Jewel variety stores well, but is susceptible to soil rot, *Streptomyces ipomoea*.

Another variety, Hernandez, which was released from the Louisiana State University Agriculture Experiment Station in 1992 (LaBonte et al., 1992) was also planted on significant acreage in key production states, namely North Carolina. Hernandez is a late developing sweetpotato, requiring 130–140 days to mature compared to 110–120 for Jewel (LaBonte et al., 1992).

Of all the varieties released in the previous 25 years, Beauregard and Covington have had the greatest impact on the United States sweetpotato industry. Beauregard is a copper skinned, orange-flesh variety that was released by the Louisiana State University Agricultural Experiment Station in 1987 (Rolston et al., 1987). It combined above average yields, good horticultural characteristics, such as above average plant production, high survivability of transplants and disease resistance into a sweetpotato that in only a few years dominated North American commercial sweetpotato market. Beauregard produced more U.S. #1 grade roots than the comparative cultivars (Centennial and Jewel) in 11 of 13 replicated trials at time of release. Beauregard is resistant to Fusarium wilt (stem rot), caused by *Fusarium oxysporum Schl.*, and is moderately resistant to soil rot (pox), caused by *Streptomyces ipomoeae* (Rolston et al., 1987). However, this variety is very susceptible to root knot nematodes, especially *Meloidigyne incongnita*. Beauregard became the industry standard only a few years after its release and has remained a dominant cultivar in most sweetpotato production areas in the United States.

Covington (Yencho et al., 2008) is another copper skinned, orange-flesh sweetpotato variety released by North Carolina State University in 2006. It has quickly established itself as the dominant variety in that state, accounting for about 65% of the acres grown. Covington has good horticultural characteristics and performs well on the sandier soil types found in North Carolina and California. It is a later maturing variety, requiring 120–130 days to reach maturity. The pack out quality of this sweetpotato in North Carolina has exceeded that traditionally experienced with Beauregard.

At present, there are over 42,000 hectares (103 thousand acres) of sweetpotato produced in the United States primarily in the states of North Carolina, Louisiana, Mississippi, California, Alabama, Arkansas, New Jersey and Texas (USDA, 2007b). Several sweetpotato varieties are currently being produced in the United States, but Beauregard and Covington are the dominant commercial cultivars.

While moist, orange-flesh varieties have dominated the U.S. market for the last 60 years; varieties have been developed with different qualities to fill important niche markets. This is especially true in California, where there is greater ethnic diversity among consumers of sweetpotatoes. The variety Garnet was instrumental in creating a red-skinned, premium yam market. In the 1980's, a purple skin, white-flesh sweetpotato variety coined Kotobuki was introduced to California. Japanese varieties have enjoyed steady growth since then, and now constitute and important marketing niche, with about 15 percent of California's production acreage (Stoddard, 2007).

Sweetpotato research and variety development is ongoing in the United States. The Louisiana State University Agricultural Experiment Station released a new variety in 2007, "Evangeline" (LaBonte et al., 2008), which has production characteristics similar to the Beauregard variety. In addition, Louisiana State University Agricultural Center, North Carolina State University and the USDA breeding laboratory in Charleston, South Carolina have productive and active breeding programs with numerous sweetpotato lines currently undergoing evaluation.

Importance of Sweetpotato in the United States

The importance of sweetpotato in the United States cannot be overstated. The crop has traditionally been used for human consumption in the United States and consumption and utilization varied considerably during the 20th Century. The changes and variability in production and consumption of the crop have primarily been influenced by the United States economy, especially during times of war or economic depression. The changing demographics of agriculture and farming, product availability, and consumer preference have also affected production, consumption and utilization of the crop.

Depression Era

The economic depression that occurred worldwide and more specifically in the United States during the 1930's had a resounding impact on industry, agriculture and the general state of the United States economy (Edmond, 1971a). Emphasis was placed on "living at home" and becoming self sufficient with respect to growing crops and sustaining individual families or communities (Edmond, 1971a). Sweetpotatoes and other subsistent vegetable crops were very important during the depression era. Sweetpotato production in the United States increased by more than 200,000 acres in the time period spanning (1930–1937), from 646,000 to 894,000 harvested acres. Yields, however, decreased by approximately 10% (Royston and Oakely, 1958). The increase in acreage accompanied by the decrease in yield is a direct reflection of ample land available for farming and limited income available for inputs such as fertilizer (Edmond, 1971a). The amount of sweetpotatoes used by farmers and the amount sold to consumers increased by approximately 13% from 1930 to 1937, while the average price paid for the crop dropped by 50% (Royston and Oakely, 1958). Per capita consumption of sweetpotatoes was at an all time high (> 25 lbs) (USDA, 2008) for the United States population, reinforcing the important niche sweetpotatoes filled in the diets of American consumers during this era.

The period following the depression saw a marked decrease in production and utilization of the crop in the United States. The decline was likely due to improved economic security in the years following the economic depression and World War II, which influenced consumer preference and ultimately per capita consumption of sweetpotatoes (Edmond, 1971a). Per capita consumption of sweetpotatoes has averaged approximately 4.5 lbs (2.0 kg) during the last thirty-five years in the United States (Table 14.1) (USDA, 2007b).

American consumers are not utilizing sweetpotatoes in their diets to the degree they did in the early 1900s, but consumption of the crop has increased slightly in recent years. The increased consumption is largely due to national advertising campaigns focused on the nutritional aspects of sweetpotatoes aimed at health conscious consumers. The sweetpotato is very high in nutritive value, and merits wider use on this account alone. Contrary to popular opinion, it is not a starchy food when baked, due to the fact that most of the starch is broken down into maltose and other soluble

Table 14.1 United States per capita consumption (lbs) of sweetpotatoes, 1970–2007

1970	**1971**	**1972**	**1973**	**1974**	**1975**	**1976**	**1977**	**1978**	**1979**
5.4	4.9	4.9	5	4.9	5.4	5.4	4.7	4.9	5.1
1980	**1981**	**1982**	**1983**	**1084**	**1985**	**1986**	**1987**	**1988**	**1989**
4.4	4.7	5.5	4.6	4.9	5.4	4.4	4.4	4.1	3.9
1990	**1991**	**1992**	**1993**	**1994**	**1995**	**1996**	**1997**	**1998**	**1999**
4.4	3.9	4.1	3.7	4.5	4.2	4.3	4.3	3.8	3.7
2000	**2001**	**2002**	**2003**	**2004**	**2005**	**2006**			
4.2	4.4	3.7	4.7	4.6	4.5	4.7			

Source: Economic Research Service/USDA Vegetables and Specialties Situation and Outlook Yearbook August 2007

sugars. Sweetpotatoes rank as one of the healthiest vegetables, because of high levels of vitamins A and C, iron, potassium, and fiber. They are also an excellent source of the vitamin A precursor, beta-carotene. One cup of the orange flesh types contains four times the recommended daily allowance of this important nutrient. The dry-flesh types contain considerably less vitamin A than the moist-fleshed types.

Major Growing Areas, Acreage, Yield and Economics, Prices to Growers

Sweetpotato production in the United States during the last 30 years was markedly reduced compared to production in the first half of the 20th century. *Producers* in the United States harvested on average 35.6 thousand hectares (88 thousand acres) per year in the eleven-year period spanning 1997–2007 (USDA, 2008). Sweetpotato production in the United States is largely concentrated in the states of North Carolina on the east coast, Louisiana and Mississippi along the Mississippi gulf coast, and California on the west coast of the United States. However, limited production occurs in several additional states. North Carolina has been the leading producer of sweetpotatoes during the last ten years, harvesting on average 36,000 acres per year during that time period (USDA, 2008). Louisiana has traditionally had the second highest acreage, with a 10-year average of 18,450 acres; however, Mississippi surpassed Louisiana in acreage in 2007 becoming the state with the second highest planted acreage (Table 14.2). United States sweetpotato production averaged 14.7 million cwt. in the years spanning 1997–2006 (Table 14.2). Sweetpotato yields realized during the last twenty years are vastly improved compared to yield estimates in the early part of the 20th century. Sweetpotato producers in the 1950s and 1960s complained that they were not receiving a "fair" market price for their crop and that it was difficult to realize a decent profit. Henry Covington, a researcher and sweetpotato breeder with North Carolina State University during the 1950–60s was of a different opinion, and his observations are still relevant in present day commercial operations. Covington (1957) stated that "Sweetpotato growers are complaining that the prices they receive are too low, and they cannot make a normal profit. The real problem is low yields. The answer to the grower's problem is to raise the yield per acre and thus cut the unit cost" (Edmond, 1971a).

To increase yields and quality, producers, have improved their cultural practices and pest management efforts, and have made greater use of virus-tested seed stock. Average yields (cwt per acre) from 1997–2006 were 9% greater than the subsequent 10-yr period.

Average total yields in California are approximately 40–45% higher than those in other major production areas (Table 14.2). Many aspects of sweetpotato production in California are different compared to production in the southeast United States. Producers in the south and along the east coast are exposed to several environmental and biological variables in any given production year, namely the weather (extreme drought or flooding) and heavy insect pressure, variables which producers

Table 14.2 Summary of major states and United States sweetpotato production: Acreage, yield, production and economics

California						
Year	Planted acres (thousand)	Harvested acres (thousand)	Yield (cwt)	Production (thousand cwt)	Price per unit U.S. dol/cwt	Value of production (thousand dollars)
1997	9.7	9.7	205	1,989	28	55,692
1998	9.7	9.7	220	2,134	25.8	55,057
1999	10	10	240	2,400	28.8	69,120
2000	10.5	10.5	250	2,625	24.1	63,263
2001	10	10	230	2,300	26.9	61,870
2002	10.4	10.4	280	2,912	23.1	67,267
2003	10.7	10.7	300	3,210	25.2	80,892
2004	11.5	11.5	280	3,220	25	80,500
2005	11.7	11.7	285	3,335	25.3	84,376
2006	12.2	12.2	305	3,721	20.1	74,792
2007	13.5	13.3	320	4,256	19	80,864
2008	14.5**	*	*	*	*	*

Table 14.2 (continued)

Louisiana						
Year	Planted acres (thousand)	Harvested acres (thousand)	Yield (cwt)	Production (thousand cwt)	Price per unit U.S. dol/cwt	Value of production (thousand dollars)
1997	21	20	170	3,400	14.4	48,960
1998	21	20	110	2,200	14.4	31,680
1999	24	23	150	3,450	14.3	49,335
2000	25	24	130	3,120	13.2	41,184
2001	24	22	140	3,080	13.3	40,964
2002	21	15	125	1,875	14.5	27,188
2003	19	18	175	3,150	20	63,000
2004	16	15.5	150	2,325	17.7	41,153
2005	18	17	145	2,465	16.2	39,933
2006	18	13.5	165	2,228	17.3	38,544
2007	16	15	195	2,925	20.3	59,378
2008	16**	*	*	*	*	*

Table 14.2 (continued)

	Mississippi					
Year	Planted acres (thousand)	Harvested acres (thousand)	Yield (cwt)	Production (thousand cwt)	Price per unit U.S. dol/cwt	Value of production (thousand dollars)
1997	8.6	8.4	130	1,092	18.5	20,202
1998	9.8	9.7	140	1,358	17.5	23,765
1999	10.5	10.3	150	1,545	21.7	33,527
2000	12.7	12.3	120	1,476	14.6	21,550
2001	16.7	16	150	2,400	13.6	32,640
2002	16	12.3	160	1,968	16.5	32,472
2003	14	13.6	175	2,380	20.8	49,504
2004	16	15.3	170	2,601	17.7	46,038
2005	17.4	17.3	180	3,114	18.3	56,986
2006	18	15.5	160	2,480	19.8	49,104
2007	20.5	20	175	3,500	19.3	67,550
2008	20**	*	*	*	*	*

Table 14.2 (continued)

North Carolina						
Year	Planted acres (thousand)	Harvested acres (thousand)	Yield (cwt)	Production (thousand cwt)	Price per unit U.S. dol/cwt	Value of production (thousand dollars)
1997	32	31	160	4,960	10.8	53,568
1998	33	32	170	5,440	11	59,840
1999	37	29	130	3,770	12	45,240
2000	38	37	150	5,550	12.3	68,265
2001	37	36	155	5,580	12	66,960
2002	40	37	130	4,810	14	67,340
2003	43	42	140	5,880	14.5	85,260
2004	45	43	160	6,880	13.5	92,880
2005	36	35	170	5,950	14.2	84,490
2006	40	39	180	7,020	16.2	113,724
2007	44	43	165	7,095	20.9	148,286
2008	47**	*	*	*	*	*

Table 14.2 (continued)

	United States					
Year	Planted acres (thousand)	Harvested acres (thousand)	Yield (cwt)	Production (thousand cwt)	Price per unit U.S. dol/cwt	Value of production (thousand dollars)
1997	85.6	82.1	162	13,327	15.8	211,177
1998	87.1	83.7	148	12,365	15.3	189,532
1999	93.7	83	147	12,221	17.6	214,754
2000	97.9	94.8	145	13,780	15.3	210,351
2001	98.3	93.6	155	14,515	15.3	222,658
2002	96.4	82.3	156	12,799	16.8	214,650
2003	95.8	92.6	172	15,891	19.2	305,448
2004	96.9	92.8	174	16,112	17.5	281,559
2005	91	88.4	178	15,730	18.1	284,103
2006	95.2	86.8	187	16,248	18.2	295,313
2007	100.6	97.5	189	18,452	20.3	373,723
2008	103.8**	*	*	*	*	*

Source: United States Department of Agriculture, National Agricultural Statistics Service: http://www.nass.usda.gov
** Projected acreage to be planted in 2008
* Data not available at time of publishing

in California are largely able to escape. In addition, producers in California receive limited rainfall each year, and as a result, manage water requirements of the crop through scheduled irrigation. Drip irrigation is the predominate form of irrigation, used on approximately 95% of acreage.

Sweetpotatoes are primarily used for human consumption in the United States. Like most vegetable crops, producers receive no direct subsidies from the government, and price is dictated in any given year by supply and demand (Edmond, 1971a). Several factors can affect the price of sweetpotatoes, including, grade, variety, time of year and transportation costs.

During the first half of the 20th century, the price of sweetpotatoes was quite volatile and fluctuated in relation to World War I, World War II and the economic depression of the 1920s and 30s. In general, the increases in pricing coincided with "war time" and the decreases with a reduction in the purchasing power of the American consumer (Edmond, 1971a). The dollar price received per cwt. from 1930–1940 in the United States, ranged between $1.36 and $1.40. After World War II, sweetpotato acreage in the United States decreased and the price received for the crop increased accordingly.

In general, the price received for the crop (dollars per cwt) increased 7–14% from 1997 through 2007 (Table 14.2). Regardless of state, prices reported to the USDA are lower than the actual price realized in a given year. This is due to the fact that the USDA reported price is reflective of all yield grades, including U.S. No. 1's, No. 2's, Jumbos, and canners, and should be interpreted as such.

Agronomic Information

Sweetpotatoes are botanically unrelated to the Irish potato (*Solanaceae* family), which includes tomatoes, peppers, eggplant, and the weedy nightshades. Sweetpotatoes are also not related to the true yam (a *true* yam is a large underground tuber in the family *Dioscoreaceae).* Native to Africa, it may vary in the size range of a white potato to enormous yams weighing 30–40 pounds and measuring as much as 3 feet in length). Irish, or "white" potatoes are *tubers*, which are thickened stems and are essentially carbohydrate storage reservoirs for the plant. Sweetpotatoes are a true root, and as a result contain much higher amounts of complex carbohydrates. The term "sweetpotato" is really a misnomer, because it implies they are potatoes that are sweet, when in reality they are as different from potatoes as carrots are. Recognizing this, the National Sweetpotato Collaborators Group and the U.S. Sweetpotato Council endorsed spelling sweetpotato as one word in 1989. Nonetheless, in the current lexicon of American English, it is still spelled with two words. Both spellings are correct.

Sweetpotatoes can be grown in various soil and climate regimes (Jansson and Ramon, 1991). Indeed, their geographic distribution ranges from Florida to southern Ontario, Canada. The crop is drought tolerant and can be grown in tropical and temperate agricultural regions that are adapted to low or high input systems with varying degrees of technology (Bouwkamp, 1985). Successful production of sweetpotatoes in the United States requires extensive inputs. Input costs are variable between geographic regions, but are considerably higher compared to the input costs of agronomic crops, such as soybean, *Glycine max* or corn, *Zea mays* in the United States. In California, inputs costs may exceed \$5,000 per acre (Stoddard et al., 2006). Similar cost of production studies conducted in Louisiana estimate production cost ranging from \$2500.00–\$3000.00 per acre (Smith et al., 2008).

In the U.S., sweetpotatoes are considered a "high risk" crop in many areas due to the potential for weather related crop failures, and the aesthetic qualities of the crop that are often negatively affected by insect feeding and disease incidence. The biology of the crop is similar in the United States as it is in other areas, but quality standards coupled with economic factors increase the risks associated with producing the crop. Minimal insect and/or disease incidence on roots can drastically reduce the marketability of the roots. Producers in the United States invest tremendous resources into cultural practices and pest management to minimize the negative effects associated with various insects, diseases, and weeds. Despite the relatively high use of technology and mechanization, sweetpotato production is considered to be more labor intensive than many other crops, especially where planting and harvesting activities are involved.

Seedbeds/Hotbeds

Sweetpotato production can be split into two distinct components: hotbeds and field production. Sweetpotato hotbeds are the nursery area that growers use to produce transplants for field production. Typically, the hotbed season begins in February or early March, when seed potatoes (small potatoes that are not sold for the market) are placed into beds to grow plants (sometimes called slips) that will later be transplanted into production fields April through July.

The sweetpotato is vegetatively propagated, and all production fields in the United States start from transplants. Thus, the hotbeds are an important and distinct operation for any grower, and must be managed as such. Each year, growers save a portion of their crop to be planted the next year. Enough seed must be saved to cover all anticipated production, as the availability of seed potatoes or plants for purchase is somewhat limited. Hotbeds have their own cultural requirements and can be very expensive to manage. The bedding process involves several steps. Often the seed potatoes are pre-sprouted in storage by increasing the storage temperature to 70–85 °F (21–29 C) and the relative humidity to 85–90%, two to three weeks prior to bedding. Plants produced from approximately 800 to 1500 lbs (365–700 kg) of seed potatoes are required to plant 1 acre (0.4 ha) of sweetpotatoes. Fifty lbs (23 kg) of seed potatoes will produce about 500 sprouts and requires 20–30 ft^2 (2–3 m^2) of bedding area (Schultheis et al., 2008). Seed potatoes are conveyed onto the soil surface mechanically or by hand (Plate 14.1), and are then covered with 2–3 inches of soil (Plate 14.2) and a plastic mulch layer (Plate 14.3). Plastic mulches are not used in

Plate 14.1 Sweetpotato bedding wagon and seed distribution in a hotbed or seed bed (Courtesy of Tara Smith, Louisiana State University Agricultural Center) (See also Plate 14 on page xxiii)

Plate 14.2 Sweetpotato bed covering machine, covering seed potatoes (Courtesy of Tara Smith, Louisiana State University Agricultural Center) (See also Plate 15 on page xxiii)

Plate 14.3 Sweetpotato mulch/plastic layer being applied to a seed bed (Courtesy of Tara Smith, Louisiana State University Agricultural Center) (See also Plate 16 on page xxiii)

Plate 14.4 Sweetpotato hotbeds in California (Courtesy of Scott Stoddard, University of California at Merced) (See also Plate 17 on page xxiv)

California; rather, clear plastic is pulled over low metal hoops to provide additional heating (Plate 14.4). The mulch layer consists of either black or clear plastic, 1.5 to 2 ml thick. The plastic is secured in place by soil and several holes, 2 inches in diameter are placed at intervals of 3–4 linear feet along each side of the bed. The mulch layer increases the soil temperature in the seedbed, which promotes earliness. The openings in the mulch layer promote ventilation and temperature regulation within the bed to control seed decay brought about by extreme high temperatures and high CO_2 levels (Boudreaux et al., 2005, Schultheis et al., 2008). Seedbeds are frequently fertilized with 300–400 lbs of a complete fertilizer, such as 13:13:13 or

Plate 14.5 Sweetpotato seed beds with transplants (Courtesy of Tara Smith, Louisiana State University Agricultural Center) (See also Plate 18 on page xxiv)

8:24:24 prior to applying the plastic mulch, or post-emergence where plastic mulch is not utilized. Six to eight weeks are required to produce transplants that are large enough to plant in production fields (Plate 14.5).

Transplanting

Sweetpotato is a perennial crop that is produced as an annual crop in the United States (Bouwkamp, 1985). Sweetpotatoes are transplanted in the spring and early summer months (April–July), with some degree of variability in time of planting between the different geographic regions. As a general rule, production in California begins earlier than other production regions in the United States. The growing period of the crop is approximately 120 days, but can be highly variable depending on cultivar and transplant spacing (Boudreaux, 2005). Most sweetpotatoes grown for commercial production and sales are planted from April 15–June 15. A percentage of the planted acreage will be harvested as seed for production the following year. (Schultheis et al., 2008).

Transplants are cut and removed from the seedbeds by hand labor or with mechanical plant cutters. Transplants in the United States most commonly consist of sprouts, which are cut from the bedded roots; however some vine cuttings are used. Sprouts are defined as entire plants, which are produced from bedded roots (Edmond, 1971b). Transplants should be cut approximately 1 inch above the soil surface, rather than pulled from beds, to reduce the potential for moving root diseases from the seed beds into the production fields (Schultheis et al., 2008).

Transplants or "slips" are harvested from the seedbeds 6–8 weeks after bedding and are then transplanted into commercial sweetpotato fields. The transplants can be planted by hand or more commonly, they are planted using mechanical transplanters (Plate 14.6). Water is usually applied after transplanting; the amount used is dependent on soil type and field moisture conditions. Mechanical transplanters contain 1–8 units, and each unit is responsible for planting one row. Transplants are most commonly planted down the row at 8–14 inch spacing, with rows on 36–48 inch centers (0.90–1.2 m); Transplants vary in length, but most are 10–14 inches in length at time of planting (Plate 14.7) (Schultheis et al., 2008). Transplants are planted at a depth of 3–4 inches and several nodes are placed underground to maximize potential root set (Boudreaux, 2005).

Plate 14.6 Sweetpotato 2-row mechanical transplanter, (Courtesy of Tara Smith, Louisiana State University Agricultural Center) (See also Plate 19 on page xxiv)

Plate 14.7 Sweetpotato transplants or slips that will be planted in a commercial production field (Courtesy of Tara Smith, Louisiana State University Agricultural Center) (See also Plate 20 on page xxv)

Soils and pH

Sweetpotatoes can be produced on a wide range of soil types (Bouwkamp, 1985), but in the United States, the majority of sweetpotatoes are produced on sandy or sandy loam soils. In general, sweetpotatoes are produced on "adapted" or excellent soil types in the United States. Favorable characteristics of adaptive soils include a level or slightly sloped field, moderate fertility, moderate to slight acidity, and adequate drainage (Edmond, 1971b). In addition, regardless of the soil texture, it is paramount to maintain good porosity and aeration and light soils are easy to manage in this regard (Bouwkamp, 1985).

In the absence of Streptomyces soil rot, soil pH in sweetpotato production fields should range from 5.5–7.0. Agricultural lime should be applied at a rate of 0.5 to 1.0 tons per acre to soils with a pH below 5.5 to effectively raise the pH to acceptable levels. Geise (1929) found lime to increase yields on acid soils and Watts and Cooper (1943) detected maximum yields at a pH range of 6.5–7.5 on Newtonia silt loam and 6.0 to 7.0 on Ruston fine sandy loam (Bouwkamp, 1985). He also indicated that vine growth, color, and vigor were superior at these ranges (Bouwkamp, 1985).

Soil rot or pox, which is caused by *Streptomyces ipomoea*, was historically a common disease in major sweetpotato production regions in the United States (Clark and Moyer, 1988). When prevalent, reducing soil pH to 5.2 or lower can significantly reduce the incidence of the disease (Bouwkamp, 1985). Hartman and Gaylor (1939) reported that 75% of storage roots were affected by the disease at a pH of 7.5 but very little infection was noted at a pH of 5.4 or less. The commercial varieties cultivated today are not as susceptible to soil rot as were older varieties such as Jewel and Centennial (Clark, personal communication). Beauregard and Covington varieties are considered moderately resistant to soil rot (Rolston et al.,

1987, Clark, personal communication); therefore producing these varieties in a low pH environment may not be necessary in the United States to control soil rot.

Fertilization

Proper fertilization is critical to realize maximum sweetpotato yield potential in the United States. Most commercial sweetpotato producers work with state extension agents or crop consultants to manage fertilizer inputs on a field-by-field basis. Fertilizer guidelines are available through state extension programs and are based on soil and tissue analyses. Without supplemental nutrients from fertilizers and/or compost, deficiency symptoms can occur (Bouwkamp, 1985).

Nitrogen, phosphorous and potassium are applied each year to commercial sweetpotato fields in the United States. Throughout much of the southeast, nitrogen is applied before transplanting or as a sidedress application 25–30 days after planting at a rate of 35–80 lbs per acre depending on cultivar, and soil type. Phosphorous is applied at a rate of 90–120 lbs and potassium is applied at a rate of 120–220 lbs per acre. Minor elements, including magnesium, calcium, zinc and boron are also applied as soil amendments or as foliar applications when needed. In California, fertilizer inputs are much higher, with recommended N rates of 125–175 pounds per acre. The nitrogen application is usually split, with about one-half being applied pre-plant from a shank injection band in the bed, and one-half applied during the growing season through the drip irrigation system (chemigation). Liquid formulations, such as urea-ammonium nitrate (UN32) and calcium nitrate (CN9 or CN17) are most common. Phosphorous and potassium is applied based on soil testing and expected yields. A response to phosphorous and potassium is not expected if the soil tests greater than 25 and 150 ppm respectively.

Irrigation

The sweetpotato is considered by many to be a drought tolerant crop, and fairly good yields have been reported in production systems suffering drought stress (Bouwkamp, 1985). Various experiments conducted from 1927 to 1979 demonstrated high sweetpotato yields under drought conditions, but also demonstrated that yields benefit from supplemental irrigation when rainfall is not adequate or where moisture distribution is erratic and unpredictable (Bouwkamp, 1985). MacGillivray (1953) stated that sweetpotatoes should not suffer moisture stress except in the last few weeks preceding harvest. Edmond (1971b) explained that sweetpotatoes in adaptive soils have well-developed fibrous root systems and a large transpiring surface. He added that if soil moisture was close to field capacity at time of transplanting, the moisture requirement for early establishment would be adequate, but that the initial moisture would be insufficient for maximum storage root development.

Early researchers working with sweetpotatoes in the first half of the 20th century agreed that irrigation, if available, should be applied to the crop when field capacity dropped below 50%. Sweetpotato yield is affected by the timing and distribution of moisture, and a lack of water can affect the quantity as well as the quality of the crop (Bouwkamp, 1985). Miller 1958, suggested that good soil moisture either from rainfall or supplemental irrigation was most critical during the first 40 days after transplanting (Kays, 1985).

Irrigation is used extensively, in sweetpotato production regions throughout the United States. In southern production regions, the variability in weather patterns makes irrigation a highly effective management tool to maximize production. Under extreme drought conditions during the production season, supplemental irrigation is applied to production fields at a rate of approximately 1 inch (2.45 cm) per week. Several types of irrigation equipment are used; including furrow irrigation with plastic polypipe, center pivots, travelling water reels, and drip tape. In the western production region of California, irrigation is required to grow the crop, as no rain occurs during the growing season. Approximately 95% of the acreage is drip irrigated, with the remainder using furrow irrigation. Water requirements vary somewhat based on time of planting, but in general the crop is irrigated every other day to match crop evapotranspiration of about two inches per week. Total water use is approximately 30–36 inches per acre (1.9–2.3 m per ha).

Virus-Tested Seed Programs

Sweetpotatoes are a vegetatively propagated crop. Roots are sprouted, and these sprouts are transplanted to the field to produce more roots. True seeds are not used in commercial production because sweetpotatoes rarely flower. An unfortunate consequence of not using true seed, however, is that viruses and mutations can accumulate in the plants, greatly diminishing both yield and quality.

Producers in the United States do save a percentage of their harvested crop each year to be used as seed in the following year. In addition, the majority of sweetpotato producers in the United States utilize virus-tested tissue culture technology and supplement their conventional seed produced on-farm with certified virus-tested foundation seed. The technology associated with virus-tested certified seed has had a tremendous positive impact on the sweetpotato industry of the United States.

First deployed at the University of California at Davis, the virus-tested sweetpotato seed program began in the 1960's in response to russet crack disease, which is caused by a strain of sweetpotato feathery mottle virus (Dangler et al., 1994). At the time, it was well known in the industry that "new" seedstock was required to prevent this disease and variety decline, which was the gradual loss of yield that occurred for many varieties. To address this problem in 1961, University of California Cooperative Extension farm advisor, Bob Scheuerman, and Extension Specialist Dennis Hall, began testing a process called meristem culture on sweetpotatoes to

Plate 14.8 Virus-tested sweetpotato plants in test tubes (Courtesy of Scott Stoddard, University of California at Merced) (See also Plate 21 on page xxv)

make clean propagation material. The procedure developed then is still in use today to provide growers with high quality seedstock. Meristem-tip culture involves aseptically removing the meristem (usually 0.5 mm long) from an apical or lateral bud of shoots produced in a greenhouse from a sprouted root. The meristems are placed in test tubes and grown out on synthetic nutrient agar to produce a new plant (Plate 14.8). After 3–4 months in culture, the plant is transplanted in the greenhouse and grown out for virus testing. To determine if the meristem-generated plants are free of virus, it is grafted onto an indicator plant (Brazilian morning glory, *Ipomoea setosa*). If the indicator plant shows no disease symptoms, then the sweetpotato plant is assumed to be virus-free. At this point, it can be propagated through cuttings and grown to produce roots for variety evaluation.

By 1966, enough plant material had been developed using this process that Scheuerman and Hall could conduct large-scale field tests. The results showed a significant yield increase using virus-tested plants as compared to plants that had not gone through meristem culture. Furthermore, there was a substantial decrease in the number of cull potatoes caused by viruses. Additional testing at North Carolina State University, Louisiana State University and other institutions since these early trials has continued to show the benefits of this technology.

Sweetpotato varieties, such as the Beauregard cultivar (Plate 14.9) have maintained their integrity as commercial cultivars largely due to the virus-tested certified seed programs. Seed selection, certification and distribution are variable among sweetpotato producing states, but in general begin with virus-tested mother plants maintained by their respective universities.

Plate 14.9 Sweetpotato commercial production field (Courtesy of Tara Smith, Louisiana State University Agricultural Center) (See also Plate 22 on page xxv)

Plate 14.10 Sweetpotato production field in California just prior to harvest (Courtesy of Scott Stoddard, University of California at Merced) (See also Plate 23 on page xxvi)

Pests and Diseases

Introduction

Many biological factors, including insects, diseases and weeds, can be limiting factors in successful sweetpotato production in the United States. Over 270 insect species and 17 mite species are known to feed on sweetpotato worldwide (Talekar, 1992). Diseases have limited sweetpotato production in the US in the past, but many of the diseases that were once significant problems have been successfully managed (Clark, personal communication). Weeds compete with sweetpotatoes for nutrients, water and sunlight and impair crop yield and quality (Smith and Miller, 2007).

Insects

Cuthbert (1967) indicated that there were over 19 insect species affecting sweetpotato production in the United States alone. Insect damage from a variety of phytophagous pests may reach 60–90% (Chalfant et al., 1990; Jansson and Ramon, 1991), and all plant parts including roots, stems, and foliage can be affected (Talekar, 1992).

Many insects reduce the quality and yield of sweetpotatoes, either by feeding directly on the storage roots or by defoliating leaves and/or boring into vines (Talekar, 1992). Aphids, whiteflies, and leafhoppers can also transmit many of the viruses known to infect sweetpotatoes (Talekar, 1992). Sweetpotato storage roots are injured by a complex of Coleopterous soil insects (Chalfant et al., 1990). Soil insects that damage the root are the most harmful because they can cause significant economic losses even in low numbers (USDA, 2001). In contrast, many foliage-feeding insects do not cause yield reductions because sweetpotato plants can often compensate for high levels of defoliation (Chalfant et al., 1990).

The sweetpotato weevil, *Cylas formicarius* Fab., is the most destructive insect affecting sweetpotato throughout tropical and subtropical production areas, including production areas located in the southern United States (Talekar, 1992). *Cylas formicarius* can attack sweetpotatoes in the field and in storage and larvae feeding on sweetpotato roots can result in major economic damage and yield loss (Chalfant et al., 1990). Larval tunneling causes the storage roots to produce terpenes, which imparts a bitter taste and leaves the sweetpotatoes unsuitable for human consumption (Uritaini et al., 1975).

Sweetpotato weevils are established in various regions in the southeastern United States (Smith and Hammond, 2006a; Mason et al., 1991). The sweetpotato weevil can also be a pest of stored sweetpotatoes in the southern United States. In addition, the sweetpotato weevil is the subject of different regulatory mandates in several states or regions in the United States and movement of sweetpotatoes from weevil-infested areas is regulated accordingly.

Education and prevention are the keys to successful weevil management. Producers are cautioned against bringing in sweetpotatoes or containers from areas where weevils are known to exist. If a threat of sweetpotato weevils exists, sex pheromone traps that are specific to sweetpotato weevils can be used to monitor plant beds, storage areas and production fields. Certain insecticides are available for use in the field and in storage facilities for controlling the sweetpotato weevil (Edmunds et al., 2008).

Damage from a variety of other soil insects that feed on sweetpotato is similar in appearance and often difficult to differentiate at harvest. For this reason, damage incurred by wireworms (*Conoderus spp.*), rootworms (*Diabrotica spp.*), and flea beetles (*Systena spp.*), is often collectively grouped into a complex referred to as the Wireworm-Diabrotica-Systena complex or WDS complex (Cuthbert, 1967).

Cucumber beetles, *Diabrotica spp.*, can be serious pests of sweetpotato (Chalfant et al., 1990). Larvae of the banded cucumber beetle, *Diabrotica balteata* (LeConte), and the spotted cucumber beetle, *Diabrotica undecimpunctata* (Barber), feed on the roots of sweetpotato (Cuthbert, 1967), resulting in unattractive scarring of the root surface (Schalk et al., 1991). Larvae of the two species are difficult to distinguish but the adults are easily recognized. The elytra (outer wings) of the banded cucumber beetle are marked with green and yellow bands and those of the spotted cucumber beetle have 11 spots on a yellow-green background (Chalfant et al., 1990). Numerous generations can develop and damage a sweetpotato crop within one season (Cuthbert, 1967).

Several species of white grubs in the genus *Pyllophaga* are serious pests of sweetpotato in the United States, particularly in Louisiana and Mississippi production areas. The most prevalent species affecting sweetpotato in the southeastern United States is *Phyllophaga ephilida* (Say). White grubs have a generation time of one or two years depending on species (Schalk et al., 1991). The larvae are the damaging stage, chewing wide gouges on the surface of sweetpotato roots (Schalk et al., 1991).

Whitefringed beetles (*Naupactus spp.*) feed on numerous plant species including sweetpotato. Larvae of these insects have a host range exceeding 380 species and damage is similar to that of other soil insects, particularly white grubs (Chalfant et al., 1990; Zehender et al., 1998). Adults of this beetle are flightless and females are parthenogenetic and can lay in excess of 3000 eggs (Young, 1939). The primary means of spread of this insect are by walking and through commercial transport (Chalfant et al., 1990).

Flea beetles, *Systena spp.* are significant insect pests of sweetpotato in southern production areas. Larval damage to sweetpotato roots caused by *Systena spp.* is similar to *Diabrotica* damage (Schalk et al., 1991). Damage from the sweetpotato flea beetle, *Chaetocnema confinis* Crotch to sweetpotato roots is easily recognizable. Larvae of this species make small winding tunnels just under the skin of the sweetpotato. Several generations can occur in one production season, as this insect completes its life cycle in about 30 days during the summer months (Cuthbert, 1967; Schalk et al., 1991).

A complex of wireworm species, including the corn wireworm, *Melanotus communis* (Gyll.), the tobacco wireworm, *Conoderus vespertinus* (F.), and the southern potato wireworm, *Conoderus falli* (Lane) will feed on sweetpotato storage roots (Chalfant and Seal, 1991) and have recently been identified as the primary insects affecting sweetpotato production in North Carolina (Abney, personal communication). In California, the main wireworm species is the sugar beet wireworm (*Limonius californicus* (Mann.). Understanding the biology of the different wireworm species, similar to other soil insects, is paramount in realizing any significant level of control.

Integrated pest management programs for sweetpotato insects have been implemented in the United States and they have the potential to improve sweetpotato insect pest management in the future (Chalfant et al., 1990). A synthetic sex pheromone has been used in managing and monitoring movement of *Cylas formicarius* (Chalfant et al., 1990). This pheromone is also a component of numerous state quarantine programs throughout the southeastern United States (Nilakhe, 1991). Identification of the attractive agents (host plant volatiles, pheromones) of other insects affecting sweetpotato, to sweetpotato fields could also be a beneficial component of sweetpotato integrated pest management.

For many years, breeding programs have selected storage roots with high levels of resistance to root insects as a method to control these pests. To date, most studies on resistance to sweetpotato weevil show little progress (Rolston et al., 1979; Mullen et al., 1985; Story et al., 1996). Recent studies by (Mao et al., 2001) indicate that storage time and production site affect resistance expression. Using cultivars with high resistance to insects could help reduce the reliance on insecticides, but the sweetpotato industry in the United States is hesitant to change from popular cultivars such as Beauregard (Schalk et al., 1991) because good horticultural characteristics, namely, (yield, color, taste, etc) have not yet been incorporated into resistant lines.

Many older insecticides currently used in sweetpotato production are under review by the U. S. Environmental Protection Agency and may not be available in the future (Schalk et al., 1991; Curtis, 2003). Consumer demand for superior quality and attractive appearance increases the need for an integrated management approach to manage sweetpotato insects (Chalfant et al., 1990). Sweetpotato pest management is evolving and continually improving and research projects continue to evaluate integrated pest management techniques (IPM) to further enhance sweetpotato agricultural production systems (Jansson and Ramon, 1991).

A multi-state (USDA-CSREES-RAMP) project conducted from 2002–2007 in North Carolina, Louisiana, Mississippi, and Alabama sought to identify the key insects and diseases affecting sweetpotato in the United States and also to identify potential management options that might reduce the use of pesticides used on the crop. Research results from the participating states identified the key insects affecting different sweetpotato production regions in the United States.

Research conducted in Louisiana and Mississippi production fields documented rootworms (banded and spotted cucumber beetles, *Diabrotica balteata* LeConte and *D. undeciumpuntata* Barber, respectively) as the most damaging soil insect pests of sweetpotatoes in Louisiana (Hammond, personal communication). Scouting and management of cucumber beetles should begin soon after transplant and extend throughout the season. An arbitrary threshold of 2 beetles per 100 sweeps is recommended when scouting for adult cucumber beetles in Louisiana (Smith et al., 2008). Using this threshold to make timely applications of preplant and foliar insecticides will potentially reduce cucumber beetle abundance and percent larval root damage (Smith and Hammond, 2006b, c). In addition, flea beetles, *Systena spp.* and the sweetpotato flea beetle, *Chaetocnema confinis*, Crotch were also identified as significant insect pests in Mississippi production fields. The sugarcane beetle, *Euetheola humilis* Burmeister has also been documented as a pest of sweetpotato in Louisiana and Mississippi. The adult stage of this insect feeds on sweetpotato storage roots, often rendering them unmarketable (Smith, 2006).

Similar studies conducted in North Carolina, identified a complex of wireworm species, including the corn wireworm, *Melanotus communis* (Gyll.), the tobacco wireworm, *Conoderus vespertinus* (F.), and the southern potato wireworm, *Conoderus falli*, Lane as the primary insects affecting sweetpotato in North Carolina. Preplant insecticides and soil barrier insecticide treatments containing chlorpyrifos and/or bifenthrin are recommended control procedures to manage wireworms in North Carolina (Abney, personal communication). Several foliar feeding insect species can also be problematic in sweetpotato production fields, including several lepidopteran species, such as the beet armyworm, *Spodoptera exigua* Hűbner, soybean loopers, *Pseudoplusia includens* Walker, and cabbage loopers, *Trichoplusia ni* Hűbner. These insects are routinely managed with labelled foliar insecticides.

In California, insects and nematodes are managed to a large extent through the use of soil fumigation. Production fields are fumigated with the chemical nematicide 1,3-dichlorpropene (Telone®) in the fall after harvest, or in the spring prior to transplanting. On-farm research has shown significant reductions in wireworm and grub damage when 1,3-D is used as compared to no fumigation (Stoddard, 2003). In areas where fumigation cannot be used, such as in buffer zones near occupied structures, chlorpyrifos (Lorsban®) and ethoprop (Mocap®) may be soil incorporated prior to planting. In season insect management may require treatment for beet armyworm, *Spodoptera exigua* Hűbner and Western Yellowstripe Armyworm, *Spodoptera praefica*, Grote, of which several insecticides are registered for use. Organic production fields make extensive use of newer, root knot nematode resistant varieties in lieu of fumigants or chemicals and biological insecticides such as *Bacillus thuringiensis* to control armyworms if required.

Diseases

Root diseases were a large problem and limited sweetpotato production in the United States to a great extent in the early 20th century. Fortunately, many of the diseases that were once significant problems have been successfully managed and in general do not cause large-scale problems for the industry. Detailed descriptions of most sweetpotato diseases are presented in the Compendium of Sweet Potato Diseases by Clark and Moyer published in 1988 and in the present book. Fusarium wilt (*Fusarium oxysporum f. spp batatas and nicotianae.)* was a limiting factor until resistant varieties were first developed in the 1950's. Since that time, however, almost all new varieties have been bred to be resistant. Resistance has also been used to successfully manage Streptomyces soil rot (pox) *Streptomyces ipomoeae* which was a limiting factor until the 1980's.

Crop rotation, disease-free mother roots, fungicides, and, cutting slips at least 2–3 cm above the soil line has significantly reduced the incidence of diseases such as black rot, *Ceratocystis fimbriata*, foot rot, *Plenodomus destruens* Harter, scurf, *Monilochaetes infuscans* Ell. & Halst. Ex Harter, charcoal rot, *Macrophomina phaseolina* (Tassi) Goid, and others that can be transmitted on seed roots. Although there is improved resistance, postharvest diseases, particularly Rhizopus soft rot, *Rhizopus spp.* can cause problems and the industry is trying to find ways to control such diseases without relying exclusively on prophylactic use of fungicides. Most states now have foundation seed programs that produce virus-tested seed. However, re-infection with viruses can occur rapidly and there is still considerable research needed to identify the viruses that reduce yield and quality and to develop means for reducing re-infection by these viruses.

The other remaining disease challenge for sweetpotato production in the U.S. is that posed by nematodes. In some production areas, root-knot nematode (*Meloidogyne incognita* Kofoid & White (Chitwood)) remains the primary nematode of concern and can cause significant reductions in yield and quality when populations are high. Control can be achieved by selecting newer, root knot nematode resistant varieties, soil fumigation, and field rotation and fallowing. The reniform nematode (*Rotylenchulus reniformis* Linford and Olveira) has also become a predominant problematic nematode in several sweetpotato-producing regions. There are no resistant varieties available for this nematode and producers must rely on fumigants or other nematicides to manage this nematode in commercial production fields (Clark and Moyer, 1988; Chris Clark, personal communication).

Weeds

The primary method of weed control in early sweetpotato production was mechanical cultivation, supplemented by hand-weeding (Welker, 1967). Currently, mechanical cultivation remains an important weed control tool with a national average of nearly three cultivations per season (Haley and Curtis, 2006). Hand-weeding is still prevalent across most of the United States to remove weed pests that have escaped other control methods, but it is not preferred due to the expense, time and labor needs associated with this task.

Rotary mowers or wick-bar wiper applicators can be used to manage weeds once a height differential occurs above the sweetpotato crop canopy. All of the weed control methods mentioned are labor and time intensive and add to the total cost of production, but all are necessary for use in weed management systems because there are few effective chemical control options available for use in sweetpotato in the United States.

Early research studies evaluating the efficacy of chemical herbicides for sweet-potatoes lead to federally approved herbicide labels for use of chloramben, diphenamide, and DCPA in sweetpotato production. Registration of napropamide (Devrinol®) and EPTC (Eptam®) followed later. EPTC is still currently used in some fields with high populations of nutsedge (*Cyperus* spp.) pressure. Research indicates that EPTC controls yellow (*Cyperus esculentus* L.) and purple (*Cyperus rotundus* L.) nutsedges up to 93% (Monks et al., 1998). However, it is not typically considered as a main component of most weed control systems because it must be applied as a pre-transplant treatment and incorporated immediately. Generally, most of these herbicides discovered before the 1980's are either no longer registered for use in sweetpotato or have to be applied at such high rates that they are not recommended.

Numerous herbicides continued to be evaluated after 1980 for use on sweetpotato with variable results on control, residual activity, and yield effects. Reports from these greenhouse and field studies indicated excellent weed control, but low to moderate crop tolerance that resulted in reduced root yield to all cultivars.

Porter (1990) conducted research to evaluate clomazone (Command®) applied after transplant for control of annual grasses and some broadleaf weeds. Research results that documented appropriate use rates and application times developed during this time-period lead to the registration of clomazone for use in sweetpotato. A survey conducted in 2005 indicated that at least 82% of all growers in Alabama, Louisiana, Mississippi, North Carolina, and Tennessee included clomazone in their weed control system (Haley and Curtis, 2006). Since the introduction of clomazone, populations of morningglory (*Ipomea* spp.), common cocklebur (*Xanthium strumarium*), and annual grasses have been reduced, but other weeds that clomazone does not control continue as pests due to ineffective or unregistered chemical control measures (Porter, 1990; Kelly et al., 2006).

The common weed species across major sweetpotato production states that need further investigation for development of effective control systems include nutsedges, pigweeds (*Amaranthus* spp.), hophornbeam copperleaf (*Acalypha ostryifolia*, Riddell), sicklepod (*Cassia obtusifolia* L.), hemp sesbania (*Sesbania exaltata*), groundcherry (*Physalis* spp.), nightshade (*Solanum* spp.), and some annual grasses (Curtis, 2003). Additional weed species that require a review of control measures include morningglory, common lambsquarters (*Chenopodium album* L.), common cocklebur, southern sandbur (*Cenchrus echinatus* L.), common ragweed (*Ambrosia artemisiifolia* L.), prickly sida (*Sida spinosa* L), and some perennial grasses (Curtis, 2003). A specific problematic weed in sweetpotato production for Louisiana is smellmelon (*Cucumis melo* L.) (Kelly et al., 2006), in Mississippi a potential weed pest that could become a problem is coast cockspur [*Echinochloa walteri* (Pursh) A. Heller], and in North Carolina new control strategies for glyphosate resistant Palmer amaranth (*Amaranthus palmeri* S. Wats.) are needed.

Recently, several herbicides have been evaluated for control of weed species that are common to most sweetpotato producing regions in the United States. Research evaluated the use of metolachlor (Dual Magnum®), dimethenamid (Outlook®), and flumioxazin (Valor®). Both metolachlor and dimethenamid have a similar mode of action in plants and control several broadleaf weeds, annual grasses, and suppression of nutsedges. Research has shown no significant phytotoxcity and excellent weed control, but root yield tended to be lower compared to other treatments. The tendency for lower root yield was probably due to an excessive use rate (Glaze and Hall, 1990).

Flumioxazin was introduced in the U.S. for use in peanuts and soybean in 2001. Research conducted in sweetpotato indicated that flumioxazin could be used in sweetpotato as a pre-transplant application (Kelly et al., 2003; Kelly et al., 2004; Main et al., 2004; Kelly and Shankle, 2005). In 2003, the U.S. Environmental Protection Agency (EPA) granted several states a Section 18 Emergency Exemption for the use of both metolachlor and flumioxazin in sweetpotato production. Metolachlor and flumioxazin remain viable weed control options for sweetpotato producers in some regions of the United States.

Currently, there are no registered herbicides in sweetpotato that provide postemergence control of nutsedge. Substantial research has shown that nutsedge can be controlled with a postmergence application of halosulfuron after the crop has been established without significant adverse effects to foliage or root yield and quality if applied at specific growth stages (Garrett et al., 2004; Shankle et al., 2005; MacRae et al., 2007; Monks et al., 1997). However, sweetpotato root quality deteriorates if applications are made at improper stages of growth.

Perennial and annual grasses can be controlled with postemergence grass herbicides such as fluazifop (Fusilade®), sethoxydim (Poast®), and clethodim (Select®). Research reports that U.S. No. 1 grade and total marketable yields were greater when a postemergent application of these herbicides were applied to control annual grasses at 4 weeks after transplant (Porter, 1993). All three of these herbicides are currently labeled and used in weed control systems by U.S. growers.

All the chemical control measures labeled for use in sweetpotatoes are species specific; therefore, growers should base their herbicide system relative to the weed species composition for certain fields. Additionally, not all chemical herbicides are registered for use in all areas where sweetpotatoes are grown, and their use may be dictated by state-specific regulations. Generally most growers will include pre-plant herbicides for early season control, cultivation and a mid-season grass herbicide, followed by mowing, handweeding, and/or wick-bar control methods mid- to late season. In California, directed sprays using glyphosate (Roundup) with shielded sprayers 4–6 weeks after transplanting has been shown to provide effective control of most troublesome weeds. This type of systems approach for weed control continues to be an important research strategy to help growers make efficient, effective, and environmentally safe weed control decisions (Main et al., 2004; Garrett et al., 2005; Garrett et al., 2007; Shankle et al., 2007).

Harvesting, Curing, Storage, Marketing

A high yielding, quality sweetpotato crop is the product of sound production practices, good pest management and careful handling (Schultheis et al., 2008). Successful field production is only the first step in achieving a profitable sweetpotato production operation. Edmond (1971) extensively reviewed early research concerning time to harvest (days after planting), harvesting methods and curing and storing of sweetpotatoes. In the United States, sweetpotatoes are harvested 110–150 days after planting; however, the days to harvest can vary considerably depending on the variety, plant spacing and weather conditions. Harvest usually begins in late July or early August and extends until November or later if necessary (Plate 14.10).

The sweetpotato crop is harvested either by hand or mechanically, using sweetpotato chain harvesters or diggers. Initial field preparation is similar for either method. If irrigated, the field is allowed to dry for 2–4 weeks before harvesting, then the foliage is removed by shredders and deviners. Numerous kinds of specialized equipment or equipment combinations may be utilized at this point, the goal being to effectively remove leaves and vines so that the harvest can proceed more efficiently.

Mechanical harvesters vary in size from 1 to 4-row equipment. The most commonly used chain harvester in Louisiana and Mississippi is the 2-row chain harvester (Plate 14.11). One 2- row harvester can harvest approximately two acres (0.80 ha) per day and requires 10–12 people to operate efficiently. The sweetpotatoes are field graded into different grade classes, including U. S. No. 1's, commercial grade, Jumbos, and canners (Table 14.3). One-row equipment can harvest approximately one acre per day and requires 5–6 people working on each harvester. One-row equipment is used extensively in California production areas. Two and 4-row bulk harvesters are also used in some areas, particularly by those producers with commercial production of 500 acres or greater. One 2-row bulk harvester can harvest close to 5

Plate 14.11 Sweetpotato harvest (2-row mechanical chain harvester) (Courtesy of Tara Smith, Louisiana State University Agricultural Center) (See also Plate 24 on page xxvi)

Table 14.3 United States Department of Agriculture sweetpotato grade standards

Grade	Weight (ounces)	Length (inches)	Diameter (inches)
U. S. Extra No. 1	≤ 18	3–9	1.75–3.25
U. S. No. 1	≤ 20	3–9	1.75–3.50
U. S. Commercial*	≤ 20	3–9	1.75–3.50
U. S. No. 2	≤ 36	3–9	1.50–3.50

*Same grade size as U. S. No. 1, with an increase tolerance for defects.
Source: http://www.ams.usda.gov/standards/sweetpot.pdf

Plate 14.12 Wooden storage bins used to harvest and store sweetpotatoes (Courtesy of Scott Stoddard, University of California at Merced) (See also Plate 25 on page xxvi)

acres (2 ha) per day and requires 11–12 people, while a 4-row bulk harvester can cover 10–12 acres (4–5 ha) per day and requires 20–22 individuals.

As the storage roots are harvested, they are graded into 20-bushel wooden or plastic bins (~ 1000 pounds; 454 kg) (Plate 14.12). The bins are moved into storage facilities, where the sweetpotatoes will undergo the curing process and will then be stored for subsequent packing and shipment. The harvested roots are quickly moved into storage facilities to prevent possible weather related damage, such as sunscald or chilling injury (Edmunds et al., 2008).

Sweetpotato producers in North Carolina prefer to harvest by hand. After the foliage is removed from the fields by shredders or deviners, the sweetpotatoes are turned up and exposed using a turning plow or shaker harvester (Plate 14.13).

Plate 14.13 Sweetpotato harvest (shaker harvester method) (Courtesy of Tara Smith, Louisiana State University Agricultural Center) (See also Plate 26 on page xxvii)

Laborers then sift through the soil, collecting the roots and sorting them into various grades. Under ideal weather and soil conditions, several acres can be harvested per day using this method.

Proper handling of sweetpotatoes begins before harvest and ends when the roots are consumed (Edmunds et al., 2008). Harvesting sweetpotatoes is a laborious task and handling and storage conditions after harvest must be addressed to safeguard against loss, such as that caused by breakdown of storage roots due to various root rots, mechanical injury, or less than ideal storage conditions. Significant losses can occur during curing, storing, and shipping. Rough handling during harvest can contribute significantly to postharvest losses by negatively affecting the cosmetic appearance of the roots or providing wounds, which can serve as a site for disease to enter the roots (Edmunds et al., 2008).

Growing season weather conditions can impact the quality of stored sweetpotatoes. Extremely dry and wet field conditions can be problematic. Drought or dry soil conditions followed by periods of heavy rain or irrigation can result in rapid root expansion and subsequent growth cracks. Growth cracks and other wounds can serve as infection sites for fungal and bacterial pathogens both before and after harvest. Saturated soil conditions caused by heavy rainfall can lead to carbon dioxide accumulation, and ultimately result in asphyxiation of roots (Edmunds et al., 2008). Sweetpotatoes have a thin skin when harvested that can easily be damaged by cuts and abrasions.

Curing

There are several positive benefits associated with the curing process of sweetpotatoes. Curing enhances the culinary or eating quality of the roots, it aids in wound healing, which reduces loss due to shrinkage and disease. Curing also sets the skin of the sweetpotato, reducing the potential for skinning during packing. Many commercial sweetpotato growers and packers in the United States have invested in curing facilities, and view proper curing as an indispensable first step in a process that allows the industry to provide a year-round supply of quality sweetpotatoes (Edmunds et al., 2008). Curing rooms are less common in California than other production areas in the United States. Instead, the crop is allowed to "field cure", where fields are not irrigated in the 2–4 weeks prior to harvest to allow the skin to set.

The curing process involves holding roots at a temperature of 85 °F/29 °C and a relative humidity of 85 to 90 percent with proper ventilation for a three- to seven-day period immediately after harvest. The duration of the curing process will vary depending on the temperature of the roots at harvest. The greater the difference between storage root temperature and 85 °F/29 °C, the longer it will take to cure. A delay of as little as 12 hours between harvest and curing has been shown to be detrimental to successful curing (Edmunds et al., 2008). The humidity during curing should be high (85 to 90 percent) but not to a level where condensation on storage walls, floors, bins, or sweetpotatoes occurs (Edmunds et al., 2008).

In addition, proper ventilation is essential for an effective curing process. Ventilation provides needed exchange of carbon dioxide and oxygen, and is necessary for heat transfer during the curing process. One-half cubic foot of outside air per bushel per day is recommended, however, injured sweetpotatoes, or those exposed to adverse weather and soil conditions prior to harvest, may require as much as 5 to 10 cubic feet of outside air per bushel per day (Edmunds et al., 2008).

Storage

Climate controlled storage facilities (Plate 14.14) are a necessity in modern sweetpotato production operations in the United States. Long term, climate controlled storage affords producers and brokers leverage in moving and marketing the crop. Sweetpotato producers with proper storage facilities can selectively move and sell their crop up to a year after harvest. Storage facilities should be designed to maintain the temperature and humidity requirements needed for long-term storage.

The primary goal of controlled storage is to maintain the quality of harvested roots by minimizing potential problems associated with physiological disorders and disease (Edmunds et al., 2008). Sweetpotato roots that are properly cured should be stored at 55 °F/13 °C and 85–90% relative humidity and with adequate ventilation. Storage roots, which are free from disease or other physiological problems will store for as long as 13 months and remain marketable under these conditions. These storage conditions were first determined in the 1920s with cultivars grown at that time and recent research has shown that these conditions are still valid for modern commercial cultivars, such as Beauregard and Covington (Edmunds et al., 2008).

Plate 14.14 Typical sweetpotato storage facility for a large commercial operation (Courtesy of Scott Stoddard, University of California at Merced) (See also Plate 27 on page xxvii)

Marketing

Sweetpotatoes can be sold during the harvest season or stored and marketed for up to a year under ideal storage conditions. Sweetpotatoes sold soon after harvest are sometimes referred to as "green" and may not be as sweet as those that have been cured. Uncured sweetpotatoes generally lack the visual appeal, extended shelf life, and superior eating quality of cured roots (Edmunds et al., 2008).

Sweetpotatoes that have undergone the initial curing process and have been stored in climate-controlled facilities for 6–8 weeks prior to being marketed are referred to as "quick cured." Harvested sweetpotatoes that have been held in common storage for 6–8 weeks are considered to be cured or "kiln dried." Both quick cured potatoes and "kiln dried" potatoes are sweeter and moister when baked compared to "green potatoes" (Boudreaux et al., 2005). The majority of sweetpotatoes that are sold to meet the increased market demand for sweetpotatoes associated with the Thanksgiving and Christmas holidays each fall from September–December are considered to be cured sweetpotatoes.

The industry adopted a standard 40-lb (18 kg) corrugated fiberboard box (Plates 14.15 and 14.16) as the standard shipping container in the late 1980's (Boudreaux et al., 2005). Several alternatives to the standard 40-lb carton exist today for fresh market sweetpotatoes, such as 10-lb (4.5 kg) specialty pack boxes, plastic bags, shrink-wrapped sweetpotatoes, and net bags. In addition bulk cardboard bins (Plate 14.17) are used to ship sweetpotatoes to processing facilities (Edmunds et al., 2008). These bins are larger and weigh more (~ 1800 lbs; 820 kg) than the wooden or plastic bins used to store and cure sweetpotatoes sold for the fresh market.

United States grade standards for sweetpotato are established and enforced by the United States Department of Agriculture (USDA, 2005). The primary marketable grades include: U. S. Extra No. 1, U. S. No. 1, U.S. No. 1 Petite, U.S. Commercial

Plate 14.15 Marketing containers (Courtesy of Tara Smith, Louisiana State University Agricultural Center and Scott Stoddard, University of California at Merced) (See also Plate 28 on page xxvii)

Plate 14.16 Marketing containers (Courtesy of Tara Smith, Louisiana State University Agricultural Center and Scott Stoddard, University of California at Merced) (See also Plate 29 on page xxviii)

Plate 14.17 Marketing containers (Courtesy of Tara Smith, Louisiana State University Agricultural Center and Scott Stoddard, University of California at Merced) (See also Plate 30 on page xxviii)

Grade, and U. S. No. 2. The U. S. No. 1 is the premium yield grade of sweetpotatoes in the United States and traditionally brings the highest price in the marketplace. The size requirements for the various grades differ considerably, on the basis of length, weight and diameter of the storage root (Table 14.3). Specifications for a U. S. No. 1 are described below and all other yield grades are described in relation to the U. S. No. 1 grade standards.

> *"U. S. No. 1 consists of sweetpotatoes of one type which are firm, fairly smooth, fairly clean, fairly well shaped, which are free from injury, internal breakdown, Black rot, other decay or wet breakdown, and free from damage caused by secondary rootlets, sprouts, cuts, bruises, scars, growth cracks, scurf, Pox (soil rot), or other diseases, wireworms, weevils, or other insects, or other means (USDA, 2005)".*

Marketing of the sweetpotato crop is handled differently in each state. The current trend in the industry is for producers to grow, pack and ship the entirety of their sweetpotato crop. Many producers choose to utilize the services of a broker in selling their crop. The broker identifies potential buyers, arranges shipments and determines the price received for a particular shipment. Sweetpotatoes grown in California are marketed and sold predominantly on the West Coast of U.S. and Canada for fresh consumption. For various ethnic and cultural reasons, the diversity of varieties used to satisfy the market is greater than in other areas of the U.S. The market for sweetpotatoes can be split into four major groups, for which various varieties are used: 1) yams, represented by varieties with copper skin and orange flesh (Beauregard, Covington); 2) red yams, with dark red skin and deep orange flesh (Diane); 3) sweetpotatoes or "Jersey Sweets", with tan skin and cream colored flesh (Golden Sweet, O'Henry); and 4) Japanese or Oriental yams, which is exclusively Koto Buki, a purple skin/white flesh sweetpotato. Because of price premiums for both #1's and mediums, the acreage planted to Koto Buki has increased significantly in recent years. While no data are available that show the percentages of each class, in general Beauregard represents 45% of the acreage grown in the state, Diane about 25%, Golden Sweet about 10%, and Koto Buki the remainder (about 20%) (C.S. Stoddard and D.R. LaBonte, 2007).

All of the major sweetpotato producing states have active grower organizations and advertising commissions, which serve to promote, advertise and support the sweetpotato industry of the United States. Individual state organizations also are members of the United States Sweetpotato Council, which is responsible for national promotion and advertising activities.

Uses of Roots and Foliage

Roots

Sweetpotato roots produced in the United States are primarily used for human consumption. Approximately 60% of the crop is sold and utilized in fresh markets, with the remainder being used for processed products such as sweetpotato chips, fries, and canned sweetpotatoes. Fresh sweetpotatoes are often marketed to large distribution companies who then move the product into grocery chains, restaurants, farmers markets and various other retail outlets. Approximately 40% of the crop harvested each fall is marketed and sold during the months of October, November and December, which corresponds with the Thanksgiving and Christmas holidays in the United States (USDA AMS, 2008). Another key marketing opportunity occurs each year in the spring, usually in March or April, coinciding with the Easter Holiday.

The amount and variety of processed sweetpotatoes has shown a significant market increase in the first part of the 21st century. The sweetpotato processing market is very important in the United States and buys substantial amounts of "off grade" and jumbo roots. Traditionally, the majority of sweetpotatoes that were sold for processing were sold to canning companies and were marketed by several companies under a variety of different labels. The canning market is still prevalent, but other processing opportunities for the crop have become increasingly popular among American consumers. Some of the different products available include: sweetpotato fries, chips, and purees. Sweetpotatoes are also being used as a key ingredient in some prepackaged juices, cookies and other flour based products, and there is some utilization as an ingredient in various pet food formulations.

Research is ongoing to determine the suitability of sweetpotato roots in the United States for use as an industrial product, such as ethanol or industrial starch. Presently, production costs are too great to realize a significant profit from these industrial uses. If high yielding, high dry matter varieties combined with new production schemes that will reduce input costs are developed, sweetpotato roots may be utilized for industrial purposes in the United States.

Foliage

Though perfectly edible, sweetpotato leaves and stems are rarely consumed in the United States. Sweetpotato greens may be available in local niche markets or in health food outlets. Most sweetpotato leaves are not utilized for any purpose and are destroyed when the storage roots are harvested each year.

Ornamentals

Ornamental sweetpotatoes, *Ipomoea batatas*, which are cultivated in flowerbeds and containers for their aesthetic quality, are grown extensively across the United States during the summer months. Ornamental sweetpotatoes are related to morning glories, but unlike their conventional counterpart, are grown strictly for their attractive foliage (Gill and Merril, 2007). Several varieties are available that offer consumers a choice of leaf attributes and growth habit. The most popular varieties are Marguerite and Blackie. Marguerite has a lime green, heart shaped leaf, while Blackie has a dark purple and deeply lobed leaf. Other varieties available include Lady Fingers, Pink Frost and Sweet Carolina Bronze.

References

Baker, J. M. 1947. Consumer preferences in sweetpotatoes. *La. Agr. Expt. Sta. Bull.* 409.

Boudreaux, J.E. 2005. Commercial Vegetable Production Recommendations. *Louisiana State University Agricultural Center,* Pub. No. 2433.

Boudreaux, J. E., Cannon, J. M., and Villordon, A. 2005. Sweetpotato, pp. 54–56. In J. E. Boudreaux (ed.) *Commercial Vegetable Production Recommendations*. LSU AgCenter Publication, No. 2433, www.lsuagcenter.com.

Bouwkamp, J. C. 1985. Introduction-part 1, pp. 3–7. In J. C. Bouwkamp (ed.) *Sweetpotato products: A Natural Resource for the Tropics*. CRC Press, Boca Raton, Florida.

Bouwkamp, J. C. 1985. Production Requirements, pp. 1–33. In J. C. Bouwkamp (ed.) *Sweetpotato Products: A Natural Resource for the Tropics*. CRC Press, Boca Raton, FL.

Bowers J. L., Harmon, S. A., and Dempsey, A. H. 1970. History of Group and Early Research, pp. 1–17. In T. P. Hernandez (ed.). *Thirty Years of Cooperative Sweetpotato Research*, 1939–1969. Baton Rouge, LA.

Chalfant, R. B. and Seal, D. R. 1991. Biology and Management of Wireworms on Sweet Potato, pp. 303–326. In Jansson & Raman (eds.) *Sweet Potato Pest Management: A Global Perspective*. Westview Press Inc., Boulder, CO.

Clark, C. A., and Moyer, J. W. 1988. *Compendium of Sweetpotato Diseases*. APS Press, St. Paul, MN.

Cuthbert, F. P. Jr. 1967. Insects affecting sweetpotatoes. *USDA Agricultural Handbook*. 329. 28 pp. Washington, D.C. *(USDA) Agricultural Research Service in Cooperation with Clemson University Agricultural Experiment Station.*

Chalfant, R. B., Jansson, R. K., Seal, D. R., and Schalk, J. M. 1990. Ecology and management of sweetpotato insects. *Annu. Rev. Entomol.* 35: 157–180.

Curtis, J. 2003. Strategic Plan for Pest Management Research and Education in Southern *Sweetpotato Production Systems*: Summary of workshops held October2002 and January, February, and August 2002. Sponsored by North Carolina State University Dept. of Horticulture, with funding from the U.S. Environmental Protection Agency, Region 4.

Dangler, J.M., R.W. Scheuerman, R.M. Campbell, and C.A. Clark. 1994. Meristem-tip culture and California's clean foundation sweetpotato program. *HortTech.* 4(3): 227–228.

Edmond, J. B. 1971a. Taxonomy, History and Industry. pp. 1–17. In J. B. Edmond and G. R. Ammerman (eds.) *Sweetpotatoes: Production, Processing and Marketing*. Avi Publishing co. Inc., Westport, CT.

Edmond, J. B. 1971b. Soils, cropping systems and commercial fertilizers. pp. 106–122. In J. B. Edmond and G. R. Ammerman (eds.) *Sweetpotatoes: Production, Processing and Marketing*. Avi Publishing co. Inc., Westport, CT.

Edmond, J. B. 1971c. Harvesting, Curing and Storing. pp. 208–246. In J. B. Edmond and G. R. Ammerman (eds.)., *Sweetpotatoes: Production, Processing and Marketing*. Avi Publishing co. Inc., Westport, CT.

Edmunds, B. A., Boyette, M., Clark, C. A., Ferrin, D., Smith, T. P., and Holmes, G. 2008. Postharvest Handling of Sweetpotatoes. *North Carolina Cooperative Extension Service*, Raleigh, NC.

Garrett, T. F., Shankle, M. W., Main, J. L. 2007. Dual magnum herbicide systems in sweetpotato. *Proc. South. Weed Sci. Soc.* 60: 139

Garrett, T. F., Shankle, M. W., Main, J. L., and Kelly, S. T. 2005. Weed control in Mississippi sweetpotato using command and valor. *Proc. South. Weed Sci. Soc.* 58: 157.

Garrett, T. F., Shankle, M. W., Kelly, S. T., and Main, J. L. 2004. Sandea rate and timing in sweetpotato. *Proc. South. Weed Sci. Soc.* 57: 136.

Gill, D., and Merril, T. 2007. *Get it Growing: Ornamental sweetpotatoes make delightful addition to summer flower garden*. Accessed February 20, 2008. *Louisiana State University Agricultural Center*, http://www.lsuagcenter.com/news_archive

Glaze, N. C., and Hall, M. R. 1990. Cultivation and herbicides for weed control in sweetpotato (*Ipomoea batatas*). *Weed Technol.* 4: 518–523.

Gray, L. C. 1933. History of Agriculture in the Southern United States to 1860. Vol. II. *Carnegie Inst. Wash. Publ.* No. 430.

Haley, J., and Curtis, J. 2006. Southern sweetpotato grower survey: report of results (2005). *NCSU, LSU, MSU and Auburn Universities*.

Hartman, J. O., and Gaylor, F. C. 1939. Soil acidity for muskmelons and sweetpotatoes on sand soil. *Proc. Am. Soc. Hortic. Sci.* 37: 847.

Jansson R. K., and Ramon, K. V. 1991. Sweetpotato pest management: A global overview. pp. 2–12, In Jansson & Raman (eds.) *Sweetpotato Pest Management: A Global Perspective*. Westview Press Inc., Boulder, CO.

Kays, S. J. 1985. The physiology of yield in the sweetpotato, pp. 80–131. In J. C. Bouwkamp (ed.) *Sweetpotato Products: A Natural Resource of the Tropics*. CRC Press, Boca Raton, FL.

Kelly, S. T., Shankle, M. W., and Miller, D. K. 2006. Efficacy and tolerance of flumioxazin on sweetpotato (*Ipomoea batatas*). *Weed Technol.* 20: 334–339.

Kelly, S. T., and Shankle, M. W. 2005. Symposia: New & potential h herbicides for vegetables & fruit crops. Development of valor in sweetpotato. *Proc. South. Weed Sci. Soc.* 58: 170.

Kelly, S. T., Cannon, J. M., Shankle, M. W. 2004. Efficacy of valor in sweetpotato. *Proc. South. Weed Sci. Soc.* 57:153.

Kelly, S. T., Carroll, H. L., Cannon, J. M., and Shankle, M. W. 2003. Sweetpotato tolerance and weed control with flumioxazin. *Proc. South. Weed Sci. Soc.* 56: 112.

LaBonte, D. R., Mulkey, W. A., Clark, C. A., Rolston, L. H., Cannon, J. M., Wilson, P. W., and St. Amand, P. C. 1992. 'Hernandez' Sweetpotato. *HortScience*. 27(4): 377.

LaBonte, D. R., Wilson, P. W., Villordon, A. Q., and Clark, C. A. 2008. 'Evangeline' Sweetpotato. *HortScience*. 43: 258–259.

Main, J. L., Shankle, M. W., and Garrett, T. F. 2004. Use of valor and command in Mississippi delta sweetpotato. *Proc. South. Weed Sci. Soc.* 57: 139.

MacRae, A. W., Monks, D. W., Batts, R. B., Thorton, A. C., and Schultheis, J. R. 2007. Sweetpotato tolerance to halosulfuron applied postemergence. *Weed Technol.* 21: 993–996.

Mao, L., Story, R. N., Hammond, A. M., and Labonte, D. R. 2001. Effect of sweet potato genotype, storage time and production site on feeding and oviposition behavior of the sweetpotato weevil, *Cylas formicarius* (Coleoptera: Apionidae). *Fla. Entomol.* 84: 259–264.

Mason, L. J., Seal, D. R., and Jansson, R. K. 1991. Response of sweetpotato weevil (Coleoptera: Apionidae) to selected insecticides. *Fla. Entomol.* 74: 350–355.

Miller, J. C. 1958. Sweetpotato Breeding and Yield Studies. *La Agric. Exp. Stn. Circ.* 41.

MacGillivray, J. H. 1953. *Vegetable Production with Special Reference to Western Crops*. Blakiston Co., New York.

Monks, D. W., Mitchem, W. E., Mills, R. J., and Greeson, C. V. 1998. Response of nutsedge and sweetpotato to EPTC and metolachlor. *Proc. South. Weed Sci. Soc.* 51: 91.

Monks, D. W., Mitchem, W. E., and Garvey, P. W. 1997. Nutsedge management in snap bean and sweetpotato. *Proc. South. Weed Sci. Soc.* 50: 79.

Mullen, M. A., Jones, A., Paterson, D. R., and Boswell, T. E. 1985. Resistance in sweet potatoes to the sweetpotato weevil, *Cylas formicarius elegantulus* (Summers). *J. Entomol. Sci.* 20: 345–350.

Nilakhe, S. S. 1991. Quarantine programs for *Cylas formicarius* in the United States, pp. 247–262, In Jansson & Raman (eds.), Sweet potato Pest Management: a global perspective. Westview, Boulder, CO.

Porter, W. C. 1993. Postemergence grass control in sweetpotatoes (*Ipomoea batatas*). *Weed Technol.* 7: 812–815.

Porter, W. C. 1990. Clomazone for weed control in sweetpotatoes (*Ipomoea batatas*). *Weed Techno.* 4: 648–651.

Rolston, L. H., Barlow, T., Hernandez, T., and Nilakhe, S. S. 1979. Field evaluation of breeding lines and cultivars of sweet potato for resistance to the sweetpotato weevil. *HortScience* 14: 634–635.

Rolston, L. H., Clark, C. A., Cannon, J. M., Randle, W. M., Riley, E. G., Wilson, P. W., and Robbins, M. L. 1987. 'Beauregard' Sweetpotato. *HortScience*. 22: 1338–1339.

Royston, R., and Oakely, M. F. 1958. Sweetpotato estimated in hundredweights by states. *U. S. Dept of Agr. Statist. Bull.* 237.

Schalk, J. M., Jones, A., Dukes, P. D., and Peterson, J. K. 1991. Approaches to the control of multiple insect problems in sweetpotato in the southern United States, pp. 283–301. In Jansson & Raman (eds.) *Sweetpotato Pest Management: A Global Perspective*. Westview Press Inc., Boulder, CO.

Shankle, M. W., Main, J. L., and Garret, T. F. 2007. Effects of cultivation on valor efficacy in sweetpotato. *Proc. South. Weed Sci. Soc.* 60: 145.

Shankle, M. W., Kelly, S. W., Garrett, T. F., and Main, J. L. 2005. Effect of halosulfuron-methyl rate and application time on sweetpotato. *Proc. South. Weed Sci. Soc.* 58: 156.

Schultheis, J.R., and L.G. Wilson. 1993. *What is the difference between a sweetpotato and a yam? Department of Hort Science, North Carolina State University Cooperative Extension.* http://www.ces.ncsu.edu/depts/hort/hil/hil-23-a.html.

Schultheis, J. R., Boudreaux, J. E., Shankle, M. W., Smith, Tara P., et al. 2008. In G. J. Holmes and J. M. Kemble (eds.) *Vegetable Crop Handbook for the Southeastern US* (Sweetpotato Section). North Carolina Vegetable Growers Association, North Carolina.

Smith, Tara P. 2008. Sweetpotato Crop Update. Louisiana Sweetpotato News. *La. State Univ. Agric. Center*. 3(1). http://www.lsuagcenter.com/en/communications/publications/newsletters/sweet_potato_news/Sweet+Potato+News.htm

Smith, Tara P., and Miller, D. K. 2007. Weed Management in Sweetpotato. *Louisiana State University Agricultural Center*, Pub. No. 3070.

Smith, Tara P., Hammond, A. M., Story, R. N., and Burris, E. 2008. Insect Pest Management in Louisiana Sweet Potatoes. *Louisiana State University Agricultural Center*, Pub. No. 2620.

Smith, Tara P., and Hammond, A. M. 2006a. Comparative susceptibility of sweetpotato weevil (Coleoptera: Brentidae) to selected insecticides. *J. Econ. Entomol.* 99: 2024–2029.

Smith, Tara P. 2006. Biology and chemical ecology of the sugarcane Beetle and integrated pest management of sweetpotato soil insects in Louisiana. *Louisiana State University Dissertation*. pp. 110.

Smith, Tara. P., and Hammond, A. M. 2006b. Evaluation of foliar insecticides for control of cucumber beetle in sweetpotato, 2005. *Arthropod Manage. Tests*. 31, http://www.entsoc.org/Protected/AMT/AMT31/resultsgood.asp

Smith, Tara. P., and Hammond, A. M. 2006c. Evaluation of granular soil insecticides for control of cucumber beetle in sweetpotato, 2004. *Arthropod Management Tests*. 31, http://www.entsoc.org/Protected/AMT/AMT31/resultsgood.asp

Stoddard, C. S., and LaBonte, D. R. 2007. *Commercial sweetpotato varieties in California and potential fit of L-01-29*. Abstract in National Sweetpotato Group progress report 2006. http://www.sweetpotatoes.org.
Stoddard, C. S., Klondky, K. M., and DeMoura, R. A. 2006. Sample costs to produce sweetpotatoes, transplant and field production in the San Joaquin Valley. *UC Davis Dept of Ag and Resource Economics*; http://coststudies.ucdavis.edu.
Stoddard, C. S. 2003. Sweetpotato Research Progress Report. *UC Cooperative Extension*, Merced, CA 95341, pp. 7–9, http://cemerced.ucdavis.edu County.
Story, R. N., Hammond, A., Murray, J., Rolston, R. H., and Labonte, D. 1996. Selection for host plant resistance in sweetpotatoes to the sweetpotato weevil, 1995, pp. 71–79. In *Sweet potato Research 1996*. LAES Mimeo series No. 117.
Talekar, N. S. 1992. Insect factors in breeding and cultivation of sweetpotato, pp. 143–153. In W. A. Hill, C. K. Bonsi and P. A. Loretan (eds.) *Sweetpotato technology for the 21st Century*. Tuskegee, AL.
Uritaini, I., Saito, T., Honda, H. and Kim, W. K. 1975. Induction of furanoterpenoids in sweetpotato roots by the larval components of the sweetpotato weevils. *Agric. Biol. Chem.* 37: 1857–1862.
(USDA) U. S. Department of Agriculture. 2001. Crop Profile for Sweet Potatoes in Louisiana, Office of Pest Management Policy & Pesticide Impact Assessment Program, February. http://pestdata.ncsu.edu/cropprofiles/docs/Lasweet potatoes.html
(USDA) U. S. Department of Agriculture. 2008. *USDA-NASS, National Agricultural Statistics Service: Vegetables*. http://www.nass.usda.gov.
(USDA) U. S. Department of Agriculture, Economic Research Service. 2007a. United States Standards for Grades of Sweetpotatoes. http://www.ers.usda.gov/Publications/vgs/Tables/Swpot.pdf
(USDA) U. S. Department of Agriculture, Economic Research Service. 2007b. *Vegetable and Melons Outlook, Sweetpotato*. http://www.ers.usda.gov/Publications/vgs/Tables/Swpot.pdf
(USDA) United States Standards for Grades of Sweet Potato. Effective April 21, 2005. http://www.ams.usda.gov/AMSv1.0/getfile?dDocName=STELPRDC5050330
Watts, V. M. and Cooper, J. R. 1943. Influence of varied soil reactions on growth and yield of vegetable crops on Newtonia silt loam and Ruston fine sandy loam soils. *Ark Agric. Exp. Stn. Bull*, No. 433.
Welker, W. V., Jr. 1967. Effect of herbicides on quality and yield of sweetpotatoes. *Weeds* 15: 112–113.
Yencho, Craig G., Pecota K. V., Schultheis, J. R., VanEsbroeck, Z., Holmes, G. J., Little, B. E., Thornton, A. C., and Truong, D. 2008. 'Covington' Sweetpotato. *HortScience*. 43: 1911–1914.
Young, H. C., and App, B. A. 1939. Biology of the white-fringed beetle (Naupactus leucoloma Boh). *USDA Bur. Entomol Plant Quarantine Bull* E-464: 22.
Zehender, G. W., Pitts, J. A., and Briggs, T. H. 1998. Management of whitefringed beetle (Coleoptera: Curculionidae) grub damage to sweetpotato with adulticide treatments. *J. Econ. Entomol*. 91: 708–714.

Chapter 15
Sweetpotato in China

L. Zhang, Q. Wang, Q. Liu and Q. Wang

History

Although several routes for the introduction of sweetpotato to China have been proposed, the leading opinion is that sweetpotato was brought first into Fujian via the sea (Ho, 1955; O'Brien, 1972; Anonymous, 1990a). It was said that an overseas Chinese businessman named Zhenlong Chen brought sweetpotato from Luzon in the Philippines to Fujian of China. It was Zhenlong Chen's son, Qinglun Chen, who presented sweetpotato, together with his explanations of the "six benefits and eight advantages" of the plant, to the governor of Fujian. The year 1594 was a famine year, and a huge area of crops was destroyed. Sweetpotato was brought to the attention of the governor. Consequently, he issued brochures on know-how of sweetpotato cultivation and ordered farmers to grow it extensively, in order to stave off famine. Sweetpotato was named "Jinshu" (golden root) at that time, because the storage roots harvested from sweetpotato saved a great number of people's lives during the famine. Although the exact time when sweetpotato was brought to Fujian is unknown, it is evident that sweetpotato must have been introduced to Fujian and grown on a small scale before being presented to the governor in the famine year 1594. The sixth generation of Zhenlong Chen was said to introduce sweetpotato from Fujian to Zhejiang, Shandong and Henan provinces of China (Ho, 1955; O'Brien, 1972; Anonymous, 1990a).

People of Zhangzhou, an important southern port of Fujian, claimed that sweetpotato was first introduced to their county and was kept as a secret for quite a long time (Ho, 1955; O'Brien, 1972). Sweetpotato was initially grown in Zhangzhou, and gradually extended northward to Quanzhou, Putian, and Changle counties. The time when sweetpotato was brought to Zhangzhou is not known yet, but the fact that sweetpotato was never called "golden root" in Zhangzhou, which was the name given by those grateful for being saved from the famine, suggests that the plant might have been grown there before the year 1594.

Q. Wang (✉)
Plant Biotechnology Laboratory, College of Horticulture, Northwest University of Agriculture and Forestry, Yangling 712100, Shaanxi, P. R. China
e-mail: qiaochunwang@nwsuaf.edu.cn

G. Loebenstein, G. Thottappilly (eds.), *The Sweetpotato*,
DOI 10.1007/978-1-4020-9475-0_15, © Springer Science+Business Media B.V. 2009

Fig. 15.1 A route map of the introduction of sweetpotato to China and extension of sweetpotato from south to north of China, as indicated by the line with arrows (Anonymous, 2003), with kind permission of Rural Industries Research and Development Corporation, Australia

It was recorded in the List of Key Events of Agriculture in China (Anonymous, 1990b) that sweetpotato was first brought from Vietnam to Dongguan of Guangdong province of China in 1582. However, detailed information is lacking.

Sweetpotato might also have been brought to China through Yunnan decades before 1594, which was recorded in Tali, a western prefecture of Yunnan near Burma, as early as the year 1563 (Ho, 1955). These data suggested that apart from a maritime route, there might have been an overland one from India and Burma. However, due to the geographical isolation of Yunnan province it is unlikely that this route was the first one for sweetpotato to arrive in China (Anonymous, 2003).

Extension of sweetpotato inside China was apparently from south to east along the coast, and from south to north through the Yangtze River and the Yellow River valleys (Anonymous, 2003). A path of the introduction of sweetpotato to China and of its extension from south to east and north inside China is outlined in Fig. 15.1 (Anonymous, 2003).

Nomenclature

Sweetpotato [*Ipomoea batatas* L. (Lam.)], a member of the family Convolvulaceae (Morning Glory), is a dicotyledonous, perennial plant producing edible storage roots (Austin, 1987). In China, nomenclature of sweetpotato differs largely from geographical regions, and is summarized in Table 15.1. Although called differently in different regions, the most common names for sweetpotato are ganshu "sweetpotato" and hongshu "red potato".

Table 15.1 Common names of sweetpotato in China. (Anonymous, 2003, with kind permission of Rural Industries Research and Development Corporation, Australia)

Chinese name	Literal translation	Region of use
Ganshu	Sweetpotato	General, throughout China
Hongshu	Red potato	General, especially in south and central China
Hongshao	Red creeper	West-central China, e.g., Henan, Sichuan
Baishu	White potato	Central China, e.g., Henan, Anhui, Beijing
Shanyu	Mountain taro	Central China, e.g., Henan, Anhui
Hongyu	Red taro	West-central China, e.g., Henan
Digua	Ground melon	North and central China, e.g., Shandong, Hebei
Fanshu	Foreign potato	Henan, Fujian, Guangdong

Importance of Sweetpotato

Sweetpotato is the fifth largest staple crop next to rice, wheat, maize and soybean in China (China's Yearbook of Agriculture, 2005), and is mainly used as food, feed and industrial materials. Historically, farmers in sweetpotato growing areas heavily depended on its cultivation, both for main income and food security. This situation started to change in the 1980s when nationally agricultural policies encouraged development of diversified agriculture. Today, although a large proportion of sweetpotato is used for industrial processing and livestock feed, sweetpotato is still a security food in the poorer areas.

Development of Sweetpotato Production in the Last 50 Years

Great changes have occurred in sweetpotato production over the last 50 years in China (Li et al., 1992). During the 1950–1960s, sweetpotato production was increased significantly to meet a rapid increase in population and shortage of food, as sweetpotato was a widely-adapted crop and gave high yield (Li et al., 1992). Total areas of sweetpotato in China reached 10.89 million ha in 1961 (Fig. 15.2). From the late 1970s, sweetpotato production decreased gradually, down to 6.2 million ha in 1985, mainly due to the changes of nationally agricultural policies, the improvement of economic situations and the development of field crops such as rice, wheat and maize. The annually decreased rate was about 6.28% during 1978 to 1985. After that, sweetpotato production has remained relatively stable between 5.5 to 6.0 million ha (Fig. 15.2). Now, the total growing areas and yields reached $\sim$ 5.5 million ha (FAO, 2005) and 106 million metric tons (FAO, 2004), respectively, which accounted for 70% and 85% of total area and yield of the world. The average yield is over 20 tons per ha, which is 1.4 times higher than the world (Fig. 15.3).

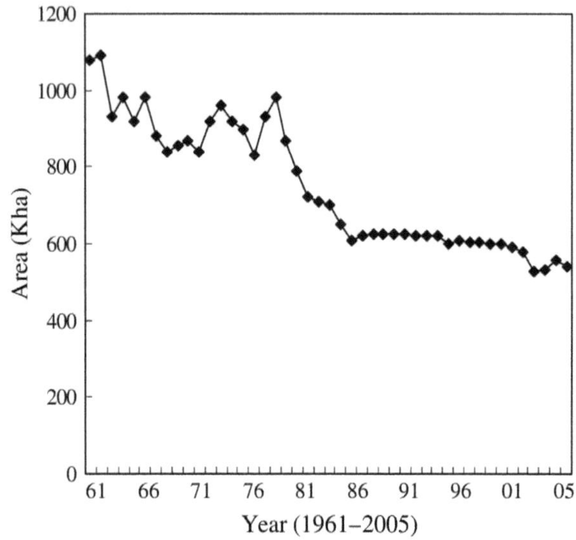

Fig. 15.2 Changes in sweetpotato growing areas in China during 1961–2005 (Data source: FAO, 1961–2005)

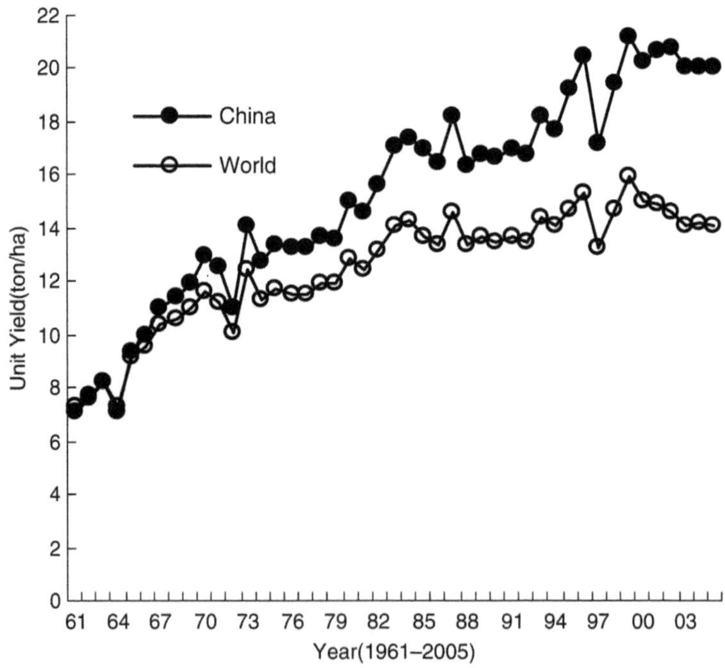

Fig. 15.3 A comparison in changes of unit yield of sweetpotato between China and the world during 1961–2005 (Data source: FAO, 1961–2005)

Five Major Regions of Sweetptato Production

Sweetpotato is grown in China, from south (Hainan) to north (Inner Mongolia) and from east (Zhejiang) to west (Tibet) with the major producing areas being concentrated in the Yellow River and the Yangtze River valleys. Five major sweetpotato areas are distinguished according to the climatic conditions and the cropping systems (Anonymous, 1984), i.e. Northern Spring Region, Yellow-Huai River Valley Spring-Summer Region, Yangtze River Valley Summer Region, Southern Summer-Autumn Region and Southern Autumn-Winter Region (Fig. 15.4). Sichuan, Henan, Chongqing, Anhui, Guangdong and Shandong were among the major producer provinces (Table 15.2, China's Yearbook of Agriculture, 2005). Climate and cropping systems in the 5 major growing regions of sweetpotato are discussed as follows:

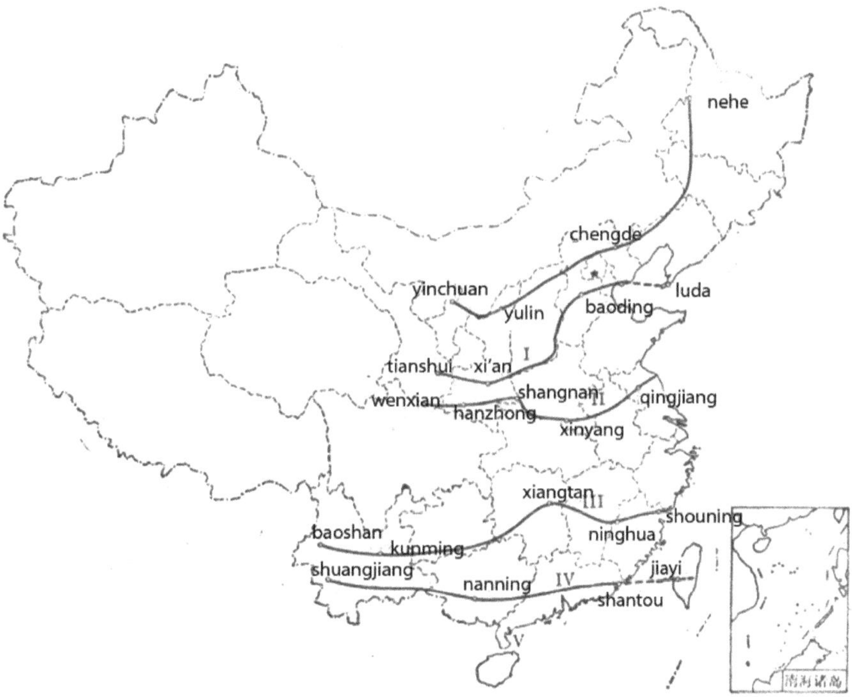

Fig. 15.4 Sketch map of 5 major sweetpotato growing regions in China. I. Northern Spring Sweetpotato Region. II. Yellow-Huai River Valley Spring-Summer sweetpotato Region. III. Yangtze River Valley Spring-Summer Sweetpotato Region. IV. Southern Summer-Autumn Sweetpotato Region. V. Southern Autumn-Winter Sweetpotato Region (Anonymous, 1984)

Table 15.2 Growing area and yield of sweetpotato in major sweetpotato producing provinces of China (China Year Book of Agriculture, 2005)

Province	Acreage (Kha)	Yield (t/ha)	Annual total yield (Kt)
Sichuan	833.2	21.20	17665
Henan	443	25.97	11505
Chongqing	410.4	21.72	8915
Anhui	349.9	17.57	6045
Guangdong	341.7	24.33	8315
Shandong	281.9	35.31	9955
Hunan	272	23.90	6500
Guangxi	256.5	12.10	3105
Fujian	237.6	24.62	5850
Yunnan	188.8	5.45	1030
Hubei	183	24.07	4405
Hebei	154.8	20.19	3125
Jiangxi	117	22.39	2620
Zhejiang	110.3	28.12	3100
Jiangsu	100.1	27.62	2765
Total	4622.1	22.18 (average)	102535

Northern Spring Region

This region is typical of humid or semi humid monsoonal, temperate and cold temperate zone climate. Generally, this region has short summer and long winter periods with large differences in day and night temperatures. Yearly average temperature is about 10.5 °C. The frost-free period is 170 days. Sunshine hour (2690 h) and rate (61%) are the highest of the 5 sweetpotato production areas. The annual rainfall (600 mm) mainly concentrates during July to August. The growing season is about 130–140 days. Spring sweetpotato production is mainly practiced and summer sweetpotato can also be found in southern part of this region, which is mainly used for production of seed tubers. Sweetpotato is planted in middle to late May and harvested in late September or early October.

Yellow-Huai River Valley Spring-Summer Region

This region has a temperate semi-humid monsoonal climate and a frost-free period of 210 days with an annual average temperature of 13.8 °C. The annual rainfall is about 760 mm, mainly from June to August. Spring is dry and short. Spring sweetpotato is planted from late April to middle May and harvested during middle to late October. Summer sweetpotato is planted from early to middle June and harvested from middle to late October. This region accounts for 40% of the total sweetpotato production in China. Shandong province in this region is the most advanced sweetpotato producer in China, with its average yield reaching 35.31 tons/ha.

Yangtze River Valley Spring-Summer Sweetpotato Region

This is the largest sweetpotato production region in China. Sichuan province is a major producer in this region. This region has a northern monsoonal subtropical humid climate, with 1800 h sunshine and annual average rainfall of about 1240 mm. The growing season is about 155 days. Sweetpotato is planted from late April to middle or late June and harvested in late October until middle November.

Southern Summer-Autumn Sweetpotato Region

This region has a monsoonal semitropical humid climate. Yearly average rainfall is about 1570 mm, and the growing season is about 120–150 days. Summer sweetpotato is widely grown and autumn sweetpotato is also grown in some parts of this region. Summer sweetpotato is planted in May, and harvested during August to October. Autumn sweetpotato is planted during middle July to early August, and harvested during late November to early December.

Southern Autumn-Winter Sweetpotato Region

This region is typical of a humid tropical monsoon climate. Two dry seasons occur when the monsoons change directions between spring and summer, and between autumn and winter. The growing reason is about 185 days with an annual average rainfall of about 1730 mm. Sweetpotato can be grown around the year. Autumn sweetpotato is planted during early July to early August, and harvested during early November to late December. Winter sweetpotato is planted in November and harvested between April and May of the following year.

Main Cultivars of Sweetpotato

"Xushu 18", a hybrid resulting from a cross of "Xindazi" × "Huabei 52–45", was bred by Institute of Agriculture of Xuzhou in 1972. This cultivar has purple peel and white-yellow flesh, and is highly resistant to sweetpotato root rot. It is grown in the northern part of China as spring sweetpotato and in Yellow-Huai Valley Spring-Summer Region. Growing areas of this cultivar cover more than 500,000 ha per year.

"Nanshu 88" was produced by a cross of "Jinzhuan 7" × "American Red" by Nanchong Institute of Agriculture, Sichuan in 1980, and mainly used for food as fresh tubers and for feed processing. This cultivar has pale red peel and pale yellow flesh, and is resistant to sweetpotato wilt disease and sweetpotato stem rot. Its growing area covers more than 1.6 million ha per year.

"Jishu 15" was bred in 2001 by Shandong Academy of Agricultural Science using a cross of "Ji 85003" × "Ji 79268". This cultivar, with red peel and light yellow flesh, is particularly suitable for starch processing. It is highly resistant to root rot and black rot, and nematodes, and also tolerant to drought and poor soil. Now, it is mainly cultivated in Shandong, with its growing areas reaching more than 200,000 ha per year.

"Jishu 98" is a hybrid crossing between *I. batatas* and *I. trifida* by Hebei Academy of Agricultural and Forestry Sciences in 2004. This cultivar has purple peel and pale yellow flesh, and is tolerant to drought, and resistant to sweetpotato black spot and root rot. It is mainly grown in Hebei on about 150,000 ha per year.

"Shangshu 19" is a hybrid by a cross of "SL-01" × "Yushu 7" by Shangqiu Institute of Agriculture, Henan. It is an early cultivar with a concentrated period of storage root formation and uniform root size. This cultivar has deep red peel and white flesh. It is the major cultivar grown in Henan and Anhui provinces, with its annually growing areas of 200,000 ha.

"Beijing 553" was bred by the former Huabei Institute of Agriculture in 1950. Peel and flesh of storage roots are brown yellow and white yellow colour, respectively. This cultivar is resistant to sweetpotato stem nematodes and black spot, but susceptible to sweetpotato root rot and *Rhizopus* soft rot (*Rhizopus stolonifer*). It is a major cultivar used for fresh table food and widely grown in the northern part of China, on about 200,000 ha per year.

"Eshu 5" was a hybrid from a cross of "CN1108-13" × "Eshu 2" by Hubei Academy of Agricultural Science in 2003. This cultivar is highly resistant to sweetpotato black rot, and resistant to sweetpotato wilt disease and *Rhizopus* soft rot (*Rhizopus stolonifer*). It has a high content of starch and is mainly used for starch processing. Its annual growing area is more than 120,000 ha.

Sichuan Academy of Agricultural Science released "Chuanshu 34", a cultivar suitable for starch processing, in 2003. This cultivar has purple peel and white flesh, and is resistant to sweetpotato black rot and medium-resistant to sweetpotato stem rot. Its annually growing area is more than 180,000 ha, mainly in Sichuan and Chongqing.

"Chuanshu 294" was bred by the Sichuan Academy of Agricultural Science in 1999. This cultivar is resistant to sweetpotato black rot and has pale red peel and white orange flesh. It is an early cultivar that can be harvested after 100 days from planting and is used for both fresh consumption and food processing. Its growing area reaches 80,000–100,000 ha per year.

"Jinshan 57" was released by Fujian Agricultural University in 1993. With its resistance to sweetpotato stem rot (*Erwinia carotovora*) and sweetpotato root rot, this cultivar is the most popular one grown in southern part of China and now its growing area covers more than 500,000 ha per year.

"Guangshu 97", bred by Guangdong Academy of Agricultural Science in 2004, is starch-rich and sweetpotato wilt disease-resistant cultivar. This cultivar is now widely grown in Guangdong and also in Guangxi, Jiangxi and Hainan provinces. The annual growing area is about 150,000–18,000 ha.

"Xiangshu 75–55" is a hybrid crossing "Xindazi" × "Hebei 67-89-419" and was released by Hunan Academy of Agricultural Science in 1999. This cultivar is resistant to sweetpotato wilt disease and stem rot, and suitable for storage.

It is widely grown in Hunan and Fijian provinces with its growing areas reaching 180,000 ha per year.

"Guangzishu 1", a hybrid of "Guangshu 95-1" × "Guangshu 88-70", was bred by Guangdong Academy of Agricultural Science in 2005. This cultivar is resistant to sweetpotato wilt disease and sweetpotato stem rot, and medium-resistant to sweetpotato root rot and sweetpotato black rot. Due to its high yields and quality of storage roots, suitability for storage and wide adaptation to various environmental conditions, it became a major cultivar in Guangdong, Fujian, Jiangxi and Hainan provinces, with increasing growing areas.

Practical Techniques for Sweetpotato Production

Propagation

Seed tubers are used as propagating materials. In general, seed tubers can be treated with hot water at 51–54 °C for 10 min, or with 50% thiophanate-methyl (1:400) or 50% carbendazim (1:500) or with the antibiotic agent 402 (1:1500–2000) for 10 min. Treated seed tubers are sown in propagation beds to produce stock shoots. When the stock shoots reach about 30 cm in length, cuttings of about 20 cm long with 5–6 nodes are taken and used for planting. For spring sweetpotato production, heated propagation beds are often employed to produce the stock shoots, while non-heated systems are used for summer and autumn sweetpotato production. Heated Kang (bed) is most often used in northern China, where temperatures in spring are too low to produce stock shoots for spring sweetpotato production. After being built, the bed is covered with 3 layers of straw and mud, with each layer of straw separated by a layer of mud. After seed tubers are placed on the bed for propagation, a fire is set to heat the bed. The heated bed can provide growers with propagating materials when temperatures are low in spring. However, the cost of heated bed is too high to be used in many sweetpotato production areas.

Planting Time, Method and Density

Time of planting is one of the most important factors affecting yield and quality of sweetpotato. Early planting promotes development of root system, resulting in early formation of storage roots and higher accumulation of starch in the tubers. In general, spring sweetpotato should be planted as early as air temperature is stable at 15 °C.

Ridge planting is the most popular method used in China. Three different ridges are used: narrow ridge, wide ridge with single row and wide ridge with double rows. Narrow ridge is widely applied in northern China and Yangtze River valley. About 45,000–54,000 and 52,500–60,000 plants/ha are used for spring and summer sweetpotato, respectively. Wide ridge with a single row is mainly adopted in western sweetpotato regions. For this system, about 52,500–57,000 plants/ha are planted. Wide ridge with double rows can be found in parts of Yangtze River valley. About 60,000 plants are planted per ha.

Although 5 planting methods are currently used for sweetpotato production in China, slant-planting is a major one. Stems of about 20 cm in length are taken from stock plants with its base (10 cm with 3 nodes) slant-planted into the soil and its top (6–10 cm) maintained above the soil.

Fertilization

Both base and top fertilizers are necessary for obtaining high yields of sweetpotato tubers with high quality. Base fertilizer, mainly composed of organic fertilizers, accounts for 80% of annually total fertilizers. About 37,000–75,000 kg/ha of base fertilizers are applied before planting. Top fertilizer is applied to sweetpotato when necessary. The time and amount of top fertilizer applied depend on growth and yield of the crops. Top application during the middle and late stage of root tuber development is generally needed for production of high yield and quality of root tubers. A mixture of 0.5% urea, 2–3% calcium superphosphate and 0.2% potassium dihydrogen phosphate is later applied to the crops once at 7 days' intervals for 2–3 times. Total amount of the mixture applied for each time is 1125–1500 kg/ha. For producing a yield of sweetpotato tubers of 37,000–52,500 tons/ha, total usage of fertilizers including base and top application should be 188, 150 and 450–600 kg/ha of nitrogen (N), phosphate (P_2O_5) and potassium (KO_2), respectively.

Control of Diseases

Sweetpotato Root Rot

Sweetpotato root rot [*Fusarium solani* (Mart.) Sacc. f. sp. *Batatas* McClure] is a serious disease widely spread in China, causing 10–50% of yield loss, and even a total loss of yields in the worst case. Disease-resistant cultivars and chemical spray are used for successful control of this disease. Disease-resistant cultivars include "Xushu 18", "Jishu 15", "Jishu 98" and "Jinshan 57".

Sweetpotato Black Rot

Sweetpotato black rot, caused by *Ceratocystis fimbriata* Ellis et Halsted, is widely distributed in China and listed as a quarantine disease. It occurs mainly in propagation beds and during storage of root tubers. Quarantine inspection and chemical spray are used for efficient control of this disease. Cultivars such as "Jishu 15", "Jishu 98", "Chuanshu 34" and "Chuanshu 294" are shown to be resistant to this disease.

Sweetpotato Wilt Disease

Sweetpotato wilt disease, a quarantine disease, is a destructive disease caused by *Ralstonia solanacearum*. The disease has been reported from Guangdong, Guangxi, Hunan, Jiangxi, Fujian and Zhejiang provinces. Yield losses range between 30 and 80%, and sometimes up to 100%. Control of this disease can be achieved through quarantine field inspection, rotation, chemical spray and usage of disease-resistant cultivars like "Eshu 5", "Guangshu 1", "Guangshu 97", "Nanshu 88" and "Xiangshu 75–55".

Sweetpotato Stem Rot

Sweetpotato stem rot (*Fusarium bulbigeum* Cooke. et Mass. Var. *batatas* Wollenw.) occurs throughout sweetpotato regions of China. Rotation, chemicals and disease-resistant cultivars are used for control of this disease. Cultivars resistant to this disease include "Guangzishu 1", "Jinshan 57", "Nanshu 88" and "Xiangshu 75–55".

Sweetpotato Scab

Sweetpotato scab (*Phaceloma batatas* Sawada) is found only in Guangdong, Guangxi and south part of Zhejiang Provinces. Infection of this disease results in reduction of starch content in root tubers. Quarantine inspection and chemicals are currently used for control of this disease.

Control of Viruses

In the late 1980s, China in collaboration with the International Potato Centre (CIP, Lima, Peru) initiated a project aiming at establishment of production and propagation system of sweetpotato virus-free plants. In this project, techniques including meristem culture, virus detection mainly by enzyme-linked immunosorbent assay (ELISA) and propagation system for virus-free stock plants were established (Fig. 15.5, Zhang et al., 1999b). Virus-free stock materials have been delivered to sweetpotato farmers for commercial production since 1994, and cultivation of virus-free plants covered 80% of sweetpotato growing areas in Shandong province in 1998 (Fuglie et al., 1999; Zhang et al., 2006). Now, virus-free plants are widely used in Shandong, Jiangsu, Henan and Guangdong provinces, covering more than 466,000 ha (Song et al., 1997; Zhang et al., 1999b and 2006; Gao et al., 2000). Cultivation of virus-free cultivars significantly increased yields ranging from 10.3% to 101.9% with an average increase of 37.9% (Fuglie et al., 1999; Zhang et al., 2006). This increased effect on yield was much more markedly with old cultivars such as "Beijing 553" and "Fengshoubai" than with new ones such as "Lushu No. 7" and "Lushu No. 8". Marketable yield (tubers > 100g) was increased by 22.2%

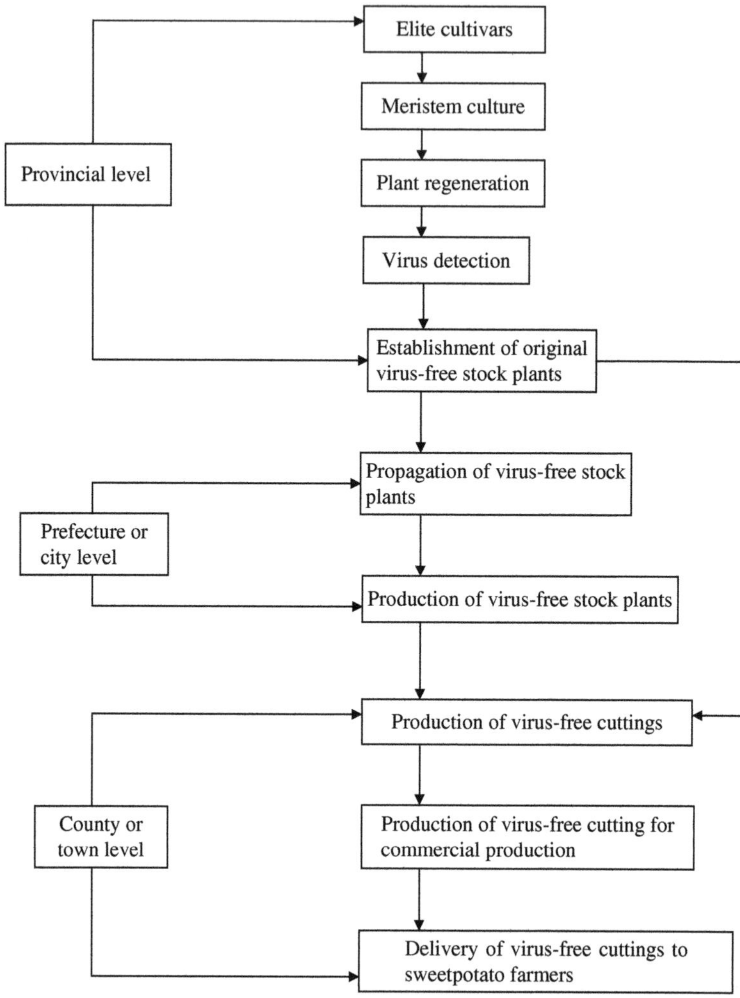

Fig. 15.5 A schematic chart of production of virus-free sweetpotato crops and their propagation system established in Shangdong Province of China (Zhang et al., 1999b)

when virus-free plants were used, as compared with virus-infected ones. A survey carried out by economists and biologists from CIP and Shandong Academy of Agricultural Science clearly showed that an annual net benefit of about $145 million was obtained by using virus-free sweetpotato seed tubers in Shandong province (Fuglie et al., 1999).

Control of Insects

Several important insects are discussed as follows.

Sweetpotato weevil

Sweetpotato weevil (*Cylas formicarius* Fab.) mainly occurs in the southern part of China, and is listed as a quarantine insect. Quarantine field inspection and chemical spray can be used for efficient control of this insect.

Alcidodes waltoni *Boheman*

Alcidodes waltoni Boheman is widely distributed in the southern part of China, including Zhejiang, Jiangxi, Fujian, Guangdong, Guangxi, Yunnan and Sichuan provinces. Adults mainly attack leaves and petioles, while larvae enter shoots, thus destroying the young plants. Rotation, capturing and killing of adults, and biological control are used for control of this insect.

Prodenia litura *Fab.*

It is found in almost all sweetpotato regions except Xinjiang and Qinghai provinces, but is most popular in regions of Yangtze River as well its south part. *Prodenia litura* damages root tubers and leaves, and even young shoots and petioles. Control methods include elimination of weeds, capturing and killing of adults, and chemical spray to kill larvae.

Protoparce convolvuli *L.*

It spreads throughout all sweetpotato regions in China. Larvae attack leaves and young shoots, destroying the whole plant. Elimination of pupa from the sweetpotato field during winter and spring seasons, capturing and killing of adults and chemical spray to kill larvae are efficient methods for control of *Protoparce convolvuli* L.

Underground Pests

Gryllus, cutworms, *Gryllotalpidae*, *Scarabaeoidea*, *Elateridae* belong to this group. Of them, the former two mainly attack stems and leaves, while the rest make damages to roots and root tubers. Killing of pupa and larva during winter, capturing of larva and adult, biological control and chemical spray are used for control of them.

*Sweetpotato Stem Nematodes (*Ditylenchus destructor *Thorne)*

Nematodes cause serious damages to sweetpotato, particularly in the northern part of China, for example, Shandong, Hebei, Beijing, and Tianjin, with increasing damage during the last years. Nematodes are listed as a quarantine disease. Nematodes cause yield losses ranging between 10 to 50%, and in some cases can totally destroy the crop. Nematode-resistant cultivars, rotation and chemicals can be used for efficient control of nematodes. Nematode-resistant cultivars included “Lushu 3”, “Yushu 10” and “Sushu 8”. However, most of these nematode-resistant cultivars currently grown in China contain low soluble substances and low starch. Rotation with maize or cotton for 3 years can largely reduce occurrence of nematode diseases. Application of Arve-mectin (1:3000) to the soil can result in about 80% of control. Treatment of propagating materials using 50% Phoxim for 20 min can also efficiently kill nematodes inhabiting inside the shoot materials. Usage of propagation beds that have never been used for propagating sweetpotato can also successfully prevent nematodes.

Harvesting, Marketing and Profitability

Harvest Time

Harvest time has an important role in determining tuber yield, storage duration and quality of stored tubers. Sweetpotato plants continue growing as long as temperature is suitable. Therefore, harvesting at premature time generally results in reduction of tuber yield, whereas a late harvesting causes freezing damages to the root tubers, which cannot be used for storage. Optimal harvesting time differs from different regions, mainly depending on the temperature of the given region. In most of sweetpotato regions except those regions of south part of China, harvesting begins when temperature drops to 15 °C and ends before temperature decreases to 12 °C.

Harvesting Method

Storage roots of sweetpotato are harvested by either hand or machine. In hilly regions, manual harvesting dominates, while harvesting by machine is popular in the plain areas. Any damages to the root tubers, especially those used for storage, must be avoided. Harvesting is generally done during morning time of sunny days so that tubers can be surface-dried by the sun and stored in the afternoon.

Storage

Cellars are most often used for storage of root tubers of sweetpotato. Various types of cellars have been built in different regions. Several important cellars are described as follows.

Arch-Cellars

Arch-cellars are made of bricks with most part built under the ground. Two types of arch-cellars are common: one with a corridor in the middle and the other with one on the side. After root tubers are placed in the cellar, the top of the cellar is covered with a layer of mud (1.5 m in thickness). Window eyes are built in each column to ensure good ventilation. Arch-cellars can be built in both hilly and plain regions.

House-Cellar

House-cellars can be built on the ground or with its half under the ground. Thickness of its wall is about 1 m. Top of it is built with three layers of straw, each layer separated by a layer of mud. Wood or straw is used to build cellar walls, which divide the cellar into storage rooms. House-cellars are common in plain regions.

Well-Cellar

An Erlenmeyer-shaped well is built under the ground with about 70 cm and 120 cm in diameter of its mouth and bottom, respectively. Height of the well is about 3–5 m. Two cellars are built at opposite directions of the bottom of the well. Each cellar is about 3 m in length, 2 m in width and 1.7 m in height. About 3500 kg of root tubers can be stored in each cellar.

Processing of Sweetpotato

Artisan farmer households and industrial factories perform processing of sweetpotato. In general, the former fulfil simple processing such as slicing and field-dried chips after harvest. These roughly processed products are sold directly to the later or transported to the markets where the middlemen collect them and then sell them to the factories for fine processing.

Food Processing

Both fresh root tubers and dried chips are used for processing to produce starch foods such as noodles, vermicelli and sheet jelly, which account for 70% of total processing products of sweetpotato. Boiled and sun-dried flavoured sweetpotato chips are becoming popular foods, which take about 10% of total processing products of sweetpotato.

China has a long history in using young stems of sweetpotato as vegetables. Sweetpotato cultivars such as "Baishu 1" and "Fushu 7–6" are particularly suitable for consuming as vegetables. Terminal young stems in 5–10 cm long are harvested once every 7–10 days, and eaten after cooking. A yield of 30–45 tons/ha can be produced, with an average price of about 1.8–2.0 Yuan (RMB)/kg at the farm gate.

Industry Processing

Starch

About 40% of total yield of root tubers are used for starch processing. Small workshops, each having about 15–20 workers, are the main productive units for rough starch processing. Starch is extracted from fresh tubers using acid method. Working duration concentrates during middle October to early December of each year when root tubers are harvested. Starch extraction in this way is quite low, about 70%. Quality of starch is poor, mainly due to old equipments and processing techniques. Starch produced by such small workshops is mainly used for producing noodles, vermicelli and sheet jelly. Recently, some medium- to large-sized companies with advanced techniques and equipments started to work on starch processing. Yield of starch from such specialized companies reaches more than 10,000 tons per year per company. Fine starch is even exported to South Korea and Japan.

Processing of Organic Products

Organic products from sweetpotato include ethanol, monosodium glutamate, citric acid, lactic acid, propanoic acid, butyric acid, butanol, oxalic acid, and amino acids. About 5–10% of total yield of sweetpotato tubers are used for this purpose. Among these products, ethanol, an energy source, seems to be the most important product and attracts great attentions of the China government. About 2.7–2.8 kg of dried sweetpotato tuber chips can produce 1 kg of fuel ethanol. Companies working on ethanol production from sweetpotato included Shandong Tielingshuguang Group, Chongqing Changlong Industrial Corporation, Sichuan Tongjiangjiuye Ltd, Hubei Jinglongquan Group and Henan Tianguan Group. At the present time, the total yield of fuel ethanol from sweetpotato is about one million tons. Projects on large-scale production of fuel ethanol from sweetpotato have already started in Henan, Hubei, Sichuan, Chongqing, Hebei, Jiangsu and Shandong provinces. It is estimated that yield of fuel ethanol from sweetpotato can reach 5 million tons in 2010.

Livestock Feed Processing

Sweetpotato stems and leaves are an important feed for pigs, cows and goats. Fresh, dried or cooked stems and leaves can be used as feed. Sweetpotato stems and leaves can be harvested 4–5 times during a growing season with a total yield of about 135 t/ha. Fresh stems or leaves are cut into pieces, mixed with wheat grain, corn flour, rice husk and water, and used as feed. Sweetpotato root tubers are also a main source for livestock feed. About 25–30% of total yield of sweetpotato are used as livestock feed, only about 20–30% of which are processed into livestock feed by companies.

Marketing

Great changes have taken place in marketing for sweetpotato since 1978, mainly due to the introduction of the reforms of economic systems. Local marketing dominated before 1978, and almost all of sweetpotatoes were sold locally in villages and towns around sweetpotato growers. After 1978, many companies and traders joined in marketing. Growers marketed sweetpotato by themselves, individually or collectively. They also sold their products to companies or traders. Fresh storage roots were sold out in relatively short distance, because they cannot be transported for long distance. Dried or chipped root tubers can be transported for long distance, and therefore, their market can be far away from the origin of production. For example, dried chips produced in Anhui province are sold to companies for processing in Beijing, Tianjin, and Shanghai in a distance more than 1,000 km. Sweetpotato chips produced in Sichuan are also sold as far as to Jilin and Heilongjiang provinces in the north of China. Recently, processed sweetpotato products such as noodles, vermicelli, snacks and defined starch are exported to Japan and South Korea.

Research

Classic Breeding

Sweetpotato breeding by crossing has long been the most important research project in China. History of sweetpotato breeding research can be divided into three stages. Breeding research before the 1950s was defined as the first stage, during which sweetpotato germplasm including a great number of local cultivars were collected, evaluated and extended to the practical usage. Introduction of elite cultivars from foreign countries such as Japan and the United States started during this stage. Cross breeding was initiated. Application of selected local cultivars, for example, cv. Yubaibai from Guangdong, significantly increased the yield up to 30%. The second stage ranged between the 1950s and the 1970s. During this stage, breeding of cultivars with high yield and resistance to diseases was targeted. Through great efforts, a huge number of new cultivars (more than 60) were bred and released to commercial productions (Sheng et al., 1987; Yuan, 1989). The most widely used cultivars included "Xushu 18", "Qingnong 2", "Fengshoubai", "Chuanshu 27" and "Nongdahong". Cultivar Xushu 18 was found highly resistant to sweetpotato root rot disease caused by *Fusarium solani*. The third stage was from the 1980s up to today. Breeding aimed at not only yield but also quality of root tubers. Breeding of cultivars suitable for variously industrial processing was also included in breeding projects. Cultivars specially used for food included "Nanshu 88", "Lushu 2" and "Zheshu 2". Cultivars with high content of starch suitable for industrial processing were "Mianfen 1", "Huaishu 3" and "Yanshu 3", while "Guangshu 62" and "Lushu 3" were the main cultivars suitable for processing of livestock feed. More recently, several cultivars were released suitable for consuming as vegetables (Cai et al., 2006; Ou et al., 2007). Cultivar Baishu 1 was one of them (Ou et al.,

Table 15.3 Institution and university involved in sweetpotato breeding research in China

Institution or university	Research subject
Anhui Acad. Agri. Sci, Hefei, Anhui	Breeding
China Agri. Uni, Beijing	Molecular biology and biotechnology
Guangdong Acad.Agri. Sci, Guangzhou, Guangdong	Breeding
Hebei Acad. Agric. Sci. Shijiazhuang, Hebei	Breeding
Henan Acad Agri. Sci., Zhengzhou, Henan	Breeding
Hubei Acad. Agri. Sci, Wuhan, Hubei	Breeding
Jiangsu Acad. Agri. Sci., Nanjing, Jiangsu	Breeding
Nanchong Prefecture Inst. of Agri. Res., Nanchong, Sichuan	Breeding
Sichuan Acad. Agric. Sci, Chengdu, Sichuan	Breeding
Shandong Acad. Agri. Sci, Jinan, Shandong	Breeding
Sweet potato Res. Center of China, Xuzhou, Jiangshu	Breeding and genebank

2007). This cultivar was bred using self-crossing of cv. Anshu 07, and is half-erect. Each plant produces 25–30 branches of stem. The stems (ca 80–100 cm) including leaves have no pubescence, are delicate and taste delicious after cooking. Analysis on nutrient components of this cultivar showed that contents of water, protein, fat, edible fibre, vitamin C and VB2 were 89%, 32g/kg, 4g/kg, 13g/kg, 400mg/kg and 1.4 mg/kg, respectively. Mineral content was also high, compared with other 24 common vegetables (Ou et al., 2005). This cultivar can also be used for production of tuber roots. With great efforts exerted in breeding projects, a great number of new cultivars suitable for food, starch, vegetable and livestock feed, respectively, are emerging every year. Main institutions and universities involved in research projects of sweetpotato breeding in China are listed in Table 15.3.

Molecular Biology and Plant Biotechnology

Over the last decade, great efforts have been made to improve sweetpotato cultivars using molecular biology and plant biotechnology methods including *in vitro* culture, plant regeneration, somatic hybridization, cell-induced mutation, genetic engineering and molecular markers (reviewed by Liu et al., 2003; Li et al., 2004, 2005; Hou et al., 2006).

Embryogenesis

In vitro plant regeneration has been achieved via organogenesis and embryogenesis using various explants including leaves (Zhou et al., 2003a, Luan et al., 2007), shoot tips (Liu et al., 1996, 1997, 2001), petioles (Zhou et al., 2003a) and shoots (Zhou et al., 2003a). Embryogenic cell suspensions have widely been used for transformation (Wang et al., 2003b; Li et al., 2005) and have their potential applications to micropropagation (Wang et al., 2003a), virus elimination (Wang et al., 2003a) and screening of mutants (Li et al., 2002; Wang et al., 2003b). For embryogenesis, shoot tips (0.5 mm) were cultured on solid MS (Murashige and Skoog, 1962) medium containing 0.2–2.0 mg/l 2,4-D to induce embryogenic callus formation at 27 °C in the dark (Liu et al., 1996, 1997, 2001). Following 6–9 weeks of culture, embryogenic calli induced were transferred to liquid MS medium supplemented with 2.0 mg/l 2,4-D and grown on a reciprocal shaker (100 rpm) at 27 °C under a 13-h photoperiod with a light intensity of 500 lux, to produce embryogenic cell suspensions. Somatic embryoes formed were transferred to solid MS medium containing 1.0 mg/l abscisic acid (ABA) for embryo maturation and germination at 27 °C under a 13-h photoperiod with a light intensity of 3000 lux. Plantlets with well-developed shoot and root system were produced in 5 weeks of culture. With this protocol, embryogenic cell suspensions and their subsequent plant regeneration were successfully obtained in more than 17 Chinese and 4 Japanese sweetpotato genotypes (Liu et al., 1996, 1997, 2001). Frequencies of embryogenic callus formation (0–76%), embryo formation (0–50%) and plant regeneration (0–100%) varied from genotypes and 2,4-D concentrations.

Resistance Breeding

Breeding of cultivars resistant to abiotic stress has brought much attention to sweetpotato breeders in China. Calli induced from leaf explants were incubated for 2 or 2.5 h with 0.5% ethylmethanesuphonate (EMS), and then cultured on a selection medium containing 200 mM NaCl (Luan et al., 2007). Salt-tolerant calli were induced to form somatic embryoes, followed by embryo germination and plant regeneration, while the control showed yellow colour and gradually died. Induction of mutants with EMS was suggested as a useful method for mutant breeding of salt-tolerant sweetpotato (Luan et al., 2007). A novel protocol was well defined using chronic irradiation to embryogenic cell suspensions (Liu et al., 2003). Embryogenic cell suspensions of sweetpotato that had been exposed to 80 Gy of gamma-ray were cultured in MS medium containing 30% PEG 6000 or 2% NaCl for selection of drought- or salt-resistant mutants, respectively (Li et al., 2002; Wang et al., 2003b; He et al., 2009). Mutants resistant to drought and salt, respectively, were successfully selected and regenerated into whole plantlets. *In vitro* and *in vivo* assays showed that salt tolerance of the mutants was significantly higher than that of the control (He et al., 2009). Breeding of drought- and salt-resistant cultivars would largely improve development of sustainable sweetpotato production, especially in hilly and mountainous regions with poor soil conditions. More recently, a mutant of callus-derived somatic embryoes induced from shoot tips of gamma ray-irradiated sweetpotato plants (cv. Kokei 14) was obtained (Wang et al., 2007b). Flesh colour of root tubers of this mutant changed from light yellow to orange, and carotenoids content of the storage root was significantly higher in the mutant than in the original plant. The yield of storage tubers from the mutant was also markedly higher than the control.

Genetic Transformation

Agrobacterium-mediated transformation of embryogenic cell suspensions has been successfully achieved and widely used in various transformation studies (Guo, et al., 2001; Luo et al., 2002; Zhai and Liu, 2003; Li et al., 2005). *A. tumefaciens* strains including LBA4404, EHA101, EHA105 and A208SE were mainly used. Using *A. tumefaciens* strain A208SE harboring the binary vector pROA93 containing the *npt* II gene and *gusA* gene, Zhai and Liu (2003) obtained approximately 48% of transformation efficiency when co-cultivated embryogenic cell suspensions were selected with 100 mg/l carbencillin and 50–75 mg/l kanamycin. Recently, an efficient *Agrobacterium*-mediated transformation of cell suspensions of sweetpotato cv. Lizixiang was reported (Yu et al., 2007). In their study, a total of 2,218 plants were regenerated from the 1,776 cell aggregates inoculated with the *A. tumefaciens* strain EHA105 harboring a binary vector pCAMBIA1301 with the *gusA* and the *hpt* II genes. Of the plantlets regenerated, 90.4% of them were transgenic as confirmed by Southern blot analysis. Such high transformation efficiency and plant regeneration of transformants exceeded all transformation studies reported so far in sweetpotato (Yu et al., 2007). Sonication-assisted *Agrobacterium*-mediated transformation (SAAT) was found to improve transformation efficiency (Wang et al., 2006). However, low transformation efficiency and genotype-dependent are still the main obstacles to production of transgenic plants.

The phytocystatin oryzacystain I (OCI), originally isolated from rice seeds (Abe et al., 1987), significantly inhibited the nymphal survival of aphids (Azzouz et al., 2005), prevented aphids from reproduction (Azzouz et al., 2005) and reduced adult weight and fecundity of aphids (Rahbé et al., 2003). Therefore, the gene encoding OCI is considered as a candidate for expression in transgenic crops and seems to be an effective method to control homopteran pests (Azzouz et al., 2005). Transformation of embryogenic cell suspensions with the OCI gene has been successfully achieved (Jiang et al., 2004; Yan et al., 2004; Li et al., 2005). Analysis using Southern blot confirmed that the plants regenerated from transformed cells were transgenic. More recently, transgenic sweetpotato plants expressing the *bar* gene for herbicide resistance were obtained (Zang et al., 2007). Transgenic plants are currently under evaluation in the field. Nevertheless, there is still a long way to go before transgenic plants will be released for commercial production of sweetpotato (Liu et al., 2003; Li et al., 2005).

Genetic Markers

Studies on sweetpotato genetic markers have been extensively reviewed in several recent publications (Liu et al., 2003; Li et al., 2005; Hou et al., 2006). Genetic linkage maps are powerful tools for the localization and map-based cloning of genes, and also for marker-assistant breeding. Genetic linkage maps of sweetpotato based on sequence-related amplified polymorphism (SRAP) (Wu et al., 2005) and amplified fragment length polymorphism (AFLP) (Pu et al., 2005) have been developed for sweetpotato using a segregating population derived from a cross between cv. Mianfen 1 and cv. Hongqi 4. Their linkage groups showed that *Ipomoea* species possessed agriculturally desirable traits such as resistances to sweetpotato weevil (*Cylas* spp.), scab [*Elsinoe batatas* (Saw.), Viegas and Jenkins] and black rot (*Ceratocystis fimbriata* Ell. Et Halst.) (Iwanaga, 1988). A better understanding of genetic diversity and relationships between sweetpotato and its wild relatives would facilitate the usage of wild *Ipomoea* genetic resources in breeding research. Using inter-simple sequence repeat (ISSR) markers and restriction site variation in four non-coding regions of chloroplast DNA, genetic diversity and relationships of ten *Ipomoea* species were established (Huan and Sun, 2000). Full length of a polyphenol oxidase gene, responsible for causing browning of sweetpotato during harvest and processing procedures, has been cloned and sequenced (Peng and Chen, 2004). Recently, a genetic linkage map with more than 2000 AFLP markers and AFLP markers closely linked to the stem nematode resistance gene have been developed for sweetpotato using a segregating population derived from a cross "Xu781" × "Xushu 18" (Jie et al., 2008).

Somatic Hybridization

As mentioned above, wild *Ipomoea* species possess some valuable resistant genes for virus and other pest diseases (Iwanaga, 1988). However, cross-incompatibility between commercial cultivars of *Ipomoea batatas* ($2n = 6x = 90$) and the wild species ($2n = 2x = 30$) of *Ipomoea* severely limited applications of cross breeding to virus- and pest-resistant cultivars. Somatic hybridization has proven to be an alternative for transfer of desired genes from wild types to cultivated plants in many species. Protocol of protoplast culture and subsequent plant regeneration has been well-established using *I. batatas* and wild *Ipomoea* species (Liu et al., 1998; Zhang et al., 1999a, 2001; Li et al., 2004; Guo et al., 2006). Protoplasts were isolated from the leaves of *in vitro* cultured *I. cairica* and *I. lacunosa*, respectively and fused with protoplasts from embryogenic cell suspensions of *I. batatas* cultivars (Zhang et al., 2001; Guo et al., 2006). Somatic hybrids were successfully obtained and regenerated into whole plants. Field performance of these somatic hybrids showed that they were fertile and able to cross with *I batatas* (Liu et al., 1998; Zhang et al., 1999a, 2001). *In vitro* somatic hybridization using wild sweetpotato species opened a new avenue for breeding of virus- and pest-resistant cultivars.

Artificial Seeds

Studies were conducted on artificial seeds using somatic embryogenic tissues or axillary buds, with some preliminary results obtained for purposes of micropropagation and germplasm storage (Tang and Li, 1994; Zhou et al., 2003b; Guo and Zhang, 2006). Attempts were made to encapsulate buds excised from virus-free *in vitro* plants, and these encapsulated buds could be used as propagating materials of virus-free plants (Zhang et al., 2004).

Post-Harvest Treatments and Processing

The major part of sweetpotato used for human food is its storage roots, which is rich in starch, dietary fibre and also vitamin A. Indeed, 100 g of cooked sweetpotatoes can provide about 11.5 mg of α-carotene or about four times the United States recommends daily allowance. The plant is also used for animal feed and for starch extraction. In addition, the stems and leaves of sweetpotato can be consumed as a green vegetable, as discussed in the above sections. Traditionally, sweetpotato was grown mainly in hilly and mountainous regions and consumed as a staple food. Post-harvest treatments and processing of sweetpotato received little attentions in the long history of sweetpotato production in China. In recent years, as the role of sweetpotato gradually shifted from staple food to industrial materials, and people of city took sweetpotato as a nutritional food or vegetable, Scientists started to pay attentions to research on post-harvest treatments and processing. Efforts have been made to prolong storage life of sweetpotato storage roots, while maintaining their quality after storage.

Low digestibility of raw starch in sweetpotato root tubers constitutes one of the constraints to feed efficiency. Digestibility depends on α-amylase activity (Dreher et al., 1984), types of starch (Zhang et al., 1993, 1995), trypsin inhibitors (proteinase inhibitors) (Zhang et al., 1998), all of which are affected by genotype (Zhang et al., 1995; 1998), storage time (Hagenimana et al., 1992, 1994) and different location (AVRDC, 1988). A systematic study was carried out on biochemical changes during storage of sweetpotato tubers (Zhang et al., 2002b). Results showed that starch content slightly decreased in most cultivars during 0–180 days of storage. Alpha-amylase activity increased during the first two months of storage, and then decreased with elongated storage time to a level similar to that at harvest. Trypsin inhibitor activity in the fresh tubers varied with genotypes from 3.9 to 21.8%. In general, there were considerable genotypic variations in digestibility, with up to 27% reduction in digestibility after 120 days in storage. This study thus provided basic, essential information on choice of sweetpotato genotype, optimum storage and processing time required by specifically industrial usage.

In order to maintain high quality of sweetpotato tubers after storage, effects of heat treatment were investigated on quality and storage life of sweetpotato (Hu and Tanaka, 2007). Tubers of sweetpotato cv. Beniotome, a Japanese cultivar that had received wound curing at 29 °C and 90–95% relative humidity for 6 day were treated in 50 °C hot water for 30 min, followed by storage under 14 °C and 90–95% relative humidity for 12 months. Results demonstrated that hot water treatment significantly inhibited sprouting and decay of sweetpotato during the storage period, and did not cause marked changes on starch properties, quality of internal components,and processing quality of the stored tubers, with less than 4% of the year-long stored tubers discarded due to spoilage. Thus, hot water treatment presented a new means for prolonging the storage life of sweetpotato with good quality and minimal loss.

Sweetpotato starch noodle (SPSN) has poor cooking quality such as dull, opaque and moderately elastic, and has high cooking loss and swelling when cooking, compared with mung bean starch noodle (MBSN). A study of Tan et al. (2006)

showed remarked differences in fine structure, molecular weight fractions, digestibility, and content and type of amylase between SPSN and MBSN. These differences resulted in a stronger distinct crystalline pattern and good cohesiveness of MBSN. This systematic study on the structure of both SPSN and MBSN offered fundamentals for improving the quality of Chinese sweetpotato starch noodle.

Glycerol, an important chemical product, has been widely used in the cosmetic, paint, tobacco, food, and pharmaceutical industries. Although other methods exist, fermentation route for efficient glycerol production is being received much attention for glycerol production. Generally, glucose is well known as a fermentation substrate for glycerol production. A preliminary study of Zhang et al (2002a) showed that sweetpotato meal might have a potential function as fermentation substrate for this purpose. Much research work still needs to be done.

Sustainable Production of Sweetpotato

Cultivation Techniques

In order to achieve high yields and quality production of sweetpotato, cultivation techniques were studied, for example, for different cultivars (Liu et al., 2004), virus-free cultivars (Cai et al., 2001), vegetable cultivars (Wang et al., 2002), and for cultivars grown in different ecological conditions (Hao and Hao, 2001; Chen and Lin, 2004; Xu et al., 2004)

Soil and Nutrition

In general, sweetpotato cultivars that produce high yields require high fertilizer supply. However, this is very difficult for poor farmers due to the high cost. Studies were performed on nutrient utilization efficiency of different sweetpotato cultivars, in order to gain useful information on selection of cultivars that produce an acceptable yield in infertile soils (Lu et al., 2003), where much of sweetpotato are being grown in China.

Plant-available phosphorus is generally low in many of sweetpotato growing soils in China due to deficiency and/or high phosphate-fixing capacities, which severely limit high yield and quality sweetpotato production. The effect of inoculation with *Arbuscular mycorrhizal* fungi (AMF) was investigated on crop productivity under small-scale farming conditions in China (Farmer et al., 2007). Inoculation with *Glomus intraradices* BEG 141 or *G. etunicatum* (HB-Bd45-Gsp4, BEG 167, BEG 168) increased yields by 10.2% and 14.0%, and tuber quality in terms of sugar or carotene contents also largely increased, compared with the control.

Pathogen Control

Whitefly (Bemisia tabaci)

Besides its being an important vector for transmission of sweetpotato viruses (Loebenstein et al., 2003) such as SPCSV, SPMMV, SPLCV, sweetpotato yellow dwarf virus (SPYDV), whitefly (*Bemisia tabaci*) has recently been recognised to be a destructive pest for agricultural production including sweetpotato in China (Ren et al., 2001), and other parts of the world (Brown et al., 1995). Although approximately 24 whitefly biotypes have been identified in the world, four of them were found in China: a B biotype, a Q biotype and two non-B/Q biotypes with the first one most widely distributed in China (Wu et al., 2003; Zhang al et. 2005). Q biotype was found only in Yunnan province and Beijing city, and non-B/Q biotypes in Shandong, Hebei, Zhejiang and Fujian provinces. More recently, a B biotype was reported in Fujian province (He et al., 2006).

Chemical control of *B. tabaci* is difficult, and meanwhile, chemical control may also seriously aggravate *B. tabaci* by reducing natural enemies (Wang et al., 2007a). Usage of biocontrol based on additional entomopathogenic fungi was tested for control of *B. tabaci* (Chen et al., 2004; Wang et al., 2004, 2005, 2007a). When applied to young sweetpotato plants, crude toxins (400 mg/l) extracted from the fungus [*Lecanicillium* (*Verticillium*) *lecanii* (Zimmermann) Gams & Zare strain V3450 and Vp28], a microparasite of *B. tabaci*, were found to significantly reduce the hatching of whitefly eggs and the subsequent survival frequency of the nymphs, and the emergence and fecundity of the progeny adults. Thus, the fungi metabolite toxins may develop into an environmentally friendly bioprotectant for effective control of *B. tabaci* (Wang et al., 2007a). Since these crude toxins at higher concentrations caused low toxicity to larva of ladybird beetles (*Delphastus catalinae*), a whitefly predator, spraying of the toxins is best avoided in field having immature stages of *D. catalinae* (Wang et al., 2005). *Encarsia bimaculata*, a parasitiod of *Bemisia tabaci*, was recently discovered in south China (Qiu and Ren, 2005).

Nematodes

Nematodes, a soil-borne pathogen, cause serious damages to sweetpotato production in China and are difficult to control. Stem nematode (*Ditylenchus dipsaci*) and root knot nematode (*Meloidogyne* spp.) are the two main nematodes attacking sweetpotato in China. Much of the interests focused on detection of molecular marker linked to nematode-resistant genes (Guo and Pan, 2002; Zhou et al., 2005a, b; Liu et al., 2005, 2006; Pan et al., 2006). Using randomly amplified polymorphic DNA (RAPD), a molecular marker was detected linked to a gene involved in resistance to the stem nematode (*Ditylenchus destructor*) of sweetpotato cv. Xu 781 (high resistant) and Xushu 18 (high susceptible) (Zhou et al., 2005a, b). A fragment marker of OPD01-700 was obtained only in cv. Xu 781. Detection of this marker in sweetpotato of high resistant (13 clones), stable resistant (10 cultivars), high susceptible (5 clones) and susceptible (8 clones), respectively, suggested that genetic marker could be used for selection of nematode-resistant cultivars. Based on the sequence of the nematode resistant gene in sugar beet, two primers were designed and used for amplifying homologous fragments from the genomic DNA of sweetpotato cv. Jinshan 25, a highly resistant cultivar to nematode (Pan et al., 2006). A special fragment of about 600 bp was obtained with its sequence homological to Hs1pro-1 in sugar beet. Thus, molecular marker would provide an efficient method for selection of nematode-resistant sweetpotato cultivars (Liu et al., 2003).

Viruses

Sweetpotato viral diseases constitute a major constraint for the sustainable sweetpotato production. Three sweetpotato viruses have long been detected since the 1950s (Zhang and Wang, 1995; Song et al., 1997): sweetpotato feathery mottle virus (SPFMV), sweetpotato latent virus (SPLV) and sweetpotato chlorotic fleck virus (SPCFV). During the last years, with the extensive introduction of sweetpotato cultivars from foreign countries to China and exchanges of sweetpotato germplasm with other countries, of the about 20 sweetpotato viruses reported in the world, 11 have been found in China (Table 15.4). SPFMV and SPLV were the most two popular viruses with their frequencies of occurrence being 20.8–100% and 2.1–90%, respectively (Zhang et al., 2006). Co-infection of SPFMV and SPLV was detected with an infection frequency at 8.9% in Shandong province (Zhang et al., 2006). Sweetpotato chlorotic stunt virus (SPCSV) was for the first time detected in Shandong province, with about 9% of incidence frequencies (Zhang et al., 2006). Fortunately, until now co-infection of SPCSV with SPFMV, resulting in development of sweetpotato virus disease (SPVD), has not been detected yet. A survey conducted in the main sweetpotato regions including Jiangsu, Sichuan, Shandong and Anhui provinces showed that sweetpotato viruses caused an average yield loss of about 20–30% (Gao et al., 2000), with the most severe case reaching 78% reported in Shandong province (Shang et al., 1996a). Virus-infected sweetpotato plants were found to be much more susceptible, than the healthy plants, to fungi *Monilochaetes infuscans* and *Ceratocystis fimbriata*, and nematodes *Pratylenchus coffeae* (Yang et al., 1998).

Further studies demonstrated that usage of seed tubers successively propagated from virus-free stock plants resulted in a decreased storage root yield (Huang et al., 2000: Wong and Chen, 2001; Zhang et al., 2006). Yield of root tubers produced from the first generation of virus-free seed tubers was similar to that from the second generation, but was significantly higher than that from the virus-infected seed tubers (Wong and Chen, 2001). However, this increased effect of yield disappeared when the virus-free seed tubers successively propagated for three years were used. Similar results were also observed by Huang et al. (2000) and Zhang et al. (2006). These data indicated that virus-free planting materials could be re-infected with time. Based on the above data, Zhang et al (2006) suggested, that virus-free planting materials needs to be renewed at least every 3 years, in order to maintain their high potential for yield.

Table 15.4 A list of sweetpotato viruses reported in China

Type of virus	Genus	Original location	Reference
C-6 virus	*Carlavirus* (a putative member)	Jiangsu and Anhui	Zhang et al., 2006
SPCFV		Jiangsu and Anhui	Zhang et al., 2006
SPCSV	*Crinivirus*	Shandong	Zhang et al., 2006
SPCV		Not specified	Gao et al., 2000
SPFMV	*Potyvirus*	Shandong, Jiangsu, Anhui and Guangdong	Zhang et al., 2006 Colient and Kummert, 1993, Colient et al., 1993, 1996, 1997, 1998; Ateka et al., 2007
SPLCV	*Begomovirus*	Not specified	Lotrakul et al., 2001
		Liaoning	Luan et al., 2006
		Not specified	Li et al., 2004
SPLV	*Potyvirus*	Shandong, Jiangsu, Anhui and Guangdong	Zhang et al., 2006, Colient et al., 1997, 1998;
		Not specified	IsHak et al., 2003
SPMMV	*Ipomovirus*	Jiangsu and Anhui	Zhang et al., 2006
SPMSV	*Potyvirus*	Jiangsu and Anhui	Zhang et al., 2006
SPVG	*Potyvirus*	Guangdong	Colient et al., 1994, 1996, 1998
SPV-2	*Potyvirus* (a tentative member)	Guangdong	Ateka et al., 2007

Virus detection was studied using the indicator plants such as *I. setosa*, *I. nil* and *Chenopodium quinoa* (Shang et al. 1996b), electron microscopic observation (Zhang et al., 1995; Yang et al., 1998) and *in situ* hybridization (Qiu et al., 1992). Identification and characteristics of tolerance of sweetpotato (*I. batatas*) and its wild relatives to viruses demonstrated that several cultivars such as "Xushu 18", "Xushu 22", "Beniazuma" and *I. cairica* originating from South America were tolerant to viruses (Li et al., 2003).

More recently, using a sweetpotato genotype CIP199004.2 infected with SPFMV and SPCSV originating from Africa, Wang and Valkonen (2008a) reported that cryotherapy of shoot tips could completely eliminate them in both single and co-infection, regardless of size of shoot tips used. Studies of Wang and Valkonen (2008b) also demonstrated that cryotherapy of shoot tips was a very efficient method for elimination of sweetpotato little leaf phytoplasma. They suggested that cryotherapy of shoot tips would provide an alternative means for elimination of sweetpotato viruses and phytoplasma, can be simultaneously used for long-term storage of sweetpotato germplasm and virus elimination and phytoplasma, and considered to be a safe method for movement of sweetpotato germplasm between regions.

Plate 15.1 **A.** "Guangshu 69", **B.** "Yanshu 5", **C.** "Guangshu 79", **D.** "Jishu 15", **E.** "Jishu 18", **F.** "Jishu 98", **G.** "Xushu 18", **H.** "Xushu 25", **I.** "Jinshan57", **J.** "Nanshu 88", **K.** "Eshu 5", **L.** "Chuanshu 34", **M.** A typical meristem used for virus elimination. **N.** Plantlet regenerated from meristem culture, **O.** Culture room of sweetpotato. **P.** Cross breeding. Bars indicate 5 cm (See also Plate 31 on page xxix)

Plate 15.2 **A.** Propagation bed. **B.** Preparation of cuttings. **C.** Ridge planting. **D.** Mulching field. **E.** Large scale production. **F.** Production of vegetable sweetpotato. **G.** Harvesting of vegetable sweetpotato. **H.** Harvesting of sweetpotato. **I.** Sun-drying of sweetpotato. **J.** Sweetpotato noodles. **K.** and **L.** Preserved Sweetpotato chips. **M.** Sweetpotato starch. **N.** Processing of Sweetpotato noodles. **O.** fresh tuber roots for market (See also Plate 32 on page xxx)

References

Abe, K., Emori, Y., Kondo, H., Suzuki, K., and Arai, S. 1987. Molecular cloning of a cysteine proteinase inhibitor of rice (Oryzacystatin). *J. Biol. Chem.* 262: 16793–16797.

Anonymous, 1984. *Sweet Potato Cultivation in China.* Shanghai Sic-Tech Press, Shanghai (in Chinese).

Anonymous, 1990a. *China's Big Encyclopedia-Agriculture.* Vol. I, p. 254 (in Chinese).

Anonymous, 1990b. *China's Big Encyclopedia-Agriculture.* Vol. II, p. 1582 (in Chinese).

Anonymous, 2003. *Select Markets for Taro, Sweet Potato and Yam.* Rural Industries Research & Development Corporation, BIRDC Publication No 03/052, Australia.

Ateka, E.M., Barg, E., Njeru, R.W., Thompson, G., and Vetten, H.J. 2007. Biological and molecular variability among geographically diverse isolates of sweet potato virus 2. *Arch. Virol.* 152: 479–488.

Austin, D.F., 1987. The taxonomy, evolution and genetic diversity of sweet potatoes and related wild species. *Exploration, Maintenance and Utilization of Sweet potato Genetic Resources.* Report of the first sweet potato planning conference, 27–59. International Potato Centre, Lima, Peru.

AVRDC, 1988. Chemistry-screening for starch digestibility of sweet potato. Progress report-1988. *Asian Vegetable Research and Development Centre (AVRDC)*, Taiwan, China, pp. 249–253.

Azzouz, H., Cherqui, A., Campan, E., Rahbe, Y., Duport, G., Jouanin, L., Kaiser, L., and Giordanengo, P. 2005. Effects of plant protease inhibitors, oryzacystatin I and soybean Bowman-Birk Inhibitor, on the aphid *Macrosiphum euphorbiae* (Homoptera, Aphididae) and its parasitoid *Aphelinus abdominalis* (Hymenoptera, Aphelinidae). *J. Insect Physiol.* 51: 75–86.

Brown, J.K., Frohlich, D.R., and Rosell, R.C. 1995. The sweet potato or silverleaf whiteflies: biotypes of *Bemisia tabaci* or a species complex. *Ann. Rev. Entomol.* 40: 511–534.

Cai, J.F., Zhang, S.Z., Li, W.F., Ma, W., Wang, X.Q., and Jiang, X.F. 2001. Growth characteristics and cultivation techniques of virus-free sweet potato. *Rain Fed Crops* 2: 53 (in Chinese).

Cai, N.T., Huang, H.K., Qou, Y.X., Zheng, X, Wu, Q.Y., Luo, W.B., and Li, G.X. 2006. Breeding of new sweet potato variety Fushu for leaf vegetable and its cultivation techniques. *J. Fujian Agri.* 21(1): 12–25 (in Chinese).

Chen, S.P., and Lin, M.Y. 2004. Good-quality, high yield and non-pollution cultural techniques of southern sweet potato. *Rain Fred Crops* 4: 223–224 (in Chinese).

Chen, Y.J., Wang, L.D., Huang, J., Lin, G.Y., Liang, Z.S., and Wu, H.W. 2004. Deterrent effect of crude extract from mycelia toxin of *Verticillium lecanii* on sweet potato whitefly (*Bemisia tabaci*). *Fujian J. Agri. Sci.* 19(4): 210–212 (in Chinese).

China's Yearbook of Agriculture, 2005. p.183 (in Chinese).

Colient, D., and Kummert, J. 1993. Identification of a sweet potato feathery mottle virus isolate from China (SPFMV-CH) by the polymerase chain reaction with degenerated primers. *J. Virol. Meth.* 45: 149–159.

Colient, D., Kummert, J., and Lepoivre, P. 1997. Evidence for the assignment of two strains of SPLV to the genus *Potyvirus* based on coat protein and 3′ non-coding region sequence data. *Virus Res.* 49: 91–100.

Colient, D., Kummert, J., Lepoivre, P., and Semal, J. 1993. Identification of distinct Potyviruses in mixedly-infected sweet potato by the polymerase chain reaction with degenerate primers. *Phytopathology* 84: 65–69.

Colient, D., Lepoivre, P., Xia, F.Z. and Kummert, J. 1996. Detection and identification of sweet potato viruses by the polymerase chain reaction. *Agro-Food-Industry* March/April: 33–34.

Colient, D., Nguyen, M., Kummert, J., Lepoivre, P., and Feng, Z.X. 1998. Differentiation among potyviruses infecting sweet potato based on genus- and virus-specific reverse transcription polymerase chain reaction. *Plant Dis.* 84: 65–69.

Dreher, M.L., Dreher, C.J., and Berry, J.W. 1984. Cultivar differences in trypsin inhibitors of sweet potato roots. *CRC Crit. Rev. Food Sci. Nutr.* 21: 47–71.

FAO, 2004. http://www.fao.org/es/ess/top/country.htm
FAO, 2005. http://www.fao.org/es/ess/top/country.htm
Farmer, F.J., Li, X., Feng, G., Zhao, B., Chatagnier, O., Gianinazzi, S., Gianinazzi-Pearson, V., and van Tuinen, D. 2007. Molecular monitoring of field-inoculated AMF to evaluate persistence in sweet potato crops in China. *Appl. Soil Ecol.* 35: 599–609.
Fuglie, K.O., Zhang, L.M., Salazar, L.F., and Walker, T.H. 1999. *Economic Impact of Virus-Free Sweet Potato Seed in Shandong Province*, China. International Potato Centre, Lima, Peru.
Gao, F., Gong, Y.F., and Zhang, P.B. 2000. Production and employment of virus-free sweet potato in China. *Crop Prot.* 19: 105–111.
Guo, F., Gong, Y.F., Lin, Z.P., and Wang, X.J. 2001. *Agrobacterium*-mediated genetic transformation of *Ipomoea batatas* and regeneration of transgenic plants. *Acta Agron. Sin.* 27(6): 751–756 (in Chinese).
Guo, J.M., Liu, Q.C., Zhai, H., and Wang, Y.P. 2006. Regeneration of plants from *Ipomoea cairica* L. protoplastts and production of somatic hybrids between *I. Cairica* L. and sweet potato *I. batatas* (L.) Lam. *Plant Cell, Tiss. Org. Cult.* 87: 321–327.
Guo, J.P., and Pan, D.R. 2002. PCR detection of nematode resistance in sweet potato. *Acta Agron. Sin.* 28(2): 321–327 (in Chinese).
Guo, X.D., and Zhang, Y.G. 2006. Preliminary study on sweet potato artificial seeds. *Jiangsu J. Agri. Sci.* 22(1): 93–94 (in Chinese).
Hagenimana, V., Vezina, L.P., and Simard, R.E. 1992. Distribution of amypase in sweet potato (*Ipomea batatas* L.) root tissue. *J. Agric. Food Chem.* 40: 1777–1783.
Hagenimana, V., Vezina, L.P., and Simard, R.E. 1994. Amylolytic activity in germinating sweet potato (*Ipomea batatas* L.). *J. Am. Soc. Hort. Sci.* 119: 313–320.
Hao, D.L., and Hao, F.W. 2001. Cultivation techniques of sweet potato in mountain areas. *Modern Agriculture.* 5: 15 (in Chinese).
He, S.Z., Han, Y.F., Wang, Y.P., Zhai, H., and Liu, Q.C. 2009. *In vitro* selection and identification of sweet potato (*Ipomoea batatas* (L.) Lam.) plants tolerant to NaCl. *Plant Cell Tiss. Org. Cult.* 96: 69–74.
He, Y.X., Yang, X.J., Weng, Q.Y., Huang, J., and Wang, L.H. 2006. Biotype identification of the populations of *Bemisia tabaci* in Fujian, China. *J. Fujian Agri. Forest. Uni.* 35(5): 486–490 (in Chinese).
Ho P-T, 1955. The introduction of American food plants into China. *Am. Anthropol.* 57: 191–201.
Hou, F.Y., Zhao, B., Wang, Q.M., Li, A.X., Zhang, H.Y., Liu, G.Z., and Zhang, L.M. 2006. Progress on molecular biology of sweet potato. *Mol. Plant Breed.* 4(1): 119–122 (in Chinese).
Hu, W.Z., and Tanaka, S.I. 2007. Effects of heat treatment on the quality and storage life of sweet potato. *J. Sci. Food Agri.* 87: 313–319 (in Chinese).
Huan, J.C., and Sun, M. 2000. Genetic diversity and relationships of sweet potato and its wild relatives in *Ipomoea* series *Batatas* (Convolvulaceae) as revealed by inter-simple sequence repeat (ISSR) and restriction analysis of chloroplast DNA. *Theor. Appl. Gen.* 100: 1050–1060.
Huang, C.X., Li, S.Z., Li, Y.X., and Gou, Y.L. 2000. Effects of generations of virus-free seed tubers on yield of sweet potato. *Shandong Agri. Sci.* 1: 23 (in Chinese).
IsHak, J.A., Kreuze, J.F., Johansson, A., Mukasa, S.B., Tairo, F., Abo El-Abbas F.M. and Valkonen, J.P.T. 2003. Some molecular characteristics of three viruses from SPVD-infected sweet potato plants in Egypt. *Arch. Virol.* 148: 2449–2460.
Iwanaga, M., 1988. Use of the wild germplasm for sweet potato breeding In: Gregory P (ed) Exploration, maintenance, and utilization of sweet potato genetic resources. International Potato Centre, Lima, Peru. pp. 199–210.
Jiang, S.J., Liu, Q.C., Zhai, H., Wu, L.S., and Wang, Y.P. 2004. Regeneration of sweet potato transgenic plants with Oryzacystatin-I (OCI) gene. *J. Agri. Biotech.* 12(1): 34–37 (in Chinese).
Jie, Q., Li Hua, Zhai, H., Wang, Y.P., and Liu, Q.C. 2008. Construction of a genetic linkage map and development of molecular markers linked to stem nematode resistance gene based on AFLP markers in sweet potato, *Ipomoea batatas* (L.) Lam. (submitted in Chinese).
Li, A.X., Liu, Q.C., Wang, Y.P., Zhai, H., Liu, B.L., and Wang, S.F. 2002. *In vitro* selection of drought- and salt-tolerant mutants in sweet potato. *J. Agri. Biotech.* 10(1): 24–28 (in Chinese).

Li, H.M., Xing, J.Y., Ma, D.F., Xie, Y.P., and Li, X. 2003. Identification and characteristics of virus tolerance of sweet potato and it wild relatives. *Crops* 3: 11–14 (in Chinese).

Li, Q., Liu, Q.C., and Ma, D.F. 2004. Protoplast culture of sweet potato. *Rain Fred Crops* 24(5): 271–274 (in Chinese).

Li, Q., Liu, Q.C., Ma, D.F., and Zhai, H. 2005. Advances, problems in genetic transformation on sweet potato. *Mol. Plant Breed.* 1: 96–106 (in Chinese).

Li, R.H., Salih, S., and Hurtt, S. 2004. Detection of Germiniviruses in sweet potato by polymerase chain reaction. *Plant Dis.* 88: 1347–1351.

Li, W.G., Wu, X.Q., Cai, H.Y., and Du, R. 1992. Sweet potato in China. In: Scott, G.J., Wiersema, S., Ferguson P.I. (eds). *Product Development for Root and Tuber Crops.* Vol I. Asia. International Potato Centre, Lima, Peru. pp.41–50.

Liu, Q.C., Lu, D.H., Ma, B., and Zhou, H.Y. 1996. Cell suspension cultures and efficient plant regeneration in sweet potato. *J. Agri. Biotech.* 4(3): 238–242 (in Chinese).

Liu, Q.C., Mi, K.X., and Zhou, H.Y. 1998. Regeneration and identification of somatic hybrids between *Ipomoea batatas* L. and *I. lacunosa. Acta Agron. Sin.* 24(25):529–535 (in Chinese).

Liu, Q.C., Mi, K.X., Lu, D.H., Zhou, H.Y., and Fu, Z. 1997. Establishment of embryogenic cell suspension cultures in sweet potato, *Ipomoea batatas* (L.) Lam. *Acta Agron. Sin.* 23(1): 22–27 (in Chinese).

Liu, Q.C., Zhai, H., and Wang, Y.P. 2003. Current status of cell engineering and molecular breeding in sweet potato. *Crops* 6:1–3 (in Chinese).

Liu, Q.C., Zhai, H., Wang, Y.P., and Zhang, D.P. 2001. Efficient plant regeneration from embryogenic suspension cultures of sweet potato. *In Vitro Cell. Dev. Biol.-Plant.* 37: 564–567.

Liu, Z.J., Li, J.G., Ying, S.L., and Niu, C.F. 2004. Characteristics of sweet potato early cultivars and techniques of two-season cultivation. *Crops* 3: 20–21 (in Chinese).

Liu, Z.S., Liu, Q.C., Zhai, H., and Wang, Y.P. 2005. Cloning of RGA related to stem nematode resistance of sweet potato (*Ipomea batatas* (L.) Lam) with modified SSAP. *Mol. Plant Breed.* 3(3): 369–374 (in Chinese).

Liu, Z.S., Liu, Q.C., Zhai, H., and Wang, Y.P. 2006. Cloning and sequence analysis of Myo inositol-1-phosphate synthase gene in sweet potato. *J. Agri. Biotech.* 21(4): 219–225 (in Chinese).

Loebenstein, G., Fuentes, S., Cohen, J., and Salazar, L.F. 2003. Sweet potato. In: Loebenstein, G., and Thottappilly G (eds), *Virus and Virus-like Diseases of Major Crops in Developing Countries*, Kluwer Academic Publishers, Dordrecht, pp. 223–248.

Lotrakul, P., Valverde, R.A., and Clark, C.A. 2001. Sweet potato leaf curl virus and related geminiviruses in sweet potato. *Acta Hort.* 583: 135–141.

Lu, G.Q., George, M.S., and Zhou, W.J. 2003. Genotype variation of sweet potato under low potassium stress. *J. Plant Nutri.* 26: 745–756.

Luan, Y.S., Zhang, J., and An, L.J. 2006. First report of sweet potato leaf curl virus in China. *Plant Dis.* 90: 1111.

Luan, Y.S., Zhang, J., Gao, X.R., and An, L.J. 2007. Mutation induced by ethlmethanesulphonate (EMS), *in vitro* screening for salt tolerance and plant regeneration of sweet potato (*Ipomoea batatas* L.). *Plant Cell, Tiss. Org. Cult.* 88: 77–81.

Luo, H.R., Zhang, Y.W., and Zhang, X.Z. 2002. Study of transformation of sweet potato with *Agrobacterium tumefaciens* and resistant calli obtained at high frequency. *J. Sichuan Uni.* 39: 21–24 (in Chinese).

Murashige, T., and Skoog F. 1962. A revised medium for rapid growth and bioassays with tobacco cell cultures. *Physiol. Plant.* 15: 473–497.

O'Brien, P. 1972. The sweet potato: its origin and dispersal. *Am. Anthropol.* 74: 342–365.

Ou, X.Q., Ren, X.J., and Li, X.H. 2007. Sweet potato variety Baishu 1. *Acta Hort. Sin.* 34: 266 (in Chinese).

Ou, X.Q., Ren, X.J., and Yang, G.T. 2005. Analysis on nutrient components in shoot-tips of new vegetable-type swee potato Baishu 1. *J. Henan Agri. Sci.* 12: 30–33 (in Chinese).

Pan, D.R., Chen, G.S., Zhou, Y.F., Guo, J.P., and Chen, J.Q. 2006. The primary study on cloning and partial sequence analysis of a gene for nematode resistance in sweet potato. *J. Fujian Agri. Forest. Uni.* 35(1): 57–59 (in Chinese).

Peng, S.Q., and Chen, S.C. 2004. Cloning and sequence analysis of sweet potato polyphenol oxidase cDNA. *J. Agri. Biotech.* 10: 241–245 (in Chinese).

Pu, Z.G., Wu, J, Tang, J., Tan, W.F., Wang, D.Y., Zhang, Z.S., and Yan, W.Z. 2005. Constructing AFLP genetic linkage map related to starch content of sweet potato. In: Ma, D.F., Liu Q.C.(eds), *Sweet potato Breeding and Industrialization in China.* China Agricultural University Press. Beijing, pp. 201–205 (in Chinese).

Qiu, B.L., and Ren, S.X. 2005. Effect of host plants on the development, survival and reproduction of *Encarisia bimaculata* (Hymenoptera:Aphelinidae), a parasitoid of *Bemisia tabaci* (Homoptera:Aleyrodidae). *Acta Entomol. Sinica* 48(3): 365–369 (in Chinese).

Qiu, B.S., Zhao, F., and Wang, X.F. 1992. cDNA probes of sweet potato feathery mottle virus (SPFMV) and its application. *Acta Microbiol. Sinica* 32(4): 242–246 (in Chinese).

Rahbé Y, Deraison, C., Bonade-Bottino, M., Girard, C., Nardon, C., and Jouanin, L. 2003. Effects of the cysteine protease inhibitor oryzacystatin (OC-I) on different aphids and reduced performance of *Myzus persicae* on OC-I expressing transgenic oil seedrape. *Plant Sci.* 164: 441–450.

Ren, S.X., Wang, Z.Z., Qiu, B.L., and Xiao, Y. 2001. The status of *Bemisica tabaci* in China and non-chemical control strategies. *Entomol. Sinica* 8(3): 279–288 (in Chinese).

Shang, Y.F., Yang, C.L., Xin, X.Q., Zhao, J.H., Li, C.S., and Luo, R.W. 1996a. Techniques of sweet potato virus-free by meristem tip culture. *Plant Prot.* 22(5): 14–16 (in Chinese).

Shang, Y.F., Yang, C.L., Zhao, J.H., and Lu, X.B. 1996b. Sweet potato virus research and application of virus-free sweet potato. *Plant Doctor* 9(4): 35–39 (in Chinese).

Sheng, J.L., Wang, Y.H., and Xue, Q.H. 1987. Current status of sweet potato research and production in China. *China Sweet Potato* 1:16–20 (in Chinese).

Song, B.F., Wang, S.W., Xie, K.Y., Yang, Y.J., Yang, C.L., Zhang, H.L., Li, R.G., Xing JY, Wu, J.Y., Guo, X.D., Meng, Q., and Zhang, L.M. 1997. Present research and development of virus-free sweet potato in China. *Sci. Agri. Sinica* 30(6): 43–48 (in Chinese).

Tan H.Z., Gu, W.Y., Zhou J.P., Wu, W.G., and Xie, Y.L. 2006. Comparative study on the starch noodle structure of sweet potato and mung bean. *J. Food. Sci.* 71: 447–455.

Tang, S.H., and Li, S.B. 1994. Studies on artificial seeds of sweet potato. *Crops*. 6: 746–750 (in Chinese).

Wang, G.L., Fang, H.J., and Li, H.Y. 2003a. Somatic embryogenesis of sweet potato and its application in virus-free propagation. *Acta Agronomica Sinica* 29(3): 345–348 (in Chinese).

Wang, K.Q., Chen, J.P., and Hu, D. 2002. Cultivation techniques for productions of sweet potato shoots. *Crop Res.* 16(4): 186–187 (in Chinese).

Wang, L.D., Huang, J., You, M.S., and Liu, B. 2004. Time-dose-mortality modelling and virulence indices for six strain of *Verticillium lacani* against sweet potato whitefly, *Bemisia tabaci* (Gennadius). *J. Appl. Ent.* 128: 494–500.

Wang, L.D., Huang, J., You, M.S., Guan, X., and Liu, B. 2005. Effects of toxins from two strains of *Verticillium lecanii* (Hyphomycetes) on bioattributes of a predatory ladybeetle, *Delphastus catalinae* (Col., Coccinellisae). *J. Appl. Ent.* 129: 32–38.

Wang, L.D., Huang, J., You, M.S., Guan, X., and Liu, B. 2007a. Toxicity and feeding deterrence of crude toxin extracts of Lecanicillium (Verticillium) lecanii (Hyphomycetes) againt sweet potato whitefly, *Bemisia tabaci* (Homoptera:Aleyrodidae). *Pest Manag. Sci.* 63: 381–387.

Wang, Q.C., and Valkonen, J.P.T. 2008a. Elimination of two viruses which interact synergistically from sweetpotato by shoot tip culture and cryotherapy. *J. Virol. Meth.* 154: 135–145.

Wang, Q.C., and Valkonen, J.P.T. 2008b. Efficient elimination of sweet potato little leaf phytoplasma by cryotherapy of shoot tips. *Plant Pathol.* 57: 338–347.

Wang, X., Zhou, Z., Li, X., Gu, X.H., and Ma, D.F. 2006. Optimization of genetic transformation using SAAT in sweet potato. *Jiangsu J. Agri. Sci.* 22(1): 14–18 (in Chinese).

Wang, Y.P., Liu, Q.C., Li, A.X., Zhai, H., Zhang, S.S., and Liu, B.L. 2003b. *In vitro* selection and identification of drought-tolerant mutants of sweet potato. *Sci. Agri. Sinica.* 36(9): 1000–1005 (in Chinese).

Wang, Y.P., Wang, F., Zhai, H., and Liu, Q.C. 2007b. Production of a useful mutant by chronic irradiation in sweet potato. *Sci. Hortic.* 111: 173–178.

Wong, D.H., and Chen, S.P. 2001. Effects of generations of virus-free seed tubers on growth and yield of sweet potato. *Rain Fed Crops*. 21(1): 29–31 (in Chinese).

Wu, J., Tan, W.F., Pu, Z.G., Wang, D.Y., Zhang, Z.S., and Yan, W.Z. 2005. Establishment of SRAP genetic linkage map and starch quantitative trait locus orientation in sweet potato. In Ma, D.F., Liu Q.C. (eds), *Sweet Potato Breeding and Industrialization in China*. China Agricultural University Press. Beijing, pp. 195–200 (in Chinese).

Wu, X.X., Li, Z.X., Hu, D.X., and Shen, Z.R. 2003. Identification of Chinese populations of *Bemisia tabaci* (Gennadius) by analyzing ribosomal ITS1 sequence. *Prog. Nat. Sci.* 13: 276–281.

Xu, Y.B., Chen, Y., and Fu, Z.G. 2004. Advance of research on drought-resistant physiology and cultivation techniques of sweet potato. *Agri. Res. Arid Areas*. 22(1): 128–131 (in Chinese).

Yan, W.Z., Wu, J., Wang, D.Y., and Tan, W.F. 2004. Introduction of rice cryteine proteinase inhibitor into sweet potato varieties. *Mol. Plant Breed.* 2: 203–207 (in Chinese).

Yang, C.L., Shang, Y.F., Zhao, J.H., and Li, C.S. 1998. Techniques and practice of production of virus-free sweet potato. *Acta Phytophylacia Sinica*. 25: 51–55 (in Chinese).

Yuan, B.Z. 1989. Fourty years of research on sweet potato in China. *China Sweet Potato*. 3: 1–5 (in Chinese).

Yu, B., Zhai, H., Wang, Y.P., Zhang, N., He, S.Z., and Liu, Q.C. 2007. Efficient *Agrobacterium tumefaciens*-mediated transformation using embryogenic suspension cultures in sweet potato, *Ipomea batabas* (L.) Lam. *Plant Cell, Tiss. Org. Cult.* 90: 265–273.

Zang, N., Zhai, H., Wang, Y.P., Yu, B., He, S.Z., and Liu, Q.C. 2007. Development of transgenic sweet potato expressing *bar* gene for herbicide resistance. *Mol. Plant Breed.* 5: 475–479 (in Chinese).

Zhai, H., and Liu, Q.C. 2003. Studies on the genetic transformation of embryogenic suspension cultures in sweet potato. *Sci. Agri. Sinica* 5: 487–491 (in Chinese).

Zhang, B.Y., Liu, Q.C., Zhai, H., and Zhou, H.Y. 1999a. Efficient regeneration of interspecific somatic hybrids between sweet potato and its wild relatives. *Sci Agri. Sinica*. 32: 23–27 (in Chinese).

Zhang, B.Y., Liu, Q.C., Zhai, H., and Zhou, H.Y. 2001. Production of fertile interspecific somatic hybrids between sweet potato and its wild relative, *Ipomoea lacunosa*. *Acta Hort.* 583: 81–85.

Zhang, D.P., Collins, W.W., and Andrade, M. 1995. Estimation of genetic variance of sweet potato. *HortSci.* 28: 348–349.

Zhang, D.P., Collins, W.W., and Andrade, M. 1998. Genotype and fertilization effects on trypsin inhibitor activity in sweet potato. *HortSci.* 33: 225–228.

Zhang, D.P., Collins, W.W., and Belding, S. 1993. Improving sweet potato starch digestibility for animal feeding. *HortSci.* 18: 325–326.

Zhang, J., Liu, D., Xie, D., Wang, Y., and Sun, Y. 2002a. Production of glycerol by fermentation using osmophilic yeast *Candida krusei* with different substrates. *Enz. Microb. Tech.* 30: 758–762.

Zhang, L.M., Wang, Q.M., Ma, D.F., and Wang, Y. 2006. The effect of major viruses and virus-free planting materials on sweet potato root yield in China. *Acta Hort.* 703: 71–77.

Zhang, L.M., Wang, Q.M., Wang, J.J., Xi, G.H., and Wang, Y.C. 1999b. Classification standard and reproduction system of virus-free sweet potato seeds. *J. Shandong Agri Sci.* 1: 24–26 (in Chinese).

Zhang, L.P., Zhang, Y.J., Zhang, W.J., Wu, Q.J., Xu, B.Y., and Chu, D. 2005. Analysis of genetic diversity among different geographical populations and determination of biotypes of *Bemisia tabaci* in China. *J. Appl. Ent.* 129: 121–128.

Zhang, X.T., Lin, S.X., and Liu, S.Y. 2004. Study on making artificial seeds of virus-elimination sweet potato. *China Seed Industry* 4: 29–30 (in Chinese).

Zhang, Z.T., Wheatley, C.C., and Corke, H. 2002b. Biochemical changes during storage of sweet potato roots differing in dry matter content. *Posth. Biol. Tech.* 24: 317–325.

Zhang, L.M., and Wang, Y.C. 1995. Progress of research and application of virus-free sweet potato in Shandong. Proc. 1st Chinese-Japanese Symposium on sweet potato and potato. Beijing Agricultural University Press. pp. 117–121 (in Chinese).

Zhou, H., Wang, X., Ma, D.F., Li, H.M., Xie, T.P., and Li, X.Y. 2005a. Genetic diversity analysis based on RAPD for resistance and susceptibility of sweet potato varieties to stem nematode. *Jiangsu J. Agri. Sci.* 21(1): 35–39 (in Chinese).

Zhou, H., Wang, X., Ma, D.F., Li, H.M., Xie, T.P., and Li, X.Y. 2005b. Identification of RAPD markers linked to stem-nematode resistant gene in sweet potato. *J. Agri.Biotech.* 13(5): 549–552 (in Chinese).

Zhou, L.Y., Gao, S.G., Bi, Y.J., and Yang, W.L. 2003a. Study on callus induction and plant regeneration of sweet potato. *Chinese Agri. Sci. Bulletin* 19(3): 61–64.

Zhou, L.Y., Gao, S.G., Bi, Y.J., Qiao, Y.K., and Shuo, Z.Q. 2003b. Study on sweet potato axillary buds of artificial seeds. *Seeds* 3: 37–38 (in Chinese).

Chapter 16
Sweetpotato in Sub-Saharan Africa

J. Low, J. Lynam, B. Lemaga, C. Crissman, I. Barker, G. Thiele, S. Namanda, C. Wheatley and M. Andrade

Historical Background

Botanically sweetpotato is called *Ipomoea batatas* (L.) Lam and belongs to the morning-glory family (*Convolvulaceae*) and originated from Latin America. The exact date of arrival of sweetpotato on the continent of Africa is unknown. Evidence indicates that slave traders brought it into the region and since its introduction it has been displacing the true yam in tropical Africa (Davidson, 1999). Given that it has been in the food system for several hundred years, whereas round, Irish, or solanum potato (*Solanum tuberosum*) was introduced in the late 1800s, sweetpotato is often falsely considered by African farmers to be an indigenous crop and is often referred to as the "local" or "traditional" potato in most Sub-Saharan African countries.

Importance of Sweetpotato in the Sub-Saharan African Region

Sweetpotato is one of the most widely grown root crops in sub-Saharan Africa (SSA), covering around 2.9 million hectares with an estimated production of 12.6 million tons of roots in 2007 (FAOSTAT, 2008 and national surveys). It is predominantly grown in small plots by poorer farmers, hence it is known as the "poor man's food" (Woolfe, 1992). However, as women predominate in sweetpotato production, "poor person's food" would be more accurate.

The crop is particularly important in countries surrounding the Great Lakes in Eastern and Central Africa; Malawi, Angola, Mozambique, and Madagascar in Southern Africa, and Nigeria in West Africa (Woolfe, 1992). It is expanding faster than any other major food crop in SSA (Fig. 16.1). It can be found from sea level up to 2500 m. Sweetpotato generates large amounts of food per unit area per unit time (Woolfe, 1992), superior to other major staples (Table 16.1). It tolerates occasional dry spells and yields even on less fertile soils in contrast to other crops such as

J. Low (✉)
International Potato Center, Sub-Saharan, Africa
e-mail: J.LOW@CGIAR.ORG

G. Loebenstein, G. Thottappilly (eds.), *The Sweetpotato*,
DOI 10.1007/978-1-4020-9475-0_16,

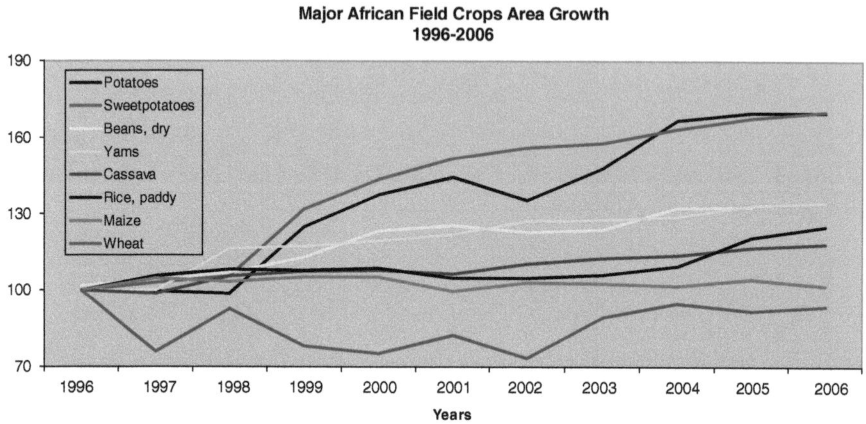

Fig. 16.1 Growth in area in Sub-Saharan Africa under major field crops: 1966–2006. Source: FAOSTAT (2008)

maize (Ewell, 1990). Compared to other crops, sweetpotato requires few inputs and relatively less labor.[1] The rapid growth of sweetpotato area in sub-Saharan Africa during the past decade is due to three major factors: (1) changes in cropping patterns, (2) unstable economies and (3) the increasing commercialization of production.

Area under sweetpotato has grown in different locales as farmers have changed cropping patterns as other traditional crops become difficult to cultivate for a variety of reasons. Perhaps the most important factor explaining change in cropping patterns affecting the area of sweetpotato is the spread of cassava and banana diseases especially in the Lake Victoria region. Numerous studies have identified the

Table 16.1 Comparative energy yields of sweetpotato and other major crops

Crop[a]	Average Tropical Yield (Tons/ Hectare)	Edible Energy Value (MJ/kg)	Proportion of Edible Energy (%)	Edible Energy per Hectare (10^3 MJ)	Mean Growth Period (Days)	Edible Energy (MJ/ha/ day)
Sweetpotato	7	4.8	88	27.2	140	194
Cassava	9	6.3	83	45.6	330	138
Yam	7	4.4	85	26.2	280	94
Banana	13	5.4	59	41.4	365	113
Rice[b]	2	14.8	70	20.8	140	149
Maize	1	15.2	100	18.8	130	145
Sorghum	<1	14.9	90	11.1	110	101
Millet	<1	15.0	100	8.2	100	82

Source: Woolfe (1992), p. 4 Notes: Based on de Vries et al. 1967.
[a] Cereals, air-dry; roots/tubers/bananas fresh.
[b] Paddy Rice.

[1] Average energy output/input ratios for rice and sweetpotato on Fijian farms were 17:1 and 60:1, respectively (Norman et al., 1984).

serious and spreading epidemic of cassava brown streak virus. There are 12 million hectares of cassava in SSA. Frequently sweetpotato and cassava are grown on the same farm or the same region. When cassava fails, farmers frequently switch to sweetpotato to substitute for the energy delivered by cassava. Another important reason driving changing cropping patterns is that cultivable land has declined in size in many countries due to increasing population. Farmers respond by growing more root and tuber crops that give higher yields per unit area than grain crops, among which sweetpotato is a significant choice.

A second important source for cropping pattern change is the decline in support for maize production. In the postcolonial era, sub-Saharan African governments seeking food self-sufficiency created a diverse set of policies that encouraged an excessive reliance on maize. During the last decade, especially in southern Africa, governments have recognized the high cost of the policies and began to dismantle some of the supports. Reduced subsidies of irrigation, fertilizer, seed and marketing costs led to declines in maize area. As farmers seek alternatives, the robust, low cost sweetpotato is often included. Soaring grain prices, resulting from the recent food crisis, have led to increased investments in inputs and staple food sectors, but the important food security role of root and tuber crops is now widely recognized.

A third important source for cropping pattern change is the impact of HIV/AIDS (Jayne et al., 2004). Now in its third decade as an epidemic, death rates in rural areas have climbed sharply. That HIV/AIDS kills those between 20 and 50 is well documented. The family and farm level impacts of the illness and death of farm family parents is being better documented. Farm families de-capitalize as savings, equipment and livestock are sold to care for the sick. In many cultures, funeral customs create additional expectations for capital expenditures to finance the ceremony. Farm families lose labor and management skills. One impact of the impoverishment, labor loss and skills loss is that the remaining farmers –typically a single parent or orphan headed households – shift to lower labor, lower cost and lower risk crops (Kennedy, 2002). Sweetpotato is a crop that answers all these needs. Rugalema (1998) showed in Bukoba District, Tanzania a shift from the intensively managed bean/coffee/banana cropping system to a simpler, extensive and low cost system based on cassava and sweetpotato.

As agriculture becomes more market-oriented, sweetpotato is one of several crops that farmers can produce to obtain cash income in addition to subsistence food security. Supply and demand factors are therefore increasingly important in determining the role sweetpotatoes will play in a more market-oriented smallholder farm sector. Markets for fresh roots and vines do exist, but with few exceptions (e.g. Uganda) are not large and, like cassava, consumption of fresh sweetpotato roots tends to decrease with urbanization, due to shifts in diets related to factors such as price, convenience and status. Enhancing the unrealized potential of sweetpotato by improving its marketing system (Ndunguru et al., 1998; Tomlins et al., 2000) and increasing demand for fresh roots and value-added sweetpotato-based products would contribute to improved productivity and well-being among rural producers, especially women, and help meet ever increasing urban food demand (Best et al., 2006; Omosa, 1997; Gakonyo, 1993).

Sweetpotato does have some "points of difference" with other root crops that justify efforts to maximize its role in rural and urban diets as well as in enhancing rural incomes and nutrition (Wheatley and Loechl, 2008):

- Orange-fleshed sweetpotato varieties (OFSP) exist that have high levels of beta-carotene, the precursor to vitamin A, in the roots. In 2005, an estimated 43 million children in SSA under 5 years of age were at risk of vitamin A deficiency (VAD) (Aguayo and Baker, 2005). The causal link between compromised vitamin A status and increased child mortality is well established.[2] Only 125 g of most OFSP varieties can supply the recommended daily allowance of vitamin A for children and non-lactating women (300–700 μg retinol activity equivalents). Most varieties in SSA are white, cream or yellow-fleshed. Hence, due to the urgency of addressing the VAD problem, sweetpotato varietal development programs have focused since the late 1990s first on adaptive testing of introduced orange-fleshed varieties, then on breeding for orange-fleshed varieties more adapted to specific agro-ecologies.
- Sweetpotato roots (orange and non-orange) are a valuable source of vitamins B, C and E. Moderate levels of iron and zinc are often found in association with high beta-carotene content.
- Vine tops have excellent micronutrient contents and adequate protein content (3–4%) (quantity and quality) for use as feed or food.
- The existence of early maturity varieties 3–4.5 month growth periods, with the potential to produce relatively high fresh and dry matter yields (of roots and vines) in less fertile conditions from sea level to 2200 m, means sweetpotato can be a component of sustainable multiple cropping systems in a broad range of agro-ecologies.
- A wide range of varieties/cultivars offering potential for many different types of utilization, including: (1) Flesh colors from white, cream, yellow, orange and purple; (2) High and low dry matter and starch contents; and (3) High and low sugar contents (sweet and non-sweet varieties).
- Unique starch properties of interest to the food industry (well-exploited already in Asia but not yet in SSA).

On the other hand, some less positive characteristics of the crop exist, but are amenable to either genetic improvement and/or post-harvest management, including:

- A lower dry matter and starch content than cassava (in general) reducing competitiveness for uses dependent on dried root chips/flour.
- Presence of anti-nutritional factors (trypsin inhibitors), complicating use of raw (either fresh or dried) roots for animal feed.

Pilot efforts in SSA have shown the potential for sweetpotato to be an excellent vehicle for creating value-added, income generating opportunities in rural villages and towns (Best et al., 2006; Wheatley et al., 1995; Kapinga et al., 2000;

---

[2] Very high mortality rates (60%) are associated with severe vitamin A deficiency and even subclinical deficiency is associated with a 23% increase in pre-schooler mortality (McGuire, 1993).

Westby et al., 2004). For value-added products to succeed, adequate farm productivity is needed to ensure low raw material costs for processors (Walker and Fuglie, 2006). Given its potential, sweetpotato has been underinvested in SSA because: (1) it is perceived as a food crop cared for by women, whose needs are often under addressed; (2) its association with "bad times" as it is heavily relied on when other crops fail; (3) its image as a crop of the poor eaten by those who cannot afford the "modern" luxuries of bread and rice; (4) lack of awareness of nutritional qualities especially of OFSP varieties and their potential contribution to combating vitamin A deficiency; (5) its bulky nature and seasonal availability which constrains market development; (6) lack of awareness and expertise concerning how to produce and market sweetpotato-based processed products and animal feeds and the potential return from these investments; (7) the difficulty of capturing production of a "piecemeal" harvest crop, and hence widespread underreporting of production in official statistics, hampering its status and resultant financial support relative to other crops, and (8) the perception that it is difficult to find appropriate resistant cultivars and agronomic practices to control major pests and disease problems. As a consequence, most under-resourced SSA national research programs have few people engaged full-time on sweetpotato research as outside funding opportunities have focused on cereals, crops with severe disease problems (e.g., cassava), and high-value crops with export potential.

Major Growing Areas, Acreage, Yields, Varietal Development and Potential Economic Impact

A. Area Under Cultivation and Yields

According to available FAO and national survey data, twenty-three countries produce 99% of all sweetpotato in SSA that is currently captured in official statistics and representative surveys (Table 16.2). Uganda and Nigeria dominate in terms of overall sweetpotato production, accounting for 33% of total reported SSA production. Per capita sweetpotato production figures are positively correlated with population density ($r = 0.75$) among these countries. This is part of a more general trend worldwide: as population increases, fertile arable land per capita diminishes and farmers turn increasingly to crops which yield a large amount of food per unit area per unit of time. Sweetpotato production per capita is over 85 kg per person in four countries: Uganda, Rwanda, Burundi, and Malawi. Seven countries (DR Congo, Ethiopia, Ghana, South Africa, Côte dÍvoire, Niger and Burkina Faso) have low production per capita levels on a national basis (less than 6 kg per capita) because sweetpotato production is concentrated in certain parts of the country and marketing channels are inadequately developed. Seven countries in Eastern and Central Africa account for half of total production, six Southern African countries 26%, and ten West African countries 21% (Fig. 16.2).

Table 16.2 Status of sweetpotato production in sub-Saharan Africa (2005–2007)

Country	Sweetpotato production ("000 Metric tons)	% of total sweetpotato grown in SSA	Sweetpotato production per capita (kgs/ person)	Population density in 2007 (persons/ sq km)	Area (Ha)	Est. yield (Tons/ ha)
Uganda	2,591	20.5	90.9	141	578,000	4.5
Nigeria*	1,578	12.5	10.9	145	464,000	3.4
Malawi*	1,146	9.1	87.5	138	159,227	7.2
Tanzania	960	7.3	24.8	42	505,000	1.9
Madagascar	874	6.9	47.8	32	123,952	7.1
Rwanda	872	6.9	93.7	346	148,996	5.9
Burundi	838	6.6	98.5	315	125,000	6.7
Kenya	769	6.1	20.8	61	70,000	11.0
Angola	689	5.4	42.2	10	143,468	4.8
Ethiopia	409	3.2	5.3	67	49,656	8.2
Mozambique*	388	3.1	19.0	25	97,000	4.0
DR Congo	228	1.7	3.6	28	45,649	5.0
Guinea	215	1.8	21.3	39	69,000	3.1
Cameroun	189	1.7	10.3	37	43,000	4.4
Mali	185	1.5	14.8	19	10,000	18.5
Zambia*	138	1.4	12.0	16	39,493	3.5
Ghana	91	0.7	4.0	97	65,250	1.4
Burkina Faso	72	0.6	4.8	51	8,550	8.4

Table 16.2 (continued)

Country	Sweetpotato production ("000 Metric tons)	% of total sweetpotato grown in SSA	Sweetpotato production per capita (kgs/ person)	Population density in 2007 (persons/ sq km)	Area (Ha)	Est. yield(Tons/ha)
Chad	65	0.5	6.0	8	25,500	2.5
Benin	56	0.4	6.2	71	11,293	5.0
South Africa	46	0.4	1.0	36	15,438	3.0
Cote d'Ivoire	45	0.4	2.2	56	20,000	2.3
Niger	44	0.3	3.1	10	2,983	14.8
Sub-Regions (No. of countries)						
East & Central (7)	6,667	52	338		1,522,301	4.4
Southern (6)	3,281	26	210		578,578	5.7
West (10)	2,540	21	84		719,576	3.5
All Countries (23)	**12,488**	**99**	**631**		**2,820,455**	**4.4**

All production and area data not marked with * are from FAOSTAT 2008 (accessed October 2008). Where variation is less than 10% across adjacent years, values are weighted: $((3 \times 2007 \text{ values}) + (2 \times 2006 \text{ values}) + (1 \times 2005 \text{ values}))/6$. Where consistent trends are apparent, most recent data (2007) are utilized. Area data for Malawi from FEWSNET (Early Warning System database); yield estimates considered to be too high (14.2 t/ha) so 50% of that value used to estimate production (7.1 t/ha). Data for Mozambique from TIA (national agricultural survey), 2005. Data for Zambia from (Central Statistical Office, 2003/2004). Data for Nigeria from the National Food Reserve Agency, 2007.

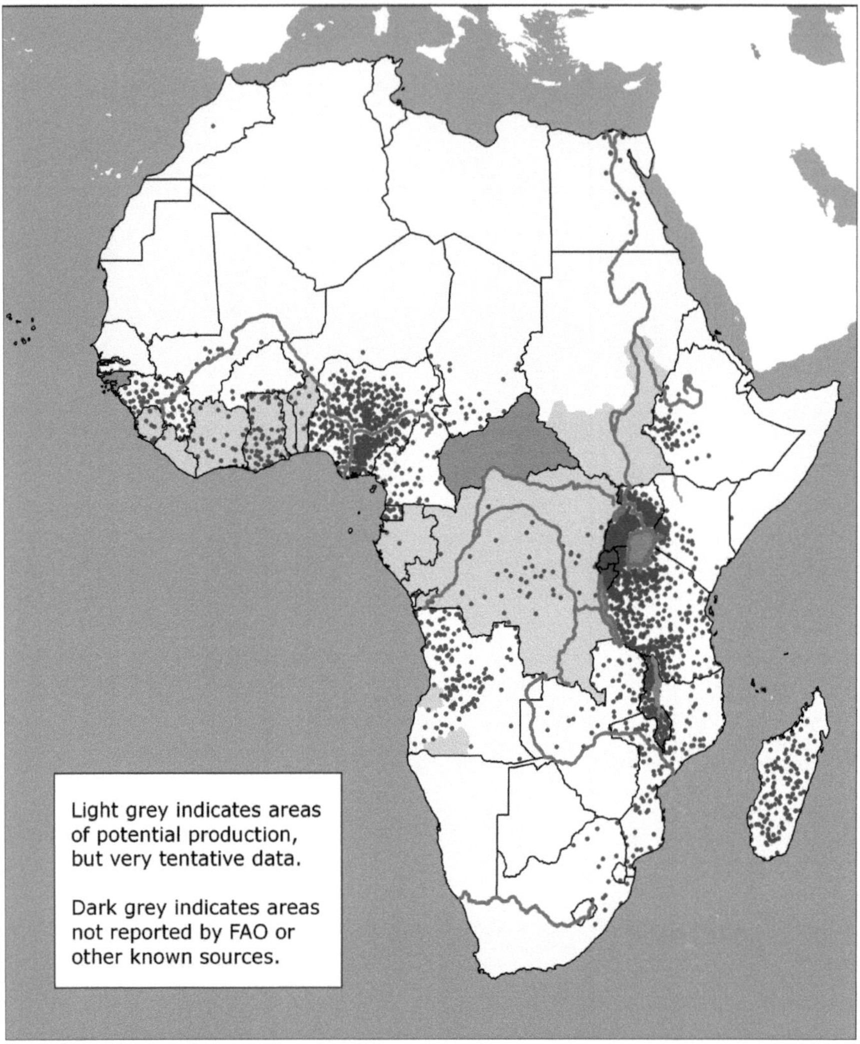

Fig. 16.2 Area under sweetpotato cultivation in sub-Saharan Africa (2007) estimates. (Each dot represents 1,000 hectares of sweetpotato cultivation)
Source: World Sweetpotato Atlas (2008).

FAO data indicate very low average sweetpotato yields (4.4 t/ha) for SSA. Given the difficulty in collecting accurate data on piecemeal harvested crops like sweetpotato and cassava, we suspect that due to underreporting actual yields on average may be higher (6 t/ha). Almost all smallholders produce sweetpotato under rainfed conditions. In contrast, commercial sweetpotato growers in South Africa who have access to irrigation, fertilizers, and credit achieve yields ranging from 50 to 70 t/ha (Niederwieser, 2004). A discussion of the factors underlying the "yield gap" between smallholders and commercial sweetpotato producers in SSA is presented in a section below.

B. Role of Sweetpotato in the Diet, Varietal Preference and Development, and the Potential Contribution of Orange-Fleshed Sweetpotato

Sweetpotatoes are principally grown in food systems in Eastern and Southern Africa dominated by root crops and bananas/plantains and secondarily grown in maize-based systems. In the root crop belt of West Africa sweetpotatoes complement the supply periods of cassava and yams, the two principal crops in those food economies. Sweetpotatoes are a strategic and flexible part of these food systems. Early maturing varieties of around three months are first to come into production to end the "hunger season" and these are often improved materials. Due to their flexible planting times and range of maturity periods farmers can manage the supply period if not constrained by an extended dry period. This helps ensure continuity of staple food supply, both for home consumption and market, so that producers can take advantage of generally higher staple food prices early in the growing season.

In countries with two rainy seasons (for example, Rwanda, Burundi and Uganda) sweetpotato is available 11 months of the year and is a primary staple. Elsewhere in SSA, where there is only one main growing season, sweetpotato is available 4–8 months of the year and it is a secondary staple consumed 2–4 times per week when in season.

With the exception of South Africa, sweetpotato production is dominated by smallholder producers. Those countries relying on sweetpotato as a food security crop tend to cultivate many varieties that have a range of maturing periods of 4–7 months and often distinct characteristics. Varieties are often mixed in the field. In SSA, white and creamed fleshed varieties predominate, and there are many popular yellow-fleshed varieties in certain countries. Adults in East and Central Africa are known to have strong preferences for floury textured varieties with dry matter contents 30% and above; those in Southern Africa also accept and often prefer varieties in the 26–30% dry matter range. In locations where commercialized sweetpotato production has emerged (e.g. Western Kenya, Eastern Uganda), distinct varietal preferences have also emerged and the number of predominant varieties has been reduced to the 2–3 demanded by the market. In the case of East Africa, many of the most highly marketed varieties are red-skinned and white-fleshed; however, during the past decade one early maturing (3 months), high dry matter (30–33%) variety SPN/O (also known as Simama, Tanzania, Kenya or Chingova depending on the country) that is white-skinned and yellow-fleshed has emerged as an important marketed variety in Uganda, Kenya, Tanzania, and Malawi.

Since the mid-1990s, the focus of adaptive testing and breeding programs in East and Southern African has been on the testing and promotion of orange-fleshed varieties. Unlike white-fleshed varieties, orange-fleshed sweetpotatoes contain Β-carotene, the major provitamin A carotenoid. Vitamin A deficiency (VAD) is widespread among young children in the developing world; globally, 127 million children under six years of age are estimated to be affected (West, 2002). Sub-Saharan Africa and India have the highest estimated prevalence rates of sub-clinical vitamin A deficiency. VAD can limit growth, weaken immunity, cause xeropthalmia leading to blindness, and increase mortality (Sommer and West, 1996).

There are two types of vitamin A available in foods: (1) preformed retinol (vitamin A itself) typically found in animal foods such as eggs, liver, and milk; and (2) provitamin A carotenoids found in plant foods such as dark green leafy vegetables, yellow and orange vegetables and fruits, and orange-fleshed sweetpotato (McLaren and Frigg, 2001).

Poor households typically cannot afford to consume the highly bioavailable animal foods on a regular basis. High rates of deficiency in the major micronutrients (vitamin A, iron, and zinc) are common among poor populations that consume plant-based diets (Hess et al., 2005). Many plant sources of vitamin A are seasonal, and after provitamin A carotenoids are absorbed into the body, they must be converted into retinol for the body to be able to make use of them. Rates of conversion vary among carotenoid containing plant foods (up to five fold) and also depend on what else is consumed at the same time (for example, fat increases absorption) and the health status of the individual (e.g. more deficient individuals absorb and/or convert at higher rates than replete individuals). Heat processing may also increase conversion rates compared to the raw product, depending on the plant matrix (Hess et al., 2005). Current guidelines recommend that provitamin A activity be expressed in Retinol Activity Equivalents (RAE). The RAE definition is based on the assumption that 16.7% of the ingested beta-carotene is absorbed and 50% is converted to retinol. This results in an average conversion factor of 12 units of beta-carotene to form 1 RAE. Many dark green leafy vegetables are probably less bioavailable than this, while palm oil is far superior (2:1 conversion factor). In contrast, the conversion factor for preformed retinol from animal sources is 1:1 and for other provitamin A carotenoids 24:1 (Institute of Medicine, 2001).

During the past five years, convincing evidence has been obtained regarding the impact of OFSP on young child vitamin A status. A South African study demonstrated that OFSP is bioavailable and efficacious in improving vitamin A status in children (Jaarsveld et al., 2005) and significant improvements in vitamin A intake and serum retinol concentrations (a proxy for vitamin A status) were obtained from an action-research study of an OFSP-based integrated agriculture-nutrition-market intervention in a very resource poor setting in Central Mozambique (Low et al., 2007a). The latter study emphasized the importance of having all three components (agriculture, nutrition and market interventions) to ensure improvement in young child vitamin A intakes and sustained adoption of the new material. A third study (Haskell et al., 2004) using the isotopic tracer deuterated retinol to estimate total vitamin A stores in 14 Bangladeshi men determined a conversion factor of 13:1 for orange-fleshed sweetpotato when it was cooked pureed with a small amount of oil.

Orange-fleshed sweetpotato as a staple food has an advantage over most vegetables in that it can supply significant amounts of vitamin A and energy simultaneously – thus helping to address both VAD and undernutrition. OFSP is an example of a *biofortified* crop in which the micronutrient status of staple foods is enhanced through plant breeding to the point where impact on micronutrient status can be achieved (Bouis, 2002). Since the poorest households typically obtained over 60% of their energy needs from food staples, this strategy is particularly suited to poor rural households that cannot access purchased fortified food products but could grow OFSP.

The intensity of the orange color reflects the amount of beta-carotene present in the sweetpotato. In most of SSA, white-fleshed varieties dominate and contain no beta-carotene. On a fresh weight basis (fwb), light orange varieties contain at least 250 RAE/100 g (30 μg/g), medium-intensity varieties at least 458 RAE/100 g (55 μg/g) and dark orange-varieties at least 833 RAE/100 g (100 μg/g). To put things into perspective, the recommended daily intake for healthy two and five year old children is 400 RAE and 500 RAE, respectively (Institute of Medicine, 2001). Depending upon the color intensity of the OFSP variety used and taking losses during cooking into account (approximately 20% through boiling) (Jaarsveld et al., 2006), 1/4 to 1 cup of boiled and mashed orange-fleshed sweetpotato meets the intake requirements of a young child.

In the late 1990s, the regional potato and sweetpotato network for Eastern and Central Africa, PRAPACE, began supporting research to on OFSP varietal development and value addition. In 2001, the International Potato Center launched the Vitamin A for Africa Initiative (VITAA) to provide a platform for bringing experts from different disciplines (especially agriculture and health) together to develop strategies and share experiences concerning the promotion of OFSP with a focus on young children and women of reproductive age (Kapinga et al., 2005). As of 2008, there are eleven African countries participating in this initiative and 38 OFSP varieties officially released (Table 16.3).

C. Potential Economic Impact of Key Interventions

Recent survey evidence indicates that the lack of sustainable seed systems is one of the key constraints to improving sweetpotato productivity in SSA. For example, a CIP survey of national agricultural research system (NARS) priorities in 2005 reported that "virus management, seed quality and supply systems" were ranked as the highest priority for future R&D against all other listed sweetpotato technologies by 91 respondents from 34 developing countries (Fuglie, 2006). CIP pro-poor research targeting further indicates that research on virus control in sweetpotato through provision of clean planting material alone could yield rates of return of between 56–84% depending on rate of adoption and adoption ceiling. The anticipated aggregate impact of the technology (assuming a *status quo* adoption ceiling) was calculated to be $74 million per annum with annual benefits to the rural poor calculated to be $49 million p.a. The maximum potential aggregate benefits and benefits to the rural poor for SSA (assuming adoption on all affected areas i.e. no adoption constraint) were calculated to be $434 million per annum and $287 million per annum, respectively (Fuglie, 2007).

Table 16.3 Status of membership in VITAA platform, and the number of sweetpotato varieties released from 2004–2007 by type, the number of Orange-fleshed Sweetpotato (OFSP) varieties likely to be released in 2008–2009

Sub-region (Country)	Member of VITAA Platform	# non-OFSP varieties released 2004–2007	# of released OFSP varieties 2004–2007	# OFSP varieties to be released in 2008–2009
East & Central				
Uganda	Yes	3	4	0
Kenya	Yes	3	4	5
Tanzania	Yes	1	0	3
Rwanda	Yes	0	4	5
Burundi	No	0	0	0
Ethiopia	Yes	4	5	3
DR Congo	No	0	2	0
Southern				
Mozambique	Yes	0	9	3
Malawi	Yes	0	0	2
Zambia	Yes	0	3	4
Madagascar	No	1	2	2
Angola	No	0	0	2
South Africa	Yes	2	3	1
West				
Ghana	Yes	2	2	4
Nigeria	Yes	0	0	2
Burkina Faso	No	0	0	2
Niger	No	0	0	2

Compelling evidence is available of the potential contribution of OFSP to improved nutrition. To evaluate potential health and economic impact, economists estimate the number of vitamin A deficiency (VAD)-related Disability-Adjusted Life Years (DALYs) that could potentially be saved through the use of biofortified sweetpotato. Results indicate that just by replacing white-fleshed varieties with orange-fleshed ones could reduce the VAD burden by 15 to 22% in 17 SSA countries where sweetpotato is widely grown (Stein et al., 2005; Fuglie and Yanggen, 2007). *Ex-ante* analysis determined that if OFSPs were adopted by one-in-six Ugandan households within 10 years of becoming available, the effort would achieve an estimated internal rate of return between 16 and 30% and yield a net present value between $23 million and $67 million (Fuglie and Yanggen, 2007).

Cultural Methods

During the past 15 years, several surveys have been conducted to better understand existing cultural practices among smallholder farmers in SSA (Bashaasha et al., 1995; Kapinga et al., 1995; INIA/SARRNET, 2003; Tewe et al., 2003). Subsequent research to improve those practices and investment in developing farmer field schools in Uganda from 2002 onwards led to the development of a useful manual that summarizes existing and recommended practices for integrated sweetpotato production and pest management (Stathers et al., 2005). Niederwieser (2004) provides useful guidelines for maximizing sweetpotato production under high and low input conditions in South Africa and provides an excellent overview of the major commercialized sweetpotato production system in SSA. These sources provide the background material for this section.

A. Smallholder Production

Most smallholder sweetpotato production occurs in rainfed systems. In systems where sweetpotato is principally grown for home consumption, priority is given to planting cereals or other cash crops at the beginning of the rains. Sweetpotato is planted 1–2 months later when sufficient moisture is still available to ensure establishment. In countries with reliable and well-distributed rainfall this approach still results in good yields. However, in Southern and many drier parts of East Africa, planting late can significantly expose sweetpotato to drought, especially at the critical time of root formation (6–8 weeks after planting) and weevil damage (as the production period extends in into the dry season). In such cases, sweetpotato production figures show considerable annual variation.

For most field work in sweetpotato, women dominate in terms of labor contribution. Men are most involved in land preparation, especially when land needs to be cleared or soils are heavy and in marketing, particularly where sweetpotato is a significant cash crop. In many areas, men also significantly contribute to weeding and harvesting, particularly when sweetpotato is intercropped.

Land Preparation

Sweetpotato is most commonly grown on mounds or ridges, and occasionally on raised beds or on the flat. Deep cultivation enhances root growth and bulking of the sweetpotato root, and production on the flat tends to be found only on sandier soils. Mounds and ridges also permit adequate drainage and can ease harvesting. Mounds are made by hoeing the soil together from the surrounding areas, and consequently are rarely in straight lines. In Uganda, mounds are approximate 100 cm wide and 60 cm tall. When mounds are used, three vine cuttings are planted singly in a triangular pattern below the tip of each mound, giving a plant population of about 33,300 plants per ha.

Farmers report that mounds are easier to make than ridges, and such systems dominate in countries like Uganda. Ridges, however, are the norm when animal traction is used and their use is increasing, as it is often the approach advocated by extension personnel. Recommended spacing varies from 70–100 cm between rows and 30 cm between plants. Raised beds are most often found in wet areas near Lake Victoria, a major sweetpotato growing zone. In this system, weeds are gathered from the surrounding area, spread on the field and then soil is pulled up to create the beds and cover the organic matter.

Traditional farmer practice varies in terms of the placement of vines on mounds and the number of vines planted per hole. In drier parts of Mozambique, it is common for 2–3 vine cuttings to be placed in a hole as farmers feel there should be a reserve plant available in case one dies. However, to maximize the use of planting material, extensionists recommended planting one per hole and replacing lost plants if they fail to establish. If single vine cuttings spaced 30 cm apart are planted on ridges, the same plant population of 33,300 plants per hectare is obtained. About 2/3 of a 25–30 cm cutting is placed below the surface at an angle.

Detailed data from 5 of 6 production zones in Tanzania (Table 16.4) serve as an example of how sweetpotato is typically grown by SSA smallholders at all altitudes, on all kinds of soils, and especially in areas where rainfall varies between 800 and 1400 mm per year. In most zones, ridges were the commonest type of seedbed, but in the Lake Zone ridges were the least common option, with the sample of farmers almost evenly divided between using mounds, raised beds, or planting on the flat.

Sources of Planting Material

Most smallholder farmers rely on their own sweetpotato fields as a source of planting material or obtain vines from their neighbors or less commonly, from extension personnel. Apical cuttings taken from disease-free mature vines make the best planting material. However, limited availability of planting material at the beginning of the rains often means that middle to lower parts of the vines are used; cuttings from the base of the vine are more likely to be infested with sweetpotato weevil.

The availability of water during the dry season, through access to lowland areas with residual or permanent moisture or access to irrigation determines the options smallholders have to draw on. In areas with dry periods lasting 4 months or longer, lack of sufficient planting material often emerges as a major constraint to expanding sweetpotato production. Namanda (2007) describes six distinct traditional practices for sourcing vines in two drier districts of Uganda (Table 16.5).

In areas where rainfall is not a major constraint, Ugandan farmers cut vines from existing fields either prior to harvest or at harvest time to plant their new field. In this study, 26% of households engaged in that method. Farmers with access to lowlands with permanent or sufficient residual moisture often obtain their vines from such plots. Lowland sites in drought prone areas are often at risk, however, of animal or pest attack and in some countries such as Mozambique are at risk of flooding if the first rains are heavy. In the Uganda study, farmers had limited access to lowland areas for maintaining their vines (only 9%). Vines maintained in the backyard often require care. In Malawi, many families place such plots near washing areas and use so-called gray water to maintain plots during the dry season.

When conditions are very difficult for maintaining vines, then SSA farmers exploit sprouts which emerge from the roots themselves. There are two sources of roots: those which just were not harvested from the field and re-sprout when the rains come (28% of cases in the Ugandan example); and those from plots which

Table 16.4 Cultural practices in five major sweetpotato producing zones of Tanzania

	Lake	Central	Eastern	Southern	Southern Highlands	Total
Sample Size	186	109	238	200	95	828
Description						
Population Density (persons/sq km)	> 50	< 50	< 50	< 50	< 50	
Number dry months	most 4–6	6–9	5–6	6–9	most 4–6	
Climate Class	Lowland semi-humid	Lowland semi-arid	Lowland Humid	Lowland semi-arid	Highland continental	
Altitude Range (m)	1000–1300	1000–1300	0–1100	0–900	1000–2500	
Cultural Practices						
% Using bed type:						
Ridges	11	90	85	93	83	72
Mounds	33	6	0	2	7	10
Raised beds	27	0	0	0	10	7
Flat	29	4	15	5	0	11
Sources of planting material (%):						
Own field	43	47	88	37	56	54
Purchase	19	28	8	2	4	12
Neighbors or others	38	25	4	61	40	34
% Intercropping	96	88	16	19	29	50
Weeding Frequency						
% Not Weeding	0	20	6	5	0	6
% Weeding once	95	70	78	50	50	69
% Using inorganic fertilizer	3	1	8	9	4	5

Source: Kapinga et al. (1995). Note that sample sizes for individual questions can vary up to 15% due to missing values.

Table 16.5 Sources of vines for smallholder sweetpotato producers in Bukedea and Kamuli districts, Uganda (2007)

Source of vines	Frequency	Percent
Sprouting roots from previously harvested fields	45	28
Sprouting roots from previously unharvested fields	15	9
Cutting from plants grown in swamps	15	9
Cutting from plants grown in backyard	35	22
Cutting from plants grown in shade	10	6
Cutting from existing plants already harvested piecemeal	42	26
Total	162	100

Source: Namanda (2007). Understanding Farmer's Traditional Knowledge in Vine Management in Bukedea and Kamuli Districts. Unpublished report.

were not harvested, with the specific purpose of obtaining sprouts from these roots for the next season (9%). Sprouting from unharvested roots begins with the onset of the rains. Hence, valuable time is lost while farmers wait to accumulate sufficient planting material and as a consequence, delayed planting (in terms of yield maximization) is the norm (Namanda, 2007).

Crop Management

Purchased fertilizer and chemical use is minimal and the use of organic manure varies depending on household livestock assets and the presence of other crops that are valued higher than sweetpotato. For example, in Tanzania only 5% of households used inorganic fertilizer on sweetpotato (Table 16.4). Monocropped plots of sweetpotato are common, but intercropping with maize, cassava, beans, pigeon pea, and occasionally other crops also occurs. Intercropping is especially common when land is scarce (e.g. Western Kenya) or labor for constructing ridges is limited (e.g. Mozambique). Practices can vary considerably within a given country as seen in the Tanzanian example in Table 16.4. In the Lake Zone, known for land shortages, intercropping is the norm. In the Eastern and Southern zones, less than 20% of households intercrop. Sweetpotato is also found grown under the shade of young perennials such as coconut and banana and under tree crops such as mango. Sweetpotato can take advantage of residual fertilizer when intercropped with fertilized maize.

Most smallholder farmers weed their sweetpotato only once (for example, Table 16.4 for Tanzania), usually within the first 2 months of planting. Many extensionists advocate a second weeding to maximize yields.

B. Commercial Production

Commercially oriented sweetpotato production, with larger amounts of land (at least one quarter of a hectare) under production as a cash crop is still limited in SSA. Two different types exist: (1) medium-scale farmers using animal traction to enable larger-scale production in Western Kenya and parts of Eastern Uganda and (2) mechanized sweetpotato production in South Africa. Each of these systems is described in this section, highlighting how they differ with smallholder production systems.

Kenyan Example

A study conducted in Western Kenya in the mid-1990s (Low, 1995) in 4 distinct agro-ecologies where sweetpotato is grown, pinpointed distinct characteristics found on farms where sweetpotato was principally grown for home consumption versus grown for cash.

These characteristics are presented in Table 16.6.

Commercially-oriented growers in Nyanza Province cultivate fewer varieties, tend to plant in pure stands, engage in progressive (harvesting one section completely at a time) or one-time harvesting of the entire field, rely heavily on oxen to plough and harvest, and employ significant amounts of hired labor. These farmers rented additional land to expand their sweetpotato production. The components of total cost for producing one acre of sweetpotato are shown in Table 16.7. At the time, average yields were 10.3 t/ha. Due to sweetpotato's bulky nature, transport and marketing costs were the major cost components.

Table 16.6 Characteristics of farms growing sweetpotatoes principally for home consumption, compared to those growing sweetpotatoes mainly for cash

Characteristics of farms growing sweetpotatoes principally for home consumption	Characteristics of farms growing sweetpotatoes principally for cash
Diversity of varieties, with differing maturation periods	1–3 varieties, all meeting qualities demanded by market (e.g. red-skinned, white-fleshed)
Piecemeal harvesting as needed (in-ground storage)	Increase in demand for early-maturing varieties
Relay and intercropping dominates pure	Progressive or one-time harvesting (still reliant on in-ground storage)
Average plot size rarely exceeds 0.5 acre	Pure stands dominate relay planting and intercropping
Never rent land strictly for sweetpotato production	Plot sizes often exceed 0.5 acre; sometimes rented
Limited male labor involvement, except for land clearing	Increased male and female labor involvement (esp. number of household members involved)
Oxen used only occasionally on heavy soils for ploughing; never for harvesting	Increased oxen use for ploughing, harvesting, and transporting to market outlets
Dominance of soils of average fertility	Intensification of pest problems, due to inadequate rotation, decrease in use of proper cultivation practices
Preference for very large tubers	Preference for medium-size tubers (easy to pack)
Planting material purchased by farmers only in areas with one annual growing season for sweetpotato	

Source: Survey of 81 sweetpotato producers in Rongo, Ndhiwa, Kabondo, and Kendu Bay in Nyanza Province, Kenya (Low, 1995).

Table 16.7 Components of total cost for 1 acre of sweetpotato production for sale in Kabondo, Kenya

	Percent of total cost
1. Land Rental	12%
2. Land Clearing	4%
3. Ploughing with Oxen	11%
4. Mound Construction	7%
5. Planting	4%
6. Single Weeding	9%
7. Harvest with Oxen	12%
8. Packing into Bags	8%
9. Transport and Marketing	33%

South African Example

Sweetpotato is considered to be a vegetable in South Africa. Commercial sweetpotato production represents less than 2% of total vegetable production in that country. The majority produced is sold to the internal fresh market; 5% is exported to Europe and the U.K. (Niederwieser, 2004).

Average commercial yields are 40 t/ha in the highly mechanized commercial sector compared to 5–10 t/ha in the rainfed smallholder sector. Clean planting material, excellent soil fertility management, and quality harvest management underlie the yield differences. The national research program, ARC-Roodeplaat, annually uses virus indexed nuclear plants of the major commercial varieties to establish mother plants and sells this clean primary material to specialized vine growers. If large roots are desired, planting densities of 33,000 plants per hectare are used; for medium sized roots (200–400 g), planting densities range from 40,000 to 45,000 plants per hectare.

Sweetpotato requires 450–600 mm water that is well distributed throughout the growing season. Key periods are at establishment (3–5 mm per day until cuttings have rooted) and 40–60 days after planting when storage roots form. Twenty-five to fifty mm per week of water is recommended during the active growing period of the crop, with water withheld 30 days prior to harvest to assure root quality. Niederwieser (2004) provides data on the amount of key nutrients sweetpotato removes from the soil, advocates soil testing for determining optimum fertilization regimes, and provides guidelines for fertilization for different yield goals. To obtain yields of 40 t/ha, farmers incorporate fertilizer prior to planting and then apply two top dressings (21 and 42 days after planting). In total, 137.3 kg of nitrogen, 80 kg of phosphorus and 106.7 kg of potassium are provided to the crop. With sweetpotato the potassium to nitrogen (K:N) ratio must be kept high to avoid excessive foliage in relation to root development. Most commercial South African farmers use herbicides to control weeds once due to their ease of application. Once the canopy is well established, further weed control is usually unnecessary (see Niederwieser, 2004 for more detailed information).

Pests and diseases and other constraints

A. Overview

Among the smallholder sector in SSA, there are five major constraints to improved productivity and incomes from sweetpotato.

1. The lack of access to virus- and pest-free "clean" planting material. Most of the local landraces and some of the introduced material are degenerated because of the sweetpotato virus disease. Yield gains of 30–60% can be obtained through use of healthy planting material (Clark and Hoy, 2006; Fuglie et al., 1999; Gibson et al., 2004; Karyeija et al., 1998). In addition, prolonged dry spells in many areas threaten the maintenance of planting material until the next growing season. Traditionally, most farmers in drought-prone areas maintain vines in valley bottoms with residual moisture, or leave roots in their fields to re-sprout when the rains return. The bottom line is that the area planted to sweetpotato is heavily constrained by limited availability of vines at the most appropriate planting times. Addressing drought through new vine conservation techniques, more drought-tolerant varieties, and small-scale dry season irrigation could radically increase sweetpotato production. In addition, insect pests facing their own food shortages during the dry season often concentrate with the first rains in vine multiplication plots with resultant high losses of planting material due to defoliators and early virus infection due to virus transmitters. Hence, locally developed and adapted pest management strategies are important to supply healthy pest and virus-free planting material to avoid early crop infestation. Moreover, as clonally propagated sweetpotato degenerates due to virus within 3–8 years, farmers must be able to periodically access clean material from primary multiplication sites typically reliant on tissue culture facilities with virus clean-up and testing capability.

The sweetpotato virus disease (SPVD) complex caused by mixed infection of *Sweetpotato feathery mottle virus* (SPFMV) and *Sweetpotato chlorotic stunt virus* (SPCSV) is, by far, the most destructive viral disease of sweetpotatoes in Africa (up to 50% in East Africa), and perhaps worldwide (Carey et al., 1999). In many cases, cultivation practices ensure that sweetpotato plants can be infected all year round. Farmer use of cuttings from their previous crop as planting material and the abundance of weed vegetation, which serves as a continually present reservoir of viruses and vectors, makes the control of the disease difficult (Karyeija et al., 1998).

Njeru et al. (2006) reported that although most farmers (73%) in Rwanda were able to identify sweetpotato virus disease (SPVD) as the most damaging disease (also confirmed by laboratory testing) the majority (65%) were not aware of what causes the disease and 53% used no control measures against it. Laboratory testing of over 300 fields in Rwanda by the same author (Njeru, personal communication in 2008) revealed that 83% of symptomatic plants and 31% of asymptomatic plants were virus infected, and with mixed infections common in symptomatic plants but not so in asymptomatic plants. Recent virus surveys in Uganda and Tanzania report varying levels of SPVD infection (Ndunguru et al., 2008; Ndunguru and Kapinga, 2008). Between 10–40% infection levels were found in Central Uganda where symptoms were recorded as being from mild to moderate. The majority of farmers practice control measures and improved varieties are widely grown. In Tanzania SPVD levels varied from 94% in the Northwest, where there was less knowledge about virus control and more local varieties grown, to 54% in the Eastern part of the lake zone (Mwanza). Limited virus testing using samples originating in this Eastern region reported less diversity of different viruses and that in particular, the mixture of viruses responsible for SPVD in Eastern Africa was not found in this particular area. In general there is a lack of current good and systematic virus survey data, particularly from Western and Southern Africa where little is known about virus incidences or the diversity of viruses present. A comprehensive virus survey has recently commenced in Rwanda, Burundi, Uganda and Congo in a CIP co-ordinated Belgian funded project (Kreuze, pers com.) Initial findings from this survey from Rwanda reinforce earlier work and indicate that 62% of farmers believe that the virus problem is increasing or at least not changing (38%). Farmers not practising any control measures amounted to 47%, 95% and 65% in Rwanda, Burundi and DR Congo, respectively.

2. Lack of improved varieties adapted to local environments. Genetic gains in terms of yield are expected to be about 20% compared to healthy local landraces (Grüneberg et al., 2004). Breeding is also the pathway for introducing quality traits (higher micronutrient content, dry matter content, sugar content, taste) that do not contribute towards improving yields but are essential for achieving other goals, such as consumer acceptance and improved diet quality.

3. Damage due to the sweetpotato weevils, particularly in drier production zones. Sweetpotato weevils are the most important pests of sweetpotato in Africa and worldwide, and production losses may often reach 60% to 100% (Stathers et al., 2003). The two species found in SSA are *Cylas puncticollis* and *Cylas brunneus*. Moderately damaged roots are unsuitable for human consumption. The search for sources of resistance to sweetpotato weevils in the crop's germplasm has not yielded reliable results and hence no conventional resistance breeding has been possible to date. Because of a lack of funding and entomological capacity, African national research programs have not been able to develop an Integrated Pest Management (IPM) program for sweetpotato to reduce major losses by sweetpotato weevils and other regional pests. That IPM can successfully work to control sweetpotato weevil was clearly demonstrated by CIP scientists and national collaborators in Cuba (Lagnaoui et al., 2000). In this collaborative IPM project the national mean damage by sweetpotato weevils was reduced from 45% to 6% on a production area of 45,000 ha within a period of 6 years. Besides using deep-rooting varieties, the major cultivation practices advocated for weevil control are: (1) crop rotation, (2) removal of infested vines, root residues, and volunteer plants, (3) use of uninfested vines (especially tips) as planting material, (4) avoiding cracks in the soil through hilling up of soil around the base of the plant, mulching, or through irrigation, (5) timing harvesting to avoid the dry season and (6) flooding for at least 48 hr after harvesting (Stathers et al., 2005). However, experience in Mozambique has indicated that smallholder farmers are only willing to adopt a significant number of cultural control practices *if* there is a market for their roots to justify their additional labor input (Low et al., 2007b).

Weevil resistance or control would bring true advantages, especially for sweetpotato production in drought-prone areas, as it would extend the possible period of in-ground storage in drier areas, reducing seasonality and improving food security. The International Potato Center is currently working with 4 countries in SSA to develop genetically modified *Bt* sweetpotato as the long-term strategy to combat this damaging pest (Ghislain, personal communication).

4. Insufficient knowledge and use of better agronomic practices. Adoption of better agronomic practices, such as site selection, planting techniques, spacing, weed control, soil fertility and water management, can substantially increase yields (more than 100%). Unfortunately, in the medium term smallholder access to fertilizers (both organic and inorganic) and irrigation is likely to remain limited. However, better management of local sources of nutrients and adoption of practices not requiring cash outlay could improve yields as much as 60% (Sevastiani et al., 2005; Niederwieser, 2004).

5. Lack of markets. Research in Mozambique and Kenya has demonstrated that farmers will substantially invest in labor-demanding technologies or technologies requiring purchased inputs to improve productivity and quality *only* when there is a market to absorb surplus root production (Low et al., 2007b; Kimathi et al., 2004). Like other roots and tubers, fresh sweetpotato roots are bulky, and hence relatively expensive to transport and good post-harvest care is essential to assure reasonable shelf life. In China and many other parts of Asia, roots are exploited as animal feed, starch, and a variety of processed products. The value-added use of sweetpotato in SSA is in its infancy. A discussion of sweetpotato marketing is provided in a section below.

B. Other Pests and Diseases

Sweetpotato is a rustic crop and while there are many diseases, which can affect sweetpotato, viruses and the sweetpotato weevil are the major ones causing economic levels of damage. Other pests and diseases of economic importance in SSA are described briefly in this section. Please refer to Skoglund and Smit (1994) for full details.

The rough sweetpotato weevil (*Bosyrus spp.*) also damages sweetpotato roots by gouging deep grooves in the surface. It is much less common and destructive than the sweetpotato weevil, but the cultural methods of control are the same.

At the beginning of the dry season in East Africa, the larvae of the sweetpotato butterfly can explode if left unchecked and cause complete defoliation. Severe outbreaks are sporadic and usually controlled with contact insecticides.

Farmers frequently complain of vertebrate pests. Near forested areas, monkeys and wild pigs can cause widespread losses and farmers have limited options for controlling them, especially in countries where firearm ownership is illegal. Domestic pigs adore sweetpotato roots and can be particularly troublesome in communities where it is common practice to allow them to roam free. Mole rats can dig through complete ridges or steal exposed roots. They can be highly destructive, at times destroying more than they eat. In certain areas, individuals knowledgeable in trapping mole rats are hired to deal with the problem. In Tanzania, a local shrub is used as a repellent. In Western Kenya, farmers try to burn cow dung mixed with pepper at the burrow entrance to smoke the mole rat out (Stathers et al., 2005).

Besides viruses, the other disease often encountered in humid environments, especially in the Central African Highlands, is the fungus *Alternaria*. The fungus survives in soil and plant debris and is spread by wind and splashing rain. It causes lesions on the foliage, especially in the base and middle sections and can lead to death of the vines. Some varieties are more resistant to *Alternaria* than others and breeding programs select for this resistance in areas with high incidence of the fungus. Use of clean planting materials and good sanitation practices are the best way to keep it under control.

C. Addressing the Yield Gap

The potential to significantly address the yield gap on farmer's fields in SSA is real. Experiments conducted in 12 East African environments with 15 improved genotypes using clean planting material yielded an average of 24.2 t/ha, compared to the average yield in the region of 5.6 t/ha (Grüneberg et al., 2004). The dissemination of clean planting material of improved varieties is capable of at least doubling current average yields under rainfed conditions from 6 to 12 t/ha. Combining such introductions with improved crop and soil fertility management practices could at least triple existing yields obtained by smallholder farmers to 18–20 t/ha. Irrigation supplied in a timely manner could also contribute to an additional 30% yield increase (Niederwieser, 2004).

Harvesting, Storage, Marketing and Profitability

A. Harvesting and Storage

In many parts of SSA there is a marked dry season and sweetpotato is only produced part of the year. Roots can be stored in the ground for an additional period but they are attacked by weevils when soil is dry and cracked. Farmers have developed in ground storage and piecemeal harvesting technology to maintain the supply of fresh sweetpotato for as long as possible (Hall et al., 1998). In Uganda this involves:

- Staggered planting, so that the crop will not all mature simultaneously
- Menu of varieties with different characteristics including maturation time to make fresh sweetpotato available over a longer period and provide roots with different post harvest characteristics e.g. yield, in ground storability and taste
- In ground storage of roots after maturity, for up to six months
- Piecemeal harvest of roots needed for immediate use

Piecemeal harvesting is an indigenous practice which may reduce weevil losses as more superficial and potentially damaged roots are harvested first (Smit, 1997). In the piecemeal system, women search the field for cracks in the mounds/ridges or exposed roots, which indicates to them that a mature root may be ready for harvesting. After removing such roots, the soil is hilled up again around the plant. When farmers are harvesting for longer periods of out-of-ground storage or sale, usually greater care is taken to avoid cutting the roots.

Damage to sweetpotato skin occurs very easily and abrasions are a major point of entry for disease pathogens. Farmers harvesting for home consumption often do not take care about avoiding damage, as immediate consumption is the intent. Roots are perishable, and unless cured or placed in stores, will likely not be marketable 1–2 weeks after harvest. Market traders in Nairobi and Kampala reported selling consignments within 3 to 4 days after arrival before rotting occurs (Omosa, 1997).

The possible duration of in-ground storage varies widely by variety with deep rooting varieties typically storing longer. Variability in maturity periods and in-ground storage capability and flexibility in planting times leads to a marked seasonality of supply with substantial price variability and deterioration in quality as the dry season progresses. Seasonality of supply creates a barrier to increasing per capita consumption and income earning possibilities both for fresh sales and for processing. There are some places, such as Rwanda, where production occurs in wetlands outside the rainy season and seasonality of supply is less marked.

If farmers could store fresh roots they could benefit from higher prices at the end of the harvest season. In practice there is little use of pits, clamps (mounds of sweetpotato sealed with earth to maintain humidity and keep out pests) or other types of stores in SSA; Malawi and Northern Nigeria are exceptions (Hall and Devereau, 2000; Tewe et al., 2003). Research by the Natural Resources Institute (NRI) on low cost storage using pits and clamps with thatched roofs showed that storage up to 4 months is possible. Stored roots are fit for home consumption but sell with a price penalty or may not be marketable because they lack the "just from the garden look" which consumers expect in fresh products (Hall and Devereau, 2000). Low cost storage was validated by the NRI in Tanzania (Rees et al., 2003). Adoption of stores for commercial use depends upon the expected price difference between the time of harvest and the moment of sale and this is variable across and within countries. A much higher price out of season was encouraging adoption of storage in Tanzania in 2004 (RIU, 2007). A cost benefit analysis of stores for home consumption in Uganda showed much higher rates of return than for any other sweetpotato enterprise (Wheatley and Loechl, 2008).

Sweetpotato roots respire during storage, but curing can reduce this. Curing sweetpotato roots at about 29 °C with high humidity for four-seven days prior to storing at 12–14 °C is used commercially in the US to heal wounds, protect against disease, reduce shrinkage and extend storage (Kemble, 2004). High ambient temperatures may mean that this type of curing is not applicable in SSA (Hall and Devereau, 2000). NRI has tested pre-harvest curing by removing sweetpotato foliage 14 days before harvesting, which reduced post harvest losses by up to 40%

(RIU, 2007). Breeding is possible to improve storability as shelf life is a varietal characteristic but it appears that cultivars that lose weight rapidly rot more (Rees et al., 1998).

B. Marketing and Profitability

Uncured sweetpotato roots are bulky and perishable. This limits the distance over which sweetpotato can be economically transported (Hall et al., 1998). Farmers in less favored locations often report that marketing sweetpotato is difficult; either markets are too distant using local transport or farmers are forced to be price takers of a sole trader serving the area. The risks of oversupply are also greater in rural locations distant from significant urban populations, as reported in Rwanda and elsewhere so that when harvests are heavy no market exists for extra production.

Production areas capable of generating surpluses tend to be relatively localized but dispersed, which leads to a lack of market integration and limits market size. Moreover, production is highly seasonal in most countries leading to marked variation in the quantity, and quality, of roots in markets and associated price swings. There is little commercial processing into chips or flour, which could be stored for year round consumption for use in stiff staple porridges, bread and cakes, or processing into fermented and dried products like *fufu* (stiff porridge). Sweetpotato consumption tends to decline as incomes rise, a change often linked with urbanization, partly because it is perceived as a "poor man's food" but also driven by the change in relative prices of root crops compared to grains in urban areas due to transport cost differentials.

The example from Kabondo in Nyanza Province Kenya cited earlier demonstrates how profitability of commercial sweetpotato production is extremely sensitive to seasonal price fluctuations and yields (Low, 1995). A farmer renting land at 800 Ksh per acre, for example, had to produce 35 bags per acre (10,300 kg/ha) in 1995 to break even, if a bag of sweetpotatoes was selling for 200 Ksh. The break-even point would drop to around 20 bags per acre when the price of sweetpotatoes rose to 300 Ksh per bag. For perspective, average prices in the major market (1995) were 300 Ksh/bag from January through April, 250 per bag during May through September, and 400 Ksh per bag from October through December. Regardless of the final price per bag, farmers on average spent 75 Ksh per bag in transport and marketing fees. Staggering planting periods and maintaining high yields were the strategies used for dealing with these price fluctuations given the lack of knowledge and unknown profitability of post-harvest fresh root storage techniques.

Discontinuous supply from relatively specialized production zones, high transaction costs and the bulky and perishable nature of the root leads to relatively high marketing costs increasing prices to urban consumers. In Kenya and Tanzania, the commercial value of sweetpotato is highest during the month of Ramadan as sweetpotato is often used to break the fasting period. In Nigeria, profits from sweetpotato can vary considerably by zone due to differences in access to markets and availability, use, and cost of inputs such as fertilizer and hired labor. One study conducted in 1995 found profit margins per hectare of 44, 57, and 105 USD in Southwest, Central, and Southeastern Nigeria, respectively (Tewe et al., 2003).

The high transport costs and tax structure of urban markets in East Africa have resulted in sweetpotato being jammed into large extended bags, often weighing over 200 kg, by traders/transporters hauling large amounts of sweetpotato to urban markets. Two to four persons are often required to load these large bags on trucks, and upon arrival, the sweetpotato is often bruised or damaged during unloading. This common practice lowers the quality of the roots and shortens shelf life.

The dispersed and seasonal nature of sweetpotato production, high costs of marketing (lack of processing opportunities), competition with other staples, the periodicity in the diet and limited consumption in towns lead to low volume or "thin" urban markets in those SSA countries where it is a secondary staple. This limits the adoption of productivity enhancing technology as additional supply leads to sharp price falls. This expectation may choke off production increases or technology adoption. These varied market problems means that there is no single critical entry point into sweetpotato value chains that would release a transformation of production and consumption. Any such transformation would have to take place across the value chain.

The promotion of orange-fleshed sweetpotato (OFSP) or diversified use into processed products or animal feed could potentially drive such a transformation of the value chain. Both strategies rely on the development and marketing of a new product. An alternative or complementary strategy would be to extend the supply period through the year, either through storage or extension of the production period.

A market transformation strategy based on the introduction and promotion of OFSP will need to be adapted to the very different market contexts for sweetpotatoes across Africa. In both West and East Africa, OFSP will have to break into markets with strong existing preferences, in these cases, for high dry matter, white or yellow-fleshed varieties. In the Southern African context, the challenge is to build market demand where sweetpotato is consumed seasonally as a secondary staple and there are no strong preferences at present, as is the case in Mozambique. This is also true in many countries emerging from conflict. Experience from pilot projects in Mozambique suggests that the second context will be less problematic for market penetration of OFSP (Low et al., 2007b).

Uses of Sweetpotato

A. Roots

Boiled and steamed roots often serve as a breakfast food or snack. In rural areas, traditional dishes often mix sweetpotato with cowpeas, coconut milk, and dark green leaves. Farmers in areas with marked dry seasons in Uganda and Tanzania sundry sweetpotato to extend the period when it may be consumed. In Uganda, roots are sliced (*amokeke*) or crushed (*inginyo*) before drying (Hall et al., 1998). *Amokeke* is reconstituted whole as a breakfast food and *inginyo* used for flour to produce *atapa*, a starchy staple. In Tanzania, roots are sliced fresh (*vichembe*) or after boiling (*matoborwa*) before drying. These products can be stored for six months in Uganda and perhaps longer in Tanzania. Attack by insects limits storage period. Artesanally dried products are mostly used for home consumption with limited commercialization, probably because they are not competitive with dried cassava chips. Since 4–5 kg of fresh roots are required to produce 1 kg of dried chips, farmers often prefer to sell in the fresh root market as the equivalent price can not be gained in the dried chip market. Slicing and drying by hand is labor intensive for processing large quantities when fresh storage would be preferable but is an option for dealing with small quantities at a time.

In Nigeria, principal sweetpotato utilization varies considerably by agro-ecology. In humid zones, sweetpotato is mixed cropped with maize and consumed as a snack, as sweetpotato flour mixed into other main dishes, and sweetpotato leaves are used for livestock feed, especially rabbits. In the sub-humid zones, sweetpotato is mixed cropped with millet and used for stiff porridge (*foofoo*), as a vegetable and in snacks. In the semi-arid north, sweetpotato is single cropped in the lowlands and is boiled for food, used to sweeten other dishes and a local fermented drink (*kanuzaki*). Dried forage is used for livestock feed in the north (Tewe et al., 2003).

Solid consumption data in urban areas for sweetpotato is rare. One study from Rwanda found that consumption of sweetpotatoes is substantially lower in urban than in rural areas and fell with increasing income confirming that its status as an inferior good in Rwandan urban areas (DeWalt, 2007). Existing evidence does indicate that its use as a breakfast and snack foods in urban settings is increasing, especially in countries facing rising cereal prices. In Zimbabwe, significant production of sweetpotato occurs in the peri-urban area surrounding the capital city of Harare. In that country, per capita sweetpotato consumption levels are higher in cities than in rural settings (Rukumi and Mutungamior, 2002). In West Africa, sweetpotato snacks are often fried and sold along the roadsides.

Some consumers report not liking to eat much sweetpotato as it can cause flatulence (due to undigested and dietary fiber). The degree to which cooking controls the flatulence varies by cultivar (Tsou and Yang, 1984) and improved techniques are needed to evaluate this negative varietal characteristic.

Regional networks such as PRAPACE for Eastern and Central Africa and SARRNET for Southern Africa have backstopped national program scientists and other stakeholders in developing a range of recipes in which sweetpotato is a primary ingredient. Uganda and Mozambique have recipe books published in English and Portuguese, respectively (Owori et al., 2007, INIA and SARRNET, 2002). Pilot efforts to link farmers to markets in Kenya (roots and flour), Uganda (roots and dried chips), and Mozambique (roots, juice, and bread) have shown that farmers respond to profitable income-earning opportunities, demonstrated by the significant expansion of sweetpotato production for market. Rural bakers substituting 38% of wheat flour (by weight) in bread buns ("*golden bread*") with boiled and mashed

OFSP from fresh roots produced economically viable products with at least 15 μg/g product of *trans*-β-carotene and hence, considered to be good sources of vitamin A (Low and van Jaarsveld, 2008). Further work is needed to improve product quality and market chain efficiencies.

Some SSA countries exhibit some of the highest urbanization rates in the world. By 2020 nearly half (46.2%) of the SSA population will be urban (UN-Habitat, 2001). Poorer rural households migrating to urban centers will carry their diverse food preferences with them and will also need good, inexpensive sources of quality food. Investments in improved infrastructure and value chain efficiency enable the expansion of sweetpotato as an urban foodstuff.

B. Foliage

Sweetpotato leaves are also consumed in many countries in SSA, with the notable exceptions of Kenya and Uganda where they are principally considered to be livestock feed. The leaves also contain significant amounts of beta-carotene, but their bioavailability is unknown and likely to be lower than the OFSP roots. In parts of Tanzania where fresh leaves are consumed, there is a strong preference for varieties with narrow leaves and deep lobes. In other parts, dried sweetpotato leaves are preferred. The traditional drying technique involves leaving fresh leaves to wither in the sun, then parboiling them for 20–30 min, removing the excess water, and subsequently sun-drying them (Kapinga et al., 1995).

Enhanced demand for sweetpotato roots could generate a secondary market for vines with income earning opportunities. The use of sweetpotato as a feed source in sub-Saharan Africa has been limited thus far and its use principally consists of vines being fed to dairy cattle and goats as supplements in highland environments. Vines from so-called dual purpose (for vines and roots) or forage varieties (vines only) are valuable as feed because of their good palatability and higher yields and crude protein content compared to alternatives such as Napier grass (Table 16.8).

In a country such as Rwanda where sweetpotato is the second largest staple crop (after banana), large quantities of sweetpotato roots are consumed and this

Table 16.8 Comparison between fresh yield, dry matter yield, and protein yield of Napier grass and sweetpotato vines

	Napier		Sweetpotato vines		
	Flat land	Rocky soil	Uganda	Kenya	Rwanda
Fresh yield (ton/ha/yr)	35	17.5	70	90	70
Dry matter content (%)	14	15	13	13	13
Dry matter yield (ton/ha/yr)	4.9	2.6	9.1	11.70	9.10
Protein yield (ton/ha/yr)	0.44	0.24	1.82	2.34	1.82

Lykuyu et al. (2007) and Peters (2008).

consumption pattern generates a large amount of sweetpotato vines that are used as feed for dairy cows. Sweetpotato vines sold in Rwandese markets are used either as planting material or feed for dairy cattle. The introduction of a zero grazing policy for cattle has led to an increase in feed prices in Rwanda. Prior to the zero grazing policy, each bundle of sweetpotato vines (15–20 kg) cost 200 Rwandese Francs (RWF), but now cost 250–300 RWF (Peters, 2008).

Because of abundant supply, Rwandan farmers have cheaper access to sweetpotato vines for feed (Table 16.9), and subsequently seem to be more familiar with the benefits of sweetpotato vines for milk production than farmers in Kenya where sweetpotato roots are consumed as a secondary staple.

Research conducted by CIP in China and Vietnam and a feasibility study in Uganda indicates that sweetpotato could also be a significant, cost-effective ingredient in pig feed in SSA (Peters, 1998; Peters, 1999; Peters et al., 2002). Until now, no major efforts have exploited this opportunity to improve low rates of weight gain among smallholder-raised pigs in SSA where around 17.6 million pigs are produced annually in the 17 major sweetpotato producing countries (FAOSTAT, 2006). The use of sweetpotato silage (fermented roots and vines) is unknown except on an experimental basis in SSA.

In conclusion, sweetpotato as a robust relatively quick maturing crop which requires few inputs is playing a significant role in securing food security in SSA. Sweetpotato is almost exclusively grown by smallholders often on small patches of land for household consumption. Its importance has steadily grown over the past decade because of increasing pressure on land, political instability and the opening of markets. The growing availability of pro-vitamin A rich orange-fleshed varieties presents an opportunity to meaningfully tackle vitamin A deficiency among vulnerable groups, especially in rural areas, and also to stimulate increased demand for sweetpotato in urban areas. A lack of year round supply is one of the constraints on further growth and this could be addressed through improved planting material supply, better management of weevils which limit storage in the ground, improvements in storage and exploiting agroecological niches where off season production is possible. In many places however, sweetpotato has been neglected in agricultural development policy and one of the principal challenges is to strengthen the evidence base to give it more visibility. The true potential of sweetpotato is of yet unleashed on the continent.

Table 16.9 The prices of sweetpotato vines for feed in Rwanda and Kenya (June 2008)

	Rwanda			Kenya	
	RWF/kg	USD/kg		Ksh/kg	USD/kg
Before zero grazing	11	0.021	When maize is available*	5	0.08
			Wet season	10	0.16
Currently	16	0.029	Dry season	20	0.32

* During maize harvest season, maize stalks are widely available and cheap, thus offering serious competition with SP vines.

Source: Peters, 2008. 545 RWF for $1 USD and 64 Ksh for $1 USD.

References

Aguayo, V.M. and Baker, S.K. 2005. Vitamin A deficiency and child survival in Sub-Saharan Africa: a reappraisal of challenges and opportunities. *Food Nutr. Bull.* 26: 348–355.

Bashaasha, B., Ewell, P.T., Mwanga, R.O.M. and Ocitti p'Obwoya, C. 1995. Sweetpotato in the farming and food systems of Uganda: A farm survey report. Lima, Peru: International Potato Center (CIP), National Agricultural Research Organization (NARO). 63p.

Best, R., Westby, A. and Ospina, B. 2006. Linking small-scale cassava and sweetpotato farmers to growth markets, experiences, lessons and challenges. *Acta Hortic.* 703: 39–46.

Bouis, H. 2002. A new tool for fighting micronutrient malnutrition. *J. Nutr.* 132 (Symposium: Plant breeding): 491S–494S.

Carey, E.E., Gibson, R.W., Fuentes, S., Machmund, M., Mwanga, R.O.M., Turyamureeba, G., Zhang, L., Ma, D., Abo El-Abbas, F., El-Bedewy, R. and Salazar, L.F. 1999. The causes and control of virus diseases of sweet potato in developing countries: Is sweet potato virus disease the main problem? Pages 241–248 in: Impact on a Changing World: CIP Program Report 1997–1998. Lima, Peru: International Potato Center.

Clark, C.A. and Hoy, M.W. 2006. Effects of common viruses on yield and quality of Beauregard sweet potato in Louisiana. *Plant Dis.* 90: 83–88.

Davidson, A. 1999. Oxford Companion to Food. New York: Oxford Uni. Press, 912p.

DeWalt, J. 2007. New Initiative will offer a fresh take on agricultural investment. Food for innovative thought website: OECD Observer N° 261. Accessed May 2007. http://www.oecdobserver.org/news/printpage.php/aid/2210/Food_for_innovative_thought.html.

Ewell, P. 1990. Sweet potatoes in Eastern and Southern Africa. Paper presented at the Workshop on Sweet Potatoes in the Food Systems of Eastern and Southern Africa. Nairobi, Kenya.

FAOSTAT. 2006. Pig stocks in sub-Saharan African countries in 2005. www. Faostat.fao.org. Last accessed 17 July 2006.

FAOSTAT. 2008. Production and area harvested statistics for sweetpotato for 2007. http://www.faostat.fao.org/site/567/default.aspx♯ancor. Last accessed 16 October 2008.

Fuglie, K.O. 2006. Priorities for sweetpotato research in developing countries: Results of a survey. Impact Enhancement Division, Lima Peru: International Potato Center, Lima, Peru.

Fuglie K. 2007. Research priority assessment for the CIP 2005–2015 Strategic Plan: Projecting impacts on poverty, employment, health and environment. Lima, Peru: International Potato Center (CIP), Lima, Peru. 105p.

Fuglie, K. and Yanggen, D. 2007. Impact of sweetpotato biofortification on vitamin A deficiency in developing countries: an *ex-ante* economic assessment. Lima, Peru: International Potato Center (CIP). Unpublished donor report. 34p.

Fuglie, K., Zhang, L., Salazar, L. and Walker, T. 1999. Economic impact of virus-free sweetpotato planting material in Shandong Province, China. Lima, Peru: International Potato Center (CIP). 27p.

Gakonyo, N. 1993. Processed sweet potato: Responding to Kenya's urban food needs. Master's thesis. Ithaca, New York: Cornell University,

Gibson, R., Aritua, V., Byamukama, E., Mpembe, I., and Kayongo, J. 2004. Control strategies for sweet potato virus disease in Africa. *Virus Res.* 100: 115–122.

Grüneberg, W.J., Abidin, E., Ndolo, P., Pereira, C.A. and Hermann, M. 2004. Variance component estimations and allocation of resources for breeding sweetpotato under East African conditions. *Plant Breed.* 123: 311–315.

Haskell, M.J., Jamil, K.M., Hassan, F., Peerson, J.M., Hossain, M.I., Fuchs, G.J. and Brown, K.H. 2004. Daily consumption of Indian spinach (*Basella alba*) or sweet potatoes has a positive effect on total body vitamin A stores in Bangadeshi men. *Am. J .Clin. Nutr.* 80: 705–714.

Hess, S.Y., Thurnham, D.I. and Hurrell, R.F. 2005. Influence of provitamin A carotenoids on iron, zinc, and vitamin A status. Washington D.C., International Food Policy Research Institute (IFPRI) and the International Center for Tropical Agriculture (CIAT).

Institute of Medicine. 2001. Dietary reference intakes for vitamin A, vitamin K, arsenic, boron, chromium, copper, iodine, iron, manganese, molybdenum, nickel, silicon, vanadium, and zinc. Washington D.C.: National Academy Press.

Instituto Nacional de Investigação Agronómica (INIA) and the Southern Africa Root Crops Research Network (SARRNET). 2002. Receitas de Batata Doce. SARRNET: Maputo, Mozambique. 12p.

Instituto Nacional de Investigação Agronómica (INIA) and the Southern Africa Root Crops Research Network (SARRNET). 2003. Cassava and sweetpotato production, processing, and marketing in Mozambique: Report of a survey conducted in June-August 2002. Maputo, Mozambique: Unpublished donor report submitted to USAID Mozambique. 115p.

Jaarsveld, P.v., Faber, M., Tanumihardjo, S.A., Nestel, P., Lombard, C.J. and Benade, A.J. 2005. Beta-carotene-rich orange-fleshed sweet potato improves the vitamin A status of primary school children assessed with the modified-relative-dose-response test. *Am. J. Clin. Nutr.* 81: 1080–1087.

Jaarsveld, P.v., Marais, D.W., Harmse, E. and Rodriguez-Amaya, D. 2006. Retention of β-carotene in boiled, mashed orange-fleshed sweet potato. *J. Food Compost. Anal.* 19: 321–329.

Hall, A., Bockett, G. and Nahdy, S. 1998. Sweetpotato Post harvest Systems in Uganda: Strategies, Constraints, and Potentials. Social Science Department Working Paper No. 1998-7. Lima, Peru: International Potato Center (CIP).

Hall, A. and Devereau, S. 2000. Low-cost storage of fresh sweet potatoes in Uganda: Lessons from participatory and on-station approaches to technology choice and adaptive testing. *Agriculture* 29: 275–282.

Jayne, T., Villarreal, M., Pingali, P. and Hemrich, G. 2004. Interactions between the agricultural sector and the HIV/AIDS pandemic: Implications for agricultural policy. East Lansing, Michigan, USA: Michigan State University International Development Paper No. 25. http://www.aec.msu.edu/agecon/fs2/index.htm.

Kapinga, R.E., Ewell, P.T., Jeremiah, S.C. and Kileo, R. 1995. Sweetpotato in Tanzanian farming and food systems: Implications for research. Lima, Peru: International Potato Center and Ministry of Agriculture Tanzania Working Paper. 47p.

Kapinga, R.E,. Rees, D., Westby, A., Ndunguru, G.T., Rwiza, E., Tomlins, K.I., Stathers. T., Jeremiah, S. and Mbilinyi, L.. 2000. Increasing the contribution of sweet potato to sustainable rural livelihoods in Tanzania. Paper presented at the Triennial Symposium of the International Society for Tropical Root Crops, Japan, 10–16 September 2000.

Karyeija, R.F., Gibson, R.W. and Valkonen, J.P.T. 1998. The significance of sweetpotato feathery mottle virus in subsistence sweetpotato production in Africa. *Plant Dis.* 82: 4–15.

Kapinga, R., Anderson, A., Crissman, C., Zhang, D., Lemaga, B. and Opio, F. 2005. Vitamin A partnership for Africa: A food-based approach to combat Vitamin A deficiency in Sub-Saharan Africa through increased utilization of orange-fleshed sweetpotato. *Chron. Horticult.* 45: 12–14.

Kemble, J. 2004. Harvesting and Curing Sweetpotatoes. Alabama Cooperative Extension System. No. ANR-1111.

Kennedy, G. 2002. Food Security in the context of urban sub-Saharan Africa Food Africa Internet Forum 31 March-11 April Internet Paper for Food Security Theme. http://foodafrica.nri.org.

Kimathi, M., Gero, R., Kirima, F. and Njoroge, S. 2004. Orange-fleshed sweet potato promotion research: Creating a suitable market environment for orange-fleshed sweetpotato and nutrient-rich baby foods. Report submitted to the International Potato Center (CIP) by the Market Access and Enterprise Development Programme of Farm Concern. 37p.

Lagnaoui, F.C., Alcázar, J. and Morales, F. 2000. A sustainable pest management strategy for the sweetpotato weevil in Cuba: A success story. *Food and Fertilizer Technology Center, Extension Bulletin* 493: 1–7.

Low, J.W. 1995. Determinants of Sweetpotato Commercialization in South Nyanza, Kenya. Paper presented at the Sixth Triennial Symposium of the International Society for Tropical Root Crops – Africa Branch. Lilongwe, Malawi, 22–28 October, 1995.

Low, J.W., Arimond, M., Osman, N., Cunguara, B., Zano, F. and Tschirley, D. 2007a. A food-based approach introducing orange-fleshed sweet potatoes increased vitamin A intake and serum retinol concentrations in young children in rural Mozambique. *J. Nutr.* 137: 1320–1327.

Low, J.W., Arimond, M., Osman, N., Cunguara, B., Zano, F. and Tschirley, D. 2007b. Ensuring supply and creating demand for a biofortified crop with a visible trait: Lessons learned from the introduction of orange-fleshed sweet potato in drought-prone areas of Mozambique. *Food Nutr. Bull.* 28: S258–S270.

Low, J. and van Jaarsveld, P. 2008. The potential contribution of bread buns fortified with B-carotene-rich sweet potato in Central Mozambique. *Food Nutr. Bull.* 29: 98–107.

Lykuyu, M., Romney, D., Ouma, R. and Sones, K.. 2007. Feeding Dairy Cattle: A manual for smallholder dairy farmers and extension workers in East Africa. ILRI. Nairobi, Kenya.

McGuire, J. 1993. Addressing Micronutrient Malnutrition. In *SCN News No. 9.* Geneva: ACC/SCN.

McLaren, D.S. and Frigg, M. 2001. Sight and Life guidebook on vitamin A in health and disease. Basel, Switzerland: Task Force Sight and Life.

Namanda, S. 2007. Understanding Farmer's Traditional Knowledge in Vine Management in Bukedea and Kamuli Districts. International Potato Center. Unpublished donor report.

Ndunguru, G., Thomson, M., Waida, R., Rwiza, E., and Westby, A. 1998. Methods for examining the relationship between quality characteristics and economic value of marketed fresh sweet potato. *Tropical Agriculture (Trinidad)* 75: 129–133.

Ndunguru, J. and Kapinga, R. 2007. Viruses and virus-like diseases affecting sweetpotato subsistence farming in southern Tanzania. *African Journal of Agricultural Research* 2(5): 232–239.

Ndunguru, J., Tumwegamire, S., Kapinga, R., Mwanga, R.O.M., and Sseruwagi, P. 2008. Unusually high populations of aphids associated with sweetpotato virus disease in south-western Uganda. *Phytoparasitica In press*

Niederwieser, J.G. (ed). 2004. Guide to Sweet Potato Production in South Africa. ARC-Roodeplaat Vegetable and Ornamental Plant Institute. Pretoria, South Africa: CPD Printers. 96p.

Njeru, R.W., Bagabe, M.C., Nsabimana, J., Ndirigue, J., Butare, L. and Kajuga, J. 2006. Farmers'knowledge on the management of sweet potato virus disease in Rwanda. *African Regional meeting on Women in Science for Food and Nutritional Security, 3–7 July 2006.* Entebbe, Uganda: Africa Conference Proceedings.

Norman, M.J.T., Pearson, C.J. and Searle, P.G.E. 1984. The Ecology of Tropical Food Crops. Cambridge: Cambridge University Press.

Omosa, M. 1997. Current and potential demand for fresh and processed sweetpotato products in Nairobi and Kisumu, Kenya. Lima, Peru: International Potato Center (CIP).

Owori, C. Lemaga, B., Mwanga, R.O.M., Namutebi, A. and Kapinga, R. 2007. Sweetpotato recipe book: Sweetpotato processed products from Eastern and Southern Africa. PRAPACE: Kampala, Uganda. 93p.

Peters, D. 1998. Setting post-harvest strategies for food security and income generation in Soroti, Uganda. Lima, Peru: International Potato Center (CIP). 28p. http://www.eseap.cipotato.org/MF-ESEAP/Fl-Library/Soroti_AHV.pdf.

Peters, D. 1999. Sweetpotato processing for poverty alleviation and rural stability in China: A case study of Yilong County in Sichuan. Lima, Peru: International Potato Center. http://www.eseap.cipotato.org/MF-ESEAP/Fl-Library/Yilong_BE.pdf.

Peters, D. 2008. Assessment of the potential of sweetpotato as livestock feed in East Africa: Rwanda, Uganda, and Kenya. Nairobi, Kenya: Unpublished report submitted to the International Potato Center (CIP), July 2008. 62p.

Peters, D., Nguyen, T. and Pham, N. 2002. Sweet potato root silage for efficient and labour-saving pig raising in Vietnam. AGRIPPA Peer Reviewed Electronic Journal. www.fao.org/docrep/article/aggripa/554_en.htm.

Rees, D., Kapinga, R., Rwiza, E., Mohammed, R. and van Oirschot, Q. 1998. The potential for extending the shelf-life of sweetpotato in East Africa through cultivar selection. *Trop. Agric. (Trinidad)* 75: 208–211.

Rees, D., Oirschot, Q.v. and Kapinga, R (eds). 2003. Sweetpotato Post-harvest Assessment: Experiences from East Africa. Chatham, U.K.: Natural Resources Institute.

RIU (Research into Use). 2007 Sweet potato boosts health and income. http://www.researchintouse.com/nrk/Topic5/Sweet%20potato%20boosts%20health%20and%20incomes%20(CPH40).htm.

Rugalema, G. 1998. Consequences of loss of labour due to HIV/AIDS in smallholder households in a Buhaya village, Bukoba District, Tanzania. Paper presented at the Regional Conference for Eastern and Southern Africa on responding to HIV/AIDS technology development needs of African smallholder agriculture, Harare, June 1998.

Rukumi, T. and Mutungamior, A. 2002. Trends in sweet potato consumption and informal trading in Harare and Chitungwiza. Development Technology Center, University of Zimbabwe. 25p.

Sevastiani, S.K., Mgonja, A., Urio, F., and Ndondi, K. 2005. Response of sweetpotato (Ipomoea batatas) to application of nitrogen and phosphorus fertilizers in the Northern highlands of Tanzania. Proceedings of the 2nd international symposium on sweetpotato and cassava (innovative technologies for commercialization). Kuala Lumpur, Malaysia, June 14–17, 2005. *Acta Horticulturae*.

Skoglund, L. and Smit, N. 1994. *Major* diseases and pests of sweetpotato in Eastern Africa. Lima, Peru: International Potato Center. 67p.

Smit, N. 1997. Integrated pest management for sweetpotato in Eastern Africa. Thesis (Ph.D.) Wageningen Agricultural University (WAU), The Netherlands. 151p.

Sommer, A. and West, K. P. 1996. Vitamin A Deficiency: Health, Survival and Vision. New York, Oxford University Press.

Stathers, T.E., Rees, D., Kabi, S., Mbilinyi, L., Smit, N., Kiozya, H., Jeremiah, S., Nyango, A. and Jeffries, D. 2003. Sweetpotato infestation by *Cylas* spp. in East Africa. 1: Cultivar differences in field infestation and the role of plant factors. *Int. J. Pest Manag*. 49: 131–140.

Stathers, T., Namanda, S., Mwanga, R.O.M., Khisa, G. and Kapinga, R. 2005. Manual for sweetpotato integrated production and pest management farmer field schools in sub-Saharan Africa. Kampala, Uganda: International Potato Center. 168p.

Stein, A.J., Meenakshi, J.V., Qaim, M., Nestel, P., Sachdev, H.P.S. and Butta, Z.A. 2005. Analyzing the health benefits of biofortified staple crops by means of the disability-adjusted life-years approach: A handbook focusing on iron, zinc, and vitamin A. Washington D.C. and Cali: International Food Policy Research Institute (IFPRI) and the International Center for Tropical Agriculture (CIAT).

Tewe, O.O., Ojeniyi, F.E. and Abu, O.A. 2003. Sweetpotato production utilization, and marketing in Nigeria. Lima, Peru: International Potato Center (CIP) and the University of Ibadan. 54p.

Tomlins, K.I., Ndunguru, G.T., Rwiza, E. and Westby, A. 2000. Post harvest handling, transport and quality of sweet potato in Tanzania. *J. Hortic. Sci. Biotechnol*. 75: 586–590.

Tsou, S.C.S. and Yang, M.H. 1984. Flatulence factors in sweet potato. *Acta Hortic*. 163: 179–186.

UN-Habitat. 2001 The State of the World's Cities, 2001. United Nations Center for Human Settlements, Nairobi, 125p.

Walker, T. and Fuglie, K. 2006. Prospects for enhancing value of crops through public-sector research: Lessons from experiences with roots and tubers. Lima, Peru: International Potato Center (CIP). 20p.

West, K. P. 2002. Extent of vitamin A deficiency among preschool children and women of reproductive age. *J. Nutr*. 132: 2857S–2866S.

Westby, A., Oirschot, Q.v., Tomlins, K., Ndunguru, G., Ngendello, T., Sanni, L., Pessey, D. and Oyewole, O. 2004. Bridging the gap between post-harvest technology development and commercialization of root and tuber crops in Africa. Invited thematic paper at the ISTRC-Africa Branch Triennial Symposium, 31 October–5 November 2004, Mombasa, Kenya.

Wheatley, C. and Loechl, C. 2008. A *C*ritical *R*eview of *S*weetpotato *P*rocessing *R*esearch *C*onducted by CIP and *P*artners in Sub-Saharan Africa. Lima, Peru: International Potato Center (CIP). Social Science Working Paper No. 2008–4.

Wheatley, C., Scott, G., Best, R. and Wiersema, S. 1995. Adding Value to Root and Tuber Crops. Cali, Colombia: International Center for Tropical Agriculture (CIAT).

Woolfe J. 1992. Sweet Potato: *A*n Untapped Resource. Cambridge: Cambridge University Press.

World Sweetpotato Atlas (WSA). 2008. International Potato Center. (http://research.cip.cgiar.org/confluence/display/WSA/Home).

Chapter 17
Sweetpotato in the Indian Sub-Continent

S. Edison, Vinayaka Hegde, T. Makeshkumar, T. Srinivas, G. Suja and G. Padmaja

Introduction

Sweetpotato, *Ipomoea batatas* (L.) Lam. is widely grown in the tropics and warm temperate regions of the world. It is the third most important tuber crop in India after potato and cassava. The root tubers are mainly used as a subsidiary food. The crop is grown in all states of, India except the states of Jammu and Kashmir, Himachal Pradesh and Sikkim. Area under this crop is declining year after year due to less importance given to this crop in the programmes of department of Horticulture/Agriculture in many states. Industrially this crop is exploited very little. Even though technologies to produce many value added products from sweetpotato are available, diversification and value addition has not gained any momentum as understood from the surveys undertaken in the production centers of the crop.

India occupies twelfth, eighth and fifth rank globally in terms of sweetpotato area, production and productivity respectively, while fifth rank in Asia in area, production and productivity respectively (Table 17.1). Sweetpotato is predominantly cultivated as rainfed crop in eastern India especially in Orissa, West Bengal, Uttar Pradesh, Bihar and Jharkhand states accounting for 77% of area and 82% of production of India (Table 17.2).

Table 17.1 Rank of India in global sweetpotato scenario

Crop	Asia	World
Area	5	12
Production	5	8
Productivity	5	5

Source: www.fao.org

Declining trends in area under sweetpotato observed in India is attributed mainly to lack of diversified uses to produce value added products, availability of food grains in sufficient quantities, changed food habits due to increased standard of

S. Edison (✉)
Central Tuber Crops Research Institute, Thiruvananthapuram, India
e-mail: ctcritvm@yahoo.com

G. Loebenstein, G. Thottappilly (eds.), *The Sweetpotato*,
DOI 10.1007/978-1-4020-9475-0_17,

Table 17.2 Sweetpotato in India

State	Area ('000 ha)	Production ('000 t)	Productivity (t ha^{-1})
Orissa	47.1	394.3	8.37
Uttar Pradesh	20.0	240.8	12.04
Bihar	4.6	63.9	13.89
Jharkhand	7.6	103.0	13.55
West Bengal	26.1	193.7	7.42
India	136.5	1211.0	8.87

Source: Agriculture, Centre for Monitoring Indian Economy, 2006

living etc. India recorded negative and significant decline in sweetpotato production. The productivity which has increased @ 0.7 per cent per annum helped in compensating the effect of area reduction on its production.

Important Features of Sweetpotato Cultivation

Important features of sweetpotato cultivation in major growing states viz., Bihar, Jharkhand and Orissa are highlighted here. Average holding size of sweetpotato farmers varies from 0.48 acres to 1.12 acres. High yielding varieties developed by Central Tuber Crops Research Institute (CTCRI) like Sree Bhadra, RS III-3, X4, Kalmegh etc. were very popular in Jharkhand and Bihar, while local varieties of sweetpotato are predominantly grown in Orissa. Sweetpotato is grown round the year in Bihar and during kharif in Jharkhand and Orissa. The main growing seasons are kharif (rainy season), September to January/February and summer (January/February to May/June) in Bihar. It is also grown as a short duration crop after devastating floods. It is grown in bheet (upland), diara (sandy land between a large river and a tributary which remains under water during the rains) and flood prone areas. The crop is usually raised as a rainfed crop, utilizing the south – west monsoon (June-August) as kharif crop and with supplemental irrigation during the north- east monsoon (October-December) as rabi season crop.

This crop is grown with the major objective of selling in the market as a vegetable. Summer crop is generally grown on chaur (low land) areas after harvesting the paddy. In some areas along the rivers, farmers plant sweetpotato mainly for commercial purpose as and when water recedes from their fields. Primary and secondary nurseries are raised before transplanting in the main field. Both ridge and furrow method and mound method of planting are practiced for planting sweetpotato. Significant differences in yields were observed among these three states. In Bihar owing to very fertile soils of gangetic belt, farmers obtained average yields of 25 t/ha. While in Orissa and Jharkhand, average yields recorded were about eight tones per ha only. Poor lateritic soils and low management practices adopted by farmers in these two states are the reasons for low yields. In Orissa, barter system was predominant in the study area. Farmers exchange sweetpotato for ragi, paddy etc.

Current Utilization Pattern of Sweetpotato

Current utilization pattern of sweetpotato is Chlorotic presented in Fig. 17.1 Of the total Sweetpotato produced, 80–85% is consumed as fresh and around 15–20% in the processed form. The common method of consumption is in the form of snack food after roasting/baking or boiling. Boiled or baked tubers are preferred with curd, milk and lassi. The poor prepare roti also from sweetpotato floor, but only a few families consume it. The majority does not prefer it because reportedly, the bread becomes too sweet. Boiled sweetpotato is also used as an adulterating material for Khoya (milk concentrate). Farmers felt that the most important factor, which limits sweetpotato consumption in Bihar is the gastric and acidity problem due to its consumption, followed by easy availability of alternate food grains and social prestige. It is believed by the local population that sweetpotato consumption along with milk and buttermilk is good for health but without these it is harmful for human health.

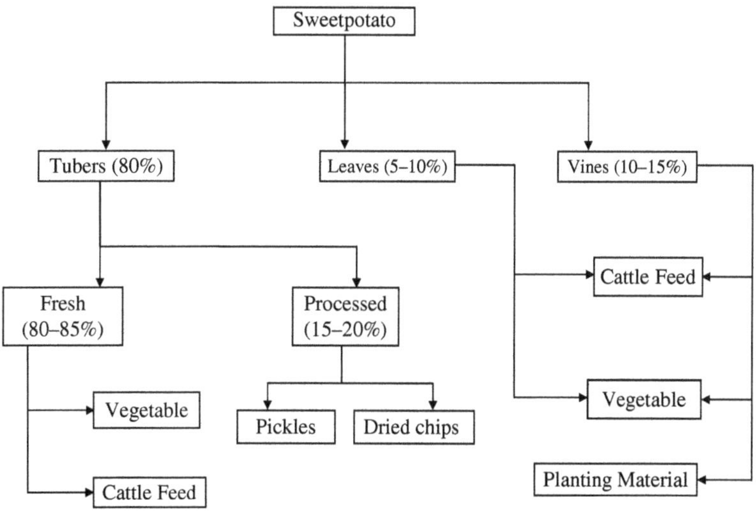

Fig. 17.1 Current utilisation pattern of sweetpotato in India

It is also occasionally used in vegetable preparations. In Bihar, fresh tubers are chipped, dried and ground into flour. This flour is mixed with wheat flour for making chapattis. On certain religious occasions when women folk are not expected to eat cereal food, sweetpotato flour is mixed with milk and sugar/jiggery and consumed. The vines are mostly fed to livestock as animal feed. About 80–90% of vines are used as animal feed and 10 to 20% are used as planting materials for the next season crop.

In Orissa, sweetpotato is predominantly cultivated by tribals. After harvesting tubers are stored for exchanging with cereals like ragi, paddy, jowar etc. under barter system during off season. So far in India, sweetpotato is not utilized for industrial purposes like starch, alcohol and liquid glucose.

Marketing of Sweetpotato

Many farmers market sweetpotato tubers by themselves at the nearby rural markets. However to a certain extent agents are also involved in the marketing of sweetpotato. They collect the tubers at the farms and take them to the neighbouring towns and cities.

Commission agents collect tubers from sweetpotato growing villages during the harvesting period and supply to wholesalers. Some wholesalers are exporting tubers in small quantities from Sri Lanka to USA. Retailers purchase tubers from wholesalers for further distribution to consumers. Some farmers sell tubers directly in the vegetable market by themselves. The marketing channel identified for sweetpotato is presented in Fig. 17.2.

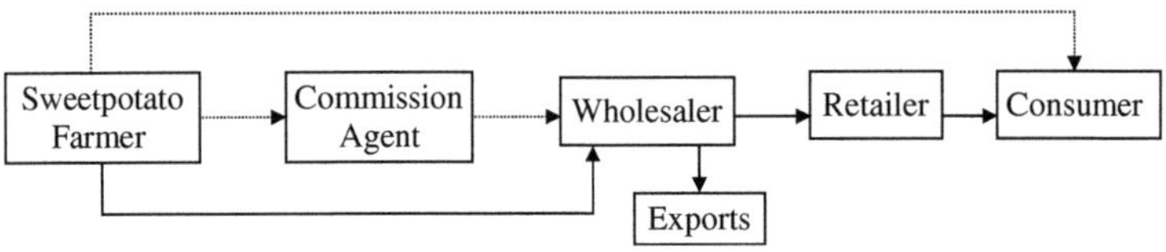

Fig. 17.2 Marketing channel for sweetpotato in India

Production Technologies

Sweetpotato has a broad agro-ecological adaptability, which is related to the genetic diversity within the cultivated species. It can be grown on a wide range of soils, but sandy loams reasonably high in organic matter with permeable subsoil are ideal. It requires a minimum of 500 mm rainfall and an annual rainfall of 700–1000 mm would be quite adequate. Short days with a low light intensity promote root development and sweetpotatoes require a day length of 11.5 hr or less to promote flowering; at 13.5 hr, flowering ceases, but yield does not appear to be affected (Nair, 2000).

Sweetpotato is a short duration crop, tolerant to a wide range of growing conditions. In India, sweetpotato is considered as a famine relief crop as it had played a pivotal role in alleviating the Bengal famine of 1942 (Nair, 2000).

Varieties

The following varieties are recommended for different states of India (Table 17.3).

Planting Season

In India, sweetpotato is grown as a rainfed crop during kharif (June–August) and as an irrigated crop during rabi (October–December). However, the major area under sweetpotato planting is during rabi season, which enjoys warm sunny days and cool

Table 17.3 Brief description of varieties recommended for different states of India

Sl. No.	Variety	Recommended region	Yield (t ha^{-1})	Duration (days)	Salient features
1	H-41	Tamil Nadu, Karnataka, Kerala	20	120–130	Semi-spreading, tuber skin reddish purple with white rind and flesh. Excellent cooking quality and low in fibre content
2	H-42	Tamil Nadu, Karnataka, Kerala	22	120–130	Semi-spreading, tuber skin purplish red and creamy flesh. Cooks easily, tastes sweet and almost free from fibre
3	VL Sakarkand-6	Hilly areas of Uttar Pradesh	20	135–140	Late maturing, spreading purple in color with light yellowish flesh. Tuber rich in carotene
4	CO-1	Tamil Nadu	28	135	Less spreading, tuber skin light pink with white flesh. Starch 24%, soft texture on cooking with high sugar content
5	Rajendra Sakarkand-5	Plains of Bihar	20	105–120	Extremely spreading, white tuber skin and greenish white flesh. suitable for two tier cropping, deep rooted suitable for fodder
6	CO-2	Tamil Nadu	32	110–120	Less spreading, leaves 10 lobed and emerging leaves pink, tuber skin light pink with white flesh. Starch 29.5%
7	CO-3	Tamil Nadu	32	110–115	Bushy habit, tuber skin light red with dark orange flesh. Carotene rich, starch 30%
8	Varsha	Konkan region of Maharastra	17	120	Semi-spreading, tubers with reddish purple skin and light yellow tuber flesh, drought tolerant
9	Sree Nandini	Kerala	20	100–105	Spreading, tubers creamy skin and white flesh. Drought tolerant, suitable for paddy fallows as a catch crop

Table 17.3 (continued)

Sl. No.	Variety	Recommended region	Yield (t ha^{-1})	Duration (days)	Salient features
10	Sree Vardhini	Kerala	20	100–105	Semi-spreading, fusiform tuber with purple skin and yellow flesh, carotene rich variety (1200 IU/100 g)
11	Samrat	Andhra Pradesh	20	110–120	Vines medium spreading, petioles violet, tubers globular with light pink skin and white flesh, medium duration
12	Rajendra Sakarkand- 43	Assam, Karnataka, Andra Pradesh, Maharashtra, Bihar	20	110–120	Medium spreading plant type, tuber skin brownish with white flesh
13	Rajendra Sakarkand-35	Assam, Bihar, Karnataka, West Bengal	25	105–110	Medium spreading, tuber, skin dull white with white flesh
14	Rajendra Sakarkand-47	Bihar, West Bengal, Uttar Pradesh	30	120–130	Spreading type, tubers purple skin with white flesh
15	Rajendra Sakarkand-92	Bihar	25	75	Spreading, red skin and white flesh, suitable for flood prone area of North Bihar
16	Kiran	Andhra Pradesh	22	110–130	Medium spreading, tuber skin red with orange flesh, early bulking
17	Sree Bhadra	Kerala, Maharashtra, Bihar, Madhya Pradesh	20	90	Semi-spreading, light pink tuber skin and creamy flesh. good cooking quality, trap crop for root knot nematode
18	Sree Rethna	Kerala	20	90–105	Spreading, purple tuber skin and orange flesh, carotene rich 3200–3500 IU/100 g, starch content 22–23%
19	Gouri	Orissa	19	110–120	Semi erect, deep orange flesh tubers, high carotene content (4.5–5.5 mg/100 g), can tolerate mid season moisture stress, suitable for both kharif and rabi season
20	Sankar	Orissa	14–16	120	Spreading, red skinned and pale yellow fleshed tubers, excellent coking quality

Table 17.3 (continued)

Sl. No.	Variety	Recommended region	Yield (t ha^{-1})	Duration (days)	Salient features
21	Sree Arun	Kerala	20–28	90	Early maturing, spreading type, pink skinned and cream fleshed tubers, good coking quality
22	Sree Varun	Kerala	20–28	90	Early maturing, spreading type, cream skinned and cream fleshed tubers, good coking quality
23	Kalinga	Orissa	16–17	105–110	Spreading plant type, tubers with purple red skin and cream flesh colour, high starch content (28%)
24	Sree Kanaka	Kerala	16–20	75–85	Short duration, cream–skinned and dark orange fleshed tubers, carotene content 8.8–10.0 mg/100 g fresh weight
25	Goutham	Orissa	18–20	105–110	Spreading, suitable for kharif and rabi seasons, white skinned and cream fleshed tubers
26	Sourin	Orissa	16	105–110	Spreading plant type, suitable for sandy loam soils, withstands mid season drought, red skinned and creamy white fleshed tubers
27	Kishan	Orissa	17	110–120	High starch (29–30%), suitable for kharif and rabi seasons, sandy loam and black sandy soils, reddish purple skin and creamy white tuber flesh

Source: Abraham et al. 2006; Palaniswami and Shirly Raichal, 2006.

nights with moderate rainfall for higher tuber yield. Surveys carried out in different regions of India on the optimum time of planting revealed that the ideal time for planting sweetpotato was late September or early October at Coimbatore, Tamil Nadu (Shanmugavelu et al., 1973), October and November at Thiruvananthapuram, Kerala (Singh and Mandai, 1976), June in the hilly areas of Tripura (Gupta and Rai, 1979) and first fortnight of September in West Bengal (Biswas et al., 1980).

A suitable cropping pattern of maize-sweetpotato-wheat and moong and maize-sweetpotato-onion has been suggested for Bihar. In Tamil Nadu, sweetpotato is followed by a cereal crop. The common rotation followed in Orissa is maize-sweetpotato-fallow and rice-sweetpotato-fallow and in West Bengal is moong-taro-sweetpotato. In Andhra Pradesh, sweetpotato follows maize and is succeeded by a vegetable crops. In Chhattisgarh/ Uttar Pradesh/ Maharashtra, vegetable cowpea-sweetpotato is the common sequence (Nair, 2000; Palaniswami and Shirly Raichal, 2006).

Planting Material

Sweetpotato is propagated by vine cuttings obtained from a nursery either raised from stored tubers or from vines of the freshly harvested crop (Onwueme and Charles, 1994). Recurrent use of vines as planting material did not result in marked variation in the yield of marketable grade tubers. However, a significant increase in the incidence of sweetpotato weevil was observed in the tubers. Vines obtained from nursery are found to be healthy and vigorous resulting in maximum tuber production (CTCRI, 1970).

Primary Nursery

In order to produce vines for planting one hectare of land, about 100 m^2 of land and 100 kg medium sized (125–150 g), weevil free tubers are required. The tubers are planted at a spacing of 20 cm on ridges formed at 60 cm apart. To ensure quick growth of vines, top dressing with urea @ 1.5 $kg/100\,m^2$ at 15 days after planting is advisable. The nursery is irrigated on alternate days for the first 10 days and once in 3 days thereafter. After 45 days of growth, the vines are cut to a length of 20–30 cm for further multiplication in the secondary nursery (CTCRI, 1987).

Secondary Nursery

Vines collected from the primary nursery are further multiplied in the secondary nursery in an area of 500 m^2 to produce enough vines for planting one hectare of land. Farmyard manure or compost is applied @ 1 kg m^{-2} and ridges are formed at a spacing of 60 cm. Vines obtained from the primary nursery or from freshly harvested crop are planted in the secondary nursery at a spacing of 20 cm on ridges. To ensure enough vegetative growth, 5 kg of urea are applied in 2 splits at 15th and 30th day after planting. The vines get ready for planting in the main field at about 40–45 days after planting.

Selection of Planting Material and Preparation of Vines

The apical cuttings are found to be the ideal to get higher tuber yield from sweetpotato (Hossain and Mondal, 1994). A vine length of 20–45 cm with at least 3–5 nodes is optimum for tuber production in different parts of India (Nair et al., 1989; Prasad, 1989). Cut vines with intact leaves are stored under shade for 24–48 hr prior to planting in the main field to promote better root initiation, early establishment of vines and higher tuber yield.

Planting Method and Spacing

Mounds, ridges, furrows and flat bed methods are being practiced in different locations (Ravindran and Nair, 1994). It is preferable to plant sweetpotatoes on mounds in areas experiencing problems of drainage. Ridges formed across the slope are recommended in sloppy areas to prevent soil erosion. Ridge method of planting gave maximum yield followed by furrow and flat methods at Bhubaneswar in Orissa (Prasad, 1989). The vines are normally planted 25–30 cm apart on ridges of height 20–25 cm made 60–75 cm apart. For Bihar, the flat method of planting and for Andhra Pradesh, Uttar Pradesh and North Eastern India the ridge method of planting are recommended.

A close spacing is generally recommended for sweetpotato to achieve maximum tuber yield. A planting distance of 30 × 60 cm within rows and 15–20 cm between plants resulted in maximum yield in different parts of India. However, when sweetpotato is planted in mounds, no specific spacing is followed and vines are planted to accommodate 3–6 cuttings per mound. CTCRI (1987) has recommended a general spacing of 60 × 20 cm for sweetpotato planted on ridges accommodating 83,000 plants in one hectare.

Planting

The cuttings are planted in the soil with both ends exposed and the middle portion buried in the soil. At IARI, New Delhi, Dayal and Sharma (1990) observed significant differences in marketable tuber yield in vertical planting, but root girth was significantly higher in horizontal planting. Planting sweetpotato vines at depth ranging from 2.5–10.0 cm in vertical method did not have any significant influence on stand establishment and final tuber yield (Ravindran and Mohankumar, 1985).

Nutritional Requirement

Organic Manuring

As sweetpotato removes appreciable quantities of plant nutrients, incorporation of considerable amount of organic manure at planting has been recommended to maintain soil productivity. Sweetpotato grown in fertile soils, generally do not receive dressings of organic manure, while soils low in organic matter content have to be supplied with organic manures at 5 or $10\,t\,ha^{-1}$ to ensure proper development of tubers. Studies conducted at CTCRI indicated that application of FYM @ $5\,t\,ha^{-1}$ resulted in higher tuber yields (Ravindran and Bala Nambisan, 1987).

Major Nutrients

Nitrogen

Sweetpotato generally responds to small doses of N application. Excessive N application results in profuse leaf production at the expense of root yield. Nitrogen deficiency is usually noticed in sandy soils and soils low in organic matter content. Sweetpotato responded to N application up to $100\,kg\,ha^{-1}$. In one of the earlier trails conducted at CTCRI (Mandal et al., 1971), significant yield increases were obtained up to $100\,kg\,ha^{-1}$ N. However, studies conducted in Kerala, Tamil Nadu and Maharashtra indicated that N @ $50\,kg\,ha^{-1}$ was sufficient for sweetpotato production (Nair and Sadanandan, 1973; Muthuswami et al., 1981 and Rajput et al., 1981 respectively). Trails conducted at Coimbatore, Tamil Nadu indicated that application of 80 kg N, 50% as basal and 50% as foliar in the form of two per cent urea at 30, 60 and 90 days after planting regulated the foliage growth and increased the tuber yield over full basal dressing and 50% basal +50% top dressing at 30 days after planting (Shanmugavelu et al., 1973). Similar observation was also made by Alexander et al. (1976) with the application of 75 kg N, 50% as basal and 50% as foliar at 35 days after planting.

Advantage of split application of N at planting and 30 days later in moderating top growth during tuber development period and thus achieving best result have been suggested by Morita (1967). On the other hand, delayed N application was unfavourable for tuber formation in sweetpotato grown on sandy loam soils (Morita, 1970).

It is a common experience that the plants utilize only 40–50% of applied N in the form of urea and the rest of the N is lost through leaching, volatilization and denitrification. Such low efficiency of utilization can be improved by modifying the urea to release N in a regulated fashion through out the growing season (Nair, 2000). Different techniques of coating urea can improve nitrogen use efficiency in the Indian farming system. Preliminary trails conducted at CTCRI (1997) indicated that the dose of N could be reduced to 37.5 kg by coating urea with cow dung and inserting into the soil as cow dung ball urea at the time of planting.

Phosphorus

Phosphorus deficiency and response to P application are most common in acid soils, especially in laterite and red soils, such as Oxisols, Ultisols, Inceptisols etc., which contain high levels of Fe and Al. Very little work has been done on the response of sweetpotato to phosphate fertilizers in India. At CTCRI, Kabeerathumma et al. (1986) reported that rock phosphate was equally effective as that of single super phosphate in direct effect, but was superior in residual effect. Since sweetpotatoes do not require very large quantities of phosphate for root development, a dose of 25–50 kg ha^{-1} is considered optimum.

Potassium

Potassium plays a major role in the translocation of photosynthates from the leaves to the roots and accelerates the process by contributing to the rapid cambial activity in the tuberous roots in which starch is stored. When K is applied, the activity of the enzyme, starch synthetase increases but when it is lacking, the enzyme activity becomes extremely low (Murata and Akazava, 1968).

Potassium deficiency is normally expected in sandy soils. Tsuno and Fujise (1968) stated that the photosynthetic activity of sweetpotato leaves should be maintained by a high level of K. Differential response to applied K has been reported from various parts of India. Soils high in available K status do not respond to added levels of K by significantly improving the tuber yield (Muthuswami et al., 1981). They also observed that between the two sources of K, schoenite had similar influence on the yield and starch content of sweetpotato as that of muriate of potash.

Based on an analysis of 392 experiments, Bao et al. (1985) concluded that K fertilization was very effective in sweetpotato. On an average, applying 70 kg ha^{-1} potash increased the tuber yield by 3.7 t. They also observed that K fertilizer increased the yield of sweetpotato by increasing the number of tubers and the ratio of large to small tubers.

Secondary Nutrients (Ca, Mg and S)

Calcium plays a major role in the water regulation of the plant, while magnesium is a constituent of chlorophyll and is, therefore essential for photosynthesis. Sulphur is a basic component of various amino acids and is required for protein synthesis. The deficiency of these nutrients is generally encountered in highly leached acid soil.

Application of CaO @ 200 kg ha^{-1} was found to be beneficial in increasing the yield and quality of the sweetpotato tubers in acid laterite soils of Kerala (Nair and Mohankumar, 1984).

Micronutrients

Introduction of more refined and complex forms of fertilizers coupled with an increase in the intensity of cropping are bound to limit the yield of crops. Therefore, a timely and precise appraisal of micronutrient deficiencies is necessary to

Table 17.4 Deficiency symptoms of various micronutrients in sweetpotato

	Symptoms	
Element	Foliar	Tuber
Manganese	Interveinal chlorosis of younger and middle leaves later changing to complete yellow	Smaller sized tubers with brownish flesh
Zinc	Occurrence of yellow spots and interveinal chlorosis, finally complete bleaching of young leaves	Brown coloration in the flesh
Copper	Young leaves turn yellow, stunted and cup shaped in appearance, finally complete bleaching of young leaves	Brown coloration in the flesh
Boron	Cessation of growth of terminal bud which becomes short in size and bunchy in appearance, finally tips start wilting and drying	No tuber formation

Source: Pillai et al. (1986)

take prompt and appropriate remedial measures to realize best yields. The deficiency symptoms in sweetpotato included under controlled conditions at CTCRI are presented in Table 17.4.

Sweetpotato responds favorably to the application of zinc. Zinc is now regarded as the third most important limiting nutrient element in crop production after N and P (Gupta, 1995). It was further reported that, Zn is essential for several enzyme systems and its deficiency retards photosynthesis and N metabolism in plants. George and Mittra (1996) reported appreciable increases in sweetpotato yield by basal as well as foliar application of Zn in the trails conducted at Kharagpur in West Bengal.

Boron is one of the essential micronutrients required for the normal growth and development of plants. Boron is needed for the development and differentiation of tissues particularly, growing tips, phloem and xylem (Sakal and Singh, 1995). It also plays an important role in sugar translocation. A combined application of ethrel (500 ppm) as vine dip and soil application of K_2O @ 90–120 kg ha^{-1} along with foliar application of boric acid @ 700 g ha^{-1} significantly increased the yield and quality of the sweetpotato tubers in Tripura. Maini et al. (1973) reported an increase in the protein content of tubers and enhancement in flowering due to the application of 10 ppm boron. Application of borax @ 10 kg ha^{-1} as basal dressing or foliar spray (0.2%) of boric acid have been recommended for overcoming the deficiency (Sakal and Singh, 1995).

Nutrient Management

Balanced application of nutrients is essential for the optimum growth and yield of sweetpotato. Nutrient recommendations for sweetpotato in different states of India are given in Table 17.5. From the above recommendations it can be concluded that conjoint application of FYM @ 5 t ha^{-1} and NPK @ 50:25:50 kg ha^{-1} is sufficient to take care of the nutrient requirement of the crop to a great extent (Ravindran

Table 17.5 Nutrient recommendations for different states of India

States	FYM (t ha^{-1}) and NPK (kg ha^{-1})
Bihar, West Bengal and Assam	FYM @ 10 t ha^{-1}, NPK @ 40–60:40:40–60
Andhra Pradesh	NPK @ 60:60:60
Karnataka	NPK @ 60:60:90
Kerala	FYM @ 5 t ha^{-1}, 50:25:50

Source: Palaniswami and Shirly Raichal (2006)

and Bala Nambisan, 1987). Full dose of P_2O_5 and K_2O and half dose of N as basal application and the remaining half N at 30 days after planting is recommended.

In sweetpotato, integrated use of *Azospirillum* and AM fungi and reduced dose of N and P fertilizers (75% and 50% of the recommended dose respectively) could maintain soil health and ensure high crop productivity (Nair et al., 2001). Trials conducted under the aegis of AICRP on tuber crops indicated that application of 2/3 recommended dose of N along with *Azospirillum* @ 2 kg ha^{-1} as vine dipping and *Azospirillum* @ 10 kg ha^{-1} as soil application is ideal for Tamil Nadu, Assam, Bihar, West Bengal and Kerala. In Andhra Pradesh the dose of fertilizer N could be further reduced to 1/3 by following the above integration with biofertilizers.

Sreelatha et al. (1999) developed equations for fertilizer recommendations to obtain targeted yields from soils with different test values based on basic data of tuber yield, nutrient uptake and soil test values.

The measures to be followed for rectifying the secondary and micronutrient deficiencies encountered in sweetpotato are given in Table 17.6.

Table 17.6 Management of secondary and micronutrients in sweetpotato

Element	Control measurers
Ca	Addition of lime, single and triple super phosphate
Mg	Incorporation of dolomitic lime or magnesium oxide in acid soils (20–50 kg ha^{-1} magnesium) or by band application of kieserite or fertilizer grade magnesium sulphate (10–40 kg ha^{-1} magnesium)
S	Application of S containing fertilizers, gypsum or elemental S or ammonium sulphate or single super phosphate
Fe	Foliar spray of chelated Fe or 1–2% ammonium ferric sulphate solution
Mn	Foliar sprays of 0.1% manganese sulfate or chelate or 2–4 kg ha^{-1} manganese, application of mulches and composts
Zn	Dipping the cuttings in 2–4% zinc sulfate for 15 min prior to planting, foliar spraying of 1–2% zinc sulphate heptahydrate solution
Cu	Foliar application of copper sulphate
B	Application of borax or other borates in the soil before planting @ 1–1.5 kg ha^{-1} on sandy soils, or up to 4 kg ha^{-1} on clayey, alkaline soils
Mo	Application of sodium molybdate or ammonium molybdate@ 0.2–0.3 kg ha^{-1}, liming the soil to raise the soil pH above 5.5 can alleviate molybdenum deficiency

Source: Susan John et al. (2006)

Weeding and Earthing Up

Sweetpotato is a quick growing crop and it covers the soil quickly and suppresses most of the weeds. About 20% reduction in tuber yield was observed in sweetpotato due to weed infestation in the early stages of growth (CTCRI, 1987). Thus, weeding may become necessary, particularly in the early stages of growth for weed control, besides improving the physical condition of the soil. The critical period of weed competition started before 14 days after planting, possibly as early as 7 DAP and continued up to 56 DAP. Critical period of crop-weed competition has been reported to be 30–45 days after planting (Nedunchezhian et al., 1998; Nair, 2000). At least one weeding and earthing up has to be done between 15 and 30 days after planting along with top dressing.

Irrigation

For proper sprouting and establishment of vines, it is appropriate to ensure sufficient moisture in the soil at the time of planting. When sufficient soil moisture is not available irrigation has to be provided on alternate days initially for the first fortnight and thereafter once in 7–10 days. A total of 12–15 irrigations are required during the entire crop period. Irrigation would generally increase the yield and improve the grade and quality of marketable tubers. In Tamil Nadu, irrigation at 60% moisture depletion level increased the tuber yield by 24% over non- irrigated crops. According to Goswami et al. (1995), 3 irrigations at tuber initiation, early tuber bulking and late tuber bulking stages were most favourable for higher number of tubers per plant, tuber bulking rate and dry matter content resulting in highest tuber yield.

Harvesting

The exact duration of the crop varies according to cultivar and the environmental condition. In North India, sweetpotato takes about 5–6 months for maturity, while it matures within 4 months in the South. The yield will increase if the crop remains in the ground longer, but the tubers become less palatable and weevil damage and rots become more noticeable with age. The maturity of tubers can be determined by cutting fresh tubers. The cut surface of the immature tubers gives a dark greenish colour, while in mature tubers the cut ends dry clearly.

Crop Protection Technologies

One of the factors limiting the production of sweetpotato in the Indian sub-continent is the infestation by pests and diseases. More than 90 insect and non-insect pests are found to infest the crop in India (Rajamma and Premkumar, 1994). Pests infest all plant parts – roots, stems, foliage, flowers, seeds and tubers. Some of them cause yield loss directly by feeding in the roots/tubers and some others indirectly by feeding the foliage or boring into the vines. Certain insects like, aphids, whiteflies and mites, in addition to feeding on the foliage, transmit diseases also. One of the reasons for the infestation by large number of pests is due to the fact that the crop can be successfully grown throughout the year.

Although a large number of insects infest the crop, only a few cause severe damage resulting in significant yield reduction. Majority of the insects are foliage feeders because of the ready availability of large number of leaves on the plant. But the crop loss caused by many of the foliage feeders is very negligible. Detailed information on the pest complex has been given by Rajamma and Pillai (1991) and Rajamma and Premkumar (1994).

Pests

Among the pests that attack the crop, the most serious and ubiquitous one is the sweetpotato weevil (SPW) (*Cylas formicarius* F.) (Palaniswami and Chattopadhyay, 2005). It causes yield loss indirectly also by damaging the vines, especially at the crown region. This pest is associated with the crop wherever it is grown and can breed and multiply throughout the year. The feeding of the sweetpotato weevil causes a characteristic terpenoid odor and a bitter cooked taste makes tubers unsuitable for consumption. A few defoliators like horned caterpillar, *Herse convolvuli* and Bihar hairy caterpillar, *Spilosoma obliqua* cause yield loss by severely defoliating the crop. Vine borer – *Omphisa anastomosalls*, though sporadic in occurrence inflict yield loss by killing the vines.

In India, on farm surveys revealed 25–45% yield loss in Orissa, 5–50% in Bihar and 4 to 50% in Kerala due to SPW (Pillai et al., 1993) while yield losses in experimental fields vary from 17–82%. The annual yield loss due to SPW alone was assessed to be to the tune of rupees 96 lakhs in Kerala (Palaniswami and Mohandas, 1994).

The study carried out on life cycle and ecology of SPW under different agro-climatic conditions has been presented in various reviews (Jayaramaiah, 1975; Rajamma, 1983). A large number of grubs, pupae and adults are seen on the attacked tubers and the tuber is completely riddled and eaten away. Infestation by the larva causes a bitter taste to the tuber due to the production of furanoterpene in the infested tubers. Harvested tuber contains different life stages of the insect and they continue the damage during storage. Weevils are also attracted to the stored tubers from the nearby fields. The weevils can survive in the plant residues left in the field after harvest, and serve as a source of inoculum. Insects are passed on to the next crop through infested planting material or by migration from other sweetpotato fields. Weeds such as *Ipomoea purpurea, I. biloba, Thumbergia* sp., in field bunds also serve as source of inoculum.

Control Measures for SPW

The concealed feeding habits of grubs and nocturnal habits of adults make the management of the pest a challenge. IPM strategies recommended by CTCRI are based on experiments and observations over a period of more than 10 years, involving mass trapping of adult weevils using a sex pheromone, agronomic practices, crop phenology and weevil life-table. The IPM package developed was a combination of weevil-free planting material, mulching with *Calophyllum inophyllum* or *Bassia latifolia* cake at 2 t ha^{-1} or mulching the plant base with leaves of *Clerodendron infortunatum* or *Chromolaena odorata* at 3 t ha^{-1}. Additional practices were reridging, mass trapping of males through sex pheromone traps, preservation of biocontrol agents, disposal of harvest residues and early harvesting. This package was found to reduce weevil damage in the field and increase the number of marketable roots. It was successfully tested in farmers' fields in 14 states of India (Palaniswami and Chattopadhyay, 2005). The synthetic sex pheromone were an efficient tool to manage population build-up of SPW (Pillai et al., 1993). Pheromone traps (1 trap/100 m^2) kept in the field throughout the cropping season and one month after harvest can reduce populations. Palaniswami and Mohandas (1994) observed the Kairomone present in the periderm of fresh sweetpotato tubers is good at trapping adult SPWs. Tarafdar and Sarkar (2005) reported that root zone application of *Beauveria bassiana* significantly reduced the intensity of root damage followed by use of sex pheromone traps of *C. formicarius* under West Bengal conditions.

Diseases

Sweetpotato is comparatively free from many serious diseases in India although the crop has been reported susceptible to more than 40 diseases from different countries. Chlorotic leaf distortion (CLD) caused by *Fusarium lateritium*, leaf spots caused by *Cercospora* spp *Alternaria* spp., *Helminthosporium euphorbias* and sweetpotato feathery mottle virus (SPFMV) are common leaf diseases; root wilt (*F. oxysporum f.* sp. *batatas*), black rot *(Ceratocystis fimbriata)*, Java block rot (*Botrydiplodia theobrmae*) and collar rot *(Sclerotium rolfsi)* on the tubers both in field and in storage were found in India. Among the diseases, stem rot/wilt caused by *Fusarium oxysporum* f. sp. *batatas* is found in Orissa, West Bengal. Uttar Pradesh and Bihar of India (Thankappan, 1994).

In general careful handling of sweetpotato tubers during harvest and transit to avoid unnecessary wounding is recommended for reducing the intensity of fungal diseases. Proper curing of the tubers and use of safe protective fungicide and improving the storage condition can also help in reducing the incidence of storage diseases.

Viruses cause economically important diseases in sweetpotato. More than 15 viruses have been reported to affect sweetpotato worldwide. In India, occurrence of sweetpotato virus disease syndrome was reported (Thankappan and Nair, 1990). Later, Kumar et al. (1991) reported the occurrence of *Sweetpotato feathery mottle virus* (SPFMV) in sweetpotato germplasm. A survey carried out in India found 10–80% viral disease incidence, depending upon age of the crop, location and varieties (Makeshkumar et al., 2001). Various types of leaf symptoms viz., ring spot, feathering, chlorotic specks, mosaic, puckering, leaf curling, leaf cupping, yellow netting and fan leaf type symptoms were observed (Jeeva et al., 2004b: Makeshkumar et al., 2001). Detection of *Sweetpotato feathery mottle virus* (SPFMV) using NCM-ELISA kits provided by CIP, Lima Peru; ISEM, DAC-ELISA and through PCR and NASH based detection (Makeshkumar et al., 2006; Jeeva et al., 2004a, b, Ganga Prasanth et al., 2006) have confirmed the wide prevalence of SPFMV ordinary strain in India. The severe form of sweetpotato virus disease (SPVD) caused by co-infection of SPFMV and *Sweetpotato chlorotic stunt virus* (SPCSV) recorded in many African countries (Karyeija et al., 2000) has not yet been observed in India. There is no report of any occurrence of tuber crack strain of SPFMV in India. Though SPFMV is the major virus prevalent in India, presence of 3 other viruses namely *Sweetpotato chlorotic fleck virus, Sweetpotato latent virus, Sweetpotato mild mottle virus* in sweetpotato germplasm collection were reported using CIP's NCM-ELISA kit (Makeshkumar et al., 2001). However, there are no further reports on occurrence of these viruses on sweetpotato. Jeeva et al. (2004b) partially purified SPFMV and prepared polyclonal antisera. The cloning and sequencing of coat protein gene of SPFMV occurring in India showed that it is more similar to SPFMV Egyptian isolate (Ganga Prasanth et al., 2006; Hegde et al., 2007a,b). The coat protein gene of the virus has been expressed in bacterial expression vector pET 32b and polyclonal antiserum specific to expressed coat protein has been produced (Hegde et al., 2007a). The recombinant antisera detects SPFMV even in symptomless sweetpotato plants using DAC-ELISA and DIBA. Recently the occurrence of *Sweetpotato leaf curl virus* was identified and based on CP gene sequencing it was found that the virus is *Sweetpotato leaf curl Georgia virus* (Makeshkumar et. al., 2007). Virus elimination in sweetpotato through meristem culture combined with thermotherapy and chemotherapy has also been standardized and virus free plants of sweetpotato have been produced (Jeeva et al., 2004a).

Nematodes

Root-knot nematodes *Meloidogyne incognita* are found in India. Many accessions showed a high degree of resistance to the nematode (Mohandas, 1994) and resistant varieties of sweetpotato could be incorporated in cropping systems where *M. incognita* is a problem. A sweetpotato variety Sree Bhadra was found to trap the root knot nematode and it can be used as a trap crop to reduce the nematode population in soil (Ramakrishnan et al., 2001).

Post Harvest Utilization

Sweetpotato can be utilized in a variety of ways in home food, in animal feed and also as a raw material for industrial products. Central Tuber Crops Research Institute has developed many technologies for the production of value added products like sweetpotato chips, candies, jams, pickels, squash, sauce, starch etc. Pigments like carotenoids and anthocyanins present in sweetpotato are natural colours and technologies are available for commercial extraction. Enzymes like β – amylase, alcoholic beverages and fermented products like citric acid, monosodium glutamate etc can be produced from sweetpotato. Both vines and roots can be used either in fresh or dried form as animal feed. Roots supply energy and vines supply protein to animals.

Sweetpotato-Various Uses (Tubers and Leaves)

Sweetpotatoes are an under exploited crops. Despite the fact that it is one the highest energy producing crops (152 MJ/ha/day), it seldom finds use as a starchy staple in countries other than China and Japan. The latter two countries have developed several value added products from sweetpotato tubers and leaves, realising its nutritional attributes. Unlike the Southeast Asian countries and India, China, which is the largest producer of sweetpotato, also utilizes a major part of its production for animal feed purposes. With the introduction of a large number of orange fleshed varieties having high β- carotene content ranging even up to 20–30 mg/100 g, there is a recent upsurge of interest in including sweetpotato roots as a part of the food based strategy to compensate vitamin A deficiency, especially in the African countries and Bangladesh.

Sweetpotato Roots as Human Food

Sweetpotato is cultivated and utilized extensively in Asia and South East Asian countries, although processed products are more common in countries like Indonesia, Philippines, Thailand, Malaysia and Vietnam. The tubers are eaten after boiling or baking and increase in sugars during these processes makes the tubers sweet after cooking. The 'dry' type of tubers will be dry and firm but mealy after cooking, while the 'moist' type will be soft and watery as well as sticky after cooking (Rao et al., 1974).

A major problem in the utilization of sweetpotato is the post harvest damage often triggered by the pathogen entry through cut surfaces. Curing at a relative humidity of 85% and temperature of 29–30 °C is often practiced in the South Asian countries and India, which facilitates wound-healing and enhances the shelf life of the roots. Studies have been made by various researchers in India to develop many value added products from sweetpotato. Sweetpotato was processed into vermicelli and its nutritional quality studied by Thirumaran and Ravindran (1992).

Sweetpotato contains water-soluble pectin, which enables its use in making jams and jellies. Due to the high starch content of sweetpotato, compared to fruits, the jam has a slightly different consistency (Truong, 1987). Jams from sweetpotato have been prepared in India (Padmaja and Premkumar, 2002). The orange-fleshed variety 'Kamala Sundari' was used for making jam with either sugar or molasses (Chaudhury, 1992). It was found that the high starch sweetpotato varieties could be incorporated only up to 60% (Padmaja and Premkumar, 2002).

Sweetpotato pickles have been made from orange-fleshed varieties in Bangladesh and Philippines and white variety in India (Chaudhury, 1992; Tan et al., 2005; Padmaja and Premkumar, 2002).

Sweetpotato based composite flours have been used in many countries for making small baked goods like cakes, cookies, biscuits, doughnuts etc. French fry type products have been prepared from roots of sweetpotato. Restructured sweetpotato sticks (RSS) are another product similar to French fries (Utomo et al., 2005).

Curd is a popular food item for the Asians and is usually made from milk by natural lactic fermentation. Sweetpotato (anthocyanin and carotene-rich varieties) was used to prepare curd having high nutritive value. Lacto-juice was made from orange-fleshed sweetpotatoes by fermenting the mash with *Lactobacillus plantarum* at $28 \pm 2\,^{\circ}$C for 48 h and expressing the juice. The sensory evaluation gave high scores for taste, aroma and texture (Panda and Ray, 2007).

Hard candies, which are favorite sweet of many contain flour and starch as texture modifiers, besides the usual ingredients like milk, fat, sugar, glucose syrup and emulsifiers. Sweetpotato was attempted as a replacer of flour and starch in candy making (Samsiah et al., 2005). Sweetpotato mash prepared from steamed/baked roots was added to the basic ingredients viz., coconut milk, sugar and glucose syrup. The basic ingredients were boiled, filtered and concentrated. Skimmed milk powder and sweetpotato mash were added, mixed and solidified with stirring. The mass was then kneaded, tempered and cut into candies. These workers obtained a high quality product from 10% steamed sweetpotato mash. Sweetpotato flour was utilized to make a 'nutrigulab jamun mix' by mixing it with refined wheat flour and milk powder. The sweet gulab jamuns, a popular dessert of India had good soft texture and taste (Padmaja and Premkumar, 2002).

Use of Roots and Vines for Animal Feeding

The roots and vines of sweetpotato are traditionally used for pig feeding in countries like China, Taiwan, Vietnam and Japan. The roots have generally low content of crude protein and fiber, when compared to the leaves, which contain as high as 11–19% crude protein (dwb) and 10–17% fiber (Bouwkamp, 1985). Sweetpotato vines are used to a greater extent than the roots for feeding cattle in China, Taiwan, Peru and Japan. Sweetpotato vine silage was prepared by Sutoh et al. (1973), who found that molasses and inorganic acids like hydrochloric acid were essential to control fermentation. Sweetpotato forage has been reported as a cheap protein source leading to good feeding efficiency in cattle. Orange-fleshed sweetpotatoes were found to improve the vitamin A content of milk (Mather et al., 1948). Fresh vines could be given up to 70 kg for a cow weighing 500 kg whilst some workers have reported that the feeding of sweetpotato roots leads to better daily weight gains in cattle, others observed slight inferior performance when fresh roots alone are fed (Mather et al., 1948).

Sweetpotatoes are not extensively used for poultry rearing in most countries, where they are produced. Feeding studies showed that chicks given 20% dried chips grew well as those fed corn diets alone (Lee and Yang, 1979). Addition of only 10% dry chip meal in layer diets and 20% in broiler diets has been suggested.

Carotene-rich sweetpotatoes were reported to result in better broiler carcass colour and egg yolk colour in layers (Garlich et al., 1974), which eliminates the need to incorporate xanthophylls in formula feeds.

Future Aspects

In India, sweetpotato will become an important crop based on the changing climate pattern and increase in labour cost. It has already proven to be a famine saver and can be harvested with in a short period. This makes the crop a choice during calamities. Any further attention on creation of processing facilities will certainly make this crop a profitable one in India.

References

Abraham, K., Edison, S., Unnikrishnan, M., Sheela, M. N., Vimala, B., Sreekumari, M. T., and Naskar, S. K. 2006. *Tuber Crops Varieties Released by the Central Tuber Crops Research Institute, Technical Bulletin Series No.24*, Central Tuber Crops Research Institute, Sreekariyam, Thiruvananthapuram, 64 pp.

Alexander, D., Sadanandan, N., and Nair, R. Y. 1976. Studies on the effect of methods of nitrogen application on mineral uptake of high yielding sweetpotato varieties. *Agric. Res. J. Kerala.* 14: 21–26.

Bao, L., De-Yin, L., and Rong-Gui, W. 1985. Response to potash fertilizers of main crops in China. Potassium in the Agricultural Systems of the Humid Tropics. *Proceedings 19th Colloquium of the International Potash Institute*, held in Bangkok. pp. 323–328.

Biswas, D., Sen. H., and Mukhopadhyay, S. K. 1980. Effect of time of planting on tuber development of sweetpotato (*Ipomoea batatas* L. Lam). *J. Root Crops.* 14: 11–15.

Bouwkamp, J. C. 1985. *Sweetpotato Products: A Natural Resource for the Tropics.* CRC Press, Boca Raton, Florida, 271pp.

Chaudhury, S. J. C. 1992. Processing of sweetpotato. *Bangladesh Agricultural Research Institute Bulletin*, pp 40.

CTCRI. 1970. *Annual Report*, Central Tuber Crops Research Institute, Sreekariyam, Thiruvananthapuram. pp. 23–24.

CTCRI. 1987. Cultural practices of sweetpotato. *Extn. Bull.*, Central Tuber Crops Research Institute, Sreekariyam, Thiruvananthapuram.

CTCRI. 1997. *Annual Report.* Central Tuber Crops Research Institute, Sreekariyam, Thiruvananthapuram. pp. 27–28.

Dayal, T. R., and Sharma, R. P. 1990. Effect of fertilizer, spacing and method of planting in sweetpotato. Paper presented in the *National Symposium. on Recent Advances in the Production and Utilization of Tropical TuberCrops.* 7–9 November 1990, Thiruvananthapuram, India.

Ganga Prasanth, Hegde, V., Makeshkumar, T., Jeeva, M. L., and Edison, S. 2006. Development of diagnostics for sweetpotato feathery mottle virus. Paper presented in *14th Triennial Symposium of the International Society for Tropical Root Crops*, 20–26 November 2006, Thiruvananthapuram, India.

Garlich, J. D., Bryant, D. M., Covington, H. M., Chamble, D. S., and Purcell, A. E. 1974. Egg yolk and broiler skin pigmentation with sweetpotato vine meal. *Poult. Sci.* 53: 692–699.

George, J., and Mittra, B. N. 1996. Effect of application of fly ash and paper factory sludge as source of micronutrient on the performance of sweetpotato. *Tropical Tuber Crops in Food Security and Nutrition.* (Eds. Balagopalan, C., Nayar, T. V. R., Sundaresan, S., Premkumar, T., and Lakshmi, K. R.). Oxford & IBH Publishing Co. Pvt. Ltd., New Delhi. pp. 338–343.

Goswami, S. B., Sen, H., and Jana, P. K 1995. Tuberization and yield potential of sweetpotato cultivars as influenced by water management practices. *J. Root Crops*. 21: 77–81.

Gupta, P. N., and Rai, M. 1979. Sweetpotato for tilla lands of Tripura. *Indian Fmg*.29: 24.

Gupta, V. K. 1995. Zinc research and agricultural production. In: *Micronutrient Research and Agricultural Production*. (Ed. Tandon, H. L. S.). Springer Netherlands, Dordrecht. pp. 132–164.

Hegde, V. Ganga Prasanth, Makeshkumar, T., Jeeva, M. L., and Edison, S. 2007a. Cloning and expression of coat protein gene of Sweetpotato Feathery Mottle Virus in *E. coli*. Paper presented in *10th International Plant Virus Epidemiology Symposium on controlling epidemics of emerging and established plant virus diseases – the way forward* held at ICRISAT, Hyderabad during 15–19 October 2007.

Hegde, V., Ganga Prasanth, Makeshkumar, T., Jeeva, M. L., and Edison, S. 2007b. Development of molecular diagnostics for sweetpotato feathery mottle virus. Paper presented *International Conference on Emerging and Re-emerging Viral Diseases in the Tropics and Subtropics* held at IARI, New Delhi during 11–14 December 2007.

Hossain, M. M., and Mondal, M. A. A. 1994. Effect of vine parts on the growth and yield of three sweetpotato varieties. *Bangladesh J. Sci. Ind. Res*. 29: 181–184.

Jayaramaiah, M. 1975. Bionomics of sweetpotato weevil *Cylas formicarius Mysore J. Agric. Sci*. 9: 99–109;

Jeeva, M. L. Balakrishanan, S., Edison, S., and Rajmohan, K. 2004a. Meristem culture and thermotherapy in the management of Sweetpotato *feathery mottle virus. J. Root Crops*. 30: 135–142.

Jeeva, M. L. Balakrishanan, S., Edison, S., Umamaheswaran, K., Makeshkumar, T. 2004b. Characterization, purification and serology of Sweetpotato feathery mottle virus in India. *J. Root Crops*. 30: 24–30.

Kabeerathumma, S., Mohankumar, B., and Potty, Y. P 1986. Efficiency of rock phosphate as a source of phosphorus to sweetpotato. *J. Indian Soc. Soil Sci*. 34: 806–809.

Karyeija, R. F., Krueze, J. F., Gibson, R. W., and Valkonen, J. P. T. 2000. Synergistic interactions of a potyvirus and a phloem-limited crinivirus in sweetpotato plants. *Virology* 269: 26–36.

Kumar, C. A. Mandal, B. B., Chandel, K. P. S. Jain, R. K., Varma, A., And Srivastava, M. 1991. Occurrence of *Sweetpotato feathery mottle virus* in germplasm of *Ipomoea batatas (L.) Lam*. in India *Curr. Sci*. 60: 321–325.

Lee, P. K., and Yang, Y. F. 1979. Nutritive value of high protein sweetpotato chips as feed ingredients for broilers. (Chinese). *J. Agric. Assoc. China* 106: 71–78.

Maini, S. B., Indira, P., and Mandai, R. C. 1973. Effect of boron, magnesium and iron on flowering, yield and quality of sweetpotato. *Madras Agric. J*. 60: 1024–1025.

Makeshkumar, T., Jeeva, M. L., Hegde, V., and Edison, S. 2006. Molecular diagnosis of sweetpotato feathery mottle disease. Poster presented in *14th Triennial Symposium of the International Society for Tropical Root Crops*, 20–26 November 2006, Thiruvananthapuram, India.

Makeshkumar, T., Sriram, S., Nair, R. R., Edison, S., And Thankappan, M. 2001 Symptomatology and characterization of sweetpotato virus diseases *J. Root Crops*. 27: 296–302.

Makeshkumar, T., Prakash Krishnan, V. S., Hegde Vinayaka, Jeeva, M. L., and Edison, S. 2007. Sweetpotato Leaf Curl Disease – a new emerging virus problem of sweetpotato in India. Paper presented in *10th International Plant Virus Epidemiology Symposium on Controlling Epidemics of Emerging and Established Plant Virus Diseases – the Way Forward* held at ICRISAT, Hyderabad during 15–19 October 2007.

Mandal, R. C., Singh, K. D., Maini, S. B., and Magoon, M. L. 1971. Response of sweetpotato to spacing and nitrogen fertilization. *Indian J. Agron*. 16: 85–87.

Mather, R. E., Linkous, W. N., and Eheart, J. F. 1948. Dehydrated sweetpotatoes as a concentrate feed for dairy cattle. *J. Dairy Sci*. 31: 569–573.

Mohandas, C. 1994. Nematodes of tropical tuber crops In *Advances in Horticulture* Vol. 8- *Tuber Crops* (Eds. Chadha, K. L., and Nayar, G. G.). MPH, New Delhi. pp. 513–522.

Morita, T. 1967. Effect of application time of nitrogenous fertilizers on the top growth, tuber formation and development of sweetpotatoes. *J. Jap. Soc. Hort. Sci*. 36: 114–121.

Morita, T. 1970. Effects of application time of nitrogenous fertilizer on the top growth and the development of the root system during the tuber formation period in sweetpotato. *Jl. Jap. Soc. Hort. Sci.* 39: 41–47.

Murata, T., and Akazava, T. 1968. Enzymic mechanism of starch synthesis in sweetpotato roots. 1.Requirement of potassium ions for starch synthatase. *Arch. Biochem. Biophys.* 126: 839–879.

Muthuswami, P., Govindaswamy, M., and Kothandaraman, G. Y. 1981. Influence of sources and levels of potash in combination with nitrogen levels of sweetpotato. *Madras Agric. J.* 68: 351–354.

Nair, G. M., and Sadanandan, N. 1973, Performance of sweetpotato varieties under varying levels of nitrogen and time of its application in red loam soils of Vellayani. *Agric. Res. J. Kerala.* 11: 14–16.

Nair, G. M., Potty, V. P., and Susan John, K. 2001. Influence of biofertilizer (*Azospirillum*) on the growth and yield of sweetpotato. *J. Root Crops.* 27: 210–213.

Nair, G. M. 2000. Cultural and manurial requirements of sweetpotato. In: *Production Technology of Tuber Crops* (Eds. Mohankumar, C. R., Nair, G. M., James George, Ravindran, C.S. and Ravi, V.), Central Tuber Crops Research Institute, Sreekariyam, Thiruvananthapuram, pp. 44–72.

Nair, G. M., Ravindran, C. S., Moorthy, S. N., and Ghosh, S. P. 1989. Sweetpotato indigenous technologies and recent advances of production, processing and utilization in India. *Proceedings International. Symposium. on Sweetpotato.* Baybay, Philippines. pp. 301–312.

Nair, P. G., and Mohankumar, B. 1984. Response of sweetpotato *(Ipomoea batatas* L.) to NPK and lime in acid laterite soil. *J. Root Crops.* 10: 17–21.

Nedunchezhian, M., Varma, S. P., and Ray, R. C. 1998. Estimation of critical period of crop weed competition in sweetpotato. *(Ipomoea batatas* L.). *Adv. Hort. Sci.* 12: 101–104.

Onwueme, I. C., and Charles, W. B. 1994. *Tropical Root and Tuber Crops: Production, Perspectives and Future Prospects.* FAO Plant Production and Protection Paper 126, Food and Agriculture Organization, pp.122–128.

Padmaja, G., and Premkumar, T. 2002. Tuber Crops Recipes, Technical Bulletin Series 36, CTCRI, Kerala, India. p. 26.

Palaniswami, M. S., and Chattopadhyay, S. 2005. Ecology based integrated management of the sweetpotato weevil in India In *Proceedings of the IInd International Symposium on Sweetpotato and Cassava: Innovative Technologies for Commercialization.* Kuala Lumpur, Malaysia, June 14–17, 2005.

Palaniswami, M. S., and Mohandas, N. 1994. Studies on attractants of adult weevil of *Cylas formicarius. J. Root Crops.* 19: 44–52

Palaniswami, M. S., and Shirly Raichal Anil. 2006. Region Specific Technologies for Tropical Tuber Crops in India, *Technical Bulletin Series No. 47,* All India Co-ordinated Research Project on Tuber Crops, Central Tuber Crops Research Institute, Sreekariyam, Thiruvananthapuram. pp. 31–39.

Panda, S., and Ray, R. C. 2007. Lactic acid fermentation of β- carotene rich sweetpotato (*Ipomoea batatas* L.) into lacto- juice. *Plant Foods Hum. Nutr.* 62: 65–70.

Pillai, K. S., Rajamma, P., and Palaniswami, M. S. 1993. New technique in the control of sweetpotato weevil using synthetic sex pheromone in India. *Int. J. Pest Manag.* 39: 84–89.

Pillai, N. G., Mohankumar, B., Kabeerathumma, S., and Nair, P. G. 1986. Deficiency symptoms of micronutrients in sweetpotato *(Ipomoea batatas* L.). *J. Root Crops.* 12: 91–95.

Prasad, M. 1989. Yield of sweetpotato as affected by planting method and vine length. J. *Root Crops.* 15: 63–64.

Rajamma, P. 1983. Biology and bionomics of .Sweetpotato weevil *Cylas formicarius* Fab. *Proceedings of the Symposium on Insect Ecology and Resource Management.* Muzaffarnagar, India. pp. 87–92.

Rajamma, P., and Pillai, K. S, 1991. Annual Report for 1990–91 CTCRI, Trivandrum.

Rajamma, P., and Premkumar, T. 1994 Pests of sweetpotato. In *Advances in Horticulture vol.8 –Tuber Crops* (Eds. Chadha, K. L., and Nayar, G. G.) Malhotra Publishing House, New Delhi. Pages: 457–484.

Rajput, S. C., Kadam, B. A., and Patil, V. K. 1981. Spacing-cum-nutritional requirement of sweetpotato. *J. Root Crops*. 7: 25–27.

Ramakrishnan, S., Viswanath, S. M., and Shenoi, M. M. 2001. Sweetpotato variety "Sree Bhadra" as a trap crop for the management of root knot nematodes in FCV tobacco nursery. *J. of root Crops*. 27: 380–382.

Rao, V. N. M., Hamann, D. D, and Humphries, E. G. 1974. Mechanical testing as a measure of kinesthetic quality of raw baked sweetpotatoes. *Trans. Am. Soc. Agric. Eng*. 17: 1187–1190.

Ravindran, C. S., and Nair, P. G. 1994. Agrotechniques and nutritional requirements of sweetpotato. In: *Advances in Horticulture, Vol 8- Tuber Crops*. (Eds. Chadha, K. L., and Nayar, G. G.) Malhotra Publishing House, New Delhi-110 064. pp. 251–269.

Ravindran, C. S., and Bala Nambisan, 1987. Effect of FYM and NPK on the yield and quality of sweetpotato. *J. Root Crops*. 13: 35–39.

Ravindran, C. S., and Mohankumar, C. R. 1985. Effect of method of land preparation and depth of planting on the yield of sweetpotato. *Proceedings National Symposium on Production and Utilization of Tropical Tuber Crops*, 27–29 November 1985, Thiruvananthapuram, India. pp. 129–133.

Sakal, R and Singh, A. P. 1995. Boron research and agricultural production. In: *Micronutrient Research and Agricultural Production*. (Ed. Tandon, H. L. S.).

Samsiah, M. S., Latifah, M. S., and Ong, C. Y. 2005. Sweetpotato candy. In: *2nd International symposium on sweetpotato and cassava*, Kuala Lumpur, Malaysia, 14–17 June, 2005. pp. 223–224.

Shanmugavelu, K. G., Thamburaj, S., and Shanmugam, A. 1973. Effect of soil and foliar application of nitrogen on the yield of sweetpotato. *Fert. News*. 18: 51–52.

Singh, K. D., and Mandai, R. C. 1976. Performance of coleus and sweetpotato in relation to seasonal variations (time of planting). *J. Root Crops*. 2: 17–22.

Sreelatha, A. K., Ravindran, C. S., and Kabeerathumma, S. 1999. Fertilizer recommendation based on soil tests for targeted yield of sweetpotato. In: *Tropical Tuber Crops: In Food Security and Nutrition* (Eds. Balagopalan, C., Nayar, T. V. R., Sundaresan, S., Premkumar, T., and Lakshmi, K. R.) Oxford & IBH Publishing Co. Pvt. Ltd., New Delhi. pp. 328–330.

Susan John, K., Suja, G., Edison, S., and Ravindran, C. S. 2006. *Nutritional Disorders of Tropical Tuber Crops. Technical Bulletin Series 48*, Central Tuber Crops Research Institute, Sreekariyam, Thiruvananthapuram, Kerala, India.74p.

Sutoh, H., Uchida, S., and Kaneda, K. 1973. Studies on silage making. XXII. The nutrient content of sweetpotato (*Ipomoea batatas* L. var. *edulis*) at the different stages and the quality of sweetpotato vine silage. (Japanese). *Sci. Rep. Fac. Agric. Okayama Univ*. 41: 61–68.

Tan, J. D., Forio, E. E., and Estoy, L. S. 2005. Production and market potential of sweetpotato flakes. In: 2nd *International Symposium on sweetpotato and cassava*, Malaysia, 14–17 June, 2005. pp. 215–216.

Tarafdar, J., and Sarkar, M. A. 2005. Managing sweetpotato weevil (*Cylas formicarius*) in West Bengal, India, by some chemicals, bioproducts and sex pheromone traps *Proceedings of the IInd Iinternational Symposium on Sweetpotato and Cassava: Innovative Technologies for Commercialization*, Kuala Lumpur, Malaysia, June 14–17, 2005.

Thankappan, M. 1994 Fungal diseases of tuber crops In *Advances in Horticulture* Vol. 8- *Tuber Crops* (Eds. Chadha, K. L., and Nayar, G. G.) MPH, New Delhi.

Thankappan, M., and Nair, N. G. 1990. New disease syndromes in sweetpotato in India *J. Root Crops*. 16: 212–215.

Thirumaran, A. S., and Ravindran, D. M. 1992. Processing of sweetpotato vermicelli in India. In: *Sweetpotato Technology for the 21st Century* (Eds. Hill, W. A., Bonsi, C. K., and Loretan, P. A.), Tuskegee university, Tuskegee, Al. pp 389–399.

Truong, V. D. 1987. New developments in processing sweetpotato for food. *Internat. Sweetpotato Symposium*, 20–26 May, 1987, Visca, Philippines.

Tsuno, Y., and Fujise, K. 1968. Studies on dry matter production of sweetpotato. *Proc. Crop Sci. Japan*. 37: 12–16.

Utomo, J. S., Che Man, Y. B., Rahman, R. A., Saad, M. S. 2005. Physical and chemical characteristics of restructured sweetpotato sticks made from three sweetpotato cultivars. In: 2nd *International Symposium on Sweetpotato and Cassava*, Kuala Lumpur, Malaysia, 14–17 June, 2005. pp. 221–222.

Chapter 18
Sweetpotato in South America

S. Fuentes and E. Chujoy

Sweetpotato [*Ipomoea batatas* (L.) Lam] is a neotropical crop grown in all subtropical and tropical regions of the world, from sea level up to 2,800 m altitude (Woolfe, 1992). Sweetpotato is a rustic crop, easy to maintain, tolerant to drought, high temperatures and humidity. As a plant of tropical origin, the sweetpotato grows better in warm regions, but new improved cultivars of high yield potential and modern cultivation techniques allow economic production in temperate regions. Its production cost is relatively low, and high returns may be obtained. It is also one of the vegetables with one of the highest production capacities in terms of energy per unit area and time (kcal/ha/day).

In South America, sweetpotato is the third most important root crop, after cassava and potato, with a production of 1,287,960 tons in 2007 (FAOSTAT, 2008). Sweetpotato production accounted for almost half (46.8%) of the production in the Americas and 1.0% of world production, which exceeded 126 million tons. Brazil, Argentina, Peru, and Paraguay were the major producers in 2005–2007, with 516,000, 350,000, 191,000 and 155,000 tons, respectively and a harvested area of 45,000, 23,000, 11,000 and 20,000 ha, respectively (Table 18.1). Peru had the highest yield with 17.2 t/ha and Guyana the lowest with 1.1 t/ha. Sweetpotato production in South America increased 10.7% in the period of 1995–1997 to 2005–2007 (Table 18.2). Among the major producers, Paraguay, Argentina and Brazil had a production change of 115.8, 14.4 and 1.8% respectively, due to increase in harvested area (Paraguay and Argentina) or yield (Brazil) (Table 18.2).

Sweetpotato is mostly grown as a mono crop, for commercial purposes in the main producing countries in South America. Commercial production often relies on modern cultivation techniques that result in higher yields, and occurs in areas closely located to main markets where the roots are destined for local use. A small proportion of roots is exported. Furthermore, sweetpotato can also be found grown as a mono crop that is used for self-consumption, often by small-scale farmers, with low input technology and lower yields. In Guyana, sweetpotato may be found intercropped with perennials such as coffee, citrus, and avocado.

S. Fuentes (✉)
International Potato Center (CIP), Apartado 1558, Lima 12, Peru
e-mail: s.fuentes@cgiar.org

G. Loebenstein, G. Thottappilly (eds.), *The Sweetpotato*,
DOI 10.1007/978-1-4020-9475-0_18,

Table 18.1 Sweetpotato production, producer price, estimated consumption, and feed use in South America

	Production 2005–2007					Estimated	Feed use	
Country	Production (tons)	Area harvested (ha)	Yield (t/ha)	Rank[a] among root crops	Producer price, 2005 US$/ton	consumption, 2003 Kg/cap/yr	Part of plant	Animal(s) fed
Brazil	516,729	45,556	11.4	3	117.2	1.4	Roots, vines	Cattle, pigs
Argentina	350,000	23,867	14.7	2	262.6	7.7	Roots, vines	Cattle, pigs
Peru	191,365	11,113	17.2	4	93.0	6.8	Roots, vines	Cattle, pigs, goats, others
Paraguay	155,343	20,663	7.5	2	100.0	18.5	Roots, vines	Cattle, pigs
Uruguay	40,000	4,233	9.5	2	136.7	15.4	Roots, vines	Cattle, pigs
Venezuela	15,469	1,486	10.4	6	424.5	0.2	Roots, vines	Livestock
Chile	9,267	1,260	7.4	2	185.4	0.5	n/a	n/a
Bolivia	7,180	1,578	4.5	4	117.0	1.1	n/a	n/a
Ecuador	3,409	1,538	2.3	4	66.3	0.2	Roots, vines	Cattle, pigs, goats
Guyana	1,333	833	1.1	2	n/a	2.3	Roots, vines	Cattle
Suriname	220	19	11.6	4	909.5	0.4	Roots, vines	Cattle

[a] Ranking of sweetpotato production compared to other root crops.

Table 18.2 Changes in sweetpotato production, harvested area, and yield in 1995–1997 and 2005–2007

Country	Production quantity (tons)			Area harvested (hectares)			Yield (t/ha)		
	1995–1997	2005–2007	Change	1995–1997	2005–2007	Change	1995–1997	2005–2007	Change
Brazil	507,435	516,729	+1.8	51,150	45,556	−10.9	9.9	11.4	+15.2
Argentina	305,880	350,000	+14.4	18,863	23,867	+26.5	16.1	14.7	−9.1
Peru	192,030	191,365	−0.3	10,893	11,113	+2.0	17.5	17.2	−1.8
Paraguay	71,984	155,343	+115.8	9,505	20,663	+117.4	7.6	7.5	−0.8
Uruguay	56,667	40,000	−29.4	6,000	4,233	−29.4	9.4	9.5	+0.9
Venezuela	10,311	15,469	+50.0	1,768	1,486	−15.9	5.8	10.4	+79.0
Chile	7,000	9,267	+32.4	1,000	1,260	+26.0	7.0	7.4	+5.1
Bolivia	12,690	7,180	−43.4	3,047	1,578	−48.2	4.2	4.5	+9.2
Ecuador	1,137	3,409	+199.8	309	1,538	+398.3	3.7	2.3	−38.2
Guyana	–	1,333	n/a	–	833	n/a	–	1.1	n/a
Suriname	349	220	−36.9	42	19	−54.4	9.7	11.6	+19.2

Sweetpotato is mostly used for human consumption as fresh product, but also, to a lesser extent, for industrial uses such as starch and flour or processed products such as the making of canned foods, or fried chips. Little attention has been given to further develop the crop for industry and feed uses, despite its potential. The increase in the world prices of petroleum and grains has now made sweetpotato an economically interesting source of bio-ethanol and starch that remains to be developed.

Improved cultivars have been released in Argentina, Brazil, Peru, Uruguay and Venezuela more often through the selection of native or introduced germplasm rather than by hybridization and systematic improvement (De la Puente et al. 2004). However, local breeding is becoming important.

Historical Background

Sweetpotato was originally domesticated in tropical America (Austin, 1988; Yen, 1974 and 1982). Based on numerical analysis of key morphological characters of sweetpotato and wild *Ipomoea* species, Austin (1988) postulated that sweetpotato originated in the region between the Yucatán Peninsula of Mexico and the Orinoco River in Venezuela. Howeve, recent molecular evidence points at Central America as the center of origin of sweetpotato (Zhang et al. 2004).

The genus *Ipomoea* comprises 600 to 700 species. The series Batatas contains 13 wild species, in addition to the sweetpotato (Austin and Huamán, 1996). Out of the 13 species, *I. trifida* was suggested to be closest to sweetpotato. The maximum variability of *I. trifida* was found between the Yucatán Peninsula of Mexico and the northwest part of South America, which overlaps with the hypothesized area of sweetpotato origin (Austin, 1988).

Materials unearthed at numerous archaeological sites in Peru representing both Inca and Pre-Inca cultures suggest that sweetpotato was one of the most important crops of this region (Ugent and Peterson, 1988). The most ancient sweetpotato remains have been found in the Chilca Canyon, 65 km southeast of Lima-Peru, 2800 m high, which date from the Neolithic period (8000–10000 B.C.) (Engel, 1970). Archaeological written evidence about the presence of sweetpotato has been found in Central America, in the region occupied by the Mayas. Diverse names in the ancient languages were given to the sweetpotato types. Thus, in Peru, the sweet and moist type was called APICHU, the drier types as KUMARA. The natives in Paraguay and Argentina called sweetpotato YETY (Montaldo, 1996).

From archaeological, linguistic and historic evidences, it is considered that the primary center of origin of sweetpotato encompasses Colombia, Ecuador and the north of Peru. Secondary centers include Mexico, Guatemala and the south of Peru (De la Puente et al. 2004). Although the oldest indications of cultivated sweetpotatoes come from Peru (Montenegro et al. 2007), a larger molecular genetic diversity seems to occur in Mexico than in Peru-Ecuador (Zhang et al. 2004). Studies on molecular diversity assessment supported the hypothesis that Central America is the primary center of diversity and most likely the center of origin of sweetpotato (Huang and Sun, 2000; Zhang et al. 2000; Zhang et al. 2004). From tropical

America, the Portuguese and Spaniards carried the sweetpotato to Europe and from there to the other continents.

Sweetpotato has different common names in South America: batata or chaco in Venezuela; batata in Argentina, Colombia and Paraguay; camote in Bolivia, Chile, Ecuador, and Peru; boniato in Uruguay; and batata-douce in Brazil (De la Puente et al. 1996).

Importance of Sweetpotato in the Region

Sweetpotato is a crop associated with low inputs that offers advantages of low cost of production, is a rustic crop that can be grown in marginal conditions and is tolerant to heat and drought but sensitive to cold temperatures (Silva and Lopes, 1995). In Argentina, Paraguay, Peru and Uruguay sweetpotato is important both as a commercial and self-consumption crop. Often producers near markets of large cities commercialize sweetpotato whereas producers in rural areas distant to markets use sweetpotato for self-consumption. Often a small volume of roots and the foliage are used for animal feed (Table 18.1), in particular in areas where pastures are limited, such as the central coast of Peru. Sweetpotato is mostly consumed as a vegetable in local dishes. It is also processed into food products such as sweets in Argentina and Uruguay, or chips in Peru that are commercialized locally.

Brazil, Argentina, Peru, Paraguay, and Uruguay concentrate around 97% of the production in South America (Table 18.1). The crop ranks second among the root crops after potato in Argentina, Uruguay and Chile, and after cassava in Paraguay and Guyana (Table 18.1). Among small-scale farmers, sweetpotato is an occasional component in their production systems. The medium (up to 25 ha) and large producers allocate a fraction of their lands to the cultivation of sweetpotato. The total sweetpotato exported by Brazil, Argentina, Paraguay, Peru and Venezuela in 2000–2005 does not exceed 2.5% of their national production (Table 18.1). Brazil is the largest sweetpotato exporter in the region with approximately 0.4% of its national production.

The annual per capita consumption in South America is low compared with those in Asia, Africa, Oceania, or the Caribbean region (FAOSTAT, 2008). In South America, the estimated per capita consumption ranges from 0.2 to 18.5 kg per year, with Paraguay and Uruguay having the highest consumption (Table 18.1). In Paraguay, sweetpotato replaces cassava, the main staple food in the country during certain periods of the year. In Uruguay and Argentina the major consumption of sweetpotato is during winter. In the northeast region of Brazil, sweetpotato becomes of great importance because it is a source of energy food in critical periods of prolonged drought. In the northern Yungas region of Peru, non-sweet type of sweetpotato replaces potato due to limited local production of the latter (Fonseca et al. 1994).

In conclusion, sweetpotato has social and economical importance for the producers because of its multiple uses both in the human and animal diet. Being a cheap and popular food, its consumption is greater in rural areas (population with less income) than in the cities (Collins, 1989).

Major Growing Areas and Acreage, Yield and Economics

Harvested Area and Yield

Sweetpotato production, harvested area and yield in 11 countries in South America for the 3-year average 1995–1997 and 2005–2007 are presented in Table 18.2. Overall, the area and yield in South America increased by 9% and 7% respectively in the ten years period, in comparison to the decrease of 2% and 3%, respectively in the world. Paraguay and Argentina had a significant increase in sweetpotato area with 117% and 26.5% correspondingly whereas Brazil and Uruguay decreased by 10.9% and 29.4% respectively (Table 18.2). In 2005–2007, Brazil, Argentina, Paraguay, Peru, and Uruguay had the largest sweetpotato area ranging from 45,556 ha to 4,233 ha, which added up to 94% of the harvested area in South America; Venezuela, Chile, Bolivia, Ecuador, Guyana and Suriname had smaller area ranging from 1,486 to 19 ha. In Argentina, sweetpotato cultivation faces competition with soybean due to increasing demand and higher international price of the latter. In Uruguay, the reduction in area resulted from an increase of performance and stagnant demand.

In 2005–2007 sweetpotato yield ranged from 17.2 (Peru) to 1.1 t/ha (Guyana) (Table 18.1), in comparison to 13.9 t/ha in the world. This wide range in yield partly reflects the level of investment in growing the crop, from a higher input technology for a commercial crop to a lower level technology for self-consumption crop with minimal use of inputs (Table 18.2). In 2005–2007 among the top 2 producers Argentina had a yield of 14.7 t/ha and Brazil 11.4 t/ha. In Brazil yield was increased 15.2% in the period 1995–1997 and 2005–2007.

Sweetpotato yield depends on the variety, environmental conditions (such as temperature and rainfall), level of production technology, occurrence of diseases and pests, and quality of planting material. The higher yield in Peru can be attributed to a modern management of the cultivation as higher input technology, use of improved varieties, control of pests, as well as to a subtropical temperature regime during the irrigated winter crop in the semiarid coast. In 1997 and 1988, high temperatures brought by El Niño provided suitable conditions for the development of high populations of whiteflies that transmitted sweetpotato virus disease to the crop of the Cañete Valley (Gutiérrez et al. 2003), the main producer area in the central coast of Peru. This resulted in yields decreasing to levels as low as 4 t/ha. In the jungle, sweetpotato has a higher yield in the cropping season of May to October, months of lesser precipitation. In 1980, the disease foot rot or "Peste Negra" caused by the fungus *Plenodomus destruens* Harter resulted in a significant drop in sweetpotato area and yield in Uruguay (Vilaró et al. 1985).

Major Growing Areas

In *Brazil*, sweetpotato is cultivated throughout the country but with greater concentration in the south and northeast regions, notably in the states of Rio Grande do

Fig. 18.1 Geographic distribution of sweetpotato in South America. Each dot represents 100 hectares harvested

Sul, Santa Catarina, Paraná, Pernambuco and Paraíba (Fig. 18.1). About 48% of the planted area is concentrated in the southern region, in the state of Rio Grande do Sul producing over 147,000 tons, with an average yield of 10.9 t/ha, which is considered very low compared with Santa Catarina and Paraná (17.8 t/ha) (Bosco et al. 2004). Among the 10 top producers states, 6 are in the northeast (Paraíba,

Sergipe, Pernambuco, Rio Grande do Norte, Bahía, and Alagoas), with 36% of the area planted of Brazil.

In *Argentina*, 95% of the production is produced in Buenos Aires (1,600 ha), Córdoba (1,400 ha), Corrientes (1,300 ha), Formosa (1,200 ha), Santa Fé (1,000 ha), Santiago del Estereo (900 ha), Chaco (700 ha), Misiones (650 ha), and Tucumán (500 ha) (Fig. 18.1) (Marti, 2008). In Santiago del Estereo, the crop is grown exclusively under irrigation. In Tucumán, sweetpotato is a profitable crop, except in the high valleys above 2000 masl.

In *Peru*, sweetpotato is grown mainly along the northern and central coast but it is also cultivated in the warm highland valleys reaching 2,800 masl and in the Amazon jungle (Fig. 18.1) (INIAA, 1990). Best quality and higher yields are obtained at elevation below 800 m (Goyas, 1989). The central coast produces 55% of the sweetpotato; the northern and southern coast 18%, the highland 22%, and the jungle 5%. The valleys of Cañete and Huaral in the Region of Lima concentrate 47% of the planted area and 56% of the production in the country. High yields exceeding 30–35 t/ha can be attained in the Cañete Valley.

In *Paraguay*, the crop is cultivated in all the areas (Fig. 18.1), with the exception of Nueva Asunción. The main producing areas are Caaguazú, Central, San Pedro and Paraguarí. While Caaguazú has the largest areas and highest production in Conception yields are highest.

In *Uruguay*, sweetpotato cultivation is widely distributed mainly in the south zone (Canelones and San José), northeast zone (Tacuarembó, Rivera, and Cerro Largo), and north zone (Salto) (Fig. 18.1). The main production is in the south, close to the main consumer market, Montevideo. The north zone is the second in importance. Production in the south and north–northeast zones are 70% and 30% of the total production, respectively. The temperature differences enable earlier planting (at least a month) in the north and northeast zones than the south zone, consequently with earlier harvests with better prices.

Economics

Sweetpotato requires low financial investment as production costs are low. The advantages of planting sweetpotato compared with other crops are: a) its low cost and low risk under diverse environmental conditions. This makes sweetpotato a quite safe crop and relatively profitable. The producer rarely loses money planting it; b) sweetpotato can be planted by small (1 to 5 ha) and medium-scale (200 ha) producers; c) it adapts well to organic production systems and of low input, and does not require high soil fertility. d) Sweetpotato allows prolonged harvest because it does not require a specific time of harvest. Compared with other crops such as rice, bananas, maize and sorghum, sweetpotato is more efficient in the amount of net energy produced per area and time unit. The disadvantages are the following: a) it requires labor for planting and harvesting. b) The demand is inelastic; sweetpotato consumption does not increase even if the price decreases.

Cultural Methods

Sweetpotato is commercially grown as a monoculture, although a small area is grown as an intercrop in Guyana and as a courtyard crop in the northern Yungas region of Peru. Also, sweetpotato is an alternating or complementing crop in some Peruvian areas (Tumbes, Arequipa, Huánuco, and Tacna), where crops as rice and banana occupy the majority of lands. In these agricultural systems, sweetpotato is an occasional crop for rotation or one of the various crops for intercropping (e.g. with banana). Sweetpotato shoots or cuttings are transplanted on ridges of 0.20 to 0.40 m height (Fig. 18.2A). Commercial producers use organic and chemical fertilization before planting and chemical fertilizers at top dressing. In areas with risk of frost, such as in Argentina and Uruguay, planting material is grown in nurseries protected with a polyethylene sheet.

In *Peru*, apical stem-cuttings 30–40 cm long from mature plants are often used as planting material. In the traditional plating material system, the stem cuttings are obtained from crops intended for root production, either from the farmer's own' crop, bought, exchanged or borrowed from other farmers, and further propagated in the field or transplanted for production. There is no dedicated planting material production. Sprouted roots are seldom used to obtain planting material. Healthy quality planting material is available from a few institutions Instituto Nacional de Innovación Agraria and Instituto Rural Valle Grande with the technical support of CIP. Depending on the level of technological input, land is prepared by agricultural machinery or animals. The stem-cutting is planted in the center of the ridge in moist soils or placed in an elbow shape in the edge of ridges in dry or sandy soils, and buried two thirds of its length. Traditional cultural practices in the Yungas region differ from that of the coast. Planting material are long vines (1.20 m), planting styles are diverse, the trailing stems are hilled up sequentially to maximize the number of edible roots produced along the vines and allow sequential harvesting over a prolonged period. The latter serves as a method of in-ground storage of the roots, as there are no sufficient and adequate means for postharvest storage of sweetpotato (Benavides et al. 1990; Fonseca et al. 1994).

In *Brazil*, sweetpotato is grown year around; stem cuttings are obtained from crops in the field (Bosco et al. 2004). Stem-cuttings are transplanted in ridges 30–40 cm height. Weeds are controlled by hand with hoe, animal or mechanical traction. In fields with high weed infestation, land is prepared two or three weeks before transplanting, during this time weeds emerge and are eliminated with no residual herbicides applied in the eve of the transplanting (Bosco et al. 2004). Hilling up is done once after the last weeding, and can be carried out mechanically before the intertwining of branches to minimize plant damage.

In *Argentina*, the field is often flood irrigated before land preparation. Stem cuttings are obtained from the spring sprout of unharvested roots from the previous crop left in ridges. To further increase planting material, producers build nurseries protected with plastic sheets where stem cuttings are produced from roots. Transplanting of shoots or cuttings is done manually and on the ridge. Weeding is done manually, 2 or 3 times until the plants cover the ground.

Fig. 18.2 (**A**) Sweetpotato after weeding and earthing up, central coast of Peru. (**B**) Variety INA 100-INIA, one of the main Peruvian varieties. (**C**) Sweetpotato affected with virus disease, central coast of Peru. (**D**) Sweetpotato roots damaged by "chacarero", a complex of whiteworms in central coast of Peru. (**E**) Control of "chacarero" by injecting liquid insecticide in the soil. (**F**) Sweetpotato roots from variety Beauregard stored in a shed and covered with polyethylene sheet in Canelones, Uruguay. (A,B,F: J.P. Molina; C: S. Fuentes; D,E: J.C. Zamudio)

Planting Days

In *Brazil*, sweetpotato is planted all year around. Most producers carry out transplanting during cloudy days or after a rain. In the center-western, southeastern, and south, planting is done from November to January during the mid-rainy season. In the Northeast, planting is done earlier at the start of the rainy season. All over the country the absence of frost allows sweetpotato producers to plant anytime, by providing that irrigation during the dry months.

In *Argentina*, there are two planting times. The early transplanting is carried out from September to October with shoots from roots raised in nurseries. Late transplanting is from November to December with cuttings from the previous crop. In Santiago del Estero, 90% of the production corresponds to this late transplanting. In Tucumán, the late transplanting can be prolonged until May, but the major percentage of the producers carry out the planting in the month of November with the arrival of the first summer rains.

In *Peru*, in the main producing areas in the central coast transplanting is done almost year around but mainly during the winter from July to September. Yield is higher during the months of high temperatures, while in cold months plant development is slow and extended. In Tumbes, on the northern coast, transplanting is at the end of the rainy period in June when the water level of the rivers descends. Here producers transplant the crop along the river edge taking advantage of the remaining humidity of the soil.

In *Paraguay*, planting is done almost year around but the most frequent planting is from March to May so harvest is from November to January when demand for sweetpotato increases. The transplanting of October to December is carried out in small plots of 0.25 ha aimed at the maintenance of genetic material.

In *Uruguay* and parts of *Argentina* the transplanting time is determined by the end of the frost risk period. In *Uruguay*, the crop is grown from September to May, with transplanting done from September until January inclusive in the north, and from October to mid-January in the south. Nurseries initiate 2 months before transplanting.

Density

In *Peru*, an estimated plant density of 55,000 plants per hectare is used in the central coast, in order to increase yield and control root size; whereas in the northern coast a lower plant density of 25,000 plants is common. Distances between rows of 0.9–1.0 m and between plants of 0.20–0.40 m are the practice. The use of new varieties with compact plant architecture of shorter stems and internodes has allowed higher plant density and yields.

In *Brazil*, *Argentina*, *Uruguay* and *Paraguay*, the distances between rows range from 0.80 to 1.0 m, and among plants 0.25 to 0.50 m.

Overall, late maturing varieties often require greater plant spacing than the early maturing varieties. Likewise, greater plant spacing is chosen for early harvesting, as this allows the development of a greater number of commercial size roots. Similarly, a crop for use in industry or animal fodder uses greater distances between rows and plants. Closer plant distances plants, are used for varieties that produce large roots.

Length of Growing Period

The vegetative growing period can vary from 3 to 6 months, depending on the variety, management, and time of planting (Table 18.3). Commercial early maturing varieties are those which grow for 3–4 months and the late maturing varieties for 5–6 months.

Varieties

Native varieties or landraces are still widely used by small-scale farmers in distant rural areas, sometimes more than one landrace is found in a plot, and many. Many landraces are only of local importance for self-consumption or subsistence agriculture. The genebank at CIP registers 2,476 landraces originating from Peru, 234 from Ecuador, 318 from Colombia, 390 from Argentina. These numbers do not represent the relative genetic diversity of sweetpotato landraces among countries, but are an example of the rich sweetpotato diversity in South America. Landraces have a range of plant characteristics such as low to high root dry matter content, white, cream, yellow, orange, brown, red to purple skin color, white, cream, yellow, orange to purple flesh color, sweet and non-sweet flesh, erect to spreading plant type.

The varieties used in commercial production can be broadly classified according to their root characteristics into low root dry matter, sweet and with orange flesh and those of high root dry matter often sweet and with white, cream, yellow or purple flesh. Further, varieties of erect and semi-erect plant growth are increasingly sought for mechanized harvest as compared to varieties of spreading plant growth. The latter varieties are preferred for both root and fodder production. Commercial varieties are selections of landraces or introductions, or result of local variety improvement through hybridization and selection. The characteristics of common varieties in Argentina, Peru and Uruguay are presented in Table 18.3. The varieties Huambachero and Huayro are landraces, Beauregard and Okinawa 100 are introductions whereas INIA Arapey, Morada INTA, INA 100-INIA are improved local varieties.

In *Argentina*, varieties with purple skin and yellow flesh such as Morada INTA and INIA Arapey are planted in the Pampeana and Cuyo regions (Buenos Aires, Cordoba, and Formosa Provinces) (Table 18.3). In the northeast region, varieties with white skin and flesh are predominant such as Okinawa 100 in Corrientes Province, or orange flesh such as Gem in Entre Ríos Province. In the northwest

Table 18.3 Characteristics of selected sweetpotato varieties in countries of South America

Country	Variety	Skin color	Flesh color	Vegetative period (months)	Plant type	Conservation	Province/Region
Argentina	Morada INTA	Purple	Yellow	5	Semi-Erect	Very good	Buenos Aires, Córdoba, Formosa
	INIA Arapey	Purple	Yellow	4	Semi-Erect	Good	Buenos Aires, Córdoba, Formosa
	Gem	Copper	Orange	4	Spreading	Good	Entre Ríos
	Okinawa 100	White	White	4	Spreading	Good	Corrientes
	Several local landraces						Corrientes, Tucumán, Misiones
Peru	INA 100-INIA	Orange	Orange	4 to 4.5	Erect	n/a	Central and North coast
	INIA 306 Huambachero	Purple	Yellow-Orange	4.5 to 5.5	Semi-Erect	n/a	Central and North coast
	Jonathan	Yellow	Orange	5	Erect	n/a	Central coast, highland
	Paramonguino Mejorado	Orange	Orange	5	Semi-Erect	n/a	North coast
	Huayro	Orange	Orange	4	Semi-Erect	n/a	North coast
	Púrpura or Lila	Red	White	4	Semi-Erect	n/a	North coast
Uruguay	INIA Arapey	Purple	Cream	3 to 4	Spreading	Good	All
	Beauregard	Bronze	Orange	3 to 4	Spreading	Very good	North litoral, South
	INIA Ayuí	Bronze	Orange	3	Erect	Regular	North litoral
	INIA Itapebí	Red	Orange	4	Spreading	Very good	North litoral
	E 9227.1	Purple	Orange	4 to 5	Semi-Erect	Very good	South

Source: H. Marti (INTA, Argentina), J.P. Molina (INIA, Peru), F. Vilaró (INIA, Uruguay)

region, landraces with white flesh coexist with others of purple skin and yellow flesh. The variety Morada INTA has a semi-erect plant type, suitable for mechanization.

In *Peru*, commercial varieties are grouped mostly in two classes: the yellow-orange type (Fig. 18.2B) and the purple skin type. Sweet varieties with purple skin and yellow-orange flesh or orange skin and flesh such as Huambachero and INA 100-INIA are predominant in the main commercial production areas of the central coast (Table 18.3). Landraces of orange skin and flesh such as Huayro and Paramonguino Mejorado are common in the northern coast. Landraces are grown all over the country, in particular on the northern coast, Yungas region and inter-Andean valleys between 1800 and 2200 masl. Landraces suitable for dual purpose (human food and animal feed) have been identified (León-Velarde, 2007; León-Velarde and Mendiburu, 2007) and would allow producers to increase their income from the sale of foliage. New sweetpotato varieties of high nutritional value suitable for consumption in fresh and export, and of high dry matter content and starch for industrial processing are being developed (INIA, 2007).

In *Uruguay*, the main variety is INIA Arapey, which is grown in 80% of the sweetpotato area (Vicente et al. 2006; Vilaró et al. 2007). This variety is also widely grown in San Pedro, a main sweetpotato area in Argentina and in two regions in southern Brazil. INIA Arapey and in lesser extent E 9227.1 (an advanced clone that has yet to be named) have a thick skin and respectively 4- and 6-month long storability (Vilaró et al. 2007).

In *Brazil*, local varieties are grown in each region and are of unknown origin. The variety Paulistinha is found in Porteirinha–Minas Gerais state; the purple skin variety Brazlandia Roxa in the Federal District region and surroundings; the purple skin, yellow flesh Rainha in the region of Anápolis–Goiás state. Most commercial varieties are of purple skin and cream flesh, but white varieties are preferred in some areas such as Rio de Janeiro, Bahia and Sergipe. The varieties are freely exchanged among farmers and are of the spreading plant type with a vegetative period of 120 days. The variety Princesa is resistance to foot rot caused by *Plenodomus destruens* and the variety Brazlandia Roxa is tolerant to West Indian sweetpotato weevil (*Euscepes postfasciatus*).

In *Paraguay*, the most cultivated varieties are called yellow, purple, and white sweetpotato.

Irrigation and Fertilization

The sweetpotato possesses a deep and ramified root system (0.75–0.90 m), which enables the plant to explore greater soil volume and to absorb water in deeper layers than the majority of the vegetables. A regime of rain of 500 mm well distributed during the cycle of the cultivation is sufficient for high productivity. In practical terms, irrigation twice per week is recommended up to 20 days after planting (DAP); once a week from 20 to 40 DAP; and every two weeks after the 40 DAP until harvesting (Bosco et al. 2004). The irrigation increases yield significantly in regions where rain is insufficient or is erratic. In non-irrigated sweetpotato fields, replanting is a common operation and is recommended when more than 10% of the cuttings do not establish.

In Brazil, sweetpotato is a rainfed crop, as rains are enough and moderately distributed throughout the year. In the northeast region, farmers cultivate sweetpotato in floodplains and river spillways. This practice is quite common in Rio Grande do Sul and in other states of the region. The use of irrigation during the dry season can increase the productivity of commercial roots by $\sim$ 35% (Oliveira and Marouelli, 1997). Organic fertilization is mostly used; on average 25 t/ha animal manure is applied. In soils previously cultivated with other vegetables, the residual fertilizer can result in a good productivity. Some producers use chemical fertilization after fertility analysis of the soil. In loam soils, 5 to 10 kg of borax is recommended. When organic material had been added, the nitrogen fertilizer is reduced to half of the recommended dose. High yields have been obtained with 200 kg/ha of fertilizer 5-20-10 in the experimental station of Ituporanga.

In Argentina, in rainfed areas (e.g. San Pedro), planting is done in November with the arrival of the first rains, whereas in the lands with irrigation (e.g. Cordoba) planting is in September. The crop is irrigated once or twice, in addition to the pre-planting irrigation, depending on the soil humidity during the summer. Usually the rainfall is sufficient for growing the crop. Chemical fertilization is used in commercial crops.

In Peru, on the coast the crop is always irrigated. "Machacado" is a pre-planting, irrigation method used to loosen the soil for easier land preparation. After planting, a light irrigation, the "enseño" irrigation, is given to ensure plant establishment. Thereon the frequency of irrigation depends on the soil type, variety, and season, and includes the irrigation after hilling up and that of preharvesting. The most frequent irrigations are carried out in the first 5–6 weeks after planting, and then irrigations diminish in the last third of the cultivation. A recommendation is to apply 2000–3000 m^3 of water per hectare per season. In the jungle, the sweetpotato crop is rainfed. Fertilization varies according to the physical and chemical characteristics of the soil, frequency of irrigation or rain, farming system and variety. Fertilization is done in places with higher technology levels, as in the Cañete Valley in the central coast where the fertilization is carried out at transplanting or hilling up (25–35 DAP), either mechanized, by animal traction or hoe by small-scale farmers. Higher potassium fertilization (e.g. 80-40-120) has often been found satisfactory. Sometimes organic material or manure at 12 to 30 t/ha is applied at land preparation or planting. Some farmers do not fertilize and use the effect of residual fertilizer from the previous crop.

In Paraguay, cattle manure is applied at doses of 10 t/ha. No chemical fertilizer is applied. The use of organic matter is more relevant in the Central zone because of its low soil fertility or long history of land use.

In Uruguay, sweetpotato is a rainfed crop (Valiente and Jeaume, 2006). In the northeastern zone, the periods of greater probability of water deficit occur between December and February. In the southeast region of Entre Ríos, the crop is irrigated when larger amounts of sweetpotato are needed to cope during the high demand months. The application of fertilizers (organic or chemical) is very variable, but in general it is scarce or nil. It is quite frequent to utilize a fertilizer solution ('starter") with the water irrigation at transplanting to enhance early root development and growth.

Pests and Other Constraints

Main sweetpotato pests (caused by insects, fungi, viruses) in selected countries in South America are listed in Table 18.4. There is no single most important pest common to all main sweet producer countries. In the field, fungal pathogens may be important in areas in Argentina and Uruguay where the crop is rainfed, roots are stored and used as planting material due to crop seasonality but are less important in Brazil and Peru where the crop is rainfed or irrigated, stem cuttings and not the roots are used as planting material, and the crop is grown year around. However fungal diseases are important during postharvest in all countries, either where roots are stored several months (Argentina and Uruguay) or commercialized shortly after harvest (Brazil and Peru). Foot rot disease, caused by *P. destruens*, is the most important fungal disease as it may damage the root either during plant growth or postharvest and can cause severe crop loss. Important pests circumscribed to certain areas are *Meloidogyne incognita*, *Anomala undulate*, and complex virus diseases. Heavy soils and in-ground storage favor the development of *Monilochaetes infuscans*.

In Argentina, major pests are sweetpotato chlorotic dwarf disease (SPCD), *P. destruens* which is important in nurseries, *Rhizopus stolonifer* in storage (Nome et al. 2008) (Table 18.4).

In Brazil, the main pests are *E. postfasciatus*, *P. destruens*, virus diseases, and soil insects (e.g. *Conoderus* sp., *Diabrotica* sp.).

In Peru, economically important and common pests are sweetpotato virus disease (Fig. 18.2C), *M. incognita*, and the "chacarero" white worm complex (Fig. 18.2D,E) (Table 18.4). Fungal diseases such as *R. stolonifer* can cause postharvest loss (INIAA, 1990; Sánchez and Vergara, 1992). *Euscepes postfasciatus* is a minor pest usually found in poorly managed or abandoned fields, although it may cause damage when harvest is delayed. Pests are controlled with chemical applications in commercial crops.

In Uruguay, common pests are *M. infuscans* and *P. destruens* during cultivation and postharvest; and *R. stolonifer*, and *Fusarium* spp during postharvest. Among insect pests, *Chaetocnema* sp., *Conoderus* spp., and *Diloboderus abderus* (Table 18.4) are common.

Integrated pest management is recommended including the use of clean planting material, pesticides, crop rotation, and resistant varieties and early harvesting, among others. Planting material selection, chemical treatment of roots for planting at storage and cuttings at transplanting are recommended to control *P. destruens*. Use of crop rotation, planting material from nurseries and field sanitation (elimination of crop residues) are recommended to control *R. stolonifer*. Weeds can be a

Table 18.4 Most important production constraints of sweetpotato in selected countries of South America (indicated with "x" in bold)

	Constraint	Argentina	Brazil	Paraguay	Peru	Uruguay	References
Insects	West Indian sweetpotato weevil: *Euscepes postfaciatus*. Coleoptera	**x**	**x**	x	x		2, 4, 5, 8, 9, 19, 25, 26, 27, 30, 31
	Ptericoptus acuminatus. Coleoptera	**x**	x	x			2, 3, 5, 6, 21
	"Pulguilla": *Chaetocnema* sp. Coleoptera	x				**x**	8, 22
	"Isoca": *Diloboderus abderus*. Coleoptera	x				**x**	(F. Vilaró, INIA-Uruguay; H. Marti, INTA-Argentina)
	Root-knot nematode: *Meloidogyne incognita*	x	x		**x**		4, 9
	Whiteworms: *Anomala undulate, A. testaceipensis*. Coleoptera				**x**		29, (J.C. Zamudio, IRVG, Peru)
	Wireworms (*Conoderus* spp). Coleoptera	x	**x**			**x**	2, 3
Fungi	"Peste negra", Foot rot, Dry rot: *Plenodomus destruens* Harter	**x**	**x**		x	**x**	3, 18, 22, 24, 25, 28, 33
	Scarf: *Monilochaetes infuscans* Harter	**x**	x		x	**x**	3, 18, 22, 24, 25, 33
	Sclerotial blight, circular spot: *Sclerotium rolfsii* Sacc.	**x**			x	x	3, 9, 18, 22, 24, 33
	Rhizopus soft rot: *Rhizopus stolonifer* (Ehrenb.) Vuill.	**x**	x	x	**x**	**x**	5, 8, 24, 25, 28, 31, 33
	Fusarium wilt, stem roft: *Fusarium oxysporum* f.sp. *batatas* W.C. Synder & H.N. Hansen	x	x	x	**x**	**x**	3, 5, 7, 8, 9, 18, 22, 24, 25, 26, 28, 31, 33
Viruses	Chlorotic dwarf disease (SPCD)	**x**					10, 24
	Sweetpotato virus disease (SPVD)		x		**x**		20
	Others (**SPFMV**, **SPCSV**, SPMMV, SPCFV, etc)	**x**	**x**		**x**	x	10, 11, 12, 13, 14, 15, 16, 17, 20, 22, 23, 24, 25, 32
Seed	Low quality	**x**	**x**	**x**	**x**	**x**	1, 3, 6, 9, 11, 18

1. Avila, 1990; **2**. Bimboni and Ruberti, 1990; **3**. Boy, 1988; **4**. Burga, 1988; **5**. Burgos, 1990; **6**. Cantos, 1990; **7**. Cardoso, 1988; **8**. Cardoso et al. 1989; **9**. Díaz, 1990; **10**. Di Feo et al. 2000; **11**. Franca, 1988; **12**. Fuentes, 1994; **13**. Fuentes et al. 1997; **14**. Fuentes et al. 1996; **15**. Fuentes et al. 1997; **16**. Fuentes and Salazar, 1992; **17**. Fuentes and Salazar, 2003; **18**. Gonzáles et al. 1989; **19**. Goyas, 1989; **20**. Gutiérrez et al. 2003; **21**. Lenscak, 1990; **22**. Marisquiera and Bettini, 1990; **23**. Nome, 1973; **24** Nome et al. 2008 **25**. Nunes, 1989; **26**. Pérez, 1998; **27**. Pletsch, 1989; **28**. Quincho et al. 1990; **29**. Reynoso, 2007; **30**. Sánchez and Vergara, 1990; **31**. Thomazelli, 1989; **32**. Untiveros et al. 2008; **33**. Vilaró, 1988.

problem in nurseries and during cultivation. With high weed infestation, herbicides can be used.

In all countries, the lack of high quality planting material is a common constraint (Avila, 1990; Bosco et al. 2004; Burgos, 1990; Cantos, 1990; Daza and Rincón, 1993; Díaz, 1990; Lenscak, 1990; Marisquiera and Bettini, 1990). Planting materials are produced by the farmer, obtained from neighbors or from commercial fields, and are usually of unknown health quality. No certified planting material production system is available. Healthy planting material is limited and produced by few agricultural institutions or producers.

Harvesting, Marketing and Profitability

Harvesting

Sweetpotato is a perennial plant that is cultivated as an annual crop and is harvested when roots reach marketable size. Harvesting time is affected by the variety, environmental conditions and can be adjusted to coincide with highest demand and market price (Silva and Lopes, 1995; Resende, 2000). In commercial cultivation, the crop is often harvested 90–180 days after transplanting. Crops for animal feed or industry can be harvested later when roots reach a higher weight. Vines are removed manually or mechanically before lifting the roots, which is done manually with hoe or mechanically (Fig. 18.3A, B). Roots are left to dry on the ridges for 30 min to 3 h (Brazil) or for 3–6 days (southern zone of Uruguay)– known as curing – in the field. Roots may be cured in a room with high relative humidity for 6–10 days before storing. Roots may be sorted by weight depending on local standards. Roots of substandard quality are often destined to industry or animal feed.

In Uruguay, sweetpotato is stored in sheds and covered with cornhusk or polyethylene sheet (Fig. 18.2F). Improved storage with ventilation systems are recommended for better postharvesting (curing and storage) management of the roots in comparison with sheds (Carballo et al. 2003).

In Peru sweetpotato is not stored but sold almost immediately, except that destined for export.

In Paraguay, farmers cultivating sweetpotato for self-consumption usually harvest the roots gradually, according to their needs throughout the year. Crops intended for sale are harvested from October to the end of January, depending on the price in the market.

Marketing

Produce is brought to markets by buyers, brokers, transporters and wholesale distributors and through them to retailers for sale to consumers and to industry.

Fig. 18.3 Mechanized (**A**) and horse power (**B**) harvesting in Peru. (**C**) 80–100 kg bags containing sweetpotato ready to be transported to the market, in central cost of Peru. (**D**) Selected and packaged roots for export in Peru. (**E**) Sweetpotato vines for animal feed. (A,C: J.P. Molina; B,D: J.C. Zamudio; E: S. Fuentes)

In Peru, the product is transported from the field to the market in jute bags of 80–100 kg (Fig. 18.3C). In Brazil, the roots are generally washed before marketing in wooden boxes containing 24–26 kg. In Uruguay, sweetpotatoes are marketed in boxes of 20–23 kg, in Argentina in bags of 20–22 kg, and in Paraguay in boxes of 55–80 kg. Even though quality standards for the classification and marketing exist, these are often not met. The marketing system is often unregulated but sweetpotato is better sorted when sold to supermarkets,

In the northeast region of Brazil, producers themselves mostly consume sweetpotatoes and only the surplus is sold in local or neighboring markets. In Argentina the main market is Buenos Aires and in Uruguay it is Montevideo. Most of the market for sweetpotato in Peru is within the greater Lima urban area (Peralta, 1992). In Paraguay, the sweetpotato market is mainly the Central Market of Asuncion.

Prices of sweetpotato fluctuate depending of the variety and quality. The seasonal price variation is related to quantities being offered. In Brazil, higher prices are obtained from January to February, in Argentina from December to January, in Peru from September to October.

In Paraguay, fresh sweetpotato are available around the year, with major production from October until January. In Tucumán, Argentina, sweetpotatoes are harvested from February to December, but the majority is sold between April and August. The low humidity in the winter and spring permit in ground storage, and roots are harvested and marketed according to demand.

Profitability

The cost of on-farm commercial production is estimated to range from $500 to $1,700 per ha, according to the level of production technology. The two main components of production cost are labor (manual and mechanized) and inputs (planting materials, fertilizers and pesticides). Estimated labor cost may represent from 45% to 70% of the production cost, and inputs from 12 to 35%. In countries such as Argentina and Uruguay, the additional cost of producing planting materials in nurseries may represent 25% of the production cost.

The profitability of sweetpotato is highly sensitive to changes in price. However, even when the price is low, producers do not lose money (Achata et al. 1990; Benavides et al. 1990). Estimated profitability may range from 15% to 80%, depending on the on-farm price, production cost and yield.

Various Uses (Roots and Foliage)

Sweetpotato is cultivated mainly for its edible roots for human consumption but its roots and vines also for animal feed. Both vines and roots may serve as planting material, and the foliage can be used for green manure.

The roots are mostly consumed fresh as boiled, baked or fried, in local dishes, soups or as a snack, in the domestic preparation of desserts, and in bakeries. Examples of typical local dishes or sweets are "ceviche" and "picarones" in Peru; "asado criollo" and "locro" in Argentina, respectively. Roots are also mashed for consumption (Fukumoto et al. 1993). Yellow sweetpotato with high beta-carotene can be boiled as an instant weaning food (Espinola et al. 1998). The starch from sweetpotato is used in the preparation of typical sweets such as "mazamorra morada" in Peru. In rural zones of Uruguay, it is often used as a substitute for bread.

In Argentina, Brazil, Peru, and Ecuador, the vines are often harvested with the roots (Table 18.1, Fig. 18.3E) and fed principally to cattle, but also to pigs (Peru) and goats (Peru and Ecuador). Vines are very commonly used for animal feeding in the coast of Peru. In Argentina and Brazil, the vines have served as an emergency cattle feed in periods of drought (Boy et al. 1988; Franca, 1988). In Peru, the vines are also used for feeding guinea pigs and rabbits. Unmarketable roots are fed generally to pigs but also to cattle (Peru and Brazil) (Burga, 1988; Franca, 1988). In Brazil, shredded-dried roots are an excellent energetic food component, which is added to the feeds of both ruminant and non-ruminant animals (Bosco et al. 2004).

Sweetpotatoes are also used as raw material in the processing of chips (Peru), sweet foods and pastes (Argentina) in local industries. Whole sweetpotato pieces in syrup are a popular dessert in Argentina and Uruguay. In Brazil, sweetpotato is mainly used in the manufacturing of sweets, known as "marron-glacé", generally sold in cans. The crystallized sweet foods are also widely consumed. In Peru, sweetpotato is processed mainly into starch, flour, fried chips, and bread; flour and starch are used in the making of sweets, desserts, and pastries (Achata et al. 1990; Peralta et al. 1992). It is a partial substitute for wheat flour in the making of sweetpotato bread (Paraguay and Peru).

Sweetpotato in Venezuela, Chile, Bolivia, Ecuador, and Colombia

Sweetpotato is a minor crop, mostly grown by small-scale farmers for self-consumption. Farmers use landraces, low input, traditional practices, and minimum land preparation and with low yields. Roots or cuttings, or both, are used as propagation material. Sweetpotatoes are cultivated often in intercropping systems. Few areas are planted as a monocrop. Sweetpotato is mostly consumed fresh. The little processing that takes place is mainly artisanal. In Colombia, cassava growers also grow sweetpotatoes; sweetpotato can be found as a perennial-like crop where varieties of the spreading type are planted at low density (2,500 plants/ha) and piecemeal harvest can continue for 3 years or more. In Chile, commercial sweetpotato cultivation has a greater specialization with preparation of nurseries and high-density planting (up to 80,000 plants/ha).

Perspectives of the Crop in the Region

No major changes in production are expected for sweetpotato in South America. There is interest in the export of fresh roots, particularly of sweet orange-fleshed varieties (Fig. 18.3D), processing of sweetpotato into flour and starch for both local and export markets, and development of new uses as health products (Yamakawa and Yoshimoto, 2002). Varieties with functional properties of high antioxidant and higher beta-carotene content are being explored by INTA–Argentina. The consumption of orange-flesh sweetpotatoes to reduce vitamin-A deficiency is being promoted by EMBRAPA in the northeast region of Brazil. Interest has also been raised for the production of ethanol from sweetpotato as a biofuel in Peru.

Concluding Remarks

In South America, sweetpotato is the third most important root and tuber crop in terms of area harvested and production, after cassava and potato. In countries with tropical and subtropical climates sweetpotato is grown year around, whereas in temperate climates the crop is grown during the warmer months. A large diversity of sweetpotato landraces is mostly grown by small-scale farmers in mixed cropping systems; these are characterized by the use of low input technology and low productivity; roots are used for self-consumption and as a food-security crop in rural areas. A few commercial varieties are grown as a monocrop using medium and high input technology, including mechanized cultivation; roots are mainly used for local consumption as fresh product in cities. The vines are widely used for animal feed in most countries. Storage of sweetpotato is almost inexistent in tropical and subtropical areas. Main production constraints include the lack of high quality healthy planting materials, and diseases caused by virus complexes and insect pests in drier areas, whereas field and postharvest fungal diseases predominate in rainfed areas. Sweetpotato is a rustic crop of low production cost that ensures profitability to producers. However its low demand and low per capita consumption restrict expansion of the crop. The processing industry is mostly aimed at local markets. Small volumes of sweetpotato are exported. There is interest in increasing exports of fresh roots, processing of flour and starch, development of processed healthy products and the production of biofuel.

Acknowledgments The authors acknowledge H. Marti and L. Di Feo of INTA, Argentina; F. Vilaró, D. Maeso and J. Paullier of INIA, Uruguay; J.P. Molina of INIA, M. Quevedo of MINAG, J.C. Zamudio of IRVG, and C. León-Velarde of CIP, Peru; J. Bosco, A. Dusi, and J.A. Buso of EMPRAPA, Brazil; F.M. Cárdenas, INIAP, Ecuador for providing information; K. Theisen for Fig. 18.1 and P. Stapleton for editing.

References

Achata, A, Fano, R., Goyas, H., Chiang, O., and Andrade, M. 1990. El Camote en el Sistema Alimentario del Perú: El Caso del Valle de Cañete. International Potato Center. Lima, Peru.

Austin, D.F. 1988. The taxonomy, evolution and genetic diversity of sweetpotatoes and related wild species. pp. 27–60 in: P. Gregory (ed.), Exploration, maintenance, and utilization of sweetpotato genetic resources. International Potato Center, Lima, Peru.

Austin, D.F., and Huamán, Z. 1996. A synopsis of *Ipomoea* (*Convolvulaceae*) in the Americas. *Taxon* 45: 3–38.

Avila, C. 1990. El cultivo de la batata en la provincia de Tucumán. In: INTA-CIP. Segundo Curso Internacional sobre el cultivo de la batata. San Pedro, Argentina. INTA, March 1990. 11pp.

Benavides, M., Prain, G., and Fonseca, C. 1990. La batata o camote, Apichu o Kumara en la conquista del Perú. Revista del INIAA 2: 44–45. Lima, Peru

Bimboni, H., and Ruberti, D. 1990. Plagas de la batata. In: INTA-CIP. Segundo Curso Internacional sobre el cultivo de la batata. San Pedro, Argentina. INTA, March 1990. 30pp.

Bosco, J., Lopes, C., and Silvestre, J. 2004. Cultura da Batata-Doce. Embrapa Hortalicas, Sistemas de Producao, 6. URL http://www.cnph.embrapa.br/sistprod/batatadoce/index.htm

Boy, A. 1988. El cultivo de la batata en Argentina. pp. 9–31 in: Centro Internacional de la Papa. Mejoramiento de la Batata (*Ipomoea batatas*) en Latinoamérica: Memorias del seminario. CIP, Lima, June 9–12, 1987.

Boy, A., Cantos, F., Fano, H., and Fernández, F. (eds.). 1988. La batata (*Ipomoea batatas* L. (Lam)) en la Argentina: Resúmenes y conclusiones del taller sobre la problemática de la producción y uso de la batata. Santiago del Estero (Argentina). June 28–29, 1988. INTA. 131pp.

Burga, J. 1988. Situación del cultivo de la batata o camote en el Perú. pp. 99–126 In: Centro Internacional de la Papa. Mejoramiento de la Batata (*Ipomoea batatas*) en Latinoamérica: Memorias del seminario. CIP, Lima, June 9–12, 1987.

Burgos, F. 1990. El cultivo de batata en Paraguay. In: INTA-CIP. Segundo Curso Internacional sobre el cultivo de la batata. San Pedro, Argentina. INTA, March 1990. 13pp.

Cantos, S. 1990. El cultivo de batata en Santiago del Estero. In: INTA-CIP. Segundo Curso Internacional sobre el cultivo de la batata. San Pedro, Argentina. INTA, March 1990. 14pp.

Cardoso, M. 1988. Producción y utilización de la batata en el Paraguay. pp. 91–97 In: *Centro Internacional de la Papa*. Mejoramiento de la Batata (*Ipomoea batatas*) en Latinoamérica: Memorias del seminario. CIP, Lima, June 9–12, 1987.

Cardoso, M. Mayeregger, T., and Stromenger, A. 1989. Informe sobre el cultivo de la batata en Paraguay. In: INTA-CIP. Primer Curso Internacional sobre cultivo de batata (*Ipomoea batatas* L.). San Pedro, Argentina. March 13–24, 1989. INTA. 1989. 9pp.

Collins, M. 1989. Economic analysis of wholesale demand for sweetpotato in Lima, Peru. Thesis for Master degree, University of Florida

Carballo, S., Telesca, J., and Cabot, M. 2003. El sistema de ventilación horizontal negativa (VHN), una alternativa para mejorar la postcosecha en boniato. Instituto Nacional de Investigación Agraria (INIA)- Reconversión y desarrollo de la granja (PREDEG), Canelones, Uruguay. URL http://www.inia.org.uy/publicaciones/documentos/lb/pol/2003/va_pos_boniato.pdf

Daza, M., and Rincón, H. 1993. Perfil tecnológico del camote (batata) en la costa central del Perú (Estudio de las zonas agroecológicas del valle de Cañete). Centro Internacional de la Papa Departamento de Ciencias Sociales. Documento de Trabajo No. 1993–1.

De la Puente, F., Austin D.F., and Diaz, J. 1996. Common names of the sweetpotato (*Ipomoea batatas*) in the America. Plant Genetic Resources. New Letter No. 106.

De la Puente, F., Reynoso, D., and Roca, W. 2004. Recursos genéticos del camote (batata o boniato) en Latino America y el Caribe. International Potato Center. 74pp.

Díaz, L. 1990. El cultivo de la batata en el Perú. In: INTA-CIP. Segundo Curso Internacional sobre el cultivo de la batata. San Pedro, Argentina. INTA, March 1990. 13pp.

Di Feo, L., Nome, S.F., Biderbost, E., Fuentes, S., and Salazar, L.F. 2000. Etiology of sweetpotato chlorotic dwarf disease in Argentina. *Plant Dis*. 84: 35–39.

Engel, F.A. 1970. Exploration of the Chilca Cayon, Peru. *Curr. Antropol*. 11: 55–58.

Espinola, N., Creed-Kanashiro, H., Ugaz, M.E., and van Hal, M. 1998. Desarrollo de un Alimento Complementario con Camote para Niños de 6 meses a 3 años. Departamento de Ciencias Sociales. Documento de Trabajo No.1998–8. International Potato Center (CIP). Lima, Peru.

FAOSTAT, 2008. FAO Statistical Database. URL http://faostat.fao.org/

Fonseca, C., Daza, M., Aguilar, C., Bezemcon, N., Benavides, M., Fano, H., Goyas, H., Prain, G., Roncal, R., and Tafur, S. 1994. El camote en los sistemas alimentarios de la Yunga Norte del Perú. Centro Internacional de la Papa. Departamento de Ciencias Sociales, Documento de Trabajo No. 1994–4.

Franca, F.H. 1988. A batata-doce no Brasil: Producao, pesquisa e difusao de tecnologia. pp. 39–56 In: Centro Internacional de la Papa. Mejoramiento de la Batata (*Ipomoea batatas*) en Latinoamérica: Memorias del seminario. CIP, Lima, June 9–12, 1987.

Fuentes, S. 1994. Preliminary identification of a sweetpotato virus (C-6). *Fitopatología* 29: 38.

Fuentes, S., Arellano, J., and Meza, M.A. 1997. Preliminary studies of a new virus, C-8, affecting sweetpotato. *Fitopatología* 32: 9–10.

Fuentes, S., Mayo, M., Jolly, C.A., Nakano, M., Querci, M., and Salazar, L.F.. 1996. A novel luteovirus from sweetpotato, sweetpotato leaf speckling virus. *Ann. appl. Biol.* 128: 491–504.

Fuentes, S., Querci, M., Salazar, L.F., and Mayo, M. 1997. Detection of SPLSV by Nucleic Acid Spot Hybridization (NASH) Test. *Fitopatología* 32: 204.

Fuentes, S., and Salazar, L.F. 1992. Identification of a new sweetpotato virus. *Fitopatología* 27: 50.

Fuentes, S., and Salazar, L.F. 2003. First report of Sweetpotato leaf curl virus in Peru. *Plant Dis.* 87: 98.

Fukumoto, M., Ugaz M.E., and Morales, E. 1993. Percepciones y actitudes sobre el uso de camote en Lima Metropolitana: Informe Final. Instituto de Investigación Nutricional. 55pp.

Gonzáles, H., Ambrosoni, S., and Vilaró, F. 1989. El cultivo de batata en Uruguay. In: INTA-CIP. Primer Curso Internacional sobre cultivo de batata (*Ipomoea batatas* L.). San Pedro, Argentina. March 13–24, 1989. INTA. 1989. 18pp.

Goyas, H. 1989. El cultivo de la batata en el Perú. In: INTA-CIP. Primer Curso Internacional sobre cultivo de batata (*Ipomoea batatas* L.). San Pedro, Argentina. March 13–24, 1989. INTA. 1989. 16pp.

Gutiérrez, D.L., Fuentes, S., and Salazar, L.F. 2003. Sweetpotato virus disease (SPVD): Distribution, incidente, and effect on sweetpotato yield in Peru. *Plant Dis.* 87: 297–302.

Huang J.C., and Sun M. 2000. Genetic diversity and relationships of sweetpotato and its wild relatives in *Ipomoea* series Batatas (Convolvulaceae) as revealed by intersimple sequence repeat (ISSR) and restriction analysis of chloroplast DNA. TAG 100: 1050–1060.

INIA, 2007. INIA investigara en la generación de nuevas variedades de camote para producción de etanol. Instituto Nacional de Investigación Agraria (INIA). Nota de Prensa 021-2007-INIA-OII-PW. http://www.inia.gob.pe/notas/nota0148.

INIAA. 1990. Memorias de la Primera Reunión Nacional de Investigadores de Camote del Perú. Instituto Nacional de Investigación Agraria y Agroindustrial (INIAA). Huaral, June 27–28. 30pp.

Lenscak, M. 1990. El cultivo de la batata en Argentina. In: INTA-CIP. Segundo Curso Internacional sobre el cultivo de la batata. San Pedro, Argentina. INTA, March 1990. 8pp.

León-Velarde, C. 2007. Manejo de batata forrajera (doble propósito); estrategias de alimentación animal. pp. 45–50 In: C. León-Velarde and R Amable (eds). Producción y uso de la batata (*Ipomoea batatas* (L.) Lam); estrategias de alimentación animal. Universidad ISA. Santiago de los Caballeros, Dominican Republic.

León-Velarde, C., and de Mendiburu, F. 2007. Variedades de batata de doble propósito. pp. 36–44 In: C. León-Velarde and R Amable (eds). Producción y uso de la batata (*Ipomoea batatas* (L.) Lam); estrategias de alimentación animal. Universidad ISA. Santiago de los Caballeros, Dominican Republic.

Marisquiera, G., and Bettini, R. 1990. Informe sobre la situación del cultivo del boniato en Uruguay. In: INTA-CIP. Segundo Curso Internacional sobre el cultivo de la batata. San Pedro, Argentina. INTA, March 1990. 25pp.

Marti, H.R. 2008 La batata: oportunidades para una hortaliza subutilizada. *Boletín Hortícola* 13: 14–17.

Montaldo, A. 1966. Manual del cultivo de la Batata (*Ipomoea batatas*). Univ. Central de Maracay, Venezuela.

Montenegro, A., Avis, C., and Weaver, A. 2007. Modeling the prehistoric arrival of the sweetpotato in Polynesia. *J. Archaeol. Sci.* 35: 355–367.

Nome, C., Nome, S.F., Di Feo, L., Martí, H., Mitidieri, I., and Pérez, B.A. 2008. Enfermedades de *Ipomoea batatas* (L.) Lam. (batata, camote). In: Nome, S.F., Docampo, D.M., Conci, L.R. (eds.) En: Atlas Fitopatológico Argentino. Vol. 1, N° 2. Septiembre 2008. ISSN 1851-8974. Córdoba, Argentina. URL: http://www.fitopatoatlas.org.ar/default.asp?hospedante=712

Nome, S.F. 1973. Sweetpotato vein mosaic in Argentina. *Phytopatho Z.* 77: 44–45

Nunes, A. 1989. O cultivo da batata doce no Rio Grande do Sul Brasil. In: INTA-CIP. Primer Curso Internacional sobre cultivo de batata (*Ipomoea batatas* L.). San Pedro, Argentina. March 13–24, 1989. INTA. 1989. 13pp.

Oliveira, C and Marouelli, W. 1997. Horticolas irrigadas na regiao dos Cerrados. VII Simposio sobre o Cerrado – Estrategias de Utilizao. 165–172pp.

Peralta, P. 1992. El camote en el Perú: producción, demanda actual y perspectivas Agroindustriales. In: G.J. Scott, J.E.Herrera, N. Espinola, M. Daza, C. Fonseca, H Fano, M. Benacides (eds.). Desarrollo de Productos de Raíces y Tubérculos. Volumen II-América Latina. Memorias del Taller sobre Procesamiento, Comercialización, y Utilización de Raíces y Tubérculos en América Latina, 8–12 Abril, 1991, en ICTA, Villa Nueva, Guatemala. CIP. Lima, Peru. 375p.

Peralta, P., Cavero, W., and Chumbe, V. 1992. Un diagnostico rápido del pan de camote en el Perú. Pp 175–194. In: G.J. Scott, J.E. Herrera, N. Espinola, M. Daza, C. Fonseca, H Fano, M. Benavides (eds.). Desarrollo de Productos de Raíces y Tubérculos. Volumen II-América Latina. Memorias del Taller sobre Procesamiento, Comercialización, y Utilización de Raíces y Tubérculos en América Latina, 8–12 Abril, 1991, en ICTA, Villa Nueva, Guatemala. CIP. Lima, Peru. 375p.

Pérez, W. 1998. *Fusarium* spp que infectan camote en el Peru. MSc thesis. Universidad Agraria La Molina, Lima, Peru. 139pp.

Pletsch, R. 1989. El cultivo de la batata en Argentina – Provincia Corrientes. In: INTA-CIP. Primer Curso Internacional sobre cultivo de batata (*Ipomoea batatas* L.). San Pedro, Argentina. March 13–24, 1989. INTA. 1989. 13pp.

Quincho, O., Otazú, V., and Elphinstone, J. 1990. Fungal Pathogen incidence in sweetpotato clones from San Ramon, Peru. *Fitopatología* 25: 9–10.

Resende, G. 2000. Caracteristicas productivas de cultivares de batata-doce en duas época de colheita, em Porteirinha MG. *Horticultura Brasileira* 18: 68–71.

Reynoso, D. 2007. La batata (*Ipomoea batatas*, Lam) generalidades del cultivo. pp. 14–25. In: C. León-Velarde and R Amable (eds). Producción y uso de la batata (*Ipomoea batatas* (L.) Lam); estrategias de alimentación animal. Universidad ISA. Santiago de los Caballeros, Dominican Republic.

Sánchez, G., and Vergara, C. 1992. Plagas del cultivo de camote. Universidad Nacional Agraria. La Molina, Lima, Peru.

Silva, J. and Lopes, C. 1995. Cultivo de batata-doce [Ipomoea batas (L.) Lam.]. Brasilia: EMBRAPA-CNPH. Technical instruction-CNPH, 7. 18pp.

Thomazelli, L. 1989. Cultivo da batata-doce no Brasil. In: INTA-CIP. Primer Curso Internacional sobre cultivo de batata (*Ipomoea batatas* L.). San Pedro, Argentina. March 13–24, 1989. INTA. 1989. 10pp.

Ugent, D., and Peterson L. 1988. Archeological remains of potato and sweetpotato in Peru. CIP Circular 16: 1–10.

Untiveros, M., Fuentes, S., and Kreuze, J.K. 2008. Molecular variability of sweetpotato feathery mottle virus and other potyviruses infecting sweetpotato in Peru. *Arch. Virol.* 153: 473–483.

Valiente, J., and Jeaume,, M.A. 2006. El cultivo de la batata. Instituto Nacional de Tecnología Agropecuaria (INTA). "Novedades Hortícolas": Centro regional Entre Ríos. Año I-No. 4. 5pp.

Vicente, E., Spina, W., Rodríguez, G., and Vilaró, F. 2006. Cultivares de boniato para el Litoral Norte. Instituto Nacional de Investigación Agropecuaria (INIA), Uruguay. Serie Actividades de Difusión Nro. 459. URL http://www.inia.org.uy/publicaciones/documentos/ad/ad_459.pdf

Vilaró, F. 1988. Batata (boniato o camote) en Uruguay. pp. 139- In: Centro Internacional de la Papa. Mejoramiento de la Batata (*Ipomoea batatas*) en Latinoamérica: Memorias del seminario. CIP, Lima, June 9–12, 1987.

Vilaró, F., Crisci, C., and Gilsanz, J. 1985. Aspectos de la producción de boniato en el Uruguay y su futuro. Estación Experimental Granjeta "Las Brujas, Canelones, Uruguay.

Vilaró, F., Rodríguez, G., Vicente, E., and Spina, W. 2007. Material de plantación: cosecha y Poscosecha. Programa Nacional de producción Hortícola, Serie de Actividades de Difusión Nro. 493. Instituto Nacional de Investigación Agropecuaria (INIA), Uruguay. URL http://www.inia.org.uy/publicaciones/documentos/ad/ad_493.pdf

Woolfe, J.A. 1992 Sweetpotato: An Untapped Food Resource. Cambridge University Press, Cambridge, 643pp.

Yamakawa, O., and Yoshimoto, M. 2002. Sweetpotato as food material with physiological functions. *Acta Horticulturae* 583: 179–185

Yen D.F. 1974. The sweetpotato and Oceania. An assay in ethnobotany. Bishop Museum Press, Hawaii, USA, Bernice P. Bishop Museum Bulletin 236.

Yen D.F. 1982. Sweetpotato in historical perspective. pp. 17–30 In: Villareal, R.L.. and Griggs T.D. (eds). Sweetpotato, *Proc. First Intern. Symp.*. AVRDC Publ. No. 82–172, Tainan, Taiwan.

Zhang, D.P., Cervantes, J., Huamán, Z., Carey, E., and Ghislain, M. 2000. Assessing genetic diversity of sweetpotato (*Ipomoea batatas* (L.) Lam.) cultivars from tropical America using AFLP. *Genet. Res. Crop Evol.* 47: 659–665.

Zhang, D.P., Rossel, G., Kriegner, A., and Hijmans, R. 2004. AFLP assessment of diversity in sweetpotato from Latin America and the Pacific region: Its implications of the dispersal of the crop. *Genet. Res. Crop Evol.* 51: 115–120.

Chapter 19
Sweetpotato in West Africa

M. Akoroda

Historical Background

West Africa (17° 15 min E, 17° 15 min W and 26° 52 min N, 15° 35 min S) with 16 countries forms the Economic Community of West African States (ECOWAS) excluding Mauritania. Mali, Burkina Faso and Niger are land-locked. The region, north of Gulf of Guinea, had a long history of slave trade, Portuguese presence, colonial exploitation of human and natural resources by European nations during 1460–1960. Her southern forests, forest-savanna transition, northern savannas agro-ecologies suit many crops using traditional techniques. However, rapid un-checked growth in human population implies greater food production. West Africa's 2000 population density was not uniform (Fig. 19.1); grew 2.5% annually to 273.5 million

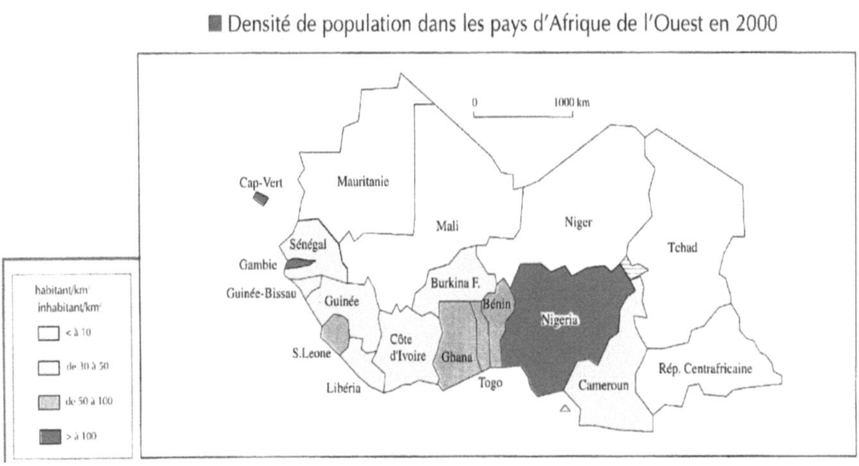

Fig. 19.1 West Africa population density in 2000

M. Akoroda (✉)
Faculty of Agriculture and Forestry, University of Ibadan, Nigeria
e-mail: m_akoroda@yahoo.com, M.AKORODA@CGIAR.ORG

G. Loebenstein, G. Thottappilly (eds.), *The Sweetpotato*,
DOI 10.1007/978-1-4020-9475-0_19, © Springer Science+Business Media B.V. 2009

inhabitants in 2006, with 51.1% in Nigeria. Population density partly influences the economy through food, feed, and industry.

Introduction of Sweetpotato to West Africa

Sweetpotato, *Ipomoea batatas* (L.) Lam is not known wild but of Tropical Central American origin. Cassava, sweetpotato, *Xanthosoma* and potato introductions from Tropical America to Africa were recent through Iberian voyages of Portugal and Spain. Christopher Columbus took sweetpotato to Spain in 1492. Portuguese brought it to their settlements or trading stations in Africa along Gulf of Guinea: Sierra Leone, Sao Tome and Principe, and Fernando Po [Bioko in Equatorial Guinea]. Sweetpotato cultivation was first mentioned in Africa in Sao Tome in 1520 (Mauny, 1953). In the 16th century, Portuguese navigators introduced sweetpotato to Africa, Europe, India, and Indonesia (Janssens, 2001); and were widely cultivated in West Africa by end of 17th Century (Leon, 1977). In the 1800s, freed slaves returning to Gulf of Guinea; brought Latin America crops. Local diet preferences dictated the share of farm resources allotted each cultivated crop.

Sweetpotato is one of over 60 common crops Table 19.1 Its spread was slowly by farmer-to-farmer diffusion involving no colonial crop dispersal agencies. Its sweetness makes it less preferred to bland starchy staples [yam, cassava, rice, cowpea, cocoyams, and acha]. Consequently, early breeding at International Institute of Tropical Agriculture (IITA) (1970–1988) screened out orange-fleshed and sweet types. New trends seek varied populations to satisfy different end-uses (Akoroda et al., 2000). Tubers contain per 100 g: 70 g water, 1.5 g protein, 0.3 g fat, 26 g carbohydrates (30.8–41.8 g starch and 3.68–10.4 g sugar), 1 g fibre, 25 mg calcium, 1 mg iron, vitamin A potency of 0–4000 international units, thiamine 0.1 mg, riboflavin 0.04 mg, nicotinamide 0.7 mg, ascorbic acid 30 mg (Tindall, 1968; Tewe et al., 2003). Africa's sweetpotato hectarage is highest in West Africa (Table 19.2). West Africa with East Africa region account for 77.4% of the continent's production.

Table 19.1 Common plants cultivated by West African farmers

Cereals: maize, rice, sorghum, millets, acha (*Digitaria exilis*), wheat
Roots and tubers: cassava, yams (10 species), sweetpotato, cocoyams (2 species), potato, carrot
Pulses and legumes: groundnut, soybean, cowpea, bambara nut, sunflower, pigeon pea, sesame,
Trees and Shrubs: oil palm, cocoa, coffee, kola, tea, cashew, coconut, rubber, shea, star apple (*Chrysophilum albidum*), Ogbono (*Irvingia gabonensis*), African pear (*Dacrodes edulis*), pepper fruit, raphia palm,
Vegetables: amaranth, telfairia, bitter leaf, celosia, corchorus, tomato, pepper, onion, garlic, roselle, cabbage, green beans, okra, watermelon, egusi, carrot, pumpkin, ginger, eggplants
Industrial: sugarcane, tobacco, kenaf, cotton
Fruits: citrus (5 species), pawpaw, banana, plantain, pineapple, guava, mango, avocado

Table 19.2 Sweetpotato production in regions of Africa (FAOSTAT, 2005)

Region (countries)	Tuber yield (kg/ha)	Harvested hectares ('000)	Production ('000 tonne)	Percent of Africa's total
West (12)	7594	1200.42	3801.44	31.6
Central (5)	4054	123.43	499.22	4.2
East (9)	7222	1065.22	5506.42	45.8
South (8)	5823	700.50	1909.90	15.9
North (2)	21474	10.58	307.51	2.6
Mean/Total	6591	3100.15	12024.48	100.0

Nigeria

More sweetpotato is produced in central humid savanna [700–1000 mm of annual rainfall] than in forests or dry savanna states (PCU, 2004). Food preferences vary between localities, northerners appreciate sweet snacks—sugarcane, sweets-toffee, date palm nuts, and sweetpotato tubers whose sweetness depends on sugar [2–9% means, 3.9–5.0% median] (IITA, 1984). Sweetpotato is a small fraction of local food-basket. Farmers' rank of its relative importance in diet base on ease of propagation and economic yield, decreasing from: yam (28), maize (34), cassava (39), cocoyam (55) to sweetpotato (69) in SE Nigeria.

Ghana

Farmers plants 65,000 ha yearly of sweetpotato that comes after cassava, and yam in importance among root crops (FAOSTAT, 2006); Grown mainly on mounds and ridges in coastal savanna and forest-savanna transition zones having two rainy seasons and two plantings in April–May (major season) and August–September (minor season); and Guinea savanna zone (one growing season with planting in June–July).

Sweetpotato has huge potential as food (boiled, fried, or roasted), shoots supplement livestock feeds, and could provide starch and flour for industries. Leaves are eaten as vegetables in Guinea and coastal savanna zones. Ghana being Vitamin A deficient (VAD); consuming orange-flesh sweetpotato (OFSP) rich in beta-carotene would help fight VAD among children.

Mali

In 1980, farmers produced 10,000 t of yams, 20,000 t of cassava and 50,000 t of sweetpotato. Malians consumed 16.8 kg sweetpotato/person in 1989, and an expected 21.4 kg in 2005 (Traore, 1994). Planting after March in some zones gives no tubers due to sweetpotato's sensibility to photoperiod. Sweetpotato is not supported by research towards supplementing cereals.

Table 19.3 Released and popular early sweetpotato varieties moderately resistant to weevil in Ghana

Cultivar name	Tuber flesh colour	Total tuber yield (t/ha)	APM	Reaction to SPVD**	Market-ability	Dry matter content (%)	Overall tuber acceptance‡
CRI-Apomuden	Reddish (deep) orange	30	Low	3	2	21.9	3
CRI-Otoo	Light orange	23	High	3	2	32.2	3
CRI-Ogyefo	White	20	Medium	3	3	40.1	3
CRI-HiStarch	Cream	18	Medium	3	2	40.0	3
Sauti	Yellow	19	Medium	3	3	40.2	4
Faara	White	22	High	4	3	36.1	4
Okumkom	White	20	Medium	3	2	30.7	2
Santom pona	Light yellow	17	Medium	3	2	34.4	2
Jukwa orange‡‡	Light orange	30	High	3	3	35.0	4

* Early – reaches maturity 4 MAP; Medium – 5 MAP; Late – 6 MAP. ** 0 – Not a problem; 1 – Susceptible; 2 – Low resistance; 3 – Moderate resistance; 4 – Resistant; 5 – High resistance; † 1 – Low; 2 – Medium; 3 – High. ‡ 1 – Bad; 2 – Fair; 3 – Good; 4 – Very good. APM: Ability to retain planting material after periods of extended drought ‡‡ – Not officially released variety, but popular in coastal savanna.

Ghana

Varieties: Farmers grow six improved white-fleshed and two OFSP varieties (Table 19.3) commonly yielding 8–10 t/ha, reaching 15–30 t/ha. Improved varieties were released in 1998 (Tables 19.4 and 19.5) after wide testing and evaluations of yield and culinary acceptances.

Planting material multiplication and distribution: Primary multiplication fields [2.5 ha] were established and 500,000 cuttings distributed to Ghanaian growers.

Ghana's Genetic Improvement

Germplasm maintenance and characterization: In 2007, germplasm of 136 accessions (6 OFSP; 89 local) were backed-up at Plant Genetic Resource Research Institute Bunso); 8 released varieties (with 2 OFSP) and 39 introduced genotypes (33 OFSP) were maintained and characterized at Fumesua.

Crossing block: Fumesua crossing block has nine flowering parents including 5 OFSPs: (Apomuden, Otoo, Beauregard, Resisto, TIS 83/0138); yellow-fleshed Sauti; and white-fleshed: Faara, HiStarch, and Excel. Major desired traits are: high dry matter, high beta carotene content, and tolerances to SPVD, weevils, and millipedes. Seeds from controlled and open-pollinated crosses are being collected for evaluation.

Multi-locational adaptability and acceptability testing: Forty OFSP and some white-fleshed varieties from CIP in 2004/05 were assessed in preliminary yield, advanced yield, and multi-locational on-farm trials. Two clones are being nominated to the National Varietal Release Committee for release after 2008 on-farm trials. A genotype x environment trial of local popular varieties, being evaluated

Table 19.4 Characteristics of sweetpotato varieties officially released on 13 Nov. 1998 in Ghana

Characteristics	TIS 8266 (Okumkon)	TIS 84/0320 (S. Pona)	TIS 3017 (Faara)	Sauti (Sauti)	Local Red	Mean
Leaf shape	Cordate	Cordate	Angulatus	Palmatus	–	–
Tuber skin colour	L. purple	Cream	D. purple	Cream	–	–
Tuber flesh colour	White	C/Y	Cream	Yellow	–	–
Tuber flesh shape	Round/ Elliptic	Long E/ Round	Long Elliptic	Long Irregular	–	–
Fresh tuber yield (t/ha) 3MAP [1996, 1997]						
Coastal savanna [CS]	8.0	6.9	5.5	5.8	3.8	–
Forest [F]	11.2	9.2	10.5	8.7	6.0	–
F/S transition [T]	5.8	4.7	4.3	3.4	1.9	–
Dry matter (%) of tuber 3MAP [1996, 1997]						
Coastal	29.0	31.5	35.3	32.5	31.1	–
Forest	32.6	29.4	33.4	34.5	33.9	–
Transition	32.8	33.8	37.0	37.3	32.8	–
Fresh tuber yield (t/ha) 4MAP [1995–1997]						
Coastal	12.5	11.1	13.8	10.0	8.5	11.2
Forest	19.9	16.7	16.9	15.4	13.5	16.5
Transition	12.0	10.2	9.3	8.5	5.9	9.2
Guinea savanna	17.8	19.9	18.9	18.7	18.3	18.7
Fresh vine yield (t/ha) 4MAP [1995–1997]						
Coastal	8.5	8.7	14.9	15.2	10.4	11.5
Forest	17.1	18.3	23.1	27.8	21.7	21.6
Transition	8.4	12.5	15.1	18.5	8.6	12.6
Guinea savanna	11.5	20.0	27.8	32.7	16.5	21.7
Fresh tuber yield (t/ha) on-farm [1996]						
Coastal	13.5	12.1	12.6	19.1	8.3	11.3
Forest	17.4	13.8	14.9	12.3	10.2	13.7
Transition	16.2	14.4	11.7	10.7	6.8	11.4
Tuber dry matter [1995, 1996]						
Coastal	29.0	36.0	37.6	35.5	37.3	36.1
Forest	32.7	31.9	34.4	36.2	36.5	34.3
Transition	30.3	35.3	36.5	38.3	36.5	35.2

S = Santom; W = white, C = cream; L = light, Y = yellow.
Source: Otoo et al. (2000).

across 11 other sub-Saharan countries within Sweetpotato Harvest-Plus Program, was established at 4 locations in 3 agro-ecologies of Ghana. The 15 genotypes include: Mayai, Gweri, K135, Zambezi, 199062.1, SPK004, K566632, Pipi, Carrot C, Ukerewe, K118, Naspot 1, Resisto, Ejumula, and a local check CRI-Apomuden.

Planting material multiplication and distribution: Primary multiplication fields [2.5 ha] were established and 500,000 cuttings distributed to Ghanaian growers.

Togo

Farmers grow sweetpotato as secondary crop to enhance household food security and diversify income sources through selling tubers. Nationally, cassava and yams are the principal tuber crops.

Table 19.5 Virus and pest infestation, sensory tests, and protein content of the first four sweetpotato varieties officially released on 13 November 1998 in Ghana

Characteristics	TIS 8266 (Okumkon)	TIS 84/0320 (S. Pona)	TIS 3017 (Faara)	Sauti (Sauti)	Local Red	Mean
Virus symptoms (1–5 scores SPVD) [1995–1997]						
1991	2.0	2.0	–	–	–	–
1995–1997	1.0	1.0	0.5	1.0	2.8	–
1998	2.0	2.0	1.5	2.5	4.0	–
Tuber attack (%) by *Cylas* sp. 4MAP [1995–1997]						
Coastal	15.9	10.9	6.0	14.5	21.3	13.7
Forest	3.0	5.7	7.2	6.1	7.1	5.8
Transition	1.1	1.1	0.0	0.6	2.8	1.1
Millipede attack (%) of tubers. [1995–1997]						
Coastal	1.7	1.7	1.8	2.3	3.8	2.3
Forest	19.4	23.4	24.4	20.4	30.8	23.7
Transition	19.0	23.6	9.6	13.8	14.4	16.1
Vine attack (%) by *Alcidodes* sp. [1995–1997]						
Coastal	9.1	2.1	1.7	1.7	3.3	3.6
Forest	0.8	0.7	0.7	0.0	4.0	1.2
Transition	1.5	4.6	0.5	0.1	9.0	3.1
Sensory test of boiled + fried tubers on-farm (T/F/C scores) 1997						
Appearance	11/12/8.5	14/13.5/19	9/17/15	38/29.5/27	0/4/18	–
Taste preference	15/22/14	17/9/17	2/18.5/8	35/26/39	5/6.5/6	–
Overall	12.5/19.6/15.4 0/11.1/8.6	–	17/18/14	4/11.6/8.2	50/20/35	–
Protein content [% dry matter]	5.1	4.8	7.0	5.3	4.2	–

S = Santom; W = white, C = cream; L = light, Y = yellow. T/F/C = Transition/Forest/Coastal.
Source: Otoo et al. (2000).

Cote d'Ivoire

Sweetpotato is grown nation-wide, with 76% of the 40 300 t from northern districts in 1985 (MINAGRA, 1990). Tubers are economically important for containing 15–25% starch, 1–2% protein, and 1–2% sugar (Onwueme and Sinha, 1991) and vitamins and minerals; providing food calories for poor populace who eat tubers boiled, fried or fufu (pounded dough); and flour could be used in bread and pastries, thus diversifying diet, and providing industries inputs. Leaves are used as vegetable. Left-over tubers unfit for food are used as feed.

Chief constraints are sweetpotato weevils, nematodes, viruses, black fungi rots affecting stored tubers, inadequate planting materials, land shortage causing restricted cultivated hectares; low yielding varieties, marketing problems, and narrow utilization.

Land shortages due to expansion of settlements reduce length of fallow for soil regeneration. Sweetpotato does not yield high in poor soils, in intercropping, or under irregular rainfall.

Tuber conservation lasts for 60 days followed by depreciation in quality and value due to sprouts and rots. Though tubers keep well at 13–16 °C at 85–90% RH (Onwueme and Sinha, 1991); such conditions are not achievable among farmers.

Commercialisation and consumption. About 10% of tubers are sold (Scott and Ewell, 1993). Difficult farm access and expensive transportation raise tuber prices and reduce profits. In some areas, its food use is restricted to hunger periods when preferred staple foods are lacking. Utilization as flour, starch, sugar, and ethanol are limited or uncommon.

Guinea

Sweetpotato is grown by small farmers as an important subsistence staple (Ibrahima Mafoulé Camara, personal communication, August 2007); more intensively in high rainfall areas than in semi-arid zones; although sweetpotato tolerates drought and diverse soils. It plays a major role during hunger period when other crops fail. It is well adapted to diverse traditional cropping system and agro-ecologies. Low sweetpotato research in national programmes puts earlier varieties at risk of being lost among farmers (Table 19.6).

These varieties entered the pool of varieties nationally and in adjacent countries but may no longer be identifiable; but continue to contribute to overall sweetpotato production. Often-times, they are planted and sold as mixtures of varieties. Usually, varieties move from one community to new localities through sales of tubers that are replanted to supplement vine supply for planting whenever farmers encounter good varieties.

Table 19.6 Sweetpotato clones previously at CRS Foulaya, Guinea 1993

Sweetpotato Variety	Origin/Source of introduction into Guinea	Crop growth cycle (month)	Mean number tubers/plant	Mean tuber diameter/plant (cm)	Fresh tuber yield (t/ha)
Burkina	Burkina Faso	2–4	–	–	–
Chinoise (1)	China	3–4	5.75	6.25	32.88
Chinoise (2)	China	3–5	–	–	–
Chinoise (3)	China	3–4	5.00	5.50	21.25
Chinoise (4)	China	3–4	–	–	–
Chinoise Bamban	China	3–5	–	–	–
CDH 30	Senegal	2–3	–	–	–
Coréenne	?	?	5.25	5.00	17.50
CIAM 8030	Cape Verde	3–4	–	–	–
Ivoirienne	Côte d'ivoire	3–5	–	–	–
Kaagbeti Wuré	Guinéa	3–4	–	–	–
Kemp	Sierra Leone	3–5	–	–	–
Molé Molé	Guinea	3–4	5.75	5.00	13.75
N'Körö	Guinea	3–5	–	–	–
Sabouya	Guinea	2–3	4.75	4.88	17.50
Sosoe Wuré	Guinea	3–5	4.00	3.50	10.75
Sweet red	USA	3–4	–	–	–

The importance of a crop-plant may be *partly* measured by efforts allotted to its research, development, improvement, and promotion. Sweetpotato research institutions and agencies in West Africa lack vigour. They are few, poorly staffed and funded relative to cassava, yam, cowpea, and cereals.

Major Growing Areas and Acreage, Yield and Economics

Sweetpotato is grown across West Africa; although Nigeria harvests 79.1% of the sweetpotato hectares. Besides high tuber yields from 5,212 ha in Sahel ecologies of Senegal, Mali, and Niger—yields average 3600 kg/ha (Table 19.7) or 10% of farmers' potential yield (36 t/ha).

Sweetpotato plots are intercropped with cassava, maize, sorghum, and vegetables or as sole crop in peri-urban holdings. As available farmland reduces, farmers tend to grow more profitable crops that require less labour and inputs, provide household food *first* and *also* generate income. Being a secondary intercrop tends to hide sweetpotato's enormous potential; thereby calling for re-education of farmers, processors, and consumers through targeted campaigns to demonstrate benefits of sweetpotato for food, feed, and industry.

Purseglove (1968) reported sweetpotato was "gradually ousting yams because of quicker return with *less work*, but yielding place to cassava due to higher yields with even lesser work". Current situation needs review by zones of each country through surveys to enable stakeholders chart future strategies.

Table 19.7 West Africa sweetpotato production statistics, 2005–2006 (FAOSTAT)

Country	2006 human population (million)	National land area (million hectares)	Fresh tuber output (tonne)	Usage per person (kg/yr)	Tuber yield [t/ha]	Total hectares harvested to the crop
Cape Verde	0.5	4	4,000	8.00	5.556	720
Senegal	12.0	197	27,000	2.25	24.281	1,112
Mauritania	3.1	1031	2,000	0.65	1.000	2,000
Gambia	1.5	11	–	–	–	–
Guinea Bissau	1.4	38	–	–	–	–
Guinea	9.8	246	60,000	6.12	6.000	10,000
Sierra Leone	5.7	72	26,000	4.56	2.476	10,500
Liberia	3.4	111	–	0.00	–	1,900
Cote d'Ivoire	19.7	322	43,000	2.18	2.150	20,000
Mali	13.9	1240	50,000	3.60	16.667	3,000
Ghana	22.6	239	90,000	3.98	1.385	65,000
Burkina Faso	13.6	274	40,864	3.00	6.917	5,908
Togo	6.3	57	3,500	0.56	1.167	3,000
Benin	8.7	113	50,018	5.75	4.477	11,171
Niger	14.4	1267	30,000	2.08	14.286	2,100
Nigeria	140.0	924	2,516,000	17.97	4.876	516,000
Total	273.5	5115	2,942,382	10.76	7.018	652,411

Best Field Practices and Cultural Methods

Field practices vary by field-size and agro-ecology. Growers' manuals stating essential steps and best-bet practices to maximise labour and inputs lack required details. Otoo et al. (2000) described sweetpotato growing, but did not quantify inputs, produce, or any specifics as to what is adjudged most suitable for each agro-ecology of Ghana. Chinaka (1983) provided no costing or inputs limits for profitable production. Both cases do not emphasize field production as an area-specific combination of quantified inputs. Thus, growers are inadequately guided on how to exploit sweetpotato's potentials.

Diversity of soils and rainfall variables superimposed on genotypic variation among sweetpotato varieties strongly require sustained research and development efforts which have been lacking for a long time. Sporadic attempts by individual scientists with little funds would only lead to under-exploitation of sweetpotato in West Africa.

Land Productivity

Acquiring or inheriting farmlands makes land selection untenable under traditional cropping systems compared to commercial enterprises. Tuber yields respond to native soil fertility. Surface soil is usually scrapped into mound, heaps or ridges into which several crops are planted. Seedbeds [20–30 cm of loose soils] facilitate tuber bulking through less resistance to penetration. Well-drained sandy loamy soils are best; whereas heavy clayey soils produce rough bad-shaped tubers.

Planting

Growers utilize planted: tubers to produce sprouts; vines from old plants; and or sprouts of varied lengths. Planting actively growing 15–30 cm apical vines with 4–6 nodes are best. Older vines establish poorly, causing missing stands in fields thus reducing yields.

Planting begins when rains start, continuing till two months before rains cease. Temperatures at planting should exceed 20 °C to spur early sprouting. Planting towards the end of rains in high rainfall areas for sweetpotato cannot tolerate excessive moisture in soil or drought without irrigation. Some varieties tolerate drought or low soil moisture; but never plant vines into dry soils. Some varieties tolerate moist soils near water ways and wetlands. Plant vines to expose two nodes and bury two nodes (7–10 cm) on *raised* seedbed to lift tubers from excessive moisture zone, as well as avoid dehydration and exposure as rains wash soil off.

Cultivars

Sweetpotato cultivars with farmers are difficult to identify because they have not been systematically characterized nationally. Small collections are maintained by research stations with little documentary and pictorial means of distinguishing them from any apparently similar types in common use. Often, new introductions enter cultivation without formal variety release. Even in marketing, cultivars are sold without labels; retailing needs only knowledge of some local varieties as to skin colour, shape and flesh colour. Variety names are as variable as their provenances; thus, one variety may possess several names according to where it was sourced.

IITA had assembled 1000 sweetpotato accessions by 1988 (IITA, 1989) including 50 non-sweet clones. Tuber yield declines when varieties are subjected to natural disease pressures; though varieties react differently to varied pest pressures. Virus-free tissue cultures stored for 6 years produced 33% higher tuber yield than stands planted with field vines (Table 19.8) but was 80% more in TIS 1499 (Table 19.9).

CRI Fumesua, Ghana assembled 136 sweetpotato genotypes, but released nine to cultivator (Otoo et al., 2000). NRCRI evaluated 358 accessions of local genotypes for tuber shape, yield, latex flow, flesh colour, and sugar content at Umudike, Nigeria (Okoli, 1980). Annual losses of germplasm by various factors are oftentimes irretrievable. Expensive *repeated* collecting expeditions under limited funding greatly retard the release of superior varieties.

There is need for descriptors for released sweetpotato varieties in West Africa that show photographs, characteristics, and field performance across diverse agroecologies of each country. Without such information, statistics of hectares cultivated; distribution flow of vines and produce would be difficult to track. National official variety release systems are not formally established in West Africa.

Table 19.8 Sweetpotato tuber yields (4MAP) from field cuttings (FC) and virus-free cuttings (TC) at Ibadan

Sweetpotato Variety	No. of Sellable Tubers/plant		Tuber weight (gram/plant)		Percent weight increase
	FC	TC	FC	TC	
TIB 11	3.8	4.6	450	770	32
TIS 2544	2.6	3.5	270	590	32
TIS 3017	1.4	1.9	330	410	8
TIS 8250	4.5	5.2	720	1100	38
TIS 70357	3.2	3.2	320	680	36

Source: IITA (1987)

Table 19.9 Effect of virus on performance of. sweetpotato (TIS, 1499)

Single plant aspect	Virus free	Virus	% Reduction
Fresh tuber yield (g)	1284	257	80.02**
Dry matter (%)	23.53	22.71	3.48 ns
Dry yield (kg)	302	59	80.46**
Fresh shoot weight (g)	743	154	79.27**
Dry shoot weight (g)	103	37	73.57**
Number of leaves	384.2	184.0	52.08 ns
Dry leaf weight/10 leaves (g)	3.31	1.33	59.81**

Source: IITA (1974)

Table 19.10 Important characteristics for selecting sweetpotato genotypes

1.	Fresh and dry shoots yield at harvest
2.	Fresh and dry tuber yield (formed within seedbeds)
3.	Tubers qualities (culinary and sensory – texture, taste/smell)
4.	Number of tubers per stand
5.	Tuber shape and size distribution
6.	Tuber flesh colour
7.	Tubers nutrients composition at optimum harvest date
8.	Tolerance to pest and disease (especially weevil and virus complex)
9.	Tuber storability in usual conditions
10.	Days to maturity for optimum tubers yield
11.	Depth of tubers at maturity (affecting harvest and weevil damage)

Sierra Leone released improved sweetpotato varieties (Dahniya et al., 1994); multiplied them at Njala for farmers in 35% of chiefdoms surveyed replacing imported potato. Few farmers (0.6%) grew sweetpotato in 1961 for market compared to 10% in 1991. Table 19.10 indicates essential breeding objectives.

Planting Density

Traditional farmers adopt varied spacing for sweetpotato in intercrop. Sole stands are 30–50 cm along rows, 80–120 cm between rows. Well-managed commercial farms seek 3:1 rectangularity (30 cm × 90 cm giving 37,037 plants/ha).

Weed Management

Sweetpotato fields should be weed-free within 6 weeks after planting (WAP) with hoeing at 3–4 WAP and 8–9 WAP (Chinaka, 1983). One spray of Glyphosate (2.4 kg a.i./ha) 4WAP; DS2 or PE Chloramben 3.4 kg a.i./kg control weeds (Unamma et al., 1984). Cost, convenience, labour availability, comparative tuber yields, and logistics appropriate to farm size should be considered before making weed control decisions.

Irrigation

Sweetpotato in West Africa is grown rain-fed. When planted some weeks before end of seasonal precipitation cease, sweetpotato matures 4–5 weeks into a dry period. Source of water and cost of labour for irrigation operations, pipes, and pumps incur new expenditures. Sweetpotato needs 530 mm of water during November to February with monthly requirements of 75, 122, 189, 145 mm respectively; but 503 mm if irrigated in South-eastern Nigeria (Chukwu, 1995).

Fertilizer

Little or no fertilizer is used by West African farmers due to poor and irregular supply and high prices. Sweetpotato responds to applied organic manures; but research on fertilizers, tuber yield, and quality for common cultivars is inadequate to guide usage. Fertilizer price and production regimes determine profits from fertilisation.

Fertilizer is applied on both ridge sides during first weeding (4–5WAP). Monitoring NPK levels in sweetpotato fields is essential in commercial farms. Rendle and Kang (1977) found P content of 0.22% in petiole of index leaves at 9 WAP indicates P is not limiting. Nitrogen levels should be low to avoid promoting shoot growth than tubers.

Organic materials should be applied *first* before supplementing with little fertilizers. Wise fertilization necessitates we consider storability and taste quality of end products, and cost-benefit ratio for each farm blocks. Rates of 125–700 kg/ha NPK (Chinaka, 1983; NRCRI, 1992; Korieocha et al., 2006; and de Geus, 1973) or 27–54 kg N, 72–120 kg P, and 72–120 kg K may be *adjusted* to satisfy local soil nutrient shortages after *soil test* of farm parcels to precisely make-up for nutrient deficiency (Table 19.11).

Most West African farmers cannot afford soil tests; thus, grading soils into low, medium, high nutrient levels using soil test kits is a first step in the right direction. Nutrients removal by each variety should be determined. Farmers should be aware of different nutrient ranges for highly productive sweetpotato yields (Table 19.12).

Length of Growing Season

Length of season depends on agro-ecological conditions of the farm site. Varieties mature early (3–4 MAP) or late (6–7MAP) according to combinations of moisture, soil fertility, weather, altitude and shading. Harvest readiness and crop life may be prolonged or shortened by the genotype-specific responses to the growth environment. Drought-tolerant varieties continue to grow actively whereas others senesce.

Table 19.11 Soil test values (STV) to guide fertilizer recommendation (***R***) for sweetpotato in Nigeria

Soil test variable	Low STV	***R***. rate (kg/ha)	Medium STV	***R***. rate (kg/ha)	High STV	***R***. rate (kg/ha)
Total NO3-N (%)	<0.10	90	0.10–0.15	45	>0.15	20
Brays P I (mg/kg)	<10	50	10–20	25	>20	0
Brays P II (mg/kg)	<15	50	15–25	25	>25	0
Exch. K (cmol/kg)	<0.15	90	0.15–0.25	48	>0.25	0
Available Zn (mg/kg)	<1.0	–	1.0–5.0	?	>5.0	–
Organic C (g/kg)	20.0	–	20.0–30.0	?	>30.0	–

Source: FFD (2002) and Chude et al. (2004).

Table 19.12 Critical levels at the 7th–9th opened leaf blade from shoot tip sampled 28 days old plants and nutrients removed by sweetpotato

Nutrient taken up by crop	Adequate nutrient range (%)	—12 t/ha Tuber	yield— Tuber/Vine	—50 t/ha Tuber	yield— Tuber/Vine
Nitrogen	4.2–5.0	26	52	110	215
Potassium	0.22–0.45	6	9	25	38
Potassium	2.6–6.0	60	90	250	376
Calcium	0.9–1.2	3.6	16	15	65
Magnesium	0.15–0.35	3	6.5	12.5	27
Sulphur	0.35–0.45	1.8	4.3	7.5	18
Chlorine	<0.9	10	18	43	75

Source: O'Sullivan et al. (1997).

Vine Removal and Sellable Tuber Yields

Dual use of field-plots to provide vines and tubers at harvest is common. Shoot removal reduces tuber yields (Table 19.13). Percentage of sellable tuber yield is an essential attribute for selecting varieties. Cultivars with many small un-sellable tubers are un-adapted to local cultural practices and or agro-ecology. Tiny un-sellable tubers constitute losses to growers; because they are sold cheaply for use in livestock feeds. Low prices make sweetpotato enterprises uncompetitive. Therefore, varieties with more sellable tubers are preferred by growers, provided consumers accept their culinary and sensory characteristics.

Table 19.13 Tuber yield (t/ha) and (percentage of sellable tubers) of improved and local sweetpotato clones according to week of vine removal

Vine pruned (at WAP)	TIS 2534	TIS 146/3092	TIS 2498	TIS 8504	Anoma (local variety)	Pruning week mean	Percent of zero pruning
0	13.0	12.6	7.7	15.6	5.4	10.9	100
	(84)	(83)	(79)	(86)	(70)	(82)	
2	14.0	9.8	6.9	13.2	5.4	9.9	90.8
	(76)	(72)	(77)	(86)	(69)	(77)	
4	10.0	9.1	8.9	13.4	4.3	9.1	83.5
	(82)	(80)	(76)	(84)	(47)	(78)	
6	9.4	6.5	5.0	11.6	3.7	7.2	66.0
	(81)	(74)	(76)	(84)	(59)	(78)	
8	7.2	4.9	4.0	8.8	2.1	5.4	49.5
	(82)	(69)	(75)	(83)	(57)	(78)	
10	6.5	5.7	3.6	7.1	3.1	5.2	47.7
	(65)	(68)	(58)	(77)	(68)	(69)	
Clone	10.0	8.1	6.0	11.6	4.0	7.9	–
mean	(79)	(77)	(75)	(84)	(63)	(76)	

Planted July 1987, Spaced 100 × 30 cm. Harvested plot: 22 m^2.
Source: 1987 Annual Report, NRCRI, Umudike, Nigeria.

Pest and Disease Management

In West Africa, sweetpotato pests and diseases occur on leaves, vines and tubers, and after harvest, on transit to store, and within market stores before sales. Two frequently cited constraints to sweetpotato production and storage are: virus and weevil (*Cylas puncticollis*). The Sweetpotato virus disease complex [SVDC] has four viruses [mosaic, mottle, feathery mottle, and vein-clearing] that depresses yields and distorts tubers, especially when repeatedly planting vines from the same source. Rossel and Thottappilly (1985) recommended varieties with good resistance to SVDC as control besides planting virus-free vines as often as possible.

Improved drought-tolerant varieties are needed for farmers in drier savanna zones with precipitation under 500 mm in West Africa (Fig. 19.2). High moisture-tolerant varieties are required for wetter localities with rainfall over 1000 mm.

Weevils threaten sweetpotato systems; attacking tubers in shallow soil or exposed from soil at maturity. Adult weevils lay eggs on stems and tubers; larvae bore into tubers; pupate on stems, and are transferred on shoots. Once established in a crop, they are difficult to control. Consequently, before planting, dip vines in insecticides. A fallow or another crop after each sweetpotato helps break cycles of weevil incidence. Eggs develop into adult weevils within 25 days; therefore, one month of careful scheduling of harvesting reduces weevil damage even if eggs have been laid on tubers. The degrees of field tolerances observed suggest there are no resistant varieties. Early harvest at 4 MAP had fewer (1–2) weevils on tubers compared to

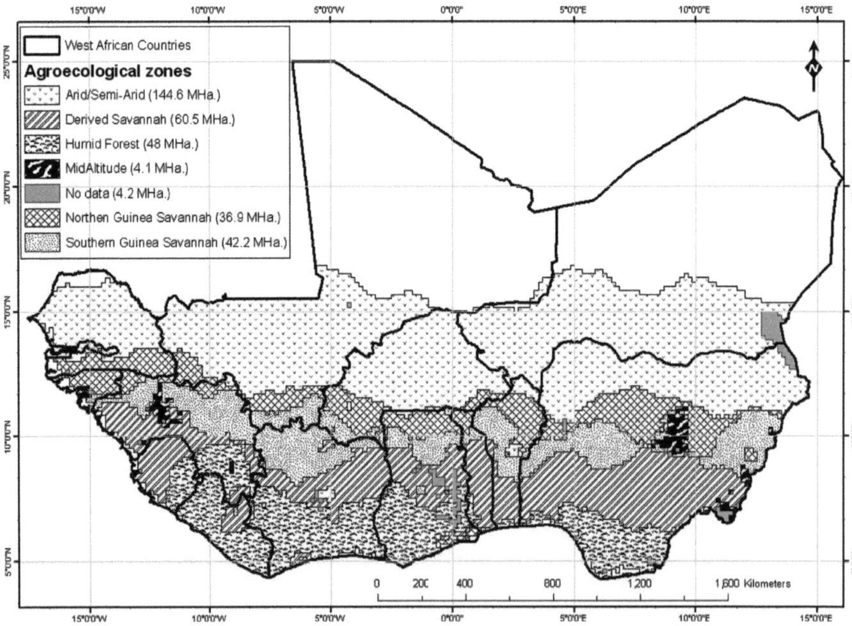

Fig. 19.2 Agro-ecological zones of West Africa suited to sweetpotato production particularly northern and southern Guinea savanna, and derived savanna zones

Table 19.14 Percentage of sweetpotato tubers rendered unmarketable due to weevil damage

Variety	Early harvest (4 MAP)	Late harvest (6 MAP)
	— Early planting (March–April) —	
TIB 4 (susceptible)	0.0	0.0
TIB 2532 (Moderately resistant)	0.0	0.0
TIS 70357 (Moderately resistant)	0.0	0.0
	— Late planting (August–September) —	
TIB 4	41.2	58.7
TIB 2532	5.0	27.5
TIS 70357	17.5	12.5
LSD (5%)	6.6	7.7

Unmarketable tubers among 20 randomly selected tubers per replicate for each variety (means of 4 replicates).
Source: Lema (1992)

delayed harvest [6 MAP] with 15–34 weevils (IITA, 1984). Deeper tubers suffer less weevil damage (IITA, 1975) suggesting re-heaping ridges helps. Adding more soil on tubers at first weeding (1MAP) introduces new expenses.

Sweetpotato farmers rarely mention nematode problems; not even associating their symptoms. In Nigeria. Fawole and Cole (2000) found *Meloidogyne incognita*, *Rotylenchus reniformis* and *Pratylenchus brachyurus* on sweetpotato tubers under mixed cropping, with maize and or cassava in holdings under 0.25 ha. Nematode damage of tuber quantity and quality depreciate food and sales; although they concluded nematodes do not seriously constrain sweetpotato production in southwestern Nigeria.

Of six species of *Cylas* feeding on sweetpotato tubers in Africa, *Cylas puncticolis* is most rampant. Late planting and harvest rendered 58.7% of tubers unmarketable through weevil damage of a susceptible variety but less so (12.5–25.5%) for moderately resistant varieties (Table 19.14). Thus, the choice of variety, appropriate planting and harvest dates minimize weevil damage.

Harvesting, Marketing, and Profitability

Harvesting. Sweetpotato tubers are harvested manually without machines in West Africa. High labour costs affect farm profits. Delaying harvest after tuber maturity, for some varieties, make tubers crack, grow too big, becoming unmarketable. Early and late varieties are harvested as they mature; evidenced by shoot senescence and drying, and sap exudation by tubers. In highlands, low temperatures delay maturity of most genotypes.

Harvesting entails cutting off shoots, carefully digging out tubers avoiding bruises, using fork shovel [from ridges sides], long wooden sticks, metal rod with flattened ends, and hoes. Under subsistence, only heaved tubers are harvested for home-use or market day. Small tubers are left to continue bulking, thereby increasing yields due to favourable source-to-sink relationships. Knowing when to harvest enables farmers obtain tubers with desirable composition (Table 19.15).

Table 19.15 Mean dry matter and starch content of fresh tubers of selected sweetpotato varieties (TIS 000A, TIS, 2534, TIS 000B) at different harvesting dates at Umudike, Nigeria

Harvest date (weeks after planting)	Leaf dry matter (%)	Stem dry matter (%)	Tuber dry matter (%)	Starch in fresh tuber (g/100 g)
6	17.47	10.74	–	–
8	15.23	8.57	–	–
10	17.33	10.60	29.60	16.85
12	18.80	15.60	33.87	23.03
14	18.73	19.53	33.87	25.56
16	20.93	18.73	33.87	21.67

Source: NRCRI (1984).

Fresh tuber yield estimates vary with growth conditions, variety, season, spacing, crop growth duration, harvest technique, and harvested area in each trial. Each factor accounts for different percentages of the total variation among tuber yields. Tuber yield range is 1–40 t/ha in 3–7 months. In Nigeria, farmers get 3–6 t/ha researchers get 30–40 t/ha in experiments; 21–38 t/ha at Ibadan, (Hill et al., 1990); 8.8–11.1 t/ha (Korieocha et al., 2006) in Umudike, and 2.4–3.5 t/ha in Asaba (Akparobi, 2006). Even delays before weighing tubers after harvest affects estimation. Until yield estimation is formally harmonised, it will continue to generate variant estimates that confuse rather than guide farmers and investors.

Tuber storage. Fresh sweetpotato tubers store poorly. In large farms, early disposal helps reduce losses to rots (wet/soft/hard/dry); sprouting; weevil damage, rodents, evaporation, and respiratory use of stored carbohydrates, skin cracks and injuries in-transit from farm to store to market; besides changes in nutritional quality. Conserving output is essential to halt tubers decline in market and culinary value during storage. About 14% of output spoils within 90 days, after Okwuowulu and Asiegbu's (2002) data were *re-analysed* and transformed to spoilage index (Table 19.16). Losses of 10–15% of weight occur 2–3 weeks after harvest

Table 19.16 Spoilage index (SI) [percentage of tubers spoilt each month] of sweetpotato tubers harvested at different months of harvest and duration of storage in baskets in barn at Nuke, Nigeria

Months of *storage*	Tuber harvest at months after *planting*			
	3	4	5	Mean
1	13.97	15.63	28.85	19.48
2	12.28	12.28	12.86	12.47
3	17.09	10.66	2.42	10.06
Mean	14.61	13.08	14.99	14.22

SI = sum of mean percentages of rots, sprouting, and weight losses excluding loss in quality caused by weevils. Mean of 3 replicates, 4 elite varieties, 3 phosphorus fertilizer levels, and 2 seasons. Baskets were covered with 2.5 mm mesh to prevent rodents.

(Table 19.16), although quality changes make tubers undesirable to buyers and consumers.

Marketing. West Africa's sweetpotato is usually sold wholesale in rural markets in baskets or sacs weighing 20–70 kg. Urban traders may contract local farmers to produce tubers, for bulking, and weekly transport in vehicles (<10 t) to meet urban demand. Tubers are brought to village markets by farmers or by middlemen traders who buy from scattered farms during piece-meal harvests that tally with local market days. Retailers sell in heaps or few tubers in containers without long-term storage.

Tubers are sold unsorted un-graded at local markets on bare ground. All sizes and shapes are mixed; sometimes including varieties with different skin colours. Traditionally preferred tastes vary from bland to very sweet according to ethnicity and age groups. Generally, larger tubers attract better prices and 20–33% of tubers are tiny or malformed. Any association of skin colour to a choice variety in a locality is based on past experience as most communities distinguish few varieties. Breeders, agronomists, and processors should manage sweetpotato systems to reduce the percentage of yield to tiny low-valued tubers.

Uniform tubers in bags or cartoons will in future be the way to market sweetpotato. The USA sweetpotato market standards have four tuber diameter grades – *No. 1* (4.4–8.9 cm, 7.6–22.9 cm long); *Canner* (2.5–4.4 cm, under 22.9 cm long); *Jumbo* (over 8.9 cm); and *Cull* (malformed or distorted tubers) (Schultheis and Jester, 2002). Sale standards are non-existent in West Africa. No sorting by standards before sale of tubers in heaps, the least diameter measured is about the combined width of the last two fingers (3–4 cm); and are as big as potato (3–5 cm thick, 5–7 cm long). There is, therefore, no way to relate prices to tuber quality characteristics as in developed economies.

Economics and Enterprise

Optimally combining production factors to achieve the highest gains is the chief aim of sweetpotato enterprises. Applying business skills with technical knowledge is critical for managing every essential step. In dynamic environments, continued changes create challenges for any sweetpotato grower, processor, or marketer. Few sweetpotato enterprises operate at large scales or business-like. Hence, high profits are unattainable appropriate inputs are never or irregularly available or if available, are expensive relative to product prices.

Economics. Production and sale of sweetpotato tubers requires that output values exceed production costs. Whole system integration should be considered before fixing prices of tubers or processed products (Table 19.17).

Varying input levels and prices affect level of profits from sweetpotato farming. In south-eastern Nigeria, input prices are ranges: seed: 400–3300 Naira/ha; fertilizer: 200–400 kg/ha; labour 1000–7000 Naira/ha, [US$1 = N83] (Tewe et al., 2003); implying no generalization of costs is possible, **local data** are needed for guiding farmers.

Table 19.17 One hectare field production cost of some arable crops in south-eastern Nigeria

Item	Sweetpotato	Cassava	Yam	Maize
1. Land clearing, packing and burning (md)	80	80	80	80
2. Ridging (md)	100	100	100	100
3. Processing of planting materials (md)	29	20	–	–
4. Planting (md)	25	30	45	35
5. Weeding (md)	40	80	80	40
6. Fertiliser application (md)	10	25	25	30
7. Roguing of weeds (md)	10	–	–	–
8. Harvesting (md)	60	75	80	50
9. Threshing (md)	–	–	–	20
10. Staking & linking (md)	–	–	90	–
11. Total man-days	345	410	500	355
12. Cost @ N100/man-day (Naira)	35,400	41,000	50,000	35,500
13. Cost of fertilizer (Naira)	1,200	1,800	1,800	1,200
14. Cost of stakes (Naira)	–	–	7,500	-
15. Cost of cassava	–	1,000	–	–
16. Cost of planting materials (Naira)	200	1,500	60,000	500
17. Cost of bags (Naira)	3, 500	–	–	–
18. Total cost of production (N)	40,300	44,300	119,400	37,200
19. Fresh tuber yield (t/ha)	7	15	10	2
20. Price (N/kg)	7	2	12	18
21. Revenue (N/ha)	49,000	30,000	120,000	36,000
22. Gross margin (N/ha)	8,700	14,300	600	1,200

Notes: md = man-day; 1.) 2–3 weedings for cassava/yam. 2.) Fertilization based on 400 and 600 kg/ha (15:15:15) for sweetpotato/maize and cassava and yam, at N150/50 kg bag. 3.) Farmer use personal and family labour. 4.)US$1 = N83.
Source: Tewe et al. (2003)

Ghana's annual production of sweetpotato is small compared to cassava and yam; though expanding in Upper East, Upper West, and Northern Regions (Adjekum, 2003). Sweetpotato's contributions to food security and exports to Burkina Faso and Togo are rising. Farmers perceive sweetpotato as profitable as its 3–6 months life fits the savanna rainfall pattern. Key issues for improving sweetpotato enterprises include: (i) screening assembled germplasm for superior varieties, widespread multiplication, and distribution of quality planting materials of new varieties; (ii) cleaning and sanitizing vines against virus; (iii) enhancing seed supply systems to assure year-round availability; (iv) promoting improved tuber processing and storage to reduce losses; (v) improving the marketing chain to more appropriately price tubers at farm gate; thereby empowering farmers to continue in profitable production.

Farm gate prices depend on: season, accessible road network to demand centres, local production costs, and labour availability. Yield estimates by farmers are difficult to verify when sold in truck-loads or harvested by field-area measured in "ropes" [acre/0.4 ha]. Traditional farmers keep no records of labour that accounts for most of the production budget. Vines are produced on-farm without agro-chemicals or fertilizers and are *not sold for much.*

Table 19.18 Returns (cedis) to resources used in production of various food crops in Ghana

Crop	Investment/ hectare	Gross revenue/ hectare	Gross margin/ hectare	Gross margin/ man-day	Gross margin/cedi invested
Cassava	1,407,038	2,837,319	1,479,156	8,913	1.08
Maize	2,907,500	4,000,000	1,292,500	26,377	0.44
Yam	10,344,090	14,000,000	3,855,910	16,270	0.37
Sweetpotato	2,390,500	9,378,240	7,174,830	49,737	3.00

[10,000 cedis = US$1 in 2008].
Source: Nurah and Ahiale (2005)

Profitability. Assessing total production cost plus the cost of capital invested in sole-crop sweetpotato is complex because few farms grow sweetpotato sole. Also profitability fluctuates periodically with prices of inputs and tubers. In Ghana, producing sweetpotato costs US$29–49/tonne with good gross margins of US$95–149/tonne (Table 19.18). However, without processing and marketing problems—storage losses constitute major risks to production (Nurah and Ahiale, 2005). Processing tubers into storable dry products and promoting manufacturing that tally with tuber production schedules would spur sweetpotato economies.

In Ghana, returns from yields (adjusted to 80% of actual on-farm tuber yields) varied with agro-ecology. Incremental net returns from existing varieties were 564,000–1,392,000 cedis in forest areas; 348,000–1,332,000 cedis in forest/savanna transition; and 192,000–636,000 cedis in coastal savanna (Otoo et al., 2000). Accounting for every expense helps determine right prices for tubers and shoots (Table 19.19).

Table 19.19 Average variable costs of sole sweetpotato crop in wet season of 2000 for Jigawa, Nasarawa, Katsina, Kebbi and Akwa Ibom States of Nigeria

Items/Operation	Cost (Naira/ha)	
Planting materials	4500	
Fertilizer	7600	
Packaging bags	1750	
Simple tools	500	
Land clearing	300	
Land cultivation	3600	
Planting	1000	
Fertilizer application	800	
Weeding/thinning	4500	
Harvesting	3000	
Bagging	250	
Transportation	1800	
Total variable costs	29600	
Fresh tuber yield (kg)	7200	
Farm gate price of tubers (Naira/kg)	8.00	
Output revenue		57,600
Enterprise gross margin		28,000
Gross margin benefit/cost ratio		0.950

(US$ = 130 Nigerian naira)
Source: PCU (2002).

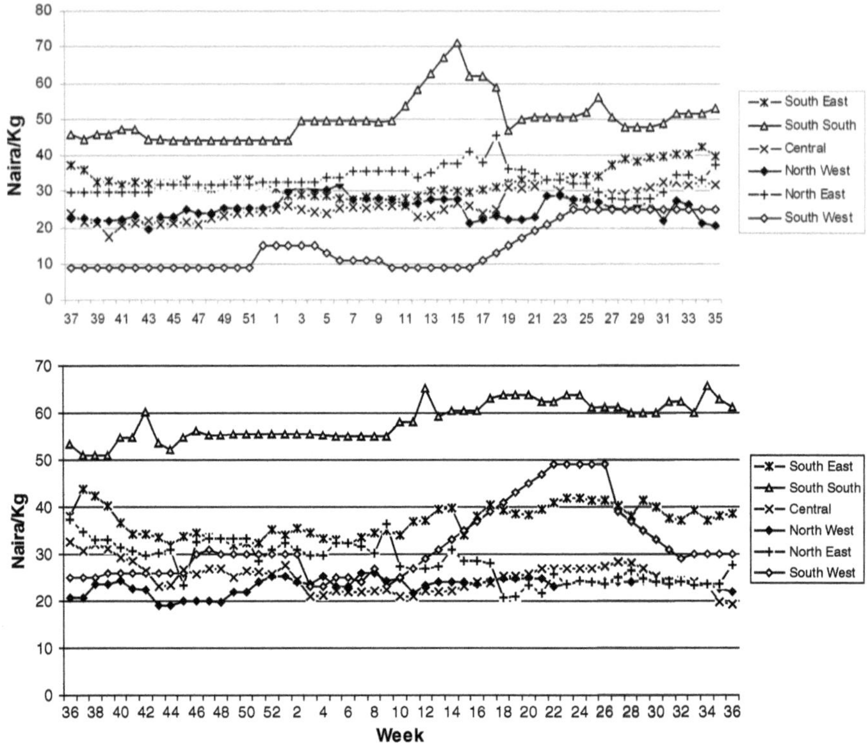

Fig. 19.3 Weekly prices of sweetpotato tubers (Naira/kg) averaged across markets in six geo-political zones of Nigeria during 2004–2006 (prepared by Paul Obasi, IITA Ibadan, 2008)

Prices of sweetpotato tubers varied periodically among geo-political zones of Nigeria (Fig. 19.3). From week 1 (1–7 January) till week 52 (25–31 December) tuber price trebled from N20/kg [N120=US$]. Exclusive use of fresh tubers will discourage higher outputs.

Various Uses of Tubers and Foliage

Every sweetpotato tuber is usable for humans or livestock. Owori et al. (2007) described 52 recipes using leaves, tubers, and derived products for Eastern and Central Africa but most are yet to enter West Africa. Tubers contain 20–35% dry matter and peels vary with shape and surface smoothness. Discounting moisture content and peels, tuber yield is 18–20% usable; although higher if peeling is not obligatory. This is an aspect requiring research as for potato where the skin need not be peeled for certain food preparations where peel fibre is incorporated.

Sweetpotato Sensory Quality

A key factor in sweetpotato selection is *sensory quality* of boiled tubers. The first annual report of IITA (1971) states: "About 10,000 clones of sweetpotato were produced from seedlings ... including varieties collected in Nigeria. It is believed that sweetpotato are **not** extensively grown and utilized for food in Tropical Africa because they are *sweet and moist when cooked*"—a true assertion even to date (2008). Some clones are less sweet, blander, and more dry-textured when boiled tasting like yam (a preferred staple).

Over 95% of sweetpotato in Osun State, Nigeria, a major producer, is for food while tiny tubers are feed to livestock (Tewe et al., 2003). Tubers are utilised boiled, baked; roasted; fried; pounded (mixed with yam) into *fufu*; or peeled chopped, sliced, parboiled, sun-dried, and stored or milled into flour sometimes mixed with cassava flour and prepared into dough. Sweetpotato flour's darkness and sweetness affecting its acceptance; but adding cassava flour dulls the sweetness and brightens the dough colour. In urban areas, sweetpotato is used as an ingredient for making meat pies.

Selecting sweetpotato varieties for food involves complex decision making including nutrients composition and density among genotypes. Processing qualities of tubers of 5 of 25 cultivars [TIS 1487, J.K. 70, Anoma, 2294 and 2291] were selected for high starch (18.5–31.3%), low moisture (61.2–73.2%), high protein (4.8–6.7%), and low fibre content (1.8–2.6%). Essential selection traits would include: dry tuber yield, acceptable sensory quality, tolerance to pest and disease in field and store, and high nutrient density per 100 g edible tubers.

Boiled tubers. In sweetpotato breeding, tubers are boiled and systematically assessed for *flesh colour*, *taste* (cortex sweetness or total free sugar content, and or its *ratio* to starch content) and *texture* (a function of density of dry matter, starch, and fibre content). Taste-texture appreciation favours staple consumed in large amounts with low sugar contents. Aikpokpodion (1998) identified 19 high yielding bland or non-sweet clones from 228 accessions (Table 19.20) and further breeding in West Africa should utilise them.

Flour. Sweetpotato flours stores longer than tubers. Utilising flour in composite flours is a desired trend for advancing processing. Diverse mixes of flours from sweetpotato, cassava, cowpea, maize, wheat, sesame, banana-plantain, groundnut paste to make juices, drinks, confectionaries or dough would raise nutrition density and better sustains dough supply.

Toasted granules. Reducing postharvest losses necessitates prompt processing of tubers into dry forms. Oduro et al. (2002) and Ellis et al. (2001) in Ghana, Sanni et al. (2001) and Meludu et al. (2003) in Nigeria processed sweetpotato tubers into toasted granules using similar steps for making *gari* from cassava. Some 152 consumers appreciated dry and water-soaked toasted granules of sweetpotato clones: Kayode, Shaba, Alphonso, GR3-25, TIS 80/004, and D.Wall (Meludu et al., 2003); it stored well in containers beyond 12 months and its vulgarisation would supplement cassava gari.

Forage and Chips. Sweetpotato shoots may be fed directly or fermented into silage; or forage and tubers for livestock may be cut, sliced, chipped, sun-dried, and bagged for addition to diets of chicken, goat, pigs, fish, and poultry. Once partially processed, their use in livestock feeding is enormous in West Africa (Tewe et al., 2003). As tuber losses to weevils have proved difficult to control cheaply; research for development should encourage prompt harvest, followed by efficient processing.

Table 19.20 Tuber characteristics of hard-textured non-sweet (bland) sweetpotato clones, Ibadan 1998

Clone	Skin/Flesh colour	Number of tubers/plant (1 × 0.5 m)	Weight (kg/) plant matter	Percent tuber dry flesh	Taste+ of tuber boiled for 15 min
TIS 3030-op-1-18	purple/cream	2	1.23	29.6	–
TIS 3030-op-1-29	purple/cream	8	1.67	38.5	–
TIS 4400-2	white/cream	6	3.25	37.3	cassava
TIS 70357-op-1-18	purple/cream	3	2.52	36.0	–
TIS 70357-op-1-78/80	purple/white	15	6.85	41.1	–
TIS 70357-op-1-79/19	purple/white	3	0.85	34.8	–
TIS 80/0140	purple/white	4	2.18	36.1	yam
TIS 82/0070 × 2532-1-43/23	purple/white	6	3.96	37.2	–
TIS 8504-op-1-80	purple/white	8	3.62	28.3	–
TIS 87/0271	cream/white	4	2.25	36.2	–
TIS 87/0319	cream/cream	4	1.40	27.4	–
TIS XDB	purple/cream	5	3.03	34.4	–
TIS XDB-12(1)	purple/cream	5	2.18	42.7	yam
AOB-25	white/cream	7	3.45	31.3	bitter
De Virousky 16/0289	purple/white	4	2.55	21.2	–
GR-3-25	purple/cream	7	3.05	39.3	–
Lima 312 (2) [Hard flesh]	purple/white	5	0.78	36.9	yam
Mogosili	cream/cream	3	5.15	34.7	yam
Unknown SPH-15	cream/white	3	0.45	38.5	yam

+ Taste may be similar to other staples
Source: Aikpokpodion (1998).

Root Storage Systems

Storage is fundamental to any sustained nutrition from sweetpotato tubers. Household storage of healthy tubers in pits (1–2 m deep) lined with dry grass is labour-intensive and impractical for large (3–7 ha) farms. Physico-chemical properties of sweetpotato flour can be altered to make them useful in food recipes by processing tubers within 7 days after harvest (Iwuoha and Nwakanma, 1999). Storage losses are low if initial tuber stock was clean and sane. Use of chemicals on tubers is not advised because of uncertain hazards of their residues to humans or livestock.

Orange-Fleshed Sweetpotato (OFSP)

The orange fleshed varieties provide a food-based approach for overcoming VAD in Africa. In Nigeria, during the 2006 flag-off of National Sensitization Workshop on Appropriate Distribution and Use of Vitamin A capsules, President Olusegun Obasanjo decried high yearly maternal mortality and 88,000 infant deaths from VAD, despite interventions by NAFDAC, UNICEF and partners (THISDAY, 2006); indicating VAD can be addressed by promoting OFSP.

Lesser use of OFSP in West Africa compared to East Africa is attributed to no promotion or use of palm oil in some areas or weak seed systems to satisfy vine supply. Some selected OFSP varieties are available for deployment to farmers. Where households cannot afford palm oil, growing OFSP is recommended.

In East and Southern Africa, sweetpotato research for development is well assisted by Centro Internacional de la Papa (CIP) offices in Africa; but less so in West Africa since 1988 when international mandate for sweetpotato passed to CIP from IITA Ibadan, Nigeria. Since 2005, CIP has collaborated more with Ghana and Nigeria through VITAA (Vitamin A for Africa) Project. Widespread cultivation of sweetpotato across West African indicates sweetpotato's great future, if well exploited for food, feed, and industry (including ethanol/bio-fuels).

A 2001–2003 food consumption and nutrition survey of Nigeria puts VAD level (based on serum retinol concentration) as 24.8% marginal [< 20 ug/dl], and 4.7% clinical [< 10 ug/dl] among children under 5 years and 13.1% deficient [<20 ug/dl] among mothers (Maziya-Dixon et al. (2004). VAD can be addressed through nation-wide campaigns for local cultivation, processing and use of OFSP.

Sweetpotato for Livestock

Tiny sweetpotato tubers are converted in chips or feed. Feeding studies indicate varying optimal levels of dried flour in livestock rations (Table 19.21). Sun-dried samples infested by micro-organisms depress pigs and poultry growth. Less chips in pig and poultry feed was recommended due to increased diarrhoea incidence at higher contents of reducing sugar and the form of their presentation. Conversely, more non-sweet tubers in rations is recommended. Factors limiting usage of

Table 19.21 Recommended sweetpotato flour inclusion level in livestock rations for optimal performance by processing method

Species		Sun-dried (Percent)	Oven-dried
Broilers:	0–4 weeks old	0	12
	5–8 weeks old	0	18
Layers:	30–40 wks old	–	10
Pigs:	Weaner	–	17
	Grower	–	17
Sheep:	Growing/Fattening	–	40

Source: Tewe (1992)

sweetpotato in feed rations include: bulkiness and requirement for rapid dehydration to avoid microbial growth, contain less proteins and dustier rations compared to maize-based rations. Overcoming these limitations, require fabricating local dryers, solar drying, adding oil or pelletizing rations to reduce impurities, using shoots to augment low tuber protein contents; and distributing non-sweet clones.

Sweetpotato for Ethanol

Sweetpotato root starch and sugars convert to ethanol through enzyme fermentation yielding about 137 l of ethanol per tonne of tubers. Short growth and high yields gives sweetpotato a comparative advantage for bio-fuels. As a secondary food crop in West Africa, use for ethanol would not adversely disrupt community food balances. Such diversified use will favour local cottage industries; generate income, besides being food reserve. If processed into high grade flours it becomes usable in pharmaceutical industry in West Africa as in Japan.

Managing the share of *national food basket* sourced from sweetpotato is important. Japan (Komaki, 2006) now has a stable contribution of sweetpotato as a ratio in relation to other foodstuffs. Sweetpotato would occupy a fair share of the food basket as a diversified diet is more healthy and sustainable in supply. Adequate statistics is quintessential for managing sweetpotato production, storage, and transportation losses. Without understanding the dynamics of seasonal marketing and consumption in West Africa, quantification of trends and proper management of flows of materials will not be feasible.

Future of Sweetpotato

Advancing sweetpotato in West Africa (Tayo, 2000) will require better funding for weak national research institutes mandated to: 1) breed desirable traits into few varieties with high shoot and tuber yields for dual use as food and feed especially in savanna zones; 2) select varieties with low sugar contents, especially if poundable into good textured dough preferred among the elderly and diabetic; 3) better manage

Table 19.22 Land area (square kilometres) under different agro-ecological zones of West Africa

West Africa country	Semi-arid/ Arid	Derived savanna	Humid forest	Mid-altitude	North Guinea savanna	South Guinea savanna	Area with no data	Country total area
Benin	1.265	2.777	0.254	0.000	3.490	3.758	0.000	11.545
Burkina Faso	20.967	0.046	0.000	0.000	4.799	1.738	0.000	27.550
Cape Verde	0.000	0.000	0.000	0.000	0.000	0.000	0.000	0.000
Cote d'ivoire	0.340	9.527	13.669	0.013	0.123	8.150	0.170	31.994
Gambia	0.839	0.000	0.000	0.000	0.039	0.000	0.810	0.959
Ghana	0.454	6.036	9.209	0.000	3.662	3.120	1.259	23.740
Guinea	0.365	7.475	4.063	1.269	2.224	8.556	0.034	23.986
Guinea Bissau	0.000	0.000	0.000	0.000	2.193	0.216	0.034	2.442
Liberia	0.000	1.430	8.063	0.000	0.000	0.068	0.000	9.562
Mali	38.691	0.000	0.000	0.000	4.198	2.853	0.033	45.776
Niger	41.368	0.000	0.000	0.000	0.000	0.000	1.231	42.599
Nigeria	26.809	25.037	9.905	2.764	11.302	13.306	1.264	90.388
Senegal	13.426	0.000	0.000	0.000	4.100	0.002	0.038	17.566
Sierra Leone	0.000	4.255	2.401	0.000	0.000	0.000	0.000	6.656
Togo	0.036	3.954	0.440	0.000	0.795	0.464	0.000	5.688
Total area	144.560	60.536	48.005	4.046	36.925	42.233	4.145	340.450

Source: Alabi (2008). Geo-Spatial Laboratory, IITA, Ibadan, Nigeria postharvest system for storing dry tuber products before spoilage (Okwuowulu and Asiegbu, 2002); 4) incorporate sweetpotato in intercropping systems to smother weeds and reduce costs; and 5) resolve policy issues that encourage neglect of research on arable staples; 6) strengthen seed systems to satisfy year-round demands.

Expanding sweetpotato hectares would be more in derived savanna and southern Guinea savanna emcompassing 10,000,000 ha in 11 countries (Table 19.22). Exploiting these zones forestalls risks posed to cereal production sensitive to unsteady rainfall in these areas. Future campaigns should focus on promoting greater cultivation and use of sweetpotato

Finally, better sweetpotato operations would involve producing efficiently, promptly processing tubers into dry storable forms for re-constitution into food-feed-based and industrial products; thus avoiding high (10–83%) post-harvest tuber losses (Nnodu, 1982). Low prices and profits discourage high investments on sweetpotato production or storage.

Acknowledgments I thank many collaborators who me sent national information; especially Dr Harrison Dapaah of Crops Research Institute Fumesua, Ghana. Mr Paul Obasi (for graphing tuber prices) and Mr Tunrayo Alabi (for agro-ecological mapping), both of IITA Ibadan Nigeria.

References

Adjekum, A., 2003. Tuber and Tuber Crops Sub-sector *Report, Root and Tuber Improvement Program*, RTIP Kumasi, Ghana and IFAD, Rome.

Aikpokpodion, P.O., 1998. Morphological characterisation and tuber quality evaluation of sweetpotato (*Ipomoea batatas* [L.] Lam.) germplasm. *MSc, University of Ibadan*, Nigeria. 70p.

Akoroda, M.O., Aikpokpodion, P., Aliyu, T.O., Fabunmi, T.O., Fatunbi, A.O. and, Olofinji, E.B., 2000. Holistic sweetpotato breeding and selection schemes: clonal trials in southwest Nigeria. *In: Proc. 5th African Potato Association Conference, Kampala*, Uganda. pp. 61–67.

Akparobi, S.O. 2006. Correlation analysis of growth parameters of sweetpotato (*Ipomoea batatas*). *In Proceedings of the 40th conference of the Agricultural Society of Nigeria*, Asumugha, G.N., Oloyede, A.O., Ikeogu, J.G., Ano, A.O., and Herbert, U. (editors) NRCRI, Umudike.16-20 October 2006. Umudike Nigeria, pp. 666–668.

Alabi, T. 2008. Personal communication. Geo-Spatial Laboratory, IITA, Ibadan, Nigeria.

Chinaka, C.C., 1983. Sweetpotato production. *Extension bulletin No. 7. AERLT, NRCRI*, Umudike, Umuahia, Nigeria. 7p.

Chude, V.O., Malgwi, W.B., Amapu, I.Y., and Ano, A.O. 2004. *Manual on Soil Fertility Assessment*. FFD/NSPFS, Abuja, Nigeria.

Chukwu, G.O., 1995. Crop irrigation water needs for sweetpotato (*Ipomoea batatas*). *Africa J. Tuber and Tuber Crops* 1: 35–38.

Dahniya, M.T., Akoroda, M.O., Alvarez, M.N., Kaindaneh, P.M., Ambe-Tumanteh, J., Okeke, J.E., and Jalloh, A., 1994. Developemnt and dissemination of appropriate root crops packages to farmers in Africa. *In: Tropical Root Crops in a Developing Economy*. Ofori F. and Hahn S.K. (editors). Proceedings of ISTRC, Accra, Ghana. pp. 2–9.

de Geus, J.G., 1973. Fertilizer guide for the tropics and subtropics. *Centre d'Etude de l'Azote, Zurich*, Switzerland. 205p.

Ellis, W.O., Oduro, I., Fianko, K., and Otoo, J.A., 2001. Quality of gari from sweetpotato varieties. *In 8th of ISTRC-AB, Symposium Proceedings*, IITA Ibadan.

FAOSTAT, 2005. Food and Agriculture Organization of the United Nations, Production Statistics Webpage. Viewed in 27 November 2007.

FAOSTAT, 2006. Sweetpotato cultivation statistics for Africa 2005. *FAO Website*.

Fawole, B., and Cole, O.A., 2000. Plant-parasitic nematodes as constraints in the production of potatoes and sweetpotatoes in Nigeria. *African Potato Association Conference Proc.* 5: 421–424. Adipala E., Nampala P. and Osiru M. (editors). NARO, Kampala, Uganda.

Tewe O.O. 1992. Biochemistry and utilization of sweetpotato (*Ipomoea batatas*) for animal feeding: implications for food security in Africa. Pages 324-327 in the Proceedings of the fifth triennial symposium of the International Society for Tropical Root Crops-Africa Branch. IITA, Ibadan, Nigeria, pp. 324–327.

FFD, 2002. Federal Fertilizer Department. Fertilizer *Use and Management Practices for Crops in Nigeria*. Aduayi EA, Chude VO, Adebusuyi, and SO Olayiiwola (editors). FMA, Abuja, Nigeria.

IITA (various years), 1971–1989. *Annual Reports and Research highlights of the International Institute of Tropical Agriculture*. IITA Ibadan, Nigeria.

Iwuoha, C.I. and Nwakanma, M.I. 1999. The physico-chemical properties of sweetpotato (*Ipomoea batatas* L. Lam) tuber flour as affected by processing condition. *Africa J. Tuber and Tuber Crops* 3: 15–18.

Janssens, M., 2001. Sweetpotato: *Ipomoea batatas* (L.) Lam. *In: Crop Production in Tropical Africa*. Raemaekers R.H. (Editor) DGIC, Brussels, Belgium. pp. 205–221.

Hill, W.A., Hortense, D., Hahn, S.K., Mulongoy, K., and Adeyeye, S.O. 1990. Sweetpotato root and biomass production with and without nitrogen fertilization. *Agronomy Journal* 82: 1120–1122.

Komaki, K., 2006. The Status of Tuber and Tuber Production, Utilization and Marketing in Japan. National Institute of Crop Science. Tsukuba, Japan. *ISTRC Trivandrum, Kerala, India.*

Korieocha, D.S., Ekeleme, F., Nwauzor, E.C., and Nwokocha, C.C., 2006. Effect of nitrogen application on the critical period for weed control in sweetpoato (*Ipomoea batatas* Lam) on an ultisol. *In*: *40th Agricultural Society of Nigeria Conference* Proceedings. Asumugha GN, Oloyede AO, Ikeogu JG, Ano AO, and Herbert U. (Editors). NRCRI, Umudike. Umudike Nigeria. pp. 663–665.

Lema, K.M., 1992. Reducing weevil dmage in sweetpotato using host-plant resistance and early planting and harvesting. *In 4th ISTRC-AB Symposium Proceedings, Kinshasa*, Zaire. pp. 345–346.

Leon, J., 1977. Origin, evolution, and early dispersal of tuber and tuber crops. *In Proceedings of 4th ISTRC symposium*, Cock J., MacIntyre R. and Graham M. (editors) CIAT, Cali, Colombia. 36pp.

Mauny, R., 1953. Notes historiques autour des principales plantes cultivees d'Afrique occidntale. *Bull. Inst.t Francaise de l'Afrique Noir* 15: 684–730.

Maziya-Dixon, B., Akinyele, I.O., Oguntona, Nokoe, S., Danusi, R.A., and Harris, E., 2004. Nigeria food consumption and nutrition survey 2001–2003. Summary. *IITA, USAID, UNICEF, and USDA. IITA*, Ibadan, Nigeria. 52+55pp.

Meludu N.T., Ajala C.G., and Akoroda M.O., 2003. Poverty alleviation through the processing of sweetpotato tubers into toasted granules and consumer preferences in Nigeria. *Africa J. Tuber and Tuber Crops* 5: 56–59.

MINAGRA, 1990. *Ministère de l'Agriculture et des Ressources Animales* (Côte d'Ivoire). Annuaire des statistiques agricoles et forestières. 176p.

Nnodu, E.C., 1982. Development of sweepotato storage system. 1982 *Annu. report, NRCRI, Umudike*, Nigeria. pp. 152–154.

NRCRI, 1984. Annual Report of the National root Crops Research Institute for 1984. Umudike, Nigeria.

NRCRI, 1994, 1992. *Annual reports, NRCRI*, Umudike, Nigeria.

Nurah, G., and Ahiale, E., 2005. Economic Sustainability of cassava and improved sweetpotato production in Ghana. *KNUST, Kumasi, Nigeria*. April 2005.

O'Sullivan, J.N., Asher, C.J., and Blamey, F.P.C., 1997. Nutrient disorders of sweetpotato. No. 48. *ACIAR Australia*, 136p.

Oduro, I., Ellis, W.O., Otoo, J.A., Akuffo, A.O., and Moses, E., 2002. Production of gari from sweetpotato tubers. *12th Symposium of ISTRC, Tsukuba*, Japan.

Okoli, O.O., 1980. The developers are coming: a case for genetic resources conservation in tuber and corms for Africa. *In Crop GeneticRresources in Africa. IITA*, Ibadan, Nigeria, pp. 150–157.

Okwuowulu, P.A., and Asiegbu, J.E., 2002. Influence of K-fertilizer treatment, variety, and age at harvest on shelf-life of sweetpotato tubers under life-shade barn. *African J. Tuber and Tuber Crops* 5: 42–45.

Onwueme, I.C., and Sinha, T.D., 1991. *Field Crop Production in Tropical Africa*, CTA, Ede, Netherlands. pp. 267–275.

Otoo, J.A., Osei, C.K., Danso, A.K., Asamoah, O.N., Gyasi-Boakye, S., Ansah, I.O.O., Berchie, J.N., Adjekum, A.A., Asante, S.N.A., and DeGraft-Johnson, A., 2000. Steps to good sweetpotato production. *RTIP, MFA and CSIR-Crops Research Institute, Kumasi*, Ghana. 16p.

Owori, C., Lemaga, B., Mwanga, R.O.M., Namutebi, and Kapinga, R. 2007. Sweetpotato recipe book: Sweetpotato processed products from Eastern and Central Africa. *ASARECA, CIP, and NARO, Kampala Uganda*. 93p.

PCU, 2002. *Farm management survey and advisory services* (FAMAS), 2000/2001 Advisory handbook. Project Coordinating Unit, FMARD, Abuja, Nigeria.

PCU, 2004. 2003 *CAYS, PCU FMARD* Abuja, Nigeria.

Purseglove, J.W. 1968. Tropical Crops: Dicotyledons. Longman, London, UK.

Rendle, C.J., and Kang, B.T., 1977. Phosphorus requirement of three sweetpotato cultivars. *In: Proceedings of 4th ISTRC symposium*, Cock J., MacIntyre R. and Graham M. (Editors). CIAT, Cali, Colombia, pp. 117–121.

Rossel, H.W., and Thottappilly, G., 1985. *Virus Diseases of Important Food Crops in Tropical Africa*. IITA, Ibadan, Nigeria.

Sanni, L.O., Ikuomola, D.P., and Sanni, S.A., 2001. Effect of length of fermentation and varieties on the qualities of sweetpotato garri. *In: Root Crops: The Small Processor and Development of Local Food Industries for Market Economy*. Proceedings of the 8th Triennial Symposium of the International Society for Tuber and Root Crops- Africa Branch (ISTRC-AB), held at Ibadan, Nigeria, pp. 208–212.

Schultheis, J.R., and Jester, W.R., 2002. Nitrogen fertilizer management in 'Beauregard' sweetpotatoes. *In: 12th ISTRC symposium, Tsukuba*, Japan. Nakatani M. and Komaki K. (Editors). pp. 188–192.

Scott, G.J., and Ewell, P., 1993. Sweetpotato in African Food Systems. *In Product Development for Root and Tuber Crops. CIP*. Lima, Peru. pp. 91–103.

Tayo, T.O., 2000. Opportunities for increased potato and sweetpotato production in West Africa. *In: African Potato Association Conference Proceedings* 5: 27–30. Adipala E., Nampala P. and Osiru M. (Editors). African Potato Association, NARO, Kampala, Uganda.

Tewe, O.O. 1992. Biochemistry and utilization of sweetpotato (*Ipomoea batatas*) for animal feeding: implications for food security in Africa. *In the Proceedings of the fifth triennial symposium of the International Society for Tropical Root Crops-Africa Branch.* IITA, Ibadan, Nigeria, pp. 324–327.

Tewe, O.O., Ojeniyi, F.E., and Abu, O.A., 2003. Sweetpotato production, utilization and marketing in Nigeria. *Social Sciences Department, Centro Internacional de la Papa (CIP)*, Lima, Peru. 53p.

THISDAY, 2006. *THISDAY newspaper*, Nigeria, 23 August 2006. Volume 11 No 4141: 8p.

Tindall, H.D. 1968. Commercial vegetable growing. Oxford Press. London. 234–237pp.

Traore, A., 1994. Situation de la production de patate douce au Mali. *In Sweetpotato situation and priority research in West and Central Africa. CIP*, Lima, Peru. pp. 21–25.

Unamma, R.P.A., Orkwor, G.C., and Igbokwe, M.C., 1984. Chemical seedbed for sweetpotato (*Ipomoea batatas* cv Dukukpuku) production. *1984Annual Report, NRCRI, Umudike*, Nigeria. pp. 245–247.

Chapter 20
Sweetpotato in Southeast Asia: Assessing the Primary Functions of a Secondary Crop

D. Campilan

Background on Southeast Asia

The Southeast Asia region encompasses the Asian mainland and surrounding islands, which are east to the Indian subcontinent and south of China. As a geographic subdivision of Asia, it includes the areas historically known as Indochina and peninsular Malaysia, and the islands extending from the Philippines in the north to Indonesia in the south.

The region has a total population of over 575 million and an average annual growth rate of 1.9% (ASEAN, 2007). About half (48.1%) depend on agriculture, cultivating a total of 448 million hectares of agricultural land (FAO, 2006).

Tropical Southeast Asia has wet and dry seasons; it is generally hot and humid all year round. Rice is the most important agricultural commodity in Southeast Asia (ASEAN, 2006), primarily because the region is endowed with extensive lowland areas suitable for cultivating the crop. Sweetpotato is generally grown in rice-based systems and in areas unsuitable for rice cultivation. It is the second most important root crop in the region, next to cassava.

Southeast Asia is the third largest sweetpotato producing region in the world, after eastern Asia (including China) and eastern Africa. Of the world's top 15 producing countries, three are in Southeast Asia – Indonesia, Vietnam and the Philippines. In 2006, the region produced 4,158,641 tons of sweetpotato (FAO, 2006).

Introduction of Sweetpotato to the Region

While questions on the introduction of sweetpotato to Southeast Asia and Oceania (Ballard et al., 2005) remain, it is widely believed that the crop was introduced through multilane transfers from the Americas. The tripartite hypothesis proposes three routes in which sweetpotato had travelled: (1) "batatas" route from the Caribbean to Europe and onwards to mainland Asia, (2) "kamote" route from Mexico to

D. Campilan (✉)
UPWARD Program, International Potato Center (CIP); c/o IRRI, DA, Metro Manila, Philippines
e-mail: d.campilan@cgiar.org

G. Loebenstein, G. Thottappilly (eds.), *The Sweetpotato*,
DOI 10.1007/978-1-4020-9475-0_20, © Springer Science+Business Media B.V. 2009

the Philippines and onward to China and east Asia, and (3) "kumara" route from Peru to central Polynesia (Yen, 1982). Recent advances in molecular analysis have provided further proof in distinguishing between these routes (Rossell et al., 2001), such as the Philippine gene pool being closer to that of Mexico and Central America (Reynoso, 2005).

Since its introduction to the region – estimated to have occurred in the early and middle part of the second millennium – sweetpotato has become a key part in the agricultural and cultural life of peoples in Southeast Asia. From West Papua (Indonesia) to the northern region of Vietnam, the crop has become a staple food for humans as well as animals, besides being an important cultural symbol for particular ethnic communities. In Southeast Asia, it is now found across agro ecologies – from the humid tropical lowlands including coastal areas to the more interior highlands (Prain, 1995).

Thousands of cultivars and landraces are grown in the field and maintained *ex situ* in various genebanks in the region (Roca, 2007). On-farm management of sweetpotato diversity is likewise commonly practiced among ethnic communities (Prain and Campilan, 1999). Sexual recombination may have given rise to new genotypes that farmers eventually conserve, while mutation might be an important source of genetic variation where incompatibility prevents sexual recombination (Reynoso, 2005). In being acknowledged as a secondary center of diversity for sweetpotato, there are now urgent calls to invest in the conservation and sustainable use of Southeast Asia's sweetpotatoes (Rao and Campilan, 2003).

Production Systems and Trends

Demonstrating its wide adaptability in Southeast Asia, sweetpotato grows under various types of production systems (Prain, 1995). First and most dominant is the lowland post-rice system, where sweetpotato is grown as a second crop after rice, or between two rice cropping seasons. This is where most of the commercial-scale production of sweetpotato is found. Second is the rainfed upland, where the crop is cultivated on a more subsistence scale along hillsides and as part of traditional slash-and-burn farming. In here, sweetpotato is usually associated with rotation systems and often grown as part of complex cropping systems, e.g. intercropped with other root crops.

Homegardens comprise the third production system for sweetpotato (Boncodin et al., 2000). Sweetpotato is grown along with other vegetables on a small piece of land adjoining the dwellings of farm households. In urban areas, sweetpotato is grown in transient, less conventional systems such as container and rooftop gardens. Besides the roots, the young leaves/tops are consumed as vegetables. A fourth production system is typically located near bodies of water – coastal areas and riverbanks. Capitalizing on the crop's ability to adapt to marginal environments with low external inputs, sweetpotato is grown in areas where rice and other food/cash crops cannot be suitably grown.

Overall, Southeast Asia's major sweetpotato-growing countries are Vietnam, Indonesia and the Philippines. These three countries account for 93% of total sweetpotato production in the region, which was over 4 million tons in 2006. Overall, Indonesia leads in terms of volume produced, while Vietnam in terms of area planted (Table 20.1).

Table 20.1 Major sweetpotato-producing countries in Southeast Asia, (FAO, 2006)

Country/Region	Area Planted (ha)	Vol Harvested (tons)
Vietnam	181,700	1, 454, 700
Indonesia	176,146	1, 851, 840
Philippines	118,829	566, 773
Southeast Asia	517,389	4, 158, 641

As FAO data revealed, there has been a decline in the region's sweetpotato production. In 2002–2006, total production decreased by 10.84%. A similar trend was observed earlier in China although in the last few years sweetpotato again picked up, as stimulated by increasing demand for animal feed and industrial uses (Fuglie and Hermann, 2004). Given Southeast Asia's potential to expand sweetpotato uses for agro-industrial markets, the same upward trend is expected to occur in the region for the coming years.

Sweetpotato in Indonesia

Sweetpotato is an important food and cash crop in the densely populated island of Java (Watson et al., 1991), and a traditional staple and feed crop in the eastern and outlying islands. Based on government statistics in 2002, a total of 170,790 hectares in the country were grown to sweetpotato. Of this, 30% was in Java Island, 15% in West Papua, and 7% each for North Sumatra and East Nusa Tenggara provinces.

In Java, sweetpotato is a common rotation crop in irrigated rice-growing areas (Braun et al., 1997). Since sweetpotato does not require as much water supply as rice, it is grown after rice or even for two cropping seasons using early maturing varieties. It is generally planted on ridges, and intercropped with maize or other annuals. Nitrogen fertilization is common, usually using surplus fertilizer from the earlier rice crop and locally available animal manure (Van de Fliert et al., 1996).

In rainfed areas of eastern Indonesia, such as the provinces of East and West Nusa Tenggara, sweetpotato is grown as cash crop and supplementary food. It thrives under dry growing conditions, on generally calcareous soil and with hardly any external inputs. Planted on mounds, it is commonly intercropped with other *palawija* (i.e. secondary food) crops in traditional slash-and-burn cultivation systems and homegardens (Minantyorini et al., 1996).

In the highlands of West Papua, sweetpotato is a subsistence food and a key source of pigfeed. It is an extremely important crop to the local communities, accounting for 90% of daily human diets and up to 100% of pigfeed (Peters, 2003).

It is grown under shifting cultivation, on individual mounds in high ridges, sometimes surrounded by deep drainage ditches. Production systems are designed to provide freshly harvested sweetpotato roots and vines year-round. Each household maintain several beds at different stages of development, and mixture of cultivars per bed. To local farmers, the diversity in cultivars per bed (i.e. which can range from 7 to 21) helps meet requirements for sustaining livelihood, ensuring food security and observing ritual practices (Prain et al., 2000). The diversity of varieties in this easternmost Indonesian province rates as second highest in the world, after Latin America (Yaku and Widyastuti, 2005).

Sweetpotato in Vietnam

Sweetpotato is grown across agro-ecological regions in Vietnam, however its dominant uses vary from the southern to northern parts of the country. It is generally grown on a small-scale to meet household needs. In hot and marginal areas along the coast, it serves as staple food and buffer during periods of rice scarcity (Hoa et al., 2000). It is only in the southern part of the country (i.e. Mekong River Delta and Southeastern region) where sweetpotato is a cash crop for the fresh market. In Central and North Vietnam, sweetpotato is grown for human and animal consumption, and in some cases strictly for vine production as fodder crop.

The country's main sweetpotato region is the Red River Delta, consisting of seven northern provinces surrounding Hanoi. It is where sweetpotato roots and vines are primarily used as pigfeed, in contrast to the typical food uses in southern provinces. This region is characterized by monsoon climate consisting of cool winter and hot summer; with clay loam as major soil type. Sweetpotato is mainly rotated with rice as winter crop.

The Northern Midland and Northern Central regions are likewise major sweetpotato producing areas in Vietnam, where sweetpotato is also primarily used as pigfeed. The Northern Midland region includes upland and mountainous areas, which are constrained by soil acidity and low soil organic matter due to erosion. In these areas, sweetpotato is grown as winter or spring crop, rotated with upland rice. Meanwhile, the Northern Central region is prone to floods and droughts, and has infertile coastal-sandy soils with some alluvial characteristics. Winter/spring sweetpotato is rotated with rice and/or leguminous crops (Peters et al., 2001).

In these three regions, local cropping systems are shifting from rice and food crops to higher value crops such as vegetables. However, farmers continue to devote part of their land to sweetpotato, as a key feed resource supporting smallholder pig production. Sweetpotato is planted on ridges, although their width and height vary across locations and seasons. Compost, consisting of animal manure and crop residues, is commonly applied together with inorganic fertilizer. In cases where farmers apply the latter, these are often in excessive doses resulting to inefficient nutrient management (Van de Fliert et al., 2001).

Sweetpotato in the Philippines

Sweetpotato is grown in three main agroecologies in the Philippines: a) as food and buffer crop in typhoon-prone regions along the Pacific coast, b) as staple food in more remote uplands, and c) as cash crop in the irrigated lowlands.

Popularly known as a crisis crop, sweetpotato provides food supply during typhoons and other natural calamities. Over a third (37%) of Philippine sweetpotato production comes from Eastern Visayas and Bicol (BAS, 2007), which are among the most typhoon-prone and economically least developed regions in the country. Sweetpotato is cultivated from low-lying coastal areas to marginal uplands, where it is proven to be a relatively hardy crop against floods and strong winds. As a short-maturing crop, it provides immediate food in post-disaster rehabilitation efforts.

Sweetpotato is also a traditional staple food and supplementary carbohydrate source in communities where rice is not widely grown, or for households unable to afford rice in their regular diet. Particularly in the northern Philippine uplands, sweetpotato is a primary food crop for indigenous communities, where it is grown on mounds in slash-and-burn farming systems (e.g. Cadelina, 1994). The lower root yield in the uplands is often due to the highly degraded soil and the shading caused by its planting between rows of taller crops and perennials (Villamayor and Amante, 2000). However in some cases, yield is intentionally sacrificed in favor of natural resource conserving benefits, e.g. cover crop, erosion control mechanism, green manure (Campilan, 1995). Additional yet smaller-scale production takes place in home gardens, where sweetpotato is grown both as root crop and vegetable, i.e. young leaves or "camote tops" (Boncodin and Gayao, 1998).

The most rapid expansion in production area has taken place in irrigated lowlands where sweetpotato is a high-value commercial crop grown after rice (Campilan, 2003). Sweetpotato demand from agroindustries, particularly from feed and food processing companies has stimulated increase in production volume and farmgate price. In Tarlac province, for example, area grown to sweetpotato reached over 22,000 hectares in 2007 (BAS), which is more than double the area in the early 2000. To support large-scale production, some sweetpotato farmers have specialized in the commercial multiplication and distribution of healthy planting materials (Laranang and Basilio, 2002). At the same time, crop residues such as vines and non-marketable roots are also used as feed resource supporting backyard beef cattle rising (Adion et al., 2007).

For large-scale production in irrigated lowlands, the main cropping season is preceded by a two-month period of growing sweetpotato for the propagation of planting materials (Data et al., 1997). Sweetpotato is grown after rice or in association with other high-value crops. In times of high market demand, farmers adjust the cropping calendar to make two sweetpotato cropping seasons possible. Compared to other production systems, large-scale sweetpotato production in irrigated lowlands involves higher input costs, with farmers investing in healthy planting materials, irrigation, fertilization and mechanization (Aganon and Bautista-Tulin, 1998).

Key Opportunities for Improving Sweetpotato Livelihoods

Efforts to help farmers' improve sweetpotato livelihoods need to consider Southeast Asia's diverse production-use systems and agroecologies associated with the crop. This diversity gives rise to a very wide range of constraints and opportunities to be potentially addressed through sweetpotato research and development.

Conservation of sweetpotato diversity. Southeast Asia's landrace/cultivar diversity is a valuable genetic resource to meet current and future livelihood needs (Roca, 2007). Of the total accessions in Southeast Asia's three main *ex situ* collections, 94% are landraces, with moderate and possibly high uniqueness rating (Table 20.2). However the potential significance and usefulness of this sweetpotato diversity are constrained by the high cost of maintaining *ex situ* collections. Seven Southeast Asian collections[1] face intermediate to important constraints in storage, regeneration and plant health (Table 20.3). In 2002, Indonesian and Philippine collections reportedly lost 100 and 574 landraces, respectively, among their registered accessions (Fuglie et al., 2002).

Table 20.2 Major sweetpotato *ex situ* collections in Southeast Asia (adapted from Roca, 2007)

	Number of accessions		
Country/Gene bank	Total	Landraces	Uniquenessrating *
Indonesia			
IABIOGRI	1,520	1,400	2.0
Philippines			
PhilRootcrops	801	771	2.0
NPRCRTC	180	170	1.5

**Uniqueness: Accessions not likely present elsewhere; occurrence of key traits and attributes (1 = low, 2 = moderate, 3 = possibly high)*

Similarly, there has been a decline in efforts for *in situ* on-farm conservation, mainly due to the lack of formal-sector institutional support and weak links with *ex situ* conservation efforts. Successful pilot efforts in *in situ* conservation efforts have been carried out in West Papua, Indonesia, yet the long-term sustainability is becoming uncertain due to the lack of national policy support for sweetpotato genetic resources conservation (Yaku and Widyastuti, 2005).

Meanwhile in the post-rice systems of Central Luzon, Philippine, there were 12 cultivars popularly grown in the mid-1990s. Cultivar diversity was maintained at on-farm, community and regional levels; households grew different varieties for different market/use destinations (Campilan and Prain, 2003). In the following years, the emerging feed-processing industry, together with the expanding fresh-roots market,

[1] Indonesian Agricultural Biotech and Genetic Resources Institute (IABIOGRI), CIP East, Southeast Asia and the Pacific (ESEAP) Office, Philippine Rootcrops Research and Training Center (PhilRootcrops), Northern Philippines Rootcrops Research and Training Center (NPRCRTC), National Plant Genetic Resources Laboratory (NPGRL), Vietnam Agricultural Sciences Institute (VASI)

Table 20.3 Major constraints in *ex situ* sweetpotato collections in Southeast Asia (adapted from Roca, 2007)

	Constraints*		
Country/Genebank	Storage	Regeneration	Plant health
Indonesia			
IABIOGRI	2	2	1
CIP-ESEAP	3	3	2
Philippines			
PhilRootcrops	2	ND	2
NPRCRTC	2	2	2
NPGRL	2	1	2
Vietnam			
VASI	2	1	1

**1 = important, 2 = intermediate, 3 = no constraint*

encouraged farmers to exploit huge market demands for a single variety, VSP-6, which met requirements of both markets. Since then, the dominance of this single variety – developed by the national breeding program – has become a threat to the sustainable *in situ* on-farm conservation of sweetpotato diversity. Only three landraces (i.e. Taiwan, Inubi and Bentong) have been proactively maintained on-farm because there is high demand for them in urban niche markets (Borromeo et al., 2005).

Crop improvement. The genetic resources used to breed new varieties released over the past 20 years have come from a relatively narrow base. Increasing the genetic diversity used to develop new hybrid crosses helps boost productivity of in-country sweetpotato breeding programs (CIP, 2007). This is especially important in efforts to breed for traits of emerging importance such as for animal feed in Vietnam, abiotic stresses in rainfed Indonesia, and agroindustrial uses in the Philippines. Data from 2002 (Table 20.4) revealed that the major varieties adopted had a wide range of target uses – from food to feed and processed products (Fuglie et al., 2002). In Vietnam for example, there was consistent preference for varieties suitable for both food and feed.

Varietal use is however subject to changing opportunities for production and utilization. Since 2002, adoption of VSP-6 in Central Luzon, Philippines reached up to 90% (Dolores et al., 2006) as farmers took advantage of increasing demand for this particular variety from both fresh and feed-processing markets. In Vietnam, adoption of new dual-purpose varieties (i.e. KB1, KL5, and K51) reached 90% of sweetpotato farmers cum pig-raisers in seven provinces, after having participated in training activities on improving feed systems for smallholder animal production (Fuglie et al., 2007).

Integrated crop management. Improved crop management can increase national average yields in Southeast Asia, which are among the lowest in the world (e.g. Rasco, 2000; Saleh and Hartojo, 2003). In the region, sweetpotato yields vary significantly between types of production systems (Table 20.5). In the Philippines, low-input subsistence to semi-commercial systems achieve yields of 10 tons/hectare

Table 20.4 Trends in varietal use in selected sweetpotato-producing areas in Southeast Asia (adapted from Fuglie et al., 2002)

Major Varieties Adopted	Estimated area covered (%)	Trend in coverage	Major uses
Indonesia – East Java/Mojokerto			
IR	45	stable	food
IR Melati	35	stable	food
Sionot	15	decreasing	sauce products
Bestak	5	decreasing	sauce products
Indonesia – West Sumatra			
Tamburin Putih	30	stable	food
Tamburin Merah	15	stable	food
Cangkuang	10	stable	food
Racih Kuning	5	stable	food
Racik Merah	5	decreasing	food
Pelo Galak	5	decreasing	food
Philippines – Central Luzon			
VSP 6	35	increasing	food, feed
Ube	22	increasing	food, feed
Bureau	21	decreasing	food, feed
Taiwan	13	stable	food
Binicol	9	stable	food
Philippines – Eastern Visayas			
PBS Sp-17	30	increasing	food, feed
PBS Sp-15	25	increasing	food, starch
PSB Sp-16	20	decreasing	food, starch
Kasapad	15	stable	food
VSP-5	10	stable	food
Vietnam – Red River Delta			
Khoai lang Hoang Long	40	–	food, feed
Khoai lang Son tim	30	–	food, feed
Tu nien	10	–	food, feed
Thai binh 3	7	–	food, feed
Vietnam – Middle North			
Khoai lang	40	–	food, feed
Buom trang	20	–	food, feed
Hong Thanh	20	–	food, feed

or less, while those in commercial high-input systems can reach as much as 30 tons/hectare (Roa, 2007).

Sweetpotato weevil (*Cylas spp.*) is the single most destructive insect pest especially during prolonged dry seasons in many parts of Southeast Asia. Weevil damage on harvestable roots of up to 70% was observed during extremely dry periods in Indonesia (Saleh and Hartojo, 2003). In Vietnam, serious weevil attack prevents farmers from producing any harvestable roots, thus they are left with only the vines useful for pigfeed (Dinh et al., 1995).

Table 20.5 Estimated crop yields (tons/hectare) of sweetpotato and cassava based on types of production system in the Philippines (adapted from Roa, 2007)

Type	Sweetpotato	Cassava
Commercial medium- to high-input	15–30	20–60
Subsistence to semi-commercial	10 or less	15 or less
National average	4.5	8

Due to inefficient fertilizer use, poor sweetpotato productivity is common in post-rice lowland systems. In Java, Indonesia, fertilizer expenditures are not significantly correlated with profitability of sweetpotato production (Van de Fliert et al., 2003). Over application of inorganic nitrogen fertilizer is common (Braun et al. 1997).

Poor quality of planting materials, mainly due to virus infection and degeneration, also causes low crop productivity. Most often, farmers obtain vine cuttings from the previous cropping season. In Java, Indonesia, farmers cannot readily recognize symptoms of viral diseases and degeneration, even if serological tests reveal seven viral pathogens, which are easily transmitted through vegetative propagation (Saleh et al., 2003).

In Central Luzon, Philippines, a virus complex with leaf-curling systems appearing as early as two weeks after planting has caused up to 100% crop loss since the late 1990s. A subsequent pilot project on the multiplication and use of pathogen-tested planting materials showed 13–20% increase in root yield (Laranang and Basilio, 2002). Still, rapid multiplication of high-quality planting materials is necessary to support larger-scale sweetpotato production. At the same time, it requires setting up institutional mechanisms certifying and monitoring quality of planting materials.

Enhancing value chains. Increased market demand for sweetpotato offers huge potential to turn the crop into a high-value commodity (Fuglie and Oates, 2003). This requires matching on-farm products with market requirements through various technological and commercial innovations.

Efforts to exploit value-adding opportunities also need to be accompanied by institutional mechanisms that enhance poor farmers' access to and participation in market chains. For example in a review of experiences with sweetpotato feed utilization for smallhoder animal production, increased efficiency in feed-to-meat conversion did not guarantee higher profit for Vietnamese and Philippine farmers (Campilan and Sister, 2008). The ultimate determinant of livelihood outcomes is the animals' market price often unilaterally set by traders. Beyond achieving better feed efficiency and weight gain for animals, farmers need improved capacity to negotiate with traders over pricing and to effectively participate in livestock market chains.

In general, there is growing recognition on potentials for value-added processing, utilization and marketing in Southeast Asia. Postharvest and marketing issues received the highest priority rating (i.e. important to very important) among the research themes suggested by sweetpotato research and development organizations in Southeast Asia (Fuglie, 2007).

R&D Challenges to Enhance Primary Functions of a Secondary Crop

Cereal staples such as rice are widely considered as primary crops and thus have been traditionally accorded highest priority in terms of research, extension and government policy. Meanwhile, other food crops are lumped under the category of "secondary crops" – they are considered of lower importance and consequently receive much less public and private sector investment in human, financial and other resources (Horton et al., 1989).

Contrary to its label as a "secondary crop", sweetpotato is utilized in Southeast Asia for multiple purposes in different agroecological and socioeconomic conditions. As in Indonesia, Vietnam and the Philippines, sweetpotato has primary functions that are critical to achieving the goals of food security, poverty alleviation and environmental conservation (Castillo, 1995; Campilan, 2002a,b).

Sweetpotato continues to be associated with resource-poor small-scale farming households in Southeast Asia. Per capita consumption of fresh sweetpotato is highest among low-income income groups (CIP, 2007), where the crop plays an important food security role especially in times of socio-economic crisis and natural disasters.

At the same time, an increasing share of sweetpotato is now being used for animal feed and various processed products for agro-industries. This promotes diversification of livelihood opportunities from the crop, while creating new value-adding market opportunities to absorb increased production.

Table 20.6 Needs and opportunities for sweetpotato R&D in Southeast Asia: a summary

Themes	Key needs and opportunities
Conservation of sweetpotato diversity	– Linking *in situ* and *ex situ* conservation
	– Rationalizing genebank collections through improved characterization and conservation methods
Crop improvement	– Increasing genetic diversity for developing hybrid crosses – Variety development and introduction for: a. high forage yield and dual-purpose food and feed b. tolerance to drought and water stresses c. high-starch and associated functional properties
Use of healthy planting materials	– Developing field-level virus detection kits
	– Strengthening community-based systems for healthy planting materials production and use – Developing mechanisms to support certification and price incentives for quality planting materials
Integrated crop management	– Integrated management of weevil (*Cylas spp.*)
	– Developing strategies for balanced, efficient fertilization – Sustainable soil management for less-favorable production systems
Enhancing value chains	– Enhancing farmers' access to markets – Optimizing use of sweetpotato and other local feed resources

Overall, sweetpotato will continue to play an even greater role in improving the livelihoods of Southeast Asia's rural poor. As such, increased investment is essential to address the key sweetpotato R&D needs and opportunities in the region (Table 20.6).

References

Adion, I. M., Valdez, M.T, Aguilar, C.J., and Basilio, C.S. 2007. *Farmer Field School on Sweetpotato-Based Cattle Fattening: Technical and Field Guides*. CIP-UPWARD, Los Baños, Laguna, Philippines. 117pp.

Aganon, C., and Bautista-Tulin, A. 1998. Participatory approaches to integrated crop management in the Philippines. *In: Sustainable Livelihood for Rural Households, Contributions from Rootcrop Agriculture*. CIP-UPWARD, Los Baños, Laguna, Philippines. 111–136.

ASEAN. 2007. Selected basic ASEAN indicators. http://www.aseansec.org/stat/Table1.pdf.

ASEAN. 2006. *ASEAN Statistical Pocketbook*. ASEAN Secretariat, Jakarta, Indonesia. 116pp.

Ballard, C., Brown, P., Bourke, R.M., and Harwood, T. 2005. The sweet potato in Oceania: a reappraisal. *Ethnology Monographs 19*. University of Pittsburg, Pittsburg PA, USA. 227pp.

BAS. 2007. *CountrySTAT Philippines*. http://countrystat.bas.gov.ph/.

Boncodin, R., Campilan, D., and Prain, G. 2000. Dynamics in tropical homegardens. *Urban Agriculture* 1(1): 10–12.

Boncodin, R., and Gayao, B. 1998. Homegardening for sustainable rootcrop research and development: case projects from the Philippines. *In: Sustainable Livelihood for Rural Households, Contributions from Rootcrop Agriculture*. CIP-UPWARD, Los Baños, Laguna, Philippines. pp. 149–161.

Borromeo, T., De Chavez, H., and Juan, Y. 2005. *Assessment of sweetpotato diversity and use in Central Luzon, Philippines*. Paper presented at the CIP-UPWARD Sweetpotaro R&D Review and Planning Workshop, 1–3 June 2005, Angeles City, Pampanga, Philippines. 23pp.

Braun, A.R, Priatna, E., Asmunati, R., Wiyanto, Widodo, Y., and Van de Fliert, E. 1997. Farmer field schools for sweetpotato in Indonesia. *In: CIP Program Report 1995–96*. CIP, Lima, Peru. pp. 198–204.

Cadelina, R.V. 1994. The sweetpotato component of Batak agriculture. *In: Local Knowledge, Global Science and Plant Genetic Resources*. CIP-UPWARD, Los Baños, Laguna, Philippines. pp. 86–96.

Campilan, D. 2002a. The importance of local knowledge in conserving crop diversity. *SciDevNet Indigenous Knowledge Policy Brief*. www.scidev.net/dossiers/indigenous_knowledge.

Campilan, D. 2002b. Challenging dominant notions on women and rootcrop livelihood. *In: Secondary Farmers of Secondary Crops? Women and Rootcrop Livelihood in the Philippines*. FAO-RAP Project. CIP-UPWARD, Los Baños, Laguna, Philippines. pp. 8–15.

Campilan, D. 2003. A livelihood systems framework for participatory agricultural research: the case of UPWARD. *In: Participatory Research and Development for Sustainable Agriculture and Natural Resource Management: A Sourcebook* (Gonsalves,J., Becker, T., Braun, A., Campilan, D., De Chavez, H., Fajber, E., Kapiriri M., Rivaca-Caminade, J., and Vernooy, R., eds). CIP-UPWARD, Los Baños, Laguna, Philippines and IDRC, Ottawa, Canada. Vol.2, pp. 212–219.

Campilan, D. 1995. *Learning to Change, Changing to Learn: Managing Natural Resources for Sustainable Agriculture in the Philippine Uplands*. Published PhD thesis. Wageningen University, Wageningen, The Netherlands. 158pp.

Campilan, D., and Sister, L. 2008. *Exploring Livelihood Outcomes of Participatory Farmer Training: The Case of Sweetpotato Feed Utilization in Vietnam and the Philippines*. Paper presented at the Rethinking Impact Workshop, 26–28 March 2008, Cali, Colombia. 11pp.

Campilan, D., and Prain, G. 2003. Farmer management of sweetpotato diversity in a changing livelihood system: a case from Central Luzon, Philippines. *In: Conservation and Sustainable Use of Agricultural Biodiversity: A Sourcebook.* Los Baños, Laguna, Philippines: CIP-UPWARD. 245–251.

Castillo, G. 1995. *Secondary Crops in Primary Functions: The Search for Systems, Synergy and Sustainability.* Working Paper Series 2. CIP-UPWARD, Los Baños, Laguna, Philippines. 42pp.

CIP. 2007. *Strengthening Our Assets, Enhancing Our Impact: The Strategic Plan for Research of the International Potato Center 2006–2016.* CIP, Lima, Peru. 75pp.

Data, E., Roa, J., and Tangonan, P. 1997. *Sweetpotato Food Systems in Central Luzon, Philippines.* UPWARD Working Paper Series No. 3. CIP-UPWARD, Los Banos, Laguna, Philippines. 42pp.

Dinh, N.V., Khang, H.L, The, N.V., Hung, H., Oanh, N.K., Ho, T.V., and Braun, A. 1995. Preliminary results on IPM application for sweetpotato at Thanhbinh, Hatay: *In: Research Results of the Faculty of Crop Science.* Hanoi Agricultural University, Hanoi, Vietnam. pp. 83–89.

FAO. 2006. *FAO statistical yearbook 2006.* http://countrystat.bas.gov.ph/

Fuglie, K. 2007. Priorities for sweetpotato research in developing countries: results of a survey. *HortScience* 42(5): 1200–12006.

Fuglie, K., Campilan, D., and Tinh, N.T. 2007. *Adoption and Impact of Innovations in Small-Holder Pig Production Through Farmer-Led Extension in Vietnam.* Impact Enhancement Division Working Paper. CIP, Lima, Peru. 23pp.

Fuglie, K., and Hermann, M. (eds). 2004. Sweetpotato post-harvest research and development in China. *Proceedings of an International Workshop, Chengdu, Sichuan, PR China, 7–8 November 2001.* CIP-ESEAP, Bogor, Indonesia. 160pp.

Fuglie, K., Mahalaya, S., and Suri, S. 2002. The economics of sweetpotato genetic resource conservation and varietal improvement in Asia. *In: Exploring the Complementarities of in situ and Ex Situ Conservation Strategies for Asian Sweetpotato Genetic Resources* (Rao, R and Campilan D, eds.). IPGRI-APO, Serdang, Malaysia. pp. 79–122.

Fuglie, K., and Oates, C. 2003. Sarch markets in Asia: opportunities and constraints. *In: From Cultivators to Consumers, Participatory Research With Various User Groups.* CIP-UPWARD, Los Baños, Laguna, Philippines. pp. 147–158.

Dolores, L.M., Rodriguez, J.D.C, and Sister, L. 2006. *Quick Survey of Sweetpotato ICM Practices in Tarlac Province, Central Luzon. Research Report.* CIP-UPWARD, Los Baños, Laguna, Philippines. 55pp.

Hoa, V.D., Loc, D.T., Ho, T.V., and Kim, H. 2000. Sweet potato in the postrice areas of Vietnam. *In: Sweet Potato in Tropical Asia* (E.T. Rasco and V. Amante, eds). PCARRD, Los Baños, Laguna, Philippines. pp. 189–230.

Horton, D., Prain, G., and Gregory, P. 1989. High-level investment for international R&D. *CIP Circular* 17(3): 1–11.

Laranang, L., and Basilio, C. 2002. Strengthening local systems for producing sweetpotato planting materials. *UPWARD Fieldnotes* 11(1–2): 8–11.

Minantyorini, Widyastuti, C.A., Djazuli, M., and Widowati, S. 1996. Local knowledge of sweetpotato in Flores, Indonesia: what next? *In: Into Action Research, Partnerships in Asian Rootcrop Research and Development.* CIP-UPWARD, Los Baños, Laguna, Philippines. pp. 11–26.

Peters, D. 2003. Poverty alleviation and food security through improving human-pig-potato systems in Papua, Indonesia. *In: Progress in Potato and Sweetpotato Research in Indonesia. Proceedings of the CIP-Indonesia Research Review Workshop, held in Bogor, Indonesia, 26–27 March 2002.* CIP-ESEAP, Bogor, Indonesia. pp. 227–237.

Peters, D., Tinh, N.T., Minh, T.T., Ton, P.H., Yen, N.T., and Hoanh, M.T. 2001. *Pig Feed Improvement Through Enhanced Use of Sweet Potato Roots and Vines in Northern and Central Vietnam.* CIP-Vietnam, Hanoi, Vietnam. 74pp.

Prain, G., and Campilan, D. 1999. Farmer maintenance of sweetpotato diversity in Asia: dominant cultivars and implications for *in situ* conservation. *In: Impact on a Changing World, CIP Program Report 1997–98.* CIP, Lima, Peru. pp. 317–328.

Prain, G., Schneider, J., and Widyastuti, C.A. 2000. Farmer maintenance of sweetpotato diversity in Irian Jaya. *In: Encouraging Diversity, The Conservation and Development of Plant Genetic*

Resources (C. Almekinders and W. de Boef, eds). Intermediate Technology, London, UK. pp. 54–59.

Prain, G. 1995. Sweetpotato in Asian production systems: an overview of UPWARD's first phase research. *In: Taking Root, Proceedings of the Third UPWARD Review and Planning Workshop*. CIP-UPWARD, Los Baños, Laguna, Philippines. pp. 1–35.

Rao, R., and Campilan, D. (eds) 2003. *Exploring the Complementarities of in situ and ex situ Conservation Strategies for Asian Sweetpotato Genetic Resources*. IPGRI-APO, Serdang, Malaysia. 202pp.

Rasco, E.T. 2000. Sweet potato production technology in tropical Asia: trends and needs for improvement. *In: Sweet Potato in Tropical Asia* (E.T. Rasco and Amante, V., eds). PCARRD, Los Baños, Laguna, Philippines. pp. 1–21.

Reynoso. D. 2005. *Sweetpotato Germplasm Characterization, Diversity Assessment and Conservation: New Strategies and Trends*. Paper presented at a special seminar at the Institute of Plant Breeding, University of the Philippines, Los Baños, Laguna, Philippines, 27 June 2005.

Roa, J. 2007. *Food Security in Fragile Lands: Philippine Cases Through the Livelihood Lens*. Published PhD thesis. Wageningen University, Wageningen, The Netherlands. 384pp.

Roca, W. 2007. *Global Strategy for ex situ Conservation of Sweetpotato Genetic Resources*. CIP, Lima, Peru. 76pp.

Rossell, G., Kriegner, A., and Zhang, D.P. 2001. From Latin America to Oceania: the historic dispersal of sweetpotato re-examined using FLP. *In: Scientist and Farmer, Partners in Research for the 21st Century. CIP Program Report 1999–2000*. CIP, Lima, Peru. pp. 315–321.

Saleh, N., Jayasinghe, U., and Rahayuningsih, S.A. 2003. Flow of sweetpotato vine cutting planting materials among farmers in East Java. *In: Progress in Potato and Sweetpotato Research in Indonesia. Proceedings of the CIP-Indonesia Research Review Workshop, held in Bogor, Indonesia, 26–27 March 2002*. CIP-ESEAP, Bogor, Indonesia. pp. 211–225.

Saleh, N., and Hartojo, K. 2003. Present status and future research in sweetpotato in Indonesia. *In: Progress in Potato and Sweetpotato Research in Indonesia. Proceedings of the CIP-Indonesia Research Review Workshop, held in Bogor, Indonesia, 26–27 March 2002*. CIP-ESEAP, Bogor, Indonesia. pp. 137–155.

Van de Fliert, E., Johnson, N., Asmunati, R., and Wiyanto. 2003. Beyond higher yields: the impact of sweetpotato integrated crop management and farmer field schools in Indonesia. *In: Progress in Potato and Sweetpotato Research in Indonesia. Proceedings of the CIP-Indonesia Research Review Workshop, held in Bogor, Indonesia, 26–27 March 2002*. CIP-ESEAP, Bogor, Indonesia. pp. 171–185.

Van de Fliert, E., Oanh, N.T.K., Son, N.T., Hoa, T.D., Thanh, P.D., Hung, L.Q., Khoang, L.H., and Lan, N.T. 2001. The sweetpotato production-postharvest use system in Vietnam: participatory needs and opportunity assessment. *In: Scientist and Farmer, Partners in Research for the 21st Century. CIP Program Report 1999–2000*. CIP, Lima, Peru. pp. 343–351.

Van de Fliert, E., Asmunati, R., Wiyanto, Widodo, Y., and Braun, A. 1996. From basic approach to tailored curriculum: participatory development of a farmer field school model for sweetpotato. *In: Into Action Research, Partnerships in Asian Rootcrop Research and Development*. CIP-UPWARD, Los Baños, Laguna, Philippines. pp. 59–74.

Villamayor, F., and Amante, V. 2000. Sweet potato in the upland agroecology of the Philippines. *In: Sweet Potato in Tropical Asia* (E.T. Rasco and Amante, V., eds). PCARRD, Los Baños, Laguna, Philippines. pp. 58–125.

Watson, G., Dimyati, A., Husni Malian, A., Bahagiawati and Wargiono, J. 1991. Sweet potato production, utilization and marketing in commercial centers of production in Java, Indonesia. *In: Sweet Potato Cultures of Asia and South Pacific. Proceedings of the 2nd Annual UPWARD International Conference*. CIP-UPWARD, Los Baños, Laguna, Philippines. pp. 361–381.

Yaku, A., and Widyastuti, C.A. 2005. Sweet potato research and development in Papua, Indonesia: a review. *In: The sweet potato in Oceania: a reappraisal* (C. Ballard, P. Brown, R.M. Bourke and T. Harwood, eds). Ethnology Monographs 19. University of Pittsburg, Pittsburg PA, USA. pp. 163–179.

Yen, D.E. 1982. Sweet potato in historical perspective. *In: Sweet Potato: Proceedings of the First International Symposium* (R.L. Villareal and T.D. Griggs, eds.). AVRDC, Taiwan. pp. 17–30.

Chapter 21
Sweetpotato in Israel

G. Loebenstein, J. Cohen and Z. Dar

History

Sweetpotatoes were first introduced to Israel around 1920 and grown at the Agricultural High School at Miqve Israel. The variety, which was best, was New Jersey with a yield of about 35 t/.ha. The Agricultural Department of the British Mandate for Palestine, as well, introduced the Puerto Rico sweetpotato cultivar. Subsequently, and during the Second World War they were grown in some home gardens instead of white potatoes (A. Cohen, personal communication) In the early fiftieth of the previous century, when the central water carrier was built and water was brought to the Northern Negev, additional cultivars were introduced (Slomnicki and Cohen, 1952), leading to a marked expansion of the crop. In those days it was thought that the sweetpotatoes would serve as raw material for starch and flour processing purposes. In the beginning yields were very high, reaching more than 60 t/ha, after a growing period of 6 months (Slomnicki and Loebenstein, 1957). However, after about 5–6 years yields declined because of virus diseases (Loebenstein and Harpaz, 1960, and see below), as farmers used for next years planting tubers or cuttings from their old fields. Sweetpotato growing for starch production by farmers was discontinued also because of shortage of water for irrigation and a subsequent increase in water pricing. In the mid 1980th a new interest in sweetpotato became apparent, for local markets but mainly for export to Europe. However, as farmers used for planting tubers or vines from their old plots, fields became heavily infected by viruses and yields averaged less than 15 t/ha. Renewed research led to the identification of the main viruses and to preparation of virus-tested planting material from meristem cultures, and commercial nurseries started to grow the virus-tested mother plants cuttings for growers. Since then swee potato acreage increased every year by 8–10% and became a profitable crop for farmers, mainly for exports.

G. Loebenstein (✉)
Agricultuarl Research Organization, Bet Dagan 50-250 Israel
e-mail: Gad-talma@barak.net.il

G. Loebenstein, G. Thottappilly (eds.), *The Sweetpotato*,
DOI 10.1007/978-1-4020-9475-0_21, © Springer Science+Business Media B.V. 2009

Growing areas and cultivation

In 2006 about 1130 ha of sweetpotato were grown, mainly in the Southern part of the country, yielding 39,000 t. Out of these 16,000 t were exported, and the rest was supplied to local markets (The Plants Production and Marketing Boards; www.plants.org.il in Hebrew). Sweetpotatoes are grown in sandy soils, loess and medium to heavy soils. The predominant variety grown is 'Georgia Jet'. Cuttings from special nurseries are planted along ridges at intervals of 30–35 cm, 75 cm apart and 30 cm high. Cuttings from one acre of nursery are sufficient to plant fields of 20 acres.. Planting starts in March and continues till June.. Fields before planting are fertilized with compost and NPK. Fields are irrigated regularly. Three weeks before harvesting irrigation is gradually reduced. Harvest begins in July and continues till January according to the size of the tubers and market demand. Most of the fields are drip- or sprinkler irrigated and given a complete PK fertilizer before planting and nitrogen during the cultivation. In sandy soils top dressing of nitrogen during the cultivation is applied at a rate of 80–120 kg per hectare. For weed control in sweetpotato it is a common practice to apply Oxadiazon before planting, incorporated by sprinklers irrigation. Tubers are harvested by lifting them mechanically after removing the vines. As Georgia Jet tubers are very susceptible to skin damage they are collected manually. Harvesting is done in October- December before soil temperatures drop below 14 °C. It is expected that with the introduction of other varieties as Beauregard, which have a more robust skin, harvesting will be more mechanized. The tubers are sprayed with a yeast solution (*Metschnikowia fruticola*). After that the tubers are held for curing for 5 days at 30 °C and 92% r.h. (Afek et al., 1998). After that the sweetpotato roots were stored at 14 °C and 90% r.h. for up to 5 months. Before marketing the tubers are sprayed with a yeast solution (*Metschnikowia fruticola*) when on the conveyor belts and sorted either according to their weight (into 6 categories from 100–150gr to 800–950gr) or a by a photo-optical device. Both devices sort about 5 tons per hour and require manual feeding of the tubers. Tubers are shipped to local markets and for exports in 6 kg cartons.

Pests and Diseases

The main insect pest of sweetpotato in Israel is the beetle *Maladera matrida*, caaled by the farmers Khomeini beetle. Adults range in length from 7 to 9 mm, they color is brown-red. Adults are active in the summer and in the spring, flying mostly in evening. They may complete two generations within a sweetpotato crop, allowing population build-up as the crop matures. They feed on the sweetpotato roots and gauge broad, shallow cavities in the storage roots. Damaged roots are not suitable for market or storage. Chemical control is difficult. The chemicals presently used are Cadusafos and Bifenthrin, applied to the soil in conjunction with overhead irrigation at a rate of 200–250 m^3/ha. Application of the pesticide starts when tubers are beginning to develop. Monthly applications are usually given and it is advisable to alternate between the two chemicals. Furthermore, as sweetpotato tubers are for consumption application of chemicals to the soil has to be terminated 6 weeks before harvest (Gol'berg et al., 1989).

Additional insect pest are the sweetpotato whitefly *Bemisia tabaci*, which can harm propagation nurseries in the autumn by infecting them with viruses. Furthermore, in the autumn sometimes sweetpotatoes can become heavily infested by whiteflies, causing secretion of honeydew on which sooty molds develop, resulting in reduced yields. The whiteflies, in case of severe outbreaks, are controlled by Bifenthrin.

Screw worms cause damage to tubers mainly in heavy soils. *Heliotis sp.* and *Spodoptera littoralis* may cause leaf and tuber damage and are controlled by Molit (Teflubenzuron) or Match (Lufenuron).

Root-knot nematodes (*.Meloidogyne*. spp.) are present mainly in light soils, causing cracks, tumors and deformations of the tubers. Control of nematodes until recently was efficient using methyl bromide, but since the use of this fumigation is now prohibited, Cadusafos is being used, but is only partially effetive.

Sweetpotato scab (Streptomyces ipomoea) causing superficial blotches on the tuber. Scab was mainly located in the coastal area. At present trials are conducted by the Extension service evaluating soil treatments.

Virus diseases: Sweetpotato feathery mottle virus, Genus *Potyvirus* (SPFMV); is the most common one. SPFMV is transmitted in a nonpersistent manner by aphids, including *Aphis gossypii*, *Myzus persicae*, and *A craccivora*. The virus can be transmitted mechanically to various *Ipomoea sp.*, as *I. batatas, I. setosa* and *I. nil* and by grafting. The virus can best be diagnosed by grafting on *I. setosa*, causing vein clearing (see Chapter 9b). Most sweetpotato cultivars infected by SPFMV alone show only mild circular spots on their leaves or light green patterns along veins, with little effects on yields. However, when infected together with the whitefly-transmitted *Sweetpotato sunken vein virus* (SPSVV) causing the Sweetpotato disease (SPVD) stunting of the plants, feathery vein clearing and yellowing of the plants are observed.

Sweetpotato sunken vein virus Genus *Crinivirus* (SPSVV). {Possible synonym: *Sweetpotato chlorotic stunt virus* (SPCSV)}. Infection of sweetpotato by SPSVV alone produced on cv. Georgia Jet mild symptoms consisting of slight yellowing of veins, with some sunken secondary veins on the upper sides of the leaves and swollen veins on their lower sides Effects on yields by SPSVV alone are minor (Milgram et al. 1996), but in complex infection with SPFMV yield losses of 50% and more are observed. SPSVV is transmitted by the whitefly *Bemisia tabaci* in a semi-persistent manner, requiring at least one hour for acquisition and infection feeding and reaching a maximum after 24 h for both of them. The virus is graft transmissible, but not by mechanical inoculation.

The virus is best being diagnosed on a pair of sweetpotato plants-one healthy, the other infected by SPFMV. On the healthy plants hardly any symptoms will become apparent, while (if carrying SPFMV) severe symptoms of SPVD will appear. The virus can also be diagnosed by immunosorbent electron microscopy (ISEM). So far ELISA does not detect the virus reliably.

Ipomeoa crinkle leaf curl virus (ICLCV). In 1992, Cohen et al. (1997) observed on *I. setosa*, grafted with scions from sweetpotato cv. Georgia Jet plants, introduced from an unknown source in North America, atypical symptoms of little leaf and crinkle symptoms. Geminate particles were observed in crude sap preparations. The virus was transmitted by *B. tabaci* in a persistent manner and by grafting, but not mechanically. The virus was transmitted to several *Ipomoea* species, including *I. hederacea, I. trifida, I. nil, I. littoralis* and *I. setosa* and induced symptoms on them, including on some cultivars of *I. batatas*, but not on cv. Georgia Jet.

Cucumber mosaic virus Genus *Cucumovirus* (CMV) severely infected sweetpotato fields in the mid-1980 ties, causing stunting, chlorosis, and yellowing of plants (Cohen and Loebenstein, 1991). Infection by CMV was dependent on the presence of SPSVV in the sweetpotato plant. Transmission of CMV, either by grafting or aphid inoculation, to healthy sweetpotato plants failed,, but was achieved when the plants carried already SPSVV.

Control of viruses. In Israel control of viruses is achieved by supplying the grower with virus-indexed propagation material. Rahan Ltd at Rosh Hanikra prepared plantlets from meristems, which were tested repeatedly at the Dept. of Virology of the Agricultural Research Organization. Commercial nurseries increased these virus-tested plants under plastic cover and supplied cuttings from them to the growers. The growers used them only for one season and planted their next year fields with fresh cuttings from the certified nurseries. This scheme resulted in almost complete disappearance of infected plants from growers' fields. Yields depended on the length of the growing period, reaching 45–60 t/ha.

Marketing and consumption. In the last decade sweetpotato consumption has increases markedly. Sweetpotatoes for both local markets and exports are sorted according to size and sent to the markets in 6 kg cartons. Attempts have been made to supply the supermarkets with clean sweetpotatoes on Styrofoam trays ready for the microwave oven. Sofar, however, this has not taken on. Consumers became aware of their nutritional value and the easiness of preparing sweetpotatoes using microwave ovens. Numerous recipes are being published in newspapers for preparing pies, chips and other dishes from sweetpotato. It appears that sweetpotatoes have now a larger share in the carbohydrate diets of local consumers.

References

Afek, U., Orenstein, J. and Nuriel, E. (1998) Increased quality and prolonged storage of sweet potatoes in Israel. *Phytoparasitica*. 26: 307–312.

Cohen, J. and Loebenstein, G. (1991) Role of a whitefly-transmitted agent in infection of sweet potato by cucumber mosaic virus. *Plant Dis*. 75: 291–292.

Cohen, J., Milgram, M., Antignus, J., Perlsman, M., Lachman, O. and Loebenstein, G. (1997) Ipomoea crinkle leaf curl caused by a whitefly–transmitted gemini-like virus. *Ann. appl. Biol.* 131: 273–282.

Gol'berg, A.M., Avigdori-Avidov, H. and Nuriel, E. (1989). Insecticide control of a white grub, *Maladera matrida*, on sweet potato. *Phytoparasitica* 17: 175–183.

Loebenstein, G. and Harpaz, I. (1960) Virus diseases of sweet potatoes in Israel. *Phytopathology* 50: 100–104

Milgram, M., Cohen, J. and Loebenstein, G. (1996) Effects of sweet potato feathery mottle virus and sweet potato sunken vein virus on sweet potato yields and rate of reinfection on virus-free planting material in Israel. *Phytoparasitica* 24: 189–193.

Slomnicki, I. and Loebenstein, G. (1957) Sweet potato growing trials on heavy soils in the Coastal Plain and loess soils in the Northern Negev. *Ktavim* 8: 65–76.

Slomnicki, I. and Cohen, A. (1952) *New Varieties of Sweet Potatoes and Their Cultivation*. Dept. Agr. Israel. Pamphlet No. 32 (in Hebrew).

Chapter 22
Sweetpotato in Oceania

R.M. Bourke

Introduction

The region in review is both large and highly diverse physically, socially and economically. It extends from the equator to about 38° south; from longitude 130° east across the International Date Line to 110° west. Within this area, sweetpotato is grown from sea level to almost 3000 m altitude.[1] Information is organised by modern political units, but their borders do not always coincide well with geographical or social boundaries and tend to reflect colonial history. The countries or other political units covered are Papua New Guinea (PNG), Indonesian Papua, Solomon Islands, Vanuatu, New Caledonia and Fiji (Melanesia); Samoa, Tonga, French Polynesia (Polynesia); Australia and New Zealand (Australasia) (Fig. 22.1).

A demographic and geographic summary for each state is: Papua New Guinea, an independent state (estimated mid 2007 population 6.3 million), which consists of the eastern half of the island of New Guinea, nearby islands in the Bismarck Archipelago and northern end of the Solomon Island chain; Indonesian Papua, a province of Indonesia (2.7 million), and composed of the western half of the island of New Guinea; Solomon Islands, an independent state (0.5 million) consisting of most islands in the Solomon chain; Vanuatu, an independent state (0.2 million), a chain of islands south-east of the Solomon Islands; New Caledonia, a territory of France (0.2 million), one main and several smaller islands; Fiji, an independent island state (0.8 million); Samoa (0.2 million) and Tonga (0.1 million), both independent island states and located in the central South Pacific; French Polynesia, a territory of France (0.3 million), and composed of many small islands in the eastern South Pacific; Australia, an independent state (21.0 million) consisting of the Australian continent and some associated islands; New Zealand, an independent state (4.1 million), composed of two larger islands and a number of smaller ones.

R.M. Bourke (✉)
Department of Human Geography, Research School of Pacific and Asian Studies,
The Australian National University, Canberra, ACT 0200, Australia
e-mail: a145250@anumail.anu.edu.au

[1] Commercial sweetpotato is grown in New Zealand as far south as Gisborne (38 °S), but in pre-European times, Maori people grew it as far south as the Banks Peninsula (43 °S). It is grown at up to 2850 m altitude in Papua New Guinea and a little higher in Papua (west New Guinea).

G. Loebenstein, G. Thottappilly (eds.), *The Sweetpotato*,
DOI 10.1007/978-1-4020-9475-0_22,

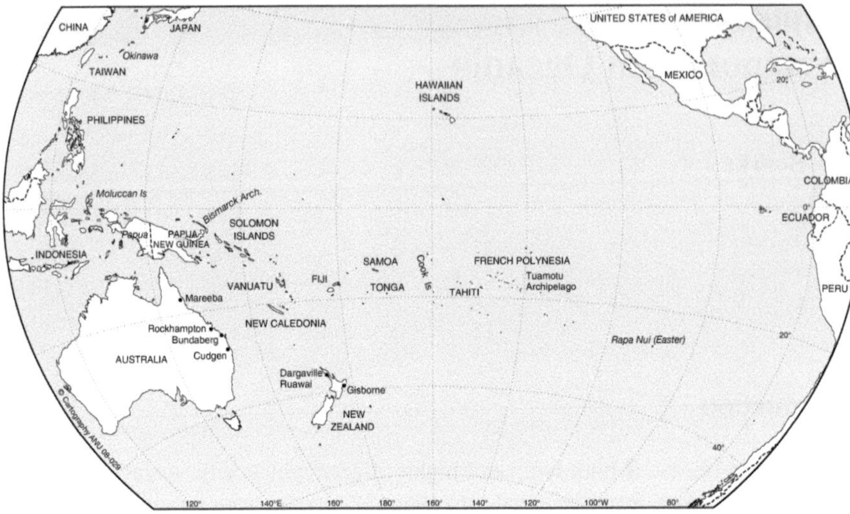

Fig. 22.1 Locations of countries and islands

Sweetpotato is particularly important in the western equatorial part of the region in Indonesian Papua, Papua New Guinea and Solomon Islands. Its significance has increased in that subregion since 1940, so that it is now the most important food crop for five million people in both the highlands and lowlands. It is of lesser significance in other Pacific Island states and in Australia and New Zealand.

More detail is given for Papua New Guinea than for other countries because of the crop's dominant role in food production and consumption, the greater availability of information and the author's experience there. The situation in Papua (west New Guinea) is similar to that on the mainland of PNG (east New Guinea) and sweetpotato production in Solomon Islands is similar to that in the Islands Region of PNG. Published production estimates for different years in various countries are recalculated here by the author for 2007 so that this is the common year for both production and population estimates.

Introduction and Adoption of Sweetpotato

Sweetpotato was introduced to the region from different sources between about 1000 AD and 1900 AD. This process of introduction is addressed here at a regional level rather than at a country level. There is considerable literature on the introduction of sweetpotato into the Pacific and, to a lesser degree, on the impact on societies there (Ballard, 2005). This summary draws heavily on Doug Yen's seminal monograph (Yen, 1974) and a number of papers in a more recent monograph, *The Sweet Potato in Oceania: A Reappraisal* (Ballard et al., 2005), in particular those by Matthew Allen, Chris Ballard, Tim Bayliss-Smith et al. (2005), Roger Green (2005) and Polly Wiessner (2005). The adoption of sweetpotato has occurred in parts of some countries, particularly in the western equatorial Pacific, during the 20th century.[2] Hence the period from its init initial introduction into the South Pacific until its widespread adoption has been about 1000 years.

The crop was probably introduced from the west coast of South America, either from Ecuador or Peru, into the southeastern Pacific by Polynesian voyagers. Other Polynesians took it to the central region of East Polynesia, in the Tuamotu Archipelago, in about 1000–1100 AD. Sweetpotato was later carried by Polynesians to three islands or groups of islands known as the apexes of the Polynesian triangle – Rapa Nui (Easter Island), Hawaii and New Zealand, all at high latitudes (Figure 1). In the case of Hawaii and Rapa Nui, the introduction of sweetpotato took place some centuries after first settlement. For New Zealand, the introduction was made by the founding inhabitants, possibly in the period 1250–1300 AD. The crop became the staple food in these three subtropical locations, while it remained a minor food in the tropical Polynesian islands. During the early period of contact between islanders and Europeans in the 18th and 19th centuries, Polynesian voyagers seem to have taken sweetpotato to Tonga, parts of Fiji, New Caledonia and Samoa.

In contrast, sweetpotato initially came to the western Pacific islands by a very different route. Plants were taken to Portugal on Columbus's first voyage across the Atlantic Ocean to the Americas, with the crop arriving in Europe in 1493. Portuguese travellers later took the crop to Africa, India and Indonesia.[3] It was present in the Moluccan Islands in east Indonesia by 1633, but was possibly there earlier. It is likely to have been taken by traders from the eastern Indonesian islands to west New Guinea, probably by the mid 17th century. Archaeological and oral history research from the Papua New Guinea highlands (east New Guinea) indicates that the crop was adopted there soon after the eruption of Long Island in 1665, that is, about 1700 AD. The adoption of sweetpotato in the highlands of east and west New Guinea in the late 17th or early 18th centuries had a profound impact on many aspects of the economic and social life of highlanders. This impact is not reviewed here but, by the time that European explorers first entered the highland valleys from the 1920s onwards, sweetpotato was by far the most important food for humans and pigs, the main domestic animal, a situation that continues to the present.

The new food crop was not carried to the islands immediately east of New Guinea until the era of direct contact with Europeans. It was introduced by European whalers, traders, explorers and missionaries to the Bismarck Archipelago of Papua New Guinea and the Solomon Islands, as well as by mission students and plantation labourers returning to their homes after living in other countries. It reached a limited number of locations after 1800 AD, but was generally introduced and adopted in the

[2] For example, sweetpotato was only widely adopted in the Sepik River basin, in the lowlands of Oro Province, in delta areas of Gulf Province and in some other locations in lowland PNG since about 1980.

[3] Sweetpotato was introduced into the Philippines in the north-west Pacific by a different route again, as Spanish travellers took it from Peru to Manila. It was then transferred from the Philippines to East Asia (Okinawa, Japan, China). Thus sweetpotato came to different parts of the Pacific via three agencies, one Polynesian and two European.

period 1870–1900. The timing and manner of the introduction of sweetpotato into Vanuatu, New Caledonia and Fiji is less certain. It is possible that it was introduced into parts of these islands from Polynesian islands to the east and spread to other places. However, in many islands in Vanuatu, it appears to be a post-European introduction in the 19th century.

The broad pattern throughout the South Pacific is the displacement of crops of Asia–Pacific origin, particularly *Colocasia* taro, but also yam (*Dioscorea* spp), banana and sago (*Metroxylon sagu*), by crops from the Americas. Sweetpotato is the most important of these American crops in the western Pacific, but cassava has become the most important staple food on many smaller islands and production continues to expand throughout the region. Other crops from the Americas that have been adopted over the past two centuries include *Xanthosoma* taro, maize and peanut. The adoption of these crops varies over time. For example, sweetpotato displaced taro as the most important food in the Solomon Island chain (including Bougainville Island in PNG) in the 1940s. The replacement of taro, yam and other Asia–Pacific crops by sweetpotato, *Xanthosoma* taro and cassava in the rest of the Papua New Guinea lowlands took place at different times and pace over the period 1945–1985.

Sweetpotato was the staple food in the north island of New Zealand by the time that European explorers first visited in the 17th century. The exact period of introduction into Australia is not known, but it was introduced by European settlers, probably in the early 19th century.

Papua New Guinea

This summary is based on the author's extensive surveys throughout Papua New Guinea (PNG) between 1970 and 2008; a comprehensive nationwide survey of agricultural systems with fieldwork conducted over a six-year period (1990–1995) (Bourke et al., 1998); and various publications including Bourke (1985, 2005a, b), Bourke and Harwood (2009) and Kokoa (2001).

Sweetpotato is the most important food in Papua New Guinea. Locally grown staples (root crops, banana and sago) provide an estimated 68% of food energy and sweetpotato provides 66% of the food energy from locally grown staple foods. Thus sweetpotato provides 45% of food energy consumed by people in PNG. No other locally grown or imported food contributes more than 9% of food energy. Most sweetpotato is consumed by people who grow it and their most important domestic animal, the pig. A small proportion, perhaps 1–2%, is sold in local or distant food markets. The tubers are the main economic product, but the foliage is fed to pigs, is sometimes used as a green manure in 'composted' mounds and is very occasionally used as a green vegetable by people.

Production was calculated as 2.9 million tonnes in 2000 and is estimated as 3.5 million tonnes in 2007. This is equivalent to 685 kg/year of sweetpotato tubers per person in rural villages. It is estimated that 500 kg of this is used for human consumption and the balance is fed to pigs, particularly in the highlands. The volume marketed is not known, but is crudely estimated as 30,000–60,000 t/year (1–2% of total production). Sweetpotato is grown in almost all rural locations in PNG and is cultivated by 99% of the rural population. It is the most important food, either on its own or with one other staple crop, for 66% of the rural population and is an important food for a further 15% of the rural population. Thus it is either *the most important* food or *an important* food for more than 80% of the rural population. No other food crop is so significant for so many people in PNG.

In the PNG highlands, sweetpotato dominates agricultural production and the area planted exceeds that for all other food or export cash crops. It is also important in many lowland and intermediate altitude locations, although a number of other staple food crops are grown in the lowlands, including banana, cassava, *Colocasia* taro, *Xanthosoma* taro, yam, coconut and sago. Sweetpotato is least important in locations subject to short-term or long-term inundation.

The area under crop is not known. Based on estimated total production of 3.5 million tonnes in 2007 and a mean yield of 14 t/ha, about 250,000 ha of sweetpotato is under cultivation in a year. There is considerable variation in tuber yield between locations and between years. A working mean yield for villagers is 13 t/ha for a 3–5 month crop in the lowlands and 15 t/ha for a 5–8 month crop in the highlands. Experimental yields are typically 15–20 t/ha in the lowlands and 20–30 t/ha in the highlands. There are indications that both experimental and village yields have declined in recent decades, but this has not been clearly established.

Using the value of rice imported from Asia, the cheapest alternative carbohydrate, it would cost around US$887 million at 2007 prices to replace the food provided by the sweetpotato produced in PNG. The value of sweetpotato sold in local food markets is not known, but is probably in the order of US$20 million per year (50,000 t at US$0.40/kg).

After sweetpotato arrived in PNG about 300 years ago, it was adopted throughout the highlands, largely replacing *Colocasia* taro, the existing staple food, probably within one or three generations. The major expansion of production in the lowlands occurred over the period 1940–2000. In 1940 it provided an estimated 41% of food energy from locally grown staple foods in PNG. This proportion increased to 45% by 1960 and to 66% by 2000.

The crop is grown from sea level up to 2700 m altitude (and occasionally as high as 2850 m). Mean annual rainfall in locations where it is grown is 1000–6500 mm/year, with 2000–3500 mm/year being the most common range. Rainfall distribution is from strongly seasonal to weakly seasonal or non-seasonal. It is grown on a wide number of soil types, ranging from light sandy loams to heavy clays with a wide range in soil fertility. It is grown on all major landforms in PNG except those subject to long-term inundation (swamps).

An estimated 5000 cultivars of sweetpotato can be found in PNG. The range of cultivars grown by different communities is 6 to 71, with a mean of 33 cultivars. Sweetpotato is generally planted throughout the year. In the limited number of locations where rainfall has a marked seasonal distribution, it is planted with other crops at the beginning of the wetter months, which is usually in October–December.

Sweetpotato is planted from cut vine portions. The range in planting density is large, with recorded densities from 15,000 to 172,000 vines per hectare. The wide range of planting densities recorded in village plots confirms experimental findings that planting density influences mean tuber size, but has limited impact on total tuber yield. Soil fertility is maintained by natural fallows. In the highlands, fallowing is not adequate and fertility is commonly enhanced by placing green manure in large mounds; planting a nitrogen fixing tree (*Casuarina oligodon*) in fallow land; or by rotating sweetpotato with peanuts (groundnuts). Inorganic fertiliser is not applied to sweetpotato in PNG. A few growers in Western Highlands and Eastern Highlands provinces plant sweetpotato after a *Solanum* potato or vegetable crop to take advantage of residual inorganic fertiliser. Occasionally villagers apply small quantities of coffee cherry pulp, fire ash or chicken manure, but this is rare.

Irrigation is rarely practised for any crop in PNG and never for sweetpotato. Rather, efforts are directed at moving excess water from the rooting zone by digging drains and planting the crop in mounds. Our nationwide surveys indicate that almost half the population who grow the crop plant it in small mounds (10–40 cm in height and 50–80 cm in diameter). A further 30% of sweetpotato producers plant it in larger mounds that are 0.4–1.2 m in height and 1–4 m in diameter. In a limited number of locations, the soil is tilled mechanically, but this is rare. Tillage and all other operations are usually done by hand, including harvesting tubers.

A large number of diseases caused by fungal, mycoplasma-like organisms, viruses, bacteria and nematodes have been recorded on sweetpotato in PNG. However, their economic impact is not known. Sweetpotato scab, caused by the fungus *Elsinoe batatas*, is possibly the most common disease but its impact on tuber yield is not clear. There are indications from recent research that virus diseases may be responsible for significant yield reductions. A range of insect pests attack the crop. The most important pest is sweetpotato weevil (*Cylas formicarius*), which causes significant tuber losses in a limited number of locations that experience a regular dry period each year or in occasional droughts, such as occurred in 1997.

Other production constraints include declining soil fertility associated with population growth and consequent pressure on land; excessive rainfall; drought; frost (above 2200 m altitude); and planting cycles. These cycles are initiated when sweetpotato is scarce in highland villages, commonly because of excessive or inadequate rainfall. This is the first of a paired food shortage. Men respond by clearing more land from fallow and, because this is more fertile than old garden land, yields are higher. Women respond by increasing the planting rate, which may be as high as high as 150 m^2 per woman per week when sweetpotato is scarce and as low as 20 m^2 per woman per week when it is particularly abundant.

The combination of crops being planted in a more fertile environment at a higher than normal planting rate results in a particularly good harvest about a year after the first food shortage. Women then reduce the area planted and men clear less land from fallow, so that a second food shortage occurs, typically about 22–24 months after the initial food shortage. The equilibrium between supply and food requirement is usually restored after that, but the cycle occasionally continues for a further two years. This pattern does not occur in the lowlands as villagers have access to other staple foods.

Greater quantities of sweetpotato could be sold in urban locations, provided that the price remains significantly less than that for imported rice. Marketing is constrained by an inefficient marketing system, poor post-harvest practices, lack of road access to the capital city Port Moresby, poorly maintained rural roads and inadequate communication between producers, middlemen and retailers. If these constrains could be reduced, there is potential for expansion of the quantity sold in urban markets.

Papua, Indonesia

This summary is based on Ploeg (2005), Yaku and Widyastuti (2005), Resosudarmo et al. (in press) and discussions with various Papua specialists, particularly Chris Ballard.[4] The broad pattern of sweetpotato production in Papua (west New Guinea) is similar to that in Papua New Guinea (east New Guinea).

Sweetpotato is the most important food in Papua. It is by far the most commonly grown and consumed food in the highlands and is also an important food in many lowland locations. As in PNG, other staple foods are also eaten in the lowlands, including locally grown rice. Sweetpotato tubers are widely fed to pigs in the highlands. Young leaves are commonly eaten in a number of locations, including in the Baliem Valley in the highlands. No estimates of overall provincial production are available. Based on a rural village population of two million in 2007, and extrapolation from the PNG data, it is crudely estimated that production in Papua is 1.4 million tonnes per year. This is equivalent to 685 kg per rural villager per year. Yields are reported to be lower than in PNG. This is consistent with the fact that soil fertility is generally lower than in east New Guinea, mainly because there is less volcanic ash in soils.

Sweetpotato is grown from sea level to about 2700 m altitude (and occasionally up to 2900 m). It is a much more important crop in the highlands than in the lowlands, particularly over the altitude range 1400–2400 m. It was introduced into west New Guinea some 300–350 years ago and probably replaced the existing staple food, taro, relatively quickly. It has been widely adopted in the lowlands over the past 60 years.

Production techniques are similar to those used in PNG and a large number of cultivars are grown. There is greater use of drained wetlands using extensive linked drains, particularly in the Baliem Valley and Paniai Lakes area in the highlands. The main pest and disease issues are similar, being weevil, sweetpotato scab and viruses. There is minimal information on the impact of pest and diseases on crop losses.

---

[4] The former Indonesian province of Papua is currently divided into two provinces (Papua in the east and West Papua in the west). However, the term 'Papua' is used to cover both provinces in this paper. Previous names for west New Guinea have included Dutch New Guinea, Irian Barat and Irian Jaya. The term 'Papua' has multiple meanings in Papua New Guinea (east New Guinea) and linguistically and hence can be confusing.

Solomon Islands

This summary draws on the author's fieldwork in Solomon Islands in 2004, Bourke et al. (2006), Jansen et al. (2006) and other literature. As in PNG and Papua, sweetpotato is the most important food in Solomon Islands. Based on extrapolation from detailed studies from Bougainville and Buka islands (the most northern islands of the Solomon chain but now part of Papua New Guinea), it has been estimated that sweetpotato provides 65% of food energy from locally grown staple foods. Most sweetpotato is consumed by subsistence producers. A small proportion, perhaps 1% of total production, is sold in local food markets. It is grown on all islands, except for several atolls, and is the most important food on all larger islands and in all provinces. No other food, locally grown or imported, is as important.

Production was estimated as 280,000 t per year in 2004 (304,000 t in 2007). This gives estimated production of about 700 kg per rural villager per year. The value can be estimated as US$71 million in 2007 by calculating the cost of replacing production using the cheapest available food, which is imported Asian rice. Sweetpotato is grown from sea level to about 500 m altitude on all soil types and on a range of landforms. Mean annual rainfall in these locations ranges from 1800 mm to 6000 mm per year.

Sweetpotato was introduced to the Solomon Islands in the mid to late 1800s, but remained a minor food crop until the 1940s, when a fungal disease of taro, caused by *Phytopthora colocasiae*, was introduced to the Solomon Island chain. The high and continuous rainfall, and consequent high humidity, resulted in the disease having a devastating impact on taro production. By the late 1940s, taro had been largely replaced by sweetpotato as the most important staple food in Solomon Islands, as well as in Bougainville and Buka islands (PNG).

A large number of cultivars are used. Sweetpotato is planted throughout the year, using cut vine portions. It is planted either in small mounds or dibbled into untilled soil. The crop is not fertilised or irrigated. Soil fertility is maintained by natural fallows, generally of low woody regrowth. Pest and disease issues are similar to those in PNG, with sweetpotato weevil, sweetpotato scab and viruses probably the most serious issues. Only hand labour is used. A number of constraints limit the quantity marketed, including few roads, insufficient and expensive inter-island shipping, poorly developed marketing chains and inadequate information for producers on market needs.

Vanuatu

This summary draws on the author's surveys of agricultural systems throughout Vanuatu in 1997–1999, Allen (2005) and Weightman (1989). Sweetpotato is grown on most islands in Vanuatu, but its overall significance is much less than in PNG, Papua or Solomon Islands. At a national level, the most important staple foods are *Colocasia* taro, banana, yam (*Dioscorea alata*), cassava and *Xanthosoma* taro. Sweetpotato, other yam species and *Alocasia* taro are less important food crops. It has become more important in locations where there is pressure on land due to high population densities, such as the central Tanna Island, in parts of central Santo Island and on Paama Island. It is also important in peri-urban villages near the capital Port Vila on Efate Island.

Sweetpotato is mainly a subsistence food in Vanuatu, although some tubers are sold in the limited number of urban markets. No production estimates are available, but production per person would be considerably less than the estimated production of about 700 kg per rural villager per year in PNG, Papua and Solomon Islands.

Sweetpotato is grown in all major environments in Vanuatu. Mean annual rainfall ranges from 1200 to over 3000 mm/year. Sweetpotato is generally not planted seasonally. Villagers on Tongoa, Epi and Ambrym islands believe that the highest and most reliable tuber yields come from plantings made in the somewhat cooler and drier months of May to August.

Other South Pacific Islands

Information on the other South Pacific islands is drawn from the author's fieldwork in these island states in the 1980s, scattered literature (such as Syed and Matalo, 1993) and discussions with various Pacific islands specialists.

Some sweetpotato is grown in other island groups in the South Pacific, including in New Caledonia, Fiji, Samoa, Tonga, Cook Islands and French Polynesia. As in Vanuatu, it is a minor food crop in New Caledonia and less important than a number of other staple foods. Cassava is the most important food crop in Fiji, with sweetpotato and *Xanthosoma* taro grown as supplementary foods. In Tonga, the most important food is cassava, followed by taro and sweetpotato. In the Cook Islands, cassava is the most important food, with dryland and wetland taro second and sweetpotato third. Similarly, sweetpotato is a minor food crop in Samoa and French Polynesia.

Australia

This summary draws on Maltby et al. (2006) and on comments from Queensland Department of Primary Industries and Fisheries (QDPIF) specialists Eric Coleman, John Maltby and Michael Hughes.

Sales of sweetpotato have grown rapidly in Australia in recent years, although it is a minor crop. Consumption has increased somewhat in recent years, partly because of a greater awareness of the value of foods with a low Glycemic Index (GI). Sweetpotato tubers have a lower GI value than *Solanum* potato tubers. From the crop's introduction to Australia in the 19th century until the early 1970s, it was predominately used as fodder for pigs and cattle. It was seen as an important feed source to sustain smallholder dairy cattle in the cool dry winter months. It probably also served as a supplementary food source for the early pioneers. Production increased in the 1980s following introduction from the United States of a number of dessert types cultivars with orange flesh.

Annual production is 50,000 t grown on over 1000 ha. Consumption of 2 kg/person/year is low compared with many Asian countries and is a tiny fraction of the amount consumed in Papua New Guinea, Papua and Solomon Islands. Mean commercial yields are 44 t/ha. Production is worth US$60 million to producers. Almost all tubers are sold in major wholesale markets or directly to large retailers in the east coast cities of Brisbane, Sydney and Melbourne.

Most (70%) of the crop is grown in the Bundaberg and Rockhampton areas in central coastal Queensland. Other growing areas are around Mareeba on the Atherton Tableland of north Queensland and Cudgen in coastal northern New South Wales (Fig. 22.1). Most producing areas have a subtropical climate, with a mild cooler winter. Mean annual rainfall ranges from 1000 mm at Bundaberg to 2000 mm at Cudgen, with more rainfall in December to March.

Four main commercial varieties are grown in Australia: Beauregard, which accounts for 95% of production; Northern Star (4%); Kestle; and WSPF. Crops are planted year round in the Bundaberg, Rockhampton and Cudgen areas and in March to June at Mareeba. The crop is grown at low altitudes and on a wide range of soil types. With the exception of Cudgen, the crops are intensively irrigated. Australia is the only country in the region in which growers intensively irrigate sweetpotato. Without irrigation most plantings would not survive the hot dry conditions.

A number of planting techniques are used, with flat cuttings planted 5 cm below the surface and maintained with daily irrigation being the recommended best technique. Mechanical planting has recently been introduced to reduce labour costs. However, approximately 60% of the crop is still hand planted in the cooler parts of the day in order to reduce transplanting losses. Both organic and inorganic fertilisers are regularly used. There is limited use of herbicides for weed control.

Mechanical cultivation is used 2–3 times early in the crop with hand removal of weeds later, especially in winter when the crop grows slowly. Beauregard is harvested between 17 and 25 weeks after planting, depending on the temperature. Harvesting is performed with machinery and tubers are not stored post harvest. The whole crop is dispatched to market in refrigerated transport and consumed within one month of harvest.

The most important pests are wireworm (family *Elateridae*) and sweetpotato weevil (*Cylas formicarius*). Insecticides are the main method for controlling these pests. Careful selection and evaluation of cultivars has almost eliminated fungal and bacterial problems. Viruses in planting material have not been controlled with breeding or selection. Until 2001 few farmers replaced their planting material on a regular basis. Since then, over 90% of farmers regularly use pathogen-tested planting material sold by QDPIF. Use of pathogen-tested planting material results in higher total yield and better shaped tubers that can be sold to supermarket chains. This one intervention has made a huge difference to the quantity of marketable tubers and has contributed greatly to the rapid growth in the popularity of sweetpotato in the Australian market.

Further improvements in marketable yield are now being attained using planting sprouts from virus free roots instead of vegetative vine from an individual sweetpotato plant. A major constraint to production lies with overcropping of farmland and the build up of pests and diseases. Continual cropping is practised due to the high capital cost of farmland and results in many farmers growing sweetpotato in the same area for many years without rotation.

New Zealand

This summary is based on Lewthwaite (1998) and discussion with Steve Lewthwaite in 2008. Sweetpotato is an important commercial crop in New Zealand. It also has cultural significance to the aboriginal Maori people, as it was their staple food prior to settlement by Europeans in the early 19th century. Following the introduction of maize and *Solanum* potato, production of sweetpotato by Maori declined, but the area planted increased from the 1950s following renewed commercial interest. While fresh domestic consumption dominates the market, the volume of processed and export products is increasing. Processed and semi-processed sweetpotato products are also imported from Australia and China.

Annual production is 18,000 t/year, which equates to mean consumption of 4 kg/person/year. This is higher than Australia, reflecting the crop's long history in New Zealand and acceptance by consumers, but is a tiny fraction of consumption levels in the western equatorial Pacific. About 1200 ha is planted to sweetpotato each year. Commercial yields are about 17 t/ha. The farm gate value is around US$11 million/year, with domestic sales worth about US$27 million/year.

About 86% of New Zealand's sweetpotato production occurs in the Dargaville–Ruawai district (36° south) on clay loam soils, while the rest is scattered about the northern part of North Island. Mean annual rainfall is about 1400 mm. Owairaka Red was released in 1954 and is the dominant variety grown (76% of production by weight). Other commercial varieties are Beauregard (15%) and Toka Toka Gold (9%). Crops are planted in November to December and harvested in March to April.

Green overwintering cover crops are incorporated into the soil prior to planting. Complete inorganic fertilisers, especially potassium, are generally applied. Fields are hand planted with sprouts cut from propagation beds. The sprouts are well watered for several days to help them establish. No other irrigation is applied. Control of weeds is by herbicides, supported by hand weeding, until the vines provide good inter-row coverage. Harvesting is generally by a tractor-drawn harvester.

The most significant economic disease is pink rot caused by the fungus *Sclerotinia sclerotiorum*. Pink rot has been an important disease in New Zealand because large areas of the national crop consist of the cultivar Toka Toka Gold, which is highly susceptible. The cultivar Beauregard is relatively resistant to pink rot and is now displacing Toka Toka Gold. The next most significant disease is scurf, caused by the fungus *Monilochaetes infuscans*, which causes largely cosmetic damage and to which there is no resistant variety. The sweetpotato weevil (*Cylas formicarius*) is not present in New Zealand.

Sweetpotato production in New Zealand is steadily increasing, sustained by research into crop management systems; the development of novel sophisticated food products and enlarged export markets.

Conclusions

Sweetpotato was first introduced into the eastern South Pacific from South America about 1000 years ago by Polynesian voyagers. Polynesians transferred it within eastern Polynesian islands and to subtropical islands Rapa Nui (Eastern Island), Hawaii and New Zealand, where it became the staple food. Polynesians also moved sweetpotato elsewhere in the central and western South Pacific in the early contact period with Europeans in the 18th and 19th century. Sweetpotato came to eastern Indonesia and New Guinea via Europe, arriving in the mid 17th or early 18th century. It was transferred to the islands east of New Guinea in the Bismarck Archipelago and Solomon Islands after 1800 and generally in the period 1870–1900. It was introduced into Australia in the early 19th century by European settlers.

The significance of the crop and the impact of its adoption vary greatly between locations in the region. It is most important in the highlands of east and west New Guinea where it dominates agricultural production and is by far the most important food consumed by people and their main domestic animal, the pig. But it is also an important crop in the lowlands there. Thus sweetpotato is the most important crop in Papua New Guinea, Indonesian Papua and Solomon Islands, where it provides about two-thirds of food energy from locally grown staple foods.[5]

Sweetpotato does not attain this level of significance in other South Pacific island countries of Vanuatu, New Caledonia, Fiji, Samoa, Tonga and French Polynesia. Nevertheless, it is widely grown in the region and is important in some localised areas. It is a minor food crop in Australia and New Zealand where it is grown for commercial sale rather than as a subsistence food. Production in Australia and New Zealand is about 1% of that in PNG. Consumption per person is somewhat greater in New Zealand than in Australia.

The volume consumed per person has increased in equatorial western Pacific (Papua, PNG and Solomon Islands) since about 1940. It has probably stabilised now as most people who can grow sweetpotato do so. The quantity grown per person elsewhere in the tropical South Pacific is less than in the western part of the region, but is still significant. Production per person is now much less in subtropical Polynesia (Rapa Nui, Hawaii and New Zealand) than before European settlement and the adoption of other foods. Production in Australia and New Zealand has increased in recent years.

[5] Note that the proportion of all food energy consumed by people is less as people obtain food energy from non-staple crops, imported foods including rice and wheat, and meat, fish and vegetable oil.

Prospects for further expansion in the region will be influenced by improvements in marketing arrangements in the less-developed economies. Everywhere greater production and consumption will follow adoption of pathogen-tested planting material with few or no viruses and improved cultivars selected for both market demand and growing conditions. The latter will include selection of cultivars that tuberise under higher temperatures as global climate continues to change.[6]

Acknowledgments Comments by area specialists Matthew Allen, Chris Ballard, Bill Clarke, Eric Coleman, Michael Hughes, Steve Lewthwaite and John Maltby are acknowledged with thanks. Editorial assistance was kindly provided by Tracy Harwood and Bryant Allen. The figure was drawn by Anthony Bright of the Cartography Unit in the Research School of Pacific and Asian Studies, The Australian National University.

References

Allen, M.G. 2005. The evidence for sweet potato in island Melanesia. In *The Sweet Potato in Oceania: A Reappraisal.* C. Ballard, P. Brown, R.M. Bourke and T. Harwood (eds). Ethnology Monographs 19, Department of Anthropology, University of Pittsburgh, Pittsburgh and Oceania Monograph 56, University of Sydney, Sydney. pp. 99–108.

Ballard, C. 2005. Still good to think with: the sweet potato in Oceania. In *The Sweet Potato in Oceania*: A Reappraisal. C. Ballard, P. Brown, R.M. Bourke and T. Harwood (eds). Ethnology Monographs 19, Department of Anthropology, University of Pittsburgh, Pittsburgh and Oceania Monograph 56, University of Sydney, Sydney. pp. 1–13.

Ballard, C., Brown, P., Bourke R.M. and Harwood T. (eds). 2005. *The Sweet Potato in Oceania: A Reappraisal.* Ethnology Monographs 19, Department of Anthropology, University of Pittsburgh, Pittsburgh and Oceania Monograph 56, University of Sydney, Sydney. 227pp.

Bayliss-Smith, T., Golson, J., Hughes, P., Blong, R. and Ambrose, W. 2005. Archaeological evidence for the Ipomoean Revolution at Kuk swamp, upper Wahgi Valley, Papua New Guinea. In *The Sweet Potato in Oceania: A Reappraisal.* C. Ballard, P. Brown, R.M. Bourke and T. Harwood (eds). Ethnology Monographs 19, Department of Anthropology, University of Pittsburgh, Pittsburgh and Oceania Monograph 56, University of Sydney, Sydney. pp. 109–120.

Bourke, R.M. 1985. Sweet potato (*Ipomoea batatas*) production and research in Papua New Guinea. *Papua New Guinea J. Agric. For. Fish.* 33(3/4):89–108.

Bourke, R.M. 2005a. Sweet potato in Papua New Guinea: the plant and people. In *The Sweet Potato in Oceania: A Reappraisal.* C. Ballard, P. Brown, R.M. Bourke and T. Harwood (eds). Ethnology Monographs 19, Department of Anthropology, University of Pittsburgh, Pittsburgh and Oceania Monograph 56, University of Sydney, Sydney. pp. 15–24.

Bourke, R.M. 2005b. The continuing Ipomoean revolution in Papua New Guinea. In *The Sweet Potato in Oceania: A Reappraisal.* C. Ballard, P. Brown, R.M. Bourke and T. Harwood (eds). Ethnology Monographs 19, Department of Anthropology, University of Pittsburgh, Pittsburgh and Oceania Monograph 56, University of Sydney, Sydney. pp. 171–179.

Bourke, R.M., Allen, B.J., Hobsbawn, P. and Conway, J. 1998. Papua New Guinea: Text Summaries. Agricultural Systems of Papua New Guinea Working Paper No 1. Two volumes. Department of Human Geography, The Australian National University, Canberra.

---

[6] Sweetpotato tuberisation is adversely affected at air temperatures of about 34 °C. Given that temperatures are near this in the equatorial part of the region year round and sometimes exceed this in the subtropics, further rises in temperature have the potential to reduce tuber yield.

Bourke, R.M. and Harwood, T. (eds). 2009. *Food and Agriculture in Papua New Guinea*. The Australian National University, Canberra.

Bourke, R.M., McGregor, A., Allen, M.G., Evans, B.R., Mullen, B.F., Pollard, A.A, Wairiu, M. and Zotalis, S. 2006. *Solomon Islands Smallholder Agriculture Study*. Volume 1. Main findings and recommendations. AusAID, Canberra. http://www.ausaid.gov.au/publications/pdf/solomon_study_vol1.pdf

Green, R. 2005. Sweet potato transfers in Polynesian prehistory. In *The Sweet Potato in Oceania: A Reappraisal*. C. Ballard, P. Brown, R.M. Bourke and T. Harwood (eds). Ethnology Monographs 19, Department of Anthropology, University of Pittsburgh, Pittsburgh and Oceania Monograph 56, University of Sydney, Sydney. pp. 43–62.

Jansen, T., Mullen, B.F., Pollard, A.A, Maemouri, R.K., Watoto, C. and Iramu, E. 2006. *Solomon Islands Smallholder Agriculture Study*. Volume 2. Subsistence Production, Livestock and Social Analysis. AusAID, Canberra. http://www.ausaid.gov.au/publications/pdf/solomon_study_vol2.pdf

Kokoa, P. 2001. Review of sweet potato diseases in PNG. In R.M. Bourke, M.G. Allen and J.G. Salisbury (eds). *Food Security for Papua New Guinea. Proceedings of the Papua New Guinea Food and Nutrition 2000 Conference*. ACIAR Proceedings No. 99. Australian Centre for International Agricultural Research, Canberra. pp. 730–735. www.aciar.gov.au/system/files/node/306/0009pr99chapter9.pdf

Lewthwaite, S.L. 1998. Commercial sweetpotato production in New Zealand: foundations for the future. In *Proceedings of the International Workshop on Sweetpotato Production System toward the 21st century*. D.R. LaBonte, M. Yamashita, and H. Mochida (eds). Kyushu National Agricultural Experiment Station, Miyakonojo, Miyazaki, Japan.

Maltby, J., Coleman, E. and Hughes, M. 2006. Sweet potato in Australia: an overview of production and research. Paper presented at ACIAR sweet potato workshop, Madang, Papua New Guinea. http://rspas.anu.edu.au/lmg/sweet_potato_workshop/

Ploeg, A. 2005. Sweet potato in the central highlands of west New Guinea. In *The Sweet Potato in Oceania: A Reappraisal*. C. Ballard, P. Brown, R.M. Bourke and T. Harwood (eds). Ethnology Monographs 19, Department of Anthropology, University of Pittsburgh, Pittsburgh and Oceania Monograph 56, University of Sydney, Sydney. pp. 149–161.

Resosudarmo, B.P., Napitupulu, L., Manning, C. and Wanggai, V. in press. Challenges for economic development in an era of political and economic change in Papua. In *Development and Environment in Eastern Indonesia: Papua, Maluku and East Nusa Tenggara*. B.P. Resosudarmo and F. Jotzo (eds). Institute of Southeast Asian Studies, Singapore.

Syed, S. and Matalo. N. 1993. *Agriculture in the Cook Islands: New Directions*. Institute of Pacific Studies, University of the South Pacific, Suva.

Yaku, A. and Widyastuti, C.A. 2005. Sweet potato research and development in Papua, Indonesia: a review. In *The Sweet Potato in Oceania: A Reappraisal*. C. Ballard, P. Brown, R.M. Bourke and T. Harwood (eds). Ethnology Monographs 19, Department of Anthropology, University of Pittsburgh, Pittsburgh and Oceania Monograph 56, University of Sydney, Sydney. pp. 163–170.

Yen, D.E. 1974. *The Sweet Potato and Oceania: An Essay in Ethnobotany*. BP Bishop Museum Bulletin 236. Bishop Museum Press, Honolulu, Hawaii.

Weightman, B. 1989. *Agriculture in Vanuatu: A Historical Review*. British Friends of Vanuatu, Cheam, Surrey.

Wiessner, P. 2005. Social, symbolic, and ritual roles of the sweet potato in Enga, from its introduction until first contact. In *The Sweet Potato in Oceania: A Reappraisal*. C. Ballard, P. Brown, R.M. Bourke and T. Harwood (eds). Ethnology Monographs 19, Department of Anthropology, University of Pittsburgh, Pittsburgh and Oceania Monograph 56, University of Sydney, Sydney. pp. 121–130.

Chapter 23
Concluding Remarks

G. Thottappilly and G. Loebenstein

Rapidly growing populations, recent total neglect of the agriculture sector by many governments, the diversion of certain food crops (maize and sorghum) for biofuel production and the frequent crop failures have aggravated food scarcity in developing countries. The poor are the most vulnerable to high and increasing prices and growing shortages of the basic food and it will lead to chronic hunger and malnutrition, and eventually to famine in extreme cases. Globally, 80–100 million children are born each year and nine out of ten of these children will be born in countries already short of food supply and the resources to grow additional food for the new mouths are lacking. However, there is a new hope if increased focus and attention are given to alternate crops such as sweetpotato as sources of energy and nutrients. Even here, persistent poverty, frequent crop failures due to unpredictable rainfall, environmental degradation, global warming, and unexpected outbreaks of pests and diseases all pose new and unprecedented challenges and opportunities in sweetpotato cultivation in the years ahead.

Root and Tuber crops are basic to the diets of millions in the tropics and subtropics where most of the world's undernourished people live. Among these crops, sweetpotatoes contain substantial levels of protein in addition to carbohydrates. They also provide substantial amounts of vitamins and minerals. In the quest for food and the struggle for human survival, the sweetpotato has historically played an important role.

Sweetpotato has inherent characteristics that limit production increases (Scott et al., 2000). The most important constraints are: 1. Technical (bulkiness/perishability, low multiplication rates, phytosanitary restrictions on the movement of germplasm, pests and diseases especially viruses), 2. Socio-economic constraints (high per unit costs as raw material, low status/stigma, supply chain linkages, small, resource poor farmers), and 3. Policy constraints (Policy neglect, absence of industry forum, weak national programs).

As scientists, our challenge is to overcome these limitations and to realise the massive potential of this crop. We must complement our research efforts with the

G. Thottappilly (✉)
Sardaya College of Engineering and Technology, Kodakara, Kerala, India
e-mail: george_thottappilly@rediffmail.com

G. Loebenstein, G. Thottappilly (eds.), *The Sweetpotato*,
DOI 10.1007/978-1-4020-9475-0_23, © Springer Science+Business Media B.V. 2009

strong commitment to help national governments to break away from policies that emphasize monocropping of cereals to most soaring food needs.

The policy makers should recognise sweetpotato for diet diversity, health and in some tropical countries for survival. What is needed is interdisciplinary research efforts worldwide, germplasm and scientific data exchange, biannual international conferences together with funding agencies and policy makers in order to create policy support and adoption by the public.

Nutrition, Taste, Versatility of Products and Utilization

Sweetpotato is a natural health food because of its high energy, dietary fibre, vitamins and mineral content (Hill et al., 1992; Padmaja, 2009). Sweetpotato has special traits and nutritional versatility in terms of its wide range of taste, texture and colour. It can be prepared in different ways – the leaves can be cooked just like other greens and the roots can be baked, boiled, fried (French fries, chips) or prepared as a salad, dessert, jam, condiment, beverages, cookies, candies etc. Sweetpotato flour can be mixed with other cereals to make bread, pancakes, and vermicelli or with other food as in a casserole or stew (Hill et al., 1992). These traits make sweetpotato comparable to other vegetables and fruits. The potential for processing is great, but it is important to minimize nutrient losses during processing. For producing sweetpotato flour, varieties are needed that are low in crude fibre and high in dietary fibre. A range of quality parameters – dryness, moisture and fibre content, 20 flavours, 15 odors and a range of non-sweet to sweet properties for about 100 advanced sweetpotato clones indicate that genetic diversity is already available for developing new cultivars with desired characteristics (McLaurin and Kays, 1992).

Sweetpotato can support more people per unit hectare than any other food crop. There is need to promote carotene-rich cultivars for children in developing countries. Since antioxidant nutrients-beta carotene (Vitamin A), ascorbic acid (Vitamin C) and tocopherol (Vitamin E) may protect against heart disease and cancer, there is great interest in sweetpotato both in developed and developing countries. Fruity products, noodles, jellied blocks, yogurt, chips and ice cream are other acceptable products. Hence cultivars for particular processing and value-added products must be properly evaluated and developed. Daily consumption of the starchy staple type of sweetpotato provides sufficient vitamin A to prevent the diseases caused by deficiency of this vitamin. An intake of about 13-gm/day of the dessert types, which are high in vitamin A, is sufficient to meet the human body's requirement (Selleck, 1982).

Sweetpotato is a low protein crop, in which crude protein content changes from 4 to 7% on a dry weight basis (Yamakawa, 1997). Marked differences are found in crude protein and individual amino acids content, thus sweetpotatos can be improved for their quantity and quality. In sweetpotato, there are currently two toxic phytochemicals: (a) stress metabolites, which cause pulmonary edema and liver and kidney damage and (b) proteinase inhibitors which cause impaired food utilization, reduced growth and pancreatic hypertrophy (Kays and Kays, 1997).

The tender shoots of sweetpotato serve as an important source of vegetable fibre and food. Studies have shown that tops contain total nutrients equivalent to beef or pork (Selleck, 1982).

High quality frozen products are needed to increase consumption in the US (Walter and Wilson, 1992). Surveys showed consumer acceptance of yogurts containing sweetpotato along with other fruits.

Much research and funds are presently allocated for developing golden rice containing vitamin-A. In sweetpotato, we have a crop with high beta-carotene content (Tsou and Hong, 1992). What is needed is to promote sweetpotato where children are suffering from Vitamin-A deficiency.

Sweetpotato is a low prestige food. It has been used as a survival food. Resistance to eating sweetpotato is both psychological and technical in nature. Co-operation in developing improved varieties; appropriate management practices and post-harvest handling and processing technology are needed to improve sweetpotato as a staple food.

Sweetpotato germplasm contains a range of colours, flavours, sweetness and textures. A variety of products have already been developed, but have limited distribution. One area of priority is to develop culturally appealing products using sweetpotato roots and greens and develop appropriate processing technology. Sweetpotato waste can be used for mushroom cultivation, fish feed or hog and steer feed, but this has to be further developed and expanded. Roots are processed into chips, which are more nutritious than corn in hog rations and more nutritious than soybeans in beef rations (Selleck, 1982).

It is important to ensure processing methods that minimize nutrient losses and to prevent post-harvest losses due to insect infestation. Furthermore it will be beneficial to develop non-food uses, e.g. fuels from starch. China produces 80–85% of the world's sweetpotatoes, which are used there mainly in the production of pork but also starch, alcohol and sugars. Strategies for marketing fresh sweetpotato to Europe, as wrapping tubers in plastic, may attract additional customers.

Sweetpotato has three kinds of colour pigments in the roots, anthocyanin, caratenoid and unidentified flavonoid (Yamakawa, 1997). Industry expects that natural colour market will become larger in place of artificials, because consumers prefer natural food ingredients to artificial ones. Anthocyanin has various important physiological functions to our health, for example, antioxidant, anti-cancer and protection against liver injury. Waste after anthocyanin extraction can be used as a material for spirit.

Crop Protection

Of over 40 species of insect pests that infest sweetpotato in various parts of the world, the sweetpotato weevil, *Cylas formicarius* and a closely related species, *C. puncticollis* are the most destructive in tropical and subtropical regions. The insect attacks sweetpotato roots in the field and in storage. Production losses can be as high 60–100% due to attack by weevils such as *C. formicarius* and *C. puncticollis* (Chalfant et al., 1990).

The weevil is difficult to control even with insecticides because of the nature of its feeding habit inside the roots. Use of weevil resistant sweetpotato cultivars appears to be more practical and the most economical method of control. The apparent inconsistency in resistance among cultivars from season to season or location to location frustrated various workers trying to develop weevil-resistant sweetpotato varieties.

Future work on the taxonomy of the sweetpotato weevil has to be done by mapping its distribution, biotyping with biochemical, karyological and molecular analysis, such as DNA fingerprinting. Use of pheromones in traps has shown great promise. Development of quality control standards for laboratory-production and field-testing protocols for synthetic pheromones should be encouraged. Also, important are the identification and synthesis of host plant volatiles – as attractants – and on biological control methods using natural enemies. A global approach of research and development of IPM strategies that take into account socio-economic and cultural factors should be further development of pheromones, other attractants and biological control methods.

In vitro screening, somaclonal variation, transformation with transgenes together with traditional breeding and molecular markers to track the new genes will bring faster progress in breeding. Host plant resistance to Reniform nematodes, Rhizopus soft rot, weevil and viruses are the priorities. The greatest challenge in breeding is to combine resistance to the important diseases and insects into commercially acceptable cultivars with high yield, root quality etc.

The major problem in developing countries is to supply the farmer with virus-tested planting material (Loebenstein et al., 2009). Breeding programs might be a future answer and such programs are in operation in Uganda, combining Sweet-potato virus disease (SPVD) resistance with desirable agronomic traits such as yield, earliness and acceptable culinary quality. Progress has been made and several cultivars were released (Turyamureeba et al., 1998). It will have to be seen if these cultivars will retain their resistance in other places where different strains of SPVD may be present. Thus, several clones that were resistant to SPFMV in CIP's tests were found to be susceptible, when Israeli (unpublished) and Ugandan isolates were tested (Karyeija et al., 1998).

Attempts have been made through genetic transformation with viral (CP) or other genes (rice cysteine –inhibitor gene) to obtain resistance against insect pests and SPVD. Although resistance to SPCSV has been genetically engineered into a sweet-potato cultivar highly resistant to SPFMV (Kreuze et al., 2009), the high levels of resistance to SPFMV breaks down following infection with SPCSV. It is hoped that future research will result in durable resistance to the "complex" infection by SPVD. In the mean time it is advisable, however, to invest more in organisational efforts to supply growers with virus-tested planting material. This approach, as practiced in Israel, Louisiana (USA) and in the Shandong province of China, resulted in stable high yields.

The chapter on sweetpotato diseases (Clark et al., 2009) summarizes our current understanding of the geographic distribution, symptomatology, pathogen biology and management of the major diseases of sweetpotatoes that occur in the plant bed, in the field (root-borne, soil-borne and foliar) and in storage. Diseases caused by fungi and bacteria can occur at any point of the sweetpotato production cycle and their prevalence is greatly influenced by geographic location and the cropping system in use. Information on diseases that occur under tropical production systems is generally lacking and represents an area for future research emphasis.

Future strategies should include *in vitro* screening, identifying sources of resistance in related species and interspecific transfer, nonpreference (antixenosis) and antibiosis. Future international efforts should concentrate on germplasm exchange, exchange of advanced breeding lines, advanced computer technology for data management, information exchange and a combination of new and emerging technologies with conventional breeding techniques

One of the greatest limitations in sweetpotato improvement is inability to transfer germplasm internationally due to diseae infections. Reliable indexing for all viruses affecting sweetpotato has not yet been developed. In certain cases biological indexing has to be applied which may require greenhouse space and time to read the results. Phytoplasma indexing is also difficult. Mixed infections of viruses are common, thus making virus indexing more difficult (Loebenstein et al., 2009).

Conventional, Wide Crosses and Molecular Breeding

Major constraints to overcome are the hexaploid nature and limited seed production of sweetpotato. Some priorities are:

1. By making germplasm available to all interested in improving protein
2. Further breeding to improve staple types and dessert types as well as types for animal feed could make a sufficient contribution in certain areas of this world.
3. There is great need for institutions to take on the responsibility of providing high quality propagules and either certified or registered seed to growers in order to maintain high quality germplasm and high levels of production.
4. Breeding for disease and insect resistance is requirement.

Protein levels of sweetpotato currently can vary from as little as 0.5% and as much as 23% on a dry-weight basis (Selleck, 1982). There is certainly potential for developing sweetpotato with high levels of protein.

Other attempts center around modification of the sweetpotato starch. Development of new cultivars with novel properties of starch is therefore one objective of sweetpotato breeding. High-amylose starches as well as amylose-free starches are in demand for their specific uses in food and other industries ranging from paper production to production of renewable, biodegradable and CO_2 neutral packing materials.

It is important to decentralize breeding strategy in response to regional needs and capabilities and international distribution and testing of elite germplasm.

There is a need to determine the distribution and effects of diseases and insects on sweetpotatoes in different parts of the world so that proper resistance can be identified at strain/biotype level. This can be achieved by enhanced communication between pathologists and entomologists worldwide. It is desirable to develop an international group on sweetpototo breeding, pests and diseases, and arrange periodic conferences to discuss and review these issues.

The challenge will be to incorporate resistance into commercially important cultivars. Further research on close relatives of sweetpotato is urgently needed and interspecific crosses should be achieved. It would be a great contribution if isolation of genes for resistance to sweetpotato weevil can be achieved. Regional research and breeding will be required to take care of the specific needs of the region.

Biotechnology

Genetic transformation offers possibilities to accelerate improvement of sweetpotato cultivars with traits that are slow or difficult to breed using conventional crossing schemes or for which no good sources have been found in sweetpotato germplasms (Kreuze et al., 2009). Sweetpotato is clonally propagated and a single superior transgenic event could be sufficient for obtaining a significantly improved cultivar for use. However, difficulties to transform and regenerate a broad range of sweetpotato genotypes with the current methods limit the approach. Novel traits cannot be introduced to all those cultivars that would make an elite variety, if the weak point such as susceptibility to viruses or weevils or the less optimal composition of starch or amino acids was not optimal. This can in some cases be adjusted by introducing the missing trait through genetic transformation. Another challenge is that the knowledge of the mechanism behind viral synergism is not yet sufficiently advanced to make a targeted approach, which could prevent severe viral diseases of sweetpotato. Thus, SPCSV eliminates effective antiviral defense and allows the co-infecting heterologous viruses to cause heavy yield reduction.

Finally, sweetpotato is genetically a challenging polyploid plant. It can hardly be considered as a model species for studies in molecular biology. Therefore, sequence data on the genomic DNA and expressed gene sequences accumulate slowly. Enhancement of molecular data and sequence information would be pivotal for improving sweetpotato by engineering its own genes and their expression (Kreuze et al., 2009).

Early crossbreeding experiments to incorporate desirable characteristics took years to obtain the desired results. But with the advent of biotechnology, such as molecular and marker-assisted selection, the end products are reached in a more organised and accelerated manner.

There is considerable interest in sweetpotato biotechnology. Also, there is great interest to grow sweetpotato in space missions. Tissue culture, disease elimination, germplasm storage and distribution are all well developed for sweetpotato. Further research is needed on molecular markers and marker assisted breeding, DNA fingerprinting and germplasm characterization, more genes and transformation of sweetpotato with these genes, and development of new products using biotechnology.

Improved germplasm storage methods and temperature regimes are to be developed to minimize the frequency of reculturing. Training is absolutely necessary to have better success with germplasm exchange in tissue culture form because many scientists are not familiar with transfer of plantlets from tubes to soil resulting in heavy losses after the plantlets are received in a particular country. A method for somatic embryo production of sweetpotato has been developed (Cantliffe, 1992).This development can lead to an automated procedure. More research is needed in this area.

Public-private partnerships are needed to share research and development costs in biotechnology. Private companies should donate more time, technology and efforts in developing appropriate sweetpotato for the region by understanding the constraints in that particular region.

It is expected that biotechnology and post-harvest processing may impact the future development of sweetpotato production. Biotechnology, including genetic engineering and molecular breeding, is providing powerful new tools, when combined with conventional breeding, hold great promise to improve sweetpotato yield, taste, nutrition, resistance to insects and diseases especially to viruses, better storability and utilization.

Agronomy

The sweetpotato is grown in many cropping systems in Africa. These systems include monoculture and intercrops with maize, cassava, banana and sorghum etc. Sweetpotato can be grown the year round, an important factor for struggling families with little real income.

Tolerance to limited water and low fertility permits production in semi-arid conditions. Acceptable yields of sweetpotato are frequently produced with residual fertility after growing rice. Because of its high level of nutrition, high productivity and low land and input requirement, it is a useful component of kitchen gardens (Selleck, 1982). It is very adaptable to systems of relay cropping, intercropping and rotation.

Sweetpotato should be changed from a catch crop. It should be raised to the status of a major crop. Right now due to storage problems, tubers are available in the market only seasonally. This should change. Year-round availability will prompt people to include sweetpotato in their daily diet. The possibility of cropping under slight irrigation should be explored. Right now it is grown in several countries only on residual moisture after rice is harvested.

Fast growing and spreading foliage of sweetpotato serves as a protective ground cover and prevent soil erosion. Also, it can help to reduce weeds in the field. It will flourish in the shade of tree crops. They are tolerant to severe weather and can be used as a relay crop in various cropping systems.

In developing countries, there is a lack of trained and dedicated people active in extension programs to help to transfer technology to farmers. Mostly, research and extension are left uncoordinated in many countries. Research is important, but without the means of transferring important information to the farmers who need it in a way that is understandable, the practical impact of technology is lost (Selleck, 1982).

Sweetpotato exhibits allelopathic defence against weeds. Biological activity of secondary compounds from sweetpotato cultivars has been demonstrated, but their particular role in allelopathy need to be further investigated (Peterson and Harrison, 1992). Methods are developed that produces sweetpotatoes hydroponically with yields of 1700 to 1800 g/plant. There is need for screening and breeding more cultivars for the development of systems for controlling experimental parameters of NASA's Controlled Ecological Life Support System (Bonsi et al., 1992).

National Policy

Our priorities should include: 1. Stimulating more demand and better utilizations 2. Developing new and low cost processing techniques 3. Genetic improvement 4. Virus-tested planting materials 5. Development of better virus diagnostics 6. Stimulating transfer of new technology 7. Value added products and 8. international exchange of germplasm

With the recent outburst of scientific and technological innovations from all over the globe, alternative and faster outcomes are possible. Proper management of science and technology is of vital importance for the improvement of the society. Shortsighted research policy, wrong priorities, inconsistent approach and lack of holistic understanding, results in poor deliverables. Scientists and technologists need to look into the changes that are required. and then develop a strategy together with the farmers in a farmer participatory crop improvement and field testing trials. The informed farmers demand informed choice of technology for implementation and the scientists could expect full support from such farmers. Hence the acceptance of research findings will depend upon the usefulness and effectiveness of the research findings. The key to success is understanding the scientific information behind the technology and proper communication and delivery of the information to decision makers. The gap between information generation, transfer to the farmers and policy development must be bridged.

Sweetpotato must play a bigger role during the next two to three decades. As scientists, our role and challenge will be to realize the massive potential of this crop. We must complement our research efforts to help national governments to change their policies from growing cereals to meet increased food needs. It should be complemented with root crops like sweetpotato.

Communication is very important these days (Gregory, 1992). While continuing our research, we should create awareness and importance of our research to the policy makers and the public alike.

We must improve the information flow about the potential of this crop to policy makers. Sweetpotato could ignite an industrial revolution in developing countries. Economic incentives to the producer are important in the development of sweetpotato. Social scientists can make an important contribution by encouraging consumers to replace vegetables with sweetpotato.

It is important that national and international bodies should increase funding for sweetpotato research. Enhanced collaboration in multidisciplinary and interdisciplinary research, exchange of germplasm as well as information and co-operative efforts on targeted problems would be most beneficial.

The goal should remain to serve the needy in developing countries by working together to enhance quantity, quality and marketability of food from sweetpotato. Luckily, the versatility of sweetpotato lends itself to a wide array of tasty, nutritious preparations in fresh and processed forms to improve the diets of people around the world.

Sweetpotato researchers worldwide should stimulate the demand for sweetpotato by making it low cost and appealing and to suit this crop into existing cropping systems (Hill et al., 1992).

Storing Sweetpotato and its Products

Under ideal storage conditions, sweetpotatoes can be stored and marketed for one year or longer. Sweetpotato is not a crop to be produced and stored for long periods. Losses of tubers after harvest are high depending on variety. It is a currently low-priced fresh commodity. As such, high expenditures on its storage will not yield profit. The better option would be to produce high yields, process the tubers soon after harvest into dry forms with better storability. Such products are eventually re-constituted into a variety of food-based and industrial products.

Let us hope that the sincere and concerted efforts of scientists all over the world will help in feeding tomorrow's hungry by growing sweetpotato. Sweetpotato has been consumed in many parts of the world for centuries, yet it remains a "survival crop" or a food for the last resort (Tsou and Villareal, 1982). People will eat it only as a snack or when they are on the verge of starvation.

Establishment of an international network by using internet for development of postharvest utilization, breeding, germplasm exchange, latest varieties available etc will be highly desirable.

In conclusion, scientists still have much to do to make sweetpotato a popular food crop. Problems associated with yield stability, postharvest loss, flower synchronization, virus diseases, weevils, management practices in various soil types and environments in the tropics remain to be solved. Loss of quality in storage and adequate supply of quality planting materials should be addressed. Effective utilization of roots and tips and their role in nutrition need further research. Attention is needed to the problems of sweetpotato trypsin inhibitor content. It is proposed that the low digestibility of raw sweetpotato could be due to the trypsin inhibitor present in the root. The appreciation of food value of the sweetpotato by the public through various educational processes is important for development. More appetizing and appealing sweetpotato dishes should be developed. This will undoubtedly help change people's attitude and encourage their acceptance of the sweetpotato. Since sweetpotato is rich in beta-carotene, vitamin C, minerals and dietary fibre, it also can be considered as a typical vegetable. As sweetpotato is high in sugar and with an attractive colour can be made into candied products and other delicacies, therefore becoming a luxury food.

Acknowledgments The authors are thankful to Dr. V. Balasubramanian for his constructive criticisms of portions of this chapter in draft form.

References

Bonsi, C.K., Hill, W.A., Mortley, D.G., Loretan, P.A., Morris, C.E. and Carliste, E.R. 1992. Growing sweetpotatoes for space mission using NFT. In: Sweetpotato Technology for the 21st Century, Walter A. Hill, Conrad K. Bonsi and Philip A. Loretan (eds), pp. 110–119, Tuskegee University, Tuskegee, Alabama, 607pp.

Cantliffe, D.J. 1992. Somatic embryos in sweetpotato. In: Sweetpotato Technology for the 21st Century, Walter A. Hill, Conrad K. Bonsi and Philip A. Loretan (eds), pp. xvii–xxv, Tuskegee University, Tuskegee, Alabama, 607pp.

Chalfant, R.B., Janson, R.K., Seal, D.R. and Schalk, J.M. 1990. Ecology and management of sweetpotato insects. Annu. Rev. Ent. 35: 157–180.

Clark, C.A., Holmes, G.J. and Ferrin, D.M. 2009. Major Fungal and Bacterial Diseases. In: Sweetpotato, G. Loebenstein and G. Thottappilly (eds), This volume, Springer Verlag, Doerdrect, The Netherlands

Gregory, P. 1992. Feeding Tomorrow's Hungry: The role of root and tuber crops. In: Sweetpotato Technology for the 21st Century, Walter A. Hill, Conrad K. Bonsi and Philip A. Loretan (eds), pp. xxvii–xxviii, Tuskegee University, Tuskegee, Alabama, 607pp.

Hill, W.A., Bonsi, C.K. and Loretan, P.A. 1992. Sweet potato research: Current status and future needs. In: Sweetpotato Technology for the 21st Century, Walter A. Hill, Conrad K. Bonsi and Philip A. Loretan (eds), pp. xvii–xxv, Tuskegee University, Tuskegee, Alabama, 607pp.

Kreuze, J.F., Valkonen, J.P.T. and Ghislain, M. 2009. Genetic Engineering. In: Sweetpotato, G. Loebenstein andG. Thottappilly (eds), This volume Springer Verlag, Dordrecht, The Netherlands.

Karyeija, R.F., Gibson, R.W. and Valkonen, J.P.T. 1998. Resistance in sweet potato virus disease (SPVD) in wild East African Ipomoea. Ann. Appl. Biol. 133: 39–44.

Kays, S.J. and Kays, S.E. 1997. Sweetpotato chemistry in relation to health. In: Proceedings of internationalworkshop on sweetpotato production system towards the 21st century, December 5–10, 1997, Miyakonojo, Japan.

Loebenstein, G., Thottappilly, G., Fuentes, S. and J. Cohen, 2009. Virus and phytoplasma diseases. In: Sweetpotato, G. Loebenstein and G. Thottappilly (eds), This volume, Springer Verlag, Dordrecht, The Netherlands.

McLaurin,W.J. and Kays, S.J. 1992. Genetic diversity in sweetpotato flavor. In: Sweetpotato Technology for the 21st Century, Walter A. Hill, Conrad K. Bonsi and Philip A. Loretan (eds), pp. 420–427, Tuskegee University, Tuskegee, Alabama, 607pp.

Padmaja, G. 2009. Uses and nutritional data of sweetpotato. In: Sweetpotato, G. Loebenstein and G. Thottappilly (eds), This volume, Springer Verlag, Dordrecht, The Netherlands.

Peterson, J.K. and Harrison, H.F. 1992. Chemical factors involved in sweet potato pest resistance and allelopathy. In: Sweetpotato Technology for the 21st Century, Walter A. Hill, Conrad K. Bonsi and Philip A. Loretan (eds), pp. 263–272, Tuskegee University, Tuskegee, Alabama, 607pp.

Scott, G.J., Best, R., Rosegrant, M. and Bokanga, M. 2000. Roots and tubers in the global food system: A vision statement to the year 2020 (including annex). A co-publication of the International Potato Center (CIP), Centro Internacional de Agricultura Tropical (CIAT), International Food Policy Research Institute (IFPRI), International Institute of Tropical Agriculture (IITA), and International Plant Genetic Resources Institute (IPRI). Peru: International Potato Center. 111pp.

Selleck, G.W. 1982. Symposium overview. In Sweet Potato: Proceedings of the first international symposium, pp. v–viii, Villareal, R.L. and Griggs, T.D. (eds), AVRDC, Tainan 74199, Taiwan.

Tsou, S.C.S. and Hong, T.L. 1992. The nutrition and utilization of sweet potato. In: Sweetpotato Technology for the 21st Century, Walter A. Hill, Conrad K. Bonsi and Philip A. Loretan (eds), pp. 359–366, Tuskegee University, Tuskegee, Alabama, 607pp.

Tsou, S.C.S. and Villareal, R.L. 1982. Resistance to eating sweet potato. In Sweet Potato: Proceedings of the first international symposium, pp. 359–366, Villareal, R.L., and Griggs, T.D. (eds), AVRDC, Tainan 74199, Taiwan.

Turyamureeba, G., Mwanga, R.O.M., Odongo, B., Ocitti p'Obwoya, C. and Carey, E.E. 1998. Application for inclusion of a crop/variety in the National Cultivar List. Submission to the Variety Release Committee for release of sweet potato varieties. The Ugandan Ministry of Agriculture, Animal Industries & Fisheries, Uganda.

Walter, Jr., W.M. and Wilson, P.W. 1992. Frozen sweetpotato products. In: Sweetpotato Technology for the 21st Century, Walter A. Hill, Conrad K. Bonsi and Philip A. Loretan (eds), pp. 400–406, Tuskegee University, Tuskegee, Alabama, 607pp.

Yamakawa, O. 1997. Development of new cultivation and utilization system for sweetpotato toward the 21st century. pp. 1–8, In: Proceedings of international workshop on sweetpotato production system towards the 21st century, December 5–10, 1997, Miyakonojo, Japan.

Index

R

S

MIX
Papier aus verantwortungsvollen Quellen
Paper from responsible sources
FSC® C105338

If you have any concerns about our products,
you can contact us on
ProductSafety@springernature.com

In case Publisher is established outside the EU,
the EU authorized representative is:
Springer Nature Customer Service Center GmbH
Europaplatz 3, 69115 Heidelberg, Germany

Printed by Libri Plureos GmbH
in Hamburg, Germany